GLOBAL WARMING
The hard science

PEARSON EDUCATION

We work with leading authors to develop the
strongest educational materials in geography,
bring cutting-edge thinking and best learning
practice to a global market.

Under a range of well-known imprints, including
Prentice Hall, we craft high quality print and
electronic publications which help readers to
understand and apply their content,
whether studying or at work.

To find out more about the complete range of our
publishing please visit us on the World Wide Web at:
www.pearsoned-ema.com

Global Warming

The Hard Science

L D DANNY HARVEY
Department of Geography
University of Toronto
100 St George Street
Toronto, Canada M5S 3G3
email: harvey@cirque.geog.utoronto.ca

PRENTICE HALL

An imprint of PEARSON EDUCATION

Harlow, England · London · New York · Reading, Massachusetts · San Francisco ·
Toronto · Don Mills, Ontario · Sydney · Tokyo · Singapore · Hong Kong · Seoul ·
Taipei · Cape Town · Madrid · Mexico City · Amsterdam · Munich · Paris · Milan

Pearson Education Limited
Edinburgh Gate
Harlow
Essex
CM20 2JE

and Associated Companies around the world

Visit us on the World Wide Web at:
www.pearsoned-ema.com

First published 2000

ISBN 0582-38167-3

British Library Cataloguing-in-Publication Data
A CIP catalogue record for this book can be obtained from the British Library.

10 9 8 7 6 5 4 3 2 1
04 03 02 01 00

Typeset in 9/11 Times by 30
Produced by Pearson Education Asia Pte Ltd.,
Printed in Singapore

To my daughter Amanda,
my genetic link with the future
of the human race,
and whose descendants will have to deal
with whatever consequences arise
from the changes in the Earth's climate
that we provoke

Contents

Preface

The focus of this book is the science of regional- and global-scale climatic change due to human activities – change that is widely referred to as "global warming". As such, this book deals with our understanding of the processes involved in translating the emissions of gases that affect climate into changes in their atmospheric concentration, in translating concentration changes into changes in the heat balance, and in translating changes in the heat balance into changes in climate and in sea level. Our primary concern will be with the so-called greenhouse gases (GHGs), which have a warming effect on climate, and on suspended liquid or solid particles in the atmosphere – known as aerosols – which have both cooling and warming effects on climate. Human activities have significantly altered the worldwide concentration of a number of important GHGs, and have significantly altered the concentrations of aerosols over vast regions of both continents and oceans.

This book is intended for individuals who want an objective, critical, and thorough introduction to the science underlying the global warming issue. It can be used as a textbook for senior undergraduate and graduate university students, by specialists in any one of the many fields covered in this book who want to enhance their understanding of what their colleagues in related fields are doing, and by the scientific advisors to government policy makers. Science writers, high school science teachers, and scientifically literate members of the general public, business organizations, and environmental groups that are interested in the global warming issue will also find this book to be a valuable resource.

The book is divided into three major parts. Part One provides background information on the workings of the climate system, on natural causes of climatic change, on the ways in which humans can influence climate, and on observed changes in the climate system during the past century, when both human and natural causes of climatic change are likely to have been important. The discussion of the climate system is centred around the major components of the system, the mass and energy flows that link them, and the physics of the so-called greenhouse effect. Part Two begins with an introduction to the hierarchy of computer models that are used to study the climate system and to project the impacts of human activities on climate. This is followed by a thorough discussion of the ways in which changes in the concentration of GHGs and aerosols alter the heat balance of the Earth. Next, the processes by which emissions of GHGs and aerosols lead to changes in their concentrations are thoroughly discussed, followed by a discussion of the methods (primarily experiments with computer simulation models) that have been used to estimate how responsive or sensitive the climate is to changes in the concentration of GHGs and aerosols. Part Two finishes with a series of chapters that discuss global-scale and regional projections of the changes in climate and associated changes in sea level. The emphasis is on the role of fundamental principles in assessing the overall credibility of such projections and in developing an appreciation of the risks of major changes in climate if unrestrained emissions of GHGs continue. Part Three discusses the science–policy interface. A glossary can be found at the end of the book which defines all terms of a technical nature, while a more quantitative discussion can be found through a series of interconnected boxes. The main text, however, can be read and understood without reference to the boxes and without any except the most trivial mathematics.

A number of other environmental issues are closely linked to the global warming issue. For example, emissions of CFCs (the chlorofluorocarbons) and related compounds are responsible for depletion of stratospheric ozone (O_3) and contribute to global warming, since these gases are powerful GHGs. The depletion of O_3 is of concern because of its role in shielding life on Earth from harmful ultraviolet radiation, but has a cooling effect on climate. The buildup of O_3 in the lower atmosphere has a number of harmful effects on plants and animals, but also a warming effect on climate. Emissions of sulphur oxides from the burning of coal, the refining of crude oil, and the smelting of certain minerals contribute to rain acid, but also have a marked regional cooling effect on climate. Loss of tropical rainforests is of great concern because of the massive species extinction that would result and the loss of valuable local ecosystem services (such as protection from floods and erosion), but the destruction of the rainforests also has implications for the global climate. All of these issues will be discussed in this book, but only to the extent that they influence climate.

The emphasis in this book is on global-scale rather than regional or local climatic change. There are two reasons for this. First, projections of regional-scale climatic change are not at all reliable, and may never be reliable. Second, this book attempts to rely as much as possible on arguments and lines of evidence that rest upon fundamental and well-established principles, that can be developed and understood based on intuitive reasoning and simple, back-of-the-envelope calculations, and that are independent of the details of any one computer simulation model. Global-scale and very broad regional-scale changes lend themselves nicely to this approach. For example, this book derives fairly strong arguments about how mid-latitude soil moisture in summer should respond in general to global warming, and discusses the competing factors that determine how strong the response will be, but one cannot say anything credible about the *magnitude* or even the *direction* of soil moisture changes at any given location. An important implication of this is that global warming can be viewed as a collective *risk* that merits a collective response among the nations of the world. There is little scientific (or moral) justification for individual nations to develop a policy response to the threat of global warming based on the anticipated climatic changes within their own jurisdiction alone.

This book stops short of assessing the impacts of projected climatic change and associated sea level rise, as that would be worthy of a whole separate book in its own right, and introduces another layer of uncertainty (depending in part on how individuals and societies respond to the change in climate). Rather, the goal of this book is to provide a clear, thorough, qualitatively rigorous, and hopefully fair discussion of the science underlying projections of climatic change. That seems to be an ambitious enough undertaking for one book.

Finally, the title of this book, *Global Warming: The Hard Science*, has a deliberate *double entendre*: the book is about "hard" science in that it deals with the physical and biological science components of the global warming issue, rather than the social science issues related to the global warming issue. The book is also about "hard" science in the sense that the system being studied is extremely complicated, making understanding anthropogenic climatic change one of the greatest challenges facing the intellectual and scientific community.

L D Danny Harvey
February 1999

Acknowledgements

I would like to thank the following people for freely providing data in electronic form for use in this book: George Boer (Plate 11), Mian Chin (Plate 8, Table 7.4), William Cooke (Fig. 7.7, Table 7.4), Curt Covey (Plate 12), Dave Etheridge (Fig. 4.1), Laurent Fairhead (Plate 11), Johann Feichter (Plates 8 and 9, Figs 7.7, 7.8, and 7.9, Tables 7.4 and 7.5), Marcel Fligge (Fig. 11.17b), Vivian Gornitz (Table 5.4), Jim Hanson and Makiko Sato (Plate 13, Figs 3.3, 3.4 and 7.5), Jim Haywood (Plate 9, Fig. 7.9), Richard Houghton (Plate 1, Fig. 8.4), Tim Johns (Plate 11), Jeff Kichl and Timothy Schneider (Plates 8, 9 and 10, Fig. 7.9, Tables 7.4 and 7.5), Joakim Langner (Plate 8), Judith Lean (Fig. 11.17b), Catherine Liousse (Fig. 7.7), Toshinobu Machida (Fig. 4.1), James Maslanik (Fig. 5.16), Gerry Meehl (Plate 11), David Parker and Phil Jacobs (Plates 4 and 5, Fig. 5.4), Mai Pham (Plate 8, Table 7.4), Alan Robock and Melissa Free (Fig. 11.17c), Makiko Sato (Fig. 11.17c), Matthias Schabel (Plate 12), Sharon Stammerjohn (Fig. 5.15), Ron Stouffer (Plate 11), Rob van Dorland (Plates 8 and 9, Fig. 7.6, Tables 7.2 and 7.4), Ian Watterson (Plate 11), Tom Wigley (Fig. 13.7). Jonathon Kenny is thanked for digging up those data in Table 7.1 that I could not find in the published literature.

I would also like to thank the following people for providing useful comments on parts of the draft text or for providing detailed answers to questions that arose in the process of writing this book: Tad Anderson, Harald Bugmann, Robert Charlson, Petr Chýlek, Bruce Douglas, William Elliot, Roger Gifford, Jonathan Gregory, Jim Haywood, Martin Heimann, Phil Jones, Thomas Karl, Tom Knutson, Maarten Krol, Richard Lindzen, Sarah Raper, Phil Rasch, Alan Robock, Brian Soden, Thomas Stocker, Ian Watterson, and Jan de Wolde.

I would like to thank Katharine Hayhoe for assistance in the preparation of many of the figures in this book, and for assistance in compiling the information needed for Tables 4.1, 4.4, 4.8, and 9.4.

Thanks are also due to Patrick Bonham who copy-edited the original manuscript for this book.

Above all, I would like to thank my wife, Connie, for her patience in enduring my virtual absence during much of the writing of this book, and my two-to-three year old daughter, Amanda, for providing comic relief.

The publishers wish to thank the following for use of the material detailed below:
Table 2.2: Lelieveld, Crutzen and Dentener (1998) Changing concentration, lifetime and climate forcing of the atmospheric methane, *Tellus* B, 50, 128–150.
Figure 2. 1: Trenbeth and Hoar, *Geophysical Research Letters*, 23, 57–60, 1996; Figure 2.14: Wang, Jacob & Logan, *Journal of Geophysical Research*, 103, 10727–10755, 1998; Figures 5.6 and 5.7: Hansen and Lebedeff, *Journal of Geophysical Research*, 92, 13345–13372, 1987; Figure 5.16: Maslanik, Serreze and Barry, *Geophysical Research Letters*, 23, 1677–1680, 1996; Figures 5.18 and 5.19: Lansez, Nicholls, Gray and Avila, *Geophysical Research Letters*, 23, 1697–1700, 1996; Figure 5.20: Lambert, *Journal of Geophysical Research*, 101, 21319–21325, 1996; Figure 5.25: Marenco, Gouget, Pages and Karcher, *Journal of Geophysical Research*, 99, 16617–16632; Figure 5.27: Novelli, Masarie and Lang, *Journal of Geophysical Research*, 103, 19015–19033, 1998; Figure 6.1: Harvey and Schneider, *Journal of Geophysical Research*, 90, 2192–2205, 1985; Figure 6.4: Harvey and Huang, *Journal of Geophysical Research* 1999; Figure 6.5: Harvey, *Global Biogeochemical Cycles*, 3, 137–153, 1989; Figure 8.4: Ciais *et al.*, *Journal of Geophysical Research*, 100, 5051–5070, 1995; Figure 8.5a: Archer, *Global Biogeochemical Cycles*, 10, 159–174, 1996; Figure 8.5b: Archer, *Global Biogeochemical Cycles*, 10, 511–526, 1996; Figure 11.10: Harvey, *Journal of Geophysical Research*, 99, 18447–18466, 1994; Figure 14.2: Schmitz, *Reviews of Geophysics*, 33, 151–173, 1995; © by the American Geophysical Union.
Figures 2.5 and 5.21: Sun and Linzden (1993) Distribution of tropical tropospheric water vapor, *Journal of the Atmospheric Sciences*, 50, 1643–1660, © 1993 AMS.
Figure 4.3: Reprinted with permission from *Nature*. Dlugokencky *et al.*, (1998) 393,447–450; Figure 5.3: Folland & Parker (1995) Correction of instrumental bias in historical sea surface temperature data. *Quarterly Journal of the Royal Meterological Society*, 121, 319–367.

Figure 5.8: Climatic Change, 31, 1995, 599–600, Marine surface temperature: observed variations and data requirements, Parker, Folland & Jackson, figure 4; Figure 8.2: *Climatic Change*, 34, 1996, 1–40, Development of a risk-hedging CO_2 emission policy Part 1, Harvey; *Climatic Change*, 34, 1996, 41–71, Development of a risk-hedging CO_2 emission policy Part II, Harvey, figure 2; Figure 5.15: *Climatic Change*, 37, 1997, 617–639, Opposing Southern Ocean climate patterns as revealed by trends in regional sea ice coverage, Stammerjohn & Smith, figure 3; Figure 5.28: *Climatic Change*, 31, 1995, 515–544, Monitoring sea level changes, Gornitz, figure 1; Figure 5.11: *Climatic Change*, 31, 1995, 349–367, On detecting long-term changes in atmospheric moisture, Elliott, figure 1, with kind permission of Kluwer Academic Publishers.

Figures 5.14 and 5.17: Nicholls, Gruza, Jouzel, Karl, Ogallo and Parker (1996) Observed climate variability and change in Houghton, Filho, Callander, Harris, Kattenberg and Maskell (eds) *Climate Change 1995: The Science of Climate Change* (Cambridge: CUP); Figure 12.2: Warrick, Le Provost, Meier, Oerlemans and Woodworth (1996) Changes in sea level, in Houghton, Filho, Callander, Harris, Kattenberg and Maskell (eds) *Climate Change 1995: The Science of Climate Change* (Cambridge: CUP) © IPCC 1996.

Figure 10.2: Mitchell, Manabe, Melshko and Tokioka (1990) Equilibrium climate change and its implications for the future, in Houghton, Jenkins and Ephraums (eds) *Climate Change: The IPCC Scientific Assessment* (Cambridge: CUP) © IPCC 1990.

Figure 5.22: Spencer and Braswell (1997) How dry is the tropical free troposphere? Implications for global warming theory. *Bulletin of the American Meteorological Society*, 78, 1097–1106; Figure 5.9: Jones, Osborn and Briffa (1997) Estimating sampling errors in large-scale temperature averages, *Journal of Climate*, 10, 2548–2568; Figure 11.11: Gordon and O'Farrell (1997) Transient climate change in the CSIRO coupled model with dynamic sea ice, *Monthly Weather Review*, 125, 875–907, © 1997 AMS.

Figure 5.26a and b: Reprinted from *Atmospheric Environment*, 23, Marenco and Said, Meridional and vertical ozone distribution in the background troposphere (70N–60S; 0–12km altitude) from scientific aircraft measurements during the Stratoz II experiment (June 1984), 185–200, © 1989; Figure 8.1: Reprinted from *Agricultural and Forest Meteorology*, 69, Idso and Idso, Plant responses to atmospheric CO_2 enrichment in the face of environmental constraints: a review of the past 10 year's research, 155–203, © 1994; Figure 10.3: Reprinted from *Paleogeography, Paleoclimatology, Paleoecology*, 121, Kheshgi and Lapenis, Estimating the accuracy of Russian paleotemperature reconstruction, 221–237, © 1996; Figure 14.3: Reprinted from *Deep Sea Research*, 28A, Aagaard, Coachman and Cormack, On the halocline of the Arctic Ocean, 529–545, © 1981 with kind permission of Elsevier Science.

Figure 6.2: Harvey (1988) a semi-analytic energy balance climate model with explicit sea ice and snow physics, *Journal of Climate*, 1, 1065–1085, © 1998 AMS.

Figure 6.3: *Climate Dynamics*, Climate sensititvity due to increased. carbon dioxide: experiments with a coupled atmosphere and ocean general circulation model, Washington and Meehl, 4, 1–38, figure 1, 1989 © Springer-Verlag.

Figures 9.1 and 9.2: Soden and Fu (1995) A satellite analysis of deep convection, upper tropospheric humidity and the greenhouse effect, *Journal of Climate*, 8, 2333–2335, © 1995 AMS.

Figure 10.4: Sato, Sellers, Randall, Schneider, Kinter, Hou and Albertazzi (1989) Effects of implementing the simple biosphere model in a general circulation model, *Journal of the Atmospheric Sciences*, 46, 2757–2782, © 1989 AMS.

Figure 11.2: *Climate Dynamics*, A zonally averaged three-basin ocean circulation model for climate studies, Hovine and Fichet, 10, 313–331, figure 3a–d, 1994 © Springer-Verlag.

Figure 11.12: Manabe, Stouffer, Spelman and Bryan (1991) Transient responses of a coupled ocean-atmosphere model to gradual changes of atmospheric CO_2, Part 1 Annual mean response, *Journal of Climate*, 4, 785–818, © 1991 AMS.

Figure 11.13: Stouffer, Manabe and Vinnokov (1994) Model assessment of the role of natural variability in three 1000-year coupled ocean-atmosphere model integrations, *Journal of Climate*, © 1994 AMS.

Figure 11.14a and b: *Climate Dynamics*, Multi-fingerprint detection and attribution analysis of greenhouse gas, greenhouse gas-plus-aerosol and solar forced climate change, Hegerl, Hasselman, Cubasch, Mitchel, Roeckna, Voss and Waszkewitz, 13, 613–634, figures 9 and 10, 1997 © Springer-Verlag.

Figure 11.16: Reprinted with permission from Tett, Mitchell, Parker and Allen (1996) Human influences of the atmospheric vertical temperature structure: detection and observations, 274, 1170–1173, figure 2.

Figure 11.18: Bryant, *Climate Process and Change*, 1997, Cambridge University Press.

Figure 12.3: Ohmura (1987) New temperature distribution maps for Greenland, *Zeitschrift für Gletscherkunde und Glazialgeologie*, 23, 1–45, figure 10, © 1987 Universitätsverlag Wagner.

Figure 12.4: Reprinted with permission from *Nature*. Oppenheimer (1998) 393, 325–332;

Figures 12.6 and 12.7: *Climate Dynamics*, The steric component of a sea level rise associated with enhanced greenhouse warming: a model study, Bryan, 12, 545–555, figure 1a and 3b, 1996 © Springer-Verlag.

Figures 13.6, 13.7 and 13.8: Reprinted with permission from *Nature*, Hoffert *et al.*, (1998) 395, 881–884 © 1998 Macmillan Magazines Ltd.

Though every effort has been made to trace the owners of copyright material, in a few cases this has proved impossible and we take this opportunity to apologise to any copyright holders whose rights may have been unwittingly infringed.

Chapter highlights

Chapter 1

- Projected climatic change under most business-as-usual scenarios will be rapid, unprecedented in recent geological history, and essentially irreversible. This is true even if the global average climatic response to a CO_2 doubling is a warming within the lower third of the commonly accepted range of 1.5–4.5 K.

Chapter 2

- The climate system consists of a number of components (the oceans, atmosphere, biosphere, cryosphere, and lithosphere) that are linked by interconnected flows of mass and energy. The major mass flows involve water, carbon, sulphur, nitrogen, and phosphorus. The mass flows are further linked through atmospheric chemistry and through a variety of feedbacks between the mass flows and climate.
- Climate can change in response to a variety of natural and anthropogenic driving forces. Irrespective of the responsiveness or sensitivity of climate to these driving forces, the natural driving forces will almost certainly be overwhelmed by the heating effect of increasing concentrations of greenhouse gases (GHGs) during the 21st century.

Chapter 3

- The so-called greenhouse effect and its effect on surface temperature arises from the reduction in upward radiation at the tropopause, not primarily from the downward emission of radiation at the surface from GHGs within the atmosphere.
- When a radiative perturbation occurs, the relevant quantity for the climatic response is the change in net upward radiation at the tropopause after allowing for stratospheric temperatures to adjust. This is referred to as the *adjusted forcing*.
- Climate sensitivity depends on a variety of radiative damping feedbacks, which determine the ease with which excess energy can be radiated to space.
- For a wide variety of radiative forcings, the climate sensitivity (the ratio of global mean temperature response to global mean forcing) is largely independent of the magnitude, nature, and spatial pattern of the forcing.
- Except for forcings with a very strong spatial variation, the latitudinal variation in the temperature response is independent of the latitudinal variation in the forcing, as the latitudinal profile is determined by climatic feedbacks that are driven by overall warming.

Chapter 4

- Direct comparison of the magnitude of natural and anthropogenic emissions of CO_2 is highly misleading because the removal processes for CO_2 molecules (uptake by plants and absorption by the oceans) are reversible.
- Direct comparison of natural and anthropogenic emissions of other GHGs and of aerosol precursors is meaningful because the removal processes for individual molecules are irreversible.
- It can be stated with complete certainty that the observed buildup of atmospheric CO_2 is due to human emissions of CO_2.

Chapter 5

· The evidence that the globally averaged climate has warmed during the preceding century is overwhelming. This warming stands out as unprecedented during the past 600 years.
· The amount of water vapour in the lower troposphere has been increasing in a manner that is broadly consistent with our expectations, given the observed warming.
· Radiosonde measurements of the amount of water vapour between altitudes of 4–8 km are significantly higher than the amounts determined from satellite sensors. The primary cause of the difference seems to be errors in the radiosonde measurements.
· The rate of decrease of stratospheric ozone and the rate of increase of tropospheric ozone both appear to have levelled off during the 1990s.
· The amount of OH in the atmosphere appears to have been modestly increasing during the past 20 years, while the CO concentration has decreased during the 1990s.
· Sea level appears to have risen during the past century at a rate of 1.8 ± 0.1 mm/year. This can be explained only if it is assumed that there is a large positive contribution from Antarctica, which is at variance with recent thinking that the Antarctic and Greenland ice mass will increase as climate warms and thereby tend to reduce the rate of sea level rise.

Chapter 6

· Computer models of various components of the climate system, or of several components coupled together, are central to much of the research today concerning the response of the climate system to human perturbations. For each climate system component, a hierarchy of models exists, ranging from quite simple to quite complex. Models at each level in these hierarchies have specific advantages and disadvantages, so no one model is ideal for the entire spectrum of research questions.

Chapter 7

· The radiative forcing due to the increase in well-mixed GHGs amounted to 2.30 ± 0.23 W m^{-2} in the early 1990s.
· Most estimates place the forcing due to the buildup of tropospheric O_3 at 0.3–0.5 W m^{-2}, but the forcing could have been as large as 0.7 W m^{-2}, with the average forcing in the northern hemisphere (NH) being about twice that in the southern hemisphere (SH).
· Recent estimates place the forcing due to the depletion of stratospheric O_3 at around –0.3 W m^{-2}, but it could have been much smaller if it is more appropriate to use the radiative perturbation at the radiative tropopause rather than at the conventionally defined tropopause.
· The direct radiative forcing of sulphate aerosols, when the effects of co-occurring black carbon aerosols are included, is close to zero in the global average, and could even be a small net heating effect.
· The indirect forcing due to sulphate aerosols, and the forcings due to carbonaceous and nitrate aerosols could, together, have been a substantial cooling effect, with the SH average forcing as large as 90% of the NH average forcing. There are enormous uncertainties in the computed indirect aerosol forcings.

Chapter 8

· The response of atmospheric CO_2 to anthropogenic emissions depends on the rate of uptake by carbon sinks, in particular the terrestrial biosphere and oceans.
· The initial, direct effect of higher atmospheric CO_2 on land plants is to stimulate the rate of photosynthesis and to improve the water-use efficiency of plants. The percentage enhancement of photosynthesis is often larger under stressed conditions (particularly low water and nutrients, or high pollution).
· The long-term effect of higher CO_2 on photosynthesis and the storage of carbon in ecosystems is quite complicated, due to a large number of interacting and often opposing factors, including biochemical changes in plants, and changes in nutrient cycling in soils, in the partitioning of plant production, in the rate of senescence, and in relative species abundance.
· In spite of these beneficial, direct effects of higher CO_2 on plants, there is widespread agreement that the terrestrial biosphere sink is limited in its capacity to absorb CO_2, and that if rapid climate change accompanied by drier continents occurs, there could be a significant pulse of carbon into the atmosphere, lasting a century or more.
· Feedbacks between climate and the size of the oceanic carbon sink, in contrast, seem to be less likely to significantly alter the buildup of atmospheric CO_2.

Chapter 9

- The responsiveness of global mean surface air temperature to a radiative heating perturbation, such as a doubling of atmospheric CO_2, depends on a number of feedbacks involving changes in the amount and distribution of water vapour in the atmosphere; changes in the vertical temperature structure in the atmosphere; changes in the amount, location, and properties of clouds; and changes in the extent and reflectivity of ice and snow.
- In the absence of any feedbacks, the global mean temperature response to a doubling of atmospheric CO_2 would be a warming of about 1.0°C.
- Direct observations of long-term trends in temperature and the total amount of water vapour in the atmosphere, of interannual variations in the amount of water vapour in the upper troposphere, and of processes governing the distribution of water vapour in the atmosphere all indicate that the water vapour feedback is positive and sufficient to increase the warming compared with the no-feedback case by at least 50%.
- Current computer simulation models simulate a global mean warming for a CO_2 doubling of 2.1°C to 4.8°C, consideration of climates during the geological past implies a warming for a CO_2 doubling of 1.4°C to 3.2°C, and the historical record of temperature change (during the past 120 years) is consistent with a warming for a CO_2 doubling of 1.0°C to perhaps 3.0°C.

Chapter 10

- The temperature response to a hypothetical, fixed increase in GHG concentrations, once the climate is fully adjusted, is characterized by greater warming at high than at low latitudes, greater warming in winter than in summer at high latitudes, greater warming of the land than of the ocean, greater warming of surface air than of the surface, and slightly greater warming by night than by day.
- An increase in CO_2 tends to cool the stratosphere, but increases in other GHGs either have very little effect on the stratosphere or tend to warm it.
- Computer simulation models, and a consideration of basic principles, indicate that there will be an increasing tendency for soils to become drier as the climate warms. The largest drying effect is likely to occur at low and middle latitudes, rather than at high latitudes.
- Simulated changes in soil moisture as the climate warms are particularly dependent on the soil moisture conditions simulated for late winter and early spring under the present climate.
- Computer simulation models indicate that the average strength of the Asian monsoon, and its interannual variability, should increase as the climate warms.
- No clear picture emerges concerning how El Niño, tropical cyclones, or mid-latitude storms should change as the climate warms, but there is general agreement that the proportion of rain as heavy rainfall should increase.
- The tendency for decreasing soil moisture in mid-latitudes as the climate warms is expected to be greater the greater the polar amplification of the temperature response. Negative feedbacks that operate at low latitudes and that reduce climate sensitivity would be associated with a greater polar amplification of the temperature response. Lower climate sensitivity in this case will not necessarily be associated with smaller impacts on soil moisture in mid-latitudes.

Chapter 11

- The time scale for warming during the transition to a warmer climate (the "transient" response) depends on both mixing of heat into the oceans and the climate sensitivity. A larger climate sensitivity results in a larger ultimate warming, but the approach to that warming is relatively slower.
- Simulations with coupled atmosphere–ocean models generally show minima in the warming in the Antarctic region and in the North Atlantic, owing to the effect of deep mixing of heat into the oceans in these regions.
- Differences in the spatial patterns of temperature change during the transient response, compared with the patterns that occur once a steady-state warming is reached (the equilibrium response), could result in quite different cloud feedbacks for transient and equilibrium climatic change. In particular, the cloud feedback could be significantly more negative during the initial transient response, producing an effective climate sensitivity during the initial transient response that is as little as one-third of the equilibrium climate sensitivity.
- The firmest basis for detection of an anthropogenic effect on climate lies in the examination of global and hemispheric mean temperature changes, rather than patterns of temperature change at continental or smaller scales. On this basis, anthropogenic effects on climate have been detected.

- Decadal and longer time-scale temperature variations are best explained assuming a climate sensitivity for CO_2 doubling of $1.0°C$ and no aerosol cooling effect, although a climate sensitivity of $2.0°C$ with 40% of the global mean heating in 1990 due to GHGs offset by aerosols also gives reasonable results. If the effective climate sensitivity during the initial transient is smaller than the equilibrium climate sensitivity, then a larger equilibrium climate sensitivity is permitted.
- Coupled 3-D atmosphere–ocean models, when driven by our best estimates of the various climatic forcing factors since 1979, and taking into account the "inertia" from the preceding buildup of GHGs, do very well in simulating the observed variation in the global mean surface, tropospheric, and stratospheric temperatures for the period 1979–1998.

Chapter 12

- For a global mean warming by 2100 of about $2.0°C$ to $3.0°C$, the rise in sea level due to thermal expansion of ocean water, melting of small glaciers and ice caps, and changes in Greenland and Antarctica, is likely to be between 25 and 70 cm.
- At a time scale of several thousand years, sea level could rise by 10–13 m, even if GHG concentrations are stabilized at no more than the equivalent of a CO_2 doubling. This would be due to a thermal expansion contribution of 0.5–3.0 m and the possible complete disappearance of the Greenland and West Antarctic ice caps.

Chapter 13

- The globally averaged temperature change serves as an index of the risks of regional impacts of climatic change.
- Adopting the globally averaged temperature change as a risk index, simple models of the climate system can be used in either forward or inverse modes.
- Under a business-as-usual scenario, where global fossil fuel CO_2 emissions increase from 6 Gt C per year at present to 20 Gt C per year in 2100, the atmospheric CO_2 concentration could more than double by 2100. When the radiative heating effect of increases in other GHGs is taken into account, the effective CO_2 increase could be more than a tripling of the pre-industrial concentration.
- For a scenario in which global fossil fuel CO_2 emissions grow from 6 Gt C per year at present to 20 Gt C per year in 2100, with less marked increases in emissions of other GHGs, the global mean surface air warming (relative to pre-industrial conditions) will reach about $2.0°C$ in 2100 for a CO_2 doubling sensitivity of $1.0°C$, and about $4.6°C$ for a doubling sensitivity of $3.0°C$.
- To limit the growth of CO_2 emissions to that given by the IPCC IS92a scenario will require developing, by 2050, a renewable energy capacity equal to the present total global energy use of 11 TW, given a continuation of the historical rate of decrease in the energy intensity of the global economy of 1% per year.
- To limit atmospheric CO_2 concentration to a doubling of the pre-industrial value will require installing 15 TW of renewable energy supply by 2050 and a continuation of historical rates of decrease in energy intensity, or no new renewable energy supply but a doubling of the historical rate of decrease in the energy intensity of the global economy to 2% per year.
- The Global Warming Potential index is a scientifically unjustifiable attempt to force the complexities of nature into a single number.

Chapter 14

- As climate warms, there is an increasing likelihood that abrupt or unexpected changes in the climate system could occur – what might be termed "surprises". Current understanding of the climate system indicates that significant reorganizations of the ocean circulation could occur, with significant impacts.
- Other potential sources of surprises include a breakdown of the vertical salinity structure in the Arctic Ocean (leading to an abrupt disappearance of any remaining pack ice), an abrupt increase in climate sensitivity, abrupt changes in precipitation patterns, or an unexpected large release of methane from methane clathrates.

List of boxes

PART 1 INTRODUCTION

CHAPTER 1

Climatic change and variability – past, present, and future

The geological record provides evidence that has been used to reconstruct changes in the Earth's climate during the past 3 billion years or so, albeit with considerably greater uncertainty for earlier geological times than for more recent geological times. Geological evidence can also be used to reconstruct past variations in the concentration of CO_2 (carbon dioxide) and CH_4 (methane), which are two important heat-trapping gases. The ice preserved in the Antarctic and Greenland ice caps contains air bubbles that were sealed off within a few hundred years of the accumulation of the snow, thereby providing samples of the atmosphere for the last 250,000 years. From these air bubbles we can measure changes in the concentrations of CO_2 and CH_4. Other, less certain evidence has been used to reconstruct the variation in the atmospheric CO_2 concentration as far back as 570 million years (e.g. Berner, 1994). Past variations in the Earth's climate and in the composition of the atmosphere provide a long-term perspective against which the human-induced changes in the atmosphere, and projected changes in the atmosphere and in climate, can be compared. In this chapter this long-term perspective is presented.

1.1 A geological perspective

Figure 1.1 shows the reconstructed variation in the average surface air temperature of the Earth, with each successive panel showing greater detail for time spans progressively closer to the present. Figure 1.1(a) gives a generalized temperature and precipitation history of the Earth during the past 3 billion years based on a variety of paleoclimatic indicators, some of them qualitative in nature. Figure 1.1(b)

shows the variation in the oxygen isotope composition of bottom-dwelling foraminifera in the global ocean during the past 70 million years, along with a rough quantitative temperature scale. The deep waters of the world's oceans decreased from about 12°C around 60 million years ago, to near 0°C at present. Figure 1.1(c) shows the variation in the oxygen isotope composition of bottom-dwelling foraminifera at Ocean Drilling Project site 677 during the past 1.2 million years; fluctuations at this time scale are largely the result of variations in the global volume of ice on land. Figure 1.1(d) shows the variation in the oxygen isotope composition of ice that accumulated on Devon Island, in the Canadian Arctic Archipelago, during the last 120,000 years. These fluctuations are related to changes in the temperature of the air masses over the ice, and a rough temperature scale is provided based on the correlation between the oxygen isotope composition of Arctic precipitation and temperature given in Johnsen et al. (1989). Finally, Fig. 1.1(e) shows the variation in the isotope ratio in Devon Island ice during the past 1000 years. From these data it can be seen that the Earth's climate changes at a wide range of time scales, with decadal-scale variations superimposed on century and millennial time-scale variations, which in turn are superimposed on variations spanning tens of thousands of years, which are superimposed on still longer time-scale variations, and so on. Major episodes of glaciation, interspersed with interglacial intervals, occurred during the pre-Cambrian period, during the Permo-Carboniferous period, and during the past 3 million years. In between these glacial–interglacial episodes have been extended periods with climates substantially warmer than during the present interglacial period. In short, the climate is a dynamic, constantly changing phenomenon.

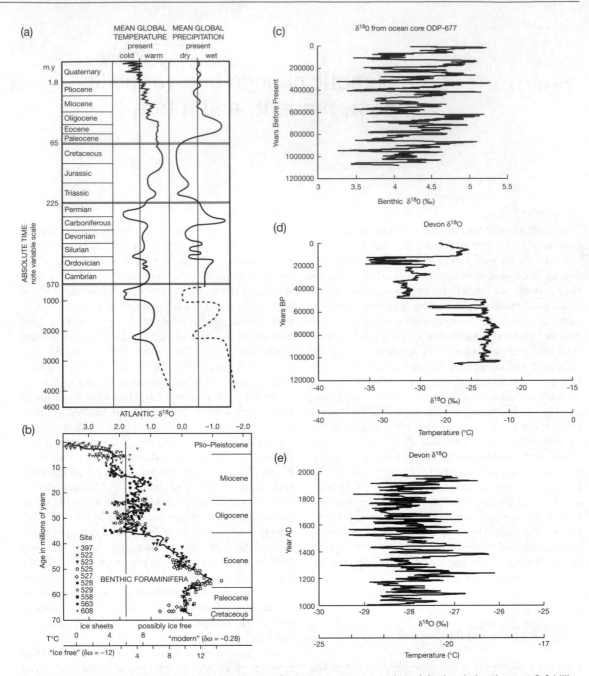

Fig. 1.1 (a) Reconstructed variation in global mean surface air temperature and precipitation during the past 3.0 billion years, from Frakes (1979). (b) Variation in the oxygen isotopic composition of deep-sea foraminifera in the global ocean during the past 70 million years, and an approximate interpretation in terms of the temperature of deep ocean water, from Miller *et al.* (1987). (c) Oxygen isotope variation of deep ocean sediments at ODP site 677 during the past 1.2 million years. (d,e) Variation in the oxygen isotope composition of ice on Devon Island during (d) the past 120,000 years, and (e) the past 1000 years, along with a very rough scale for temperature fluctuations. Data for (c) were obtained from the paleoclimatic data web site: ftp://ftp.ngdc.noaa.gov/paleo/paleocean/sedimentfiles/complete/odp-677.csv, while data for (d) and (e) were obtained from ftp://ftp.ngdc.noaa.gov/paleo/polar/devon/devon data/d72del.200, and ftp://ftp.ngdc.noaa.gov/paleo/polar/devon/devon data/d7273del.5yr, respectively.

Figure 1.2 shows the variation in atmospheric CO_2 concentration over the last 570 million years as inferred from computer models that simulate the geological time-scale flows of carbon. There is no direct record of atmospheric CO_2 concentration this far back, but measurements of quantities that would have been affected by changes in the carbon cycle, such as the chemical composition of oceanic sediments, can be used as a rough check on these models. According to the results shown in Fig. 1.2, the atmospheric CO_2 concentration has varied significantly over periods of tens to hundreds of millions of years, reaching concentrations of 4–16 times the preindustrial concentration of 280 ppmv (parts per million by volume), albeit with a large uncertainty range. Figure 1.3 shows the variation in atmospheric CO_2 and CH_4 concentration over the past 160,000 years as directly measured in air bubbles trapped in Antarctic ice. Also shown on Fig. 1.3 are the present concentrations of CO_2 and CH_4, and a continuation of the upward trend during the next few centuries as a result of human emissions. During the last 160,000 years, CO_2 underwent natural variations in concentration from as low as 180 ppmv to as high as 300 ppmv, while CH_4 varied between 0.3 and 0.7 ppmv. In contrast, human activities during the past 200 years have increased the CO_2 concentration to over 360 ppmv and the CH_4 concentration to over 1.7 ppmv.

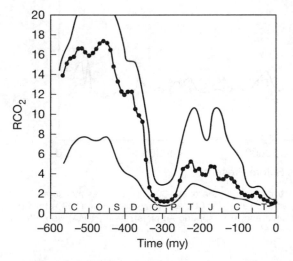

Fig. 1.2 Reconstructed variation in atmospheric CO_2 concentration during the past 570 million years. Reproduced from Berner (1994).

Furthermore, the rates of change during the last 200 years far exceed the rates of change that occurred naturally during the preceding 160,000 years.

1.2   Future prospects and the scientific basis of concern

Not only are the recent, human-induced changes in the atmospheric concentration of CO_2 and CH_4 unprecedented in speed and magnitude, but "business-as-usual" (BAU) projections (i.e., assuming no effort to restrain emissions) indicate that far larger changes will occur during the 21st century and beyond. Figure 1.4 shows a closeup of the BAU projections of atmospheric CO_2 and CH_4 concentration that were shown as a sharp spike at the right-hand edge of Fig. 1.3. The emissions assumed for this projection are discussed in Chapter 4 but are not at all excessive. As seen from this figure, the atmospheric CO_2 concentration could reach six times the preindustrial concentration, while CH_4 could reach four times the pre-industrial concentration. Significant increases in the concentrations of other GHGs are also projected. On the basis of verifiable laboratory measurements, these concentration increases will lead to a significant trapping of heat, and on the basis of very fundamental physical principles and observational evidence, this heat trapping will almost certainly lead to changes in climate that are significant from a human and ecological point of view. Also shown in Fig. 1.4 are projections of the change in global mean temperature due to business-as-usual buildup of GHGs. These projections were made assuming that the eventual climatic warming for a doubling of atmospheric CO_2 only (ΔT_{2x}) is 1.5°C or 3.0°C, which is not excessive either. Nevertheless, by the year 2100, the global average temperature could very well have increased by 2.5–4.0°C above the late 19th century value, and an eventual warming of 3.5–7.0°C occurs. Inclusion of the cooling effect of aerosols might reduce the global mean warming by 0.5–1.0°C and significantly alter regional weather patterns. By comparison, the difference between the last ice age and the present interglacial period is estimated to have been only 4–5°C, but the post-glacial warming occurred over a period of about 10,000 years.

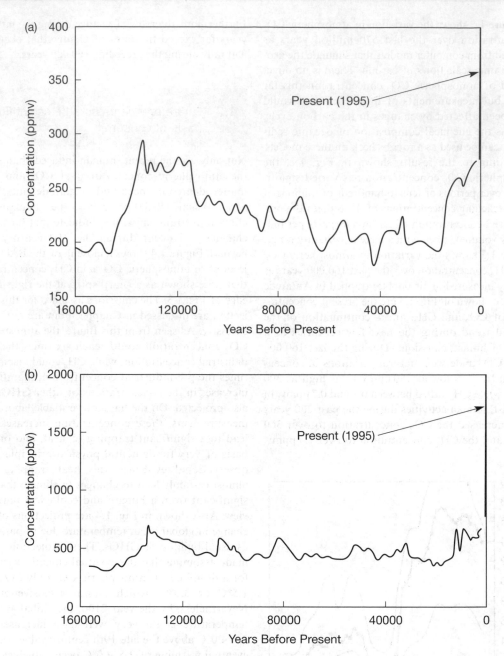

Fig. 1.3 Variation in atmospheric (a) CO_2 and (b) CH_4 concentration during the past 160,000 years, as measured in air bubbles trapped in Antarctic ice. Based on data in Barnola *et al.* (1991) for CO_2 and Chapellaz *et al.* (1990) for CH_4. Both datasets were obtained from the National Oceanographic and Atmospheric Administration (NOAA) web site at http://www.ngdc.noaa.gov/paleo.

Scientific concern over the buildup of GHGs in the atmosphere is not based on some simple-minded extrapolation of past trends in concentrations or in temperatures. Rather, it is based on a step-by-step, process-based assessment of the sequence of events and feedbacks leading from emissions of GHGs to climatic change and the associated impacts. The changes in the atmosphere's composition provoked by human activities are unprecedented in recent geological history, extremely rapid, and – in the case of the CO_2 buildup – essentially irreversible.

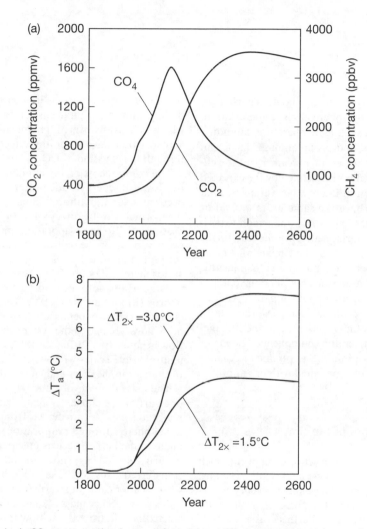

Fig. 1.4 (a) Atmospheric CO_2 concentration (ppmv, left scale) and CH_4 (ppbv, right scale) as projected to the year 2100 according to a plausible business-as-usual emission scenario. (b) Change in global mean surface air temperature corresponding to the business-as-usual CO_2 and CH_4 concentrations shown in (a). Results in (b) are shown assuming eventual global mean temperature increases for a doubling of atmospheric CO_2 concentration of 1.5°C and 3.0°C.

CHAPTER 2

The climate system and climatic change

The term "weather" refers to the day-to-day changes in the state of the atmosphere at a specific location. It includes variables such as temperature, humidity, windiness, cloudiness, and precipitation. Climate can be defined in part as the average weather, but the climate of a given region is also defined by the expected year-to-year variation in the weather variables. That is, two regions that have the same mean annual or seasonal temperatures, but where one region experiences much greater variation from one year to the next, have different climates. The climate of a given region, however, depends on much more than just the atmosphere. Rather, it depends on all of the components that interact together to form part of the *climate system*. This chapter identifies the major components of the climate system and the linkages and interactions between the components. The major natural and anthropogenic causes of climatic change are then introduced and compared with one another.

2.1 Components of the climate system

A system can be defined as a set of components such that each component influences, and is influenced by, all the other components. The climate system consists of the atmosphere, oceans, biosphere, cryosphere (ice and snow), and lithosphere (Earth's crust). Each of these components influences, and is influenced by, all of the others, so they form part of a single system. The sun, in contrast, is not part of the climate system because the climate cannot affect the sun. Rather, the sun is said to be an *external forcing*. Volcanic eruptions, although originating from inside the Earth and its atmospheric envelope, are also an external forcing because they are external in a system sense – they influence, but are not influenced by, the climate

system. Volcanic eruptions influence climate through the injection of sulphur gases into the stratosphere, which are transformed chemically into sulphate aerosols that have a cooling effect on climate, and through the emission of CO_2.

The components of the climate system are linked by flows of energy and matter. The energy flows occur as solar and infrared radiation, as sensible heat (heat which can be directly felt or sensed), as latent heat (related to the evaporation and condensation of water vapour, or the freezing and melting of ice), and through the transfer of momentum between the atmosphere and ocean. The major mass flows involve water, carbon, sulphur, and nutrients such as phosphorus (P) and nitrate (NO_3^-). The behaviour of the climate system depends on how these energy and mass flows change as the system changes, on how the flows themselves influence the system, and on the multiple time scales with which the system responds to changes in the mass and energy flows.

The mass flows involve the transfer of matter from one part of the climate system to another. For example, water can be transferred from the ocean to the atmosphere (through evaporation), from the atmosphere to the land surface as precipitation, can be temporarily stored as snow or ice, and eventually transferred back to the ocean. The different places where matter can temporarily reside are referred to as "reservoirs". Since mass is conserved (it can be neither created nor destroyed), the flow from one reservoir to another and eventually back to the original reservoir constitutes a cycle. Thus, the water, carbon, sulphur, phosphorus, and nitrogen mass flows are referred to as the hydrological cycle, the carbon cycle, and so on.

In the following sections of this chapter, the main features of the energy and mass flows in the climate system are described, with an emphasis on the interactions between the energy and mass flows or between different mass flows, and on how the energy and mass flows are affected by and can affect climate. The discussion that follows focuses almost entirely on natural flows and processes, with only passing reference – where appropriate – to anthropogenic emissions. The latter are discussed extensively and compared with the natural flows in Chapter 4.

2.2   Energy flows

The major energy flows within the climate system are illustrated in Fig. 2.1. Energy from the sun (solar energy) occurs at a variety of wavelengths, ranging from ultraviolet (0.1 0.4 μm) and visible (0.4–0.7 μm) to near-infrared (0.7–4.0 μm). Much of the ultraviolet radiation is absorbed by ozone (O_3) in the stratosphere and, to a lesser extent, in the troposphere (the region from the surface to the base of the stratosphere). Some (a few percent) of the visible and near infrared radiation is reflected back to space by aerosol particles, and an even smaller fraction is absorbed by aerosols. The major types of aerosol are sea salts (which form particles when spray from the oceans evaporates), dust, soot (black carbon) aerosols directly emitted from the burning of biomass, organic carbon aerosols that are either produced by photochemical reactions involving gases emitted by plants (e.g. terpenes) or directly emitted, and sulphate particles that form from emissions of sulphur compounds by bacteria in the ocean, land plants, soils, and volcanoes. Table 2.1 gives recent estimates of the total mass of major types of aerosols in the atmosphere. Water vapour, which is overwhelmingly concentrated in the troposphere, absorbs

a portion of the incoming near-infrared radiation. A portion of the radiation that reaches the surface is absorbed, while the rest is reflected back up. The fraction of incident radiation that is reflected is called the *albedo* and is denoted by the symbol α. Some of the reflected radiation escapes to space, but some is absorbed by water vapour, aerosols, and ozone, and some is reflected back to the surface to be either absorbed or re-reflected. Clouds exert an enormous influence on the flow of solar energy, by typically reflecting 40–80% of the incident radiation back to space and by absorbing 5–15%.

The land surface, oceans, clouds, and atmosphere emit their own radiation, in the infrared part of the spectrum (wavelengths from 4.0 to 50.0 μm). They also absorb some or all of the incident (or impinging) infrared radiation. The absorption and emission of radiation depends on the inherent ability of a substance to emit (the *emissivity*), with emission also depending on – and increasing with – the temperature of the emitting object. The Earth's surface is comparatively warm and has close to the maximum possible emissivity of 1.0, so the global mean surface emission is comparatively large – about 390 W m^{-2} (watts per square metre) according to Fig. 2.1. The atmosphere absorbs a portion of the infrared radiation emitted from the surface, and replaces the absorbed emission with its own emission. However, since the atmosphere is generally colder than the Earth's surface, the amount of infrared radiation that the atmosphere re-emits is smaller than the amount it absorbs, as seen in Fig. 2.1. The net result is to reduce the amount of infrared radiation that escapes to space. The atmosphere also emits some infrared radiation back to the Earth's surface.

Table 2.1 Estimated present-day loading (dry mass) of different aerosol types in the global atmosphere, as given by Andreae (1995). The total loadings are also given as average column loadings in order to permit comparison with the purely anthropogenic loading estimates given in Chapter 7 (Table 7.4).

Aerosol type	Atmospheric loading	Global average column loading
Mineral dust	16.4×10^{12} g	32.2 mg/m^2
Sea salts	3.6×10^{12} g	7.0 mg/m^2
Natural and anthropogenic sulphates[a]	3.3×10^{12} g	6.5 mg/m^2
Products from the burning of biomass by humans	1.8×10^{12} g	3.4 mg/m^2
Products from the oxidation of naturally emitted organic compounds	1.1×10^{12} g	2.1 mg/m^2
Natural and anthropogenic nitrates	0.6×10^{12} g	1.3 mg/m^2

[a] Sulphate aerosols occur either as dilute sulphuric acid (75% H_2SO_4, 25% H_2O) or as ammonium sulphate ((NH_4)$_2SO_4$). Only the mass of SO_4 is given here. The sulphate aerosol mass is further increased by the fact that sulphate aerosols absorb water vapour to an extent that depends on the atmospheric relative humidity.

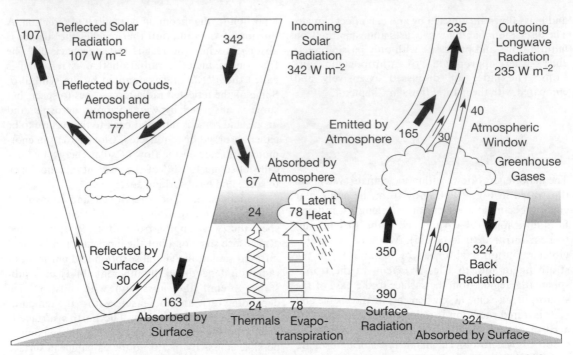

Fig. 2.1 Global mean energy flows between the surface and atmosphere. Reproduced from Trenberth *et al.* (1996).

Since radiation – whether solar or infrared – is a form of energy, the reduction in the loss of infrared radiation to space by the atmosphere tends to make the climate warmer than it would be otherwise. This is the so-called "greenhouse effect" although, as explained in Box 2.1, the term is a misnomer since a greenhouse is warm largely for different reasons. The main gases responsible for the greenhouse effect are those gases that are capable of absorbing and re-emitting infrared radiation. The most important of these is water vapour, followed by CO_2, O_3, CH_4, and N_2O (nitrous oxide). These are all naturally occurring gases, so the greenhouse effect is a naturally occurring phenomenon. Clouds are also effective greenhouse agents, as they absorb all of the radiation emitted from the Earth's surface below (except for thin, wispy clouds that are known as cirrus clouds) but re-emit a considerably smaller amount of radiation owing to the fact that the tops of clouds are generally much colder than the Earth's surface.

Also shown in Fig. 2.1 are the flows of sensible and latent heat from the surface to the atmosphere. The Earth's surface on average absorbs about $100\,W\,m^{-2}$ more solar + infrared radiation than its emission of infrared radiation, while the atmosphere emits about $100\,W\,m^{-2}$ more radiation than it absorbs. About 80%

of the excess radiative energy at the surface is used to evaporate water rather than raising the surface temperature. When the water vapour condenses in the atmosphere, enough heat is released to offset about

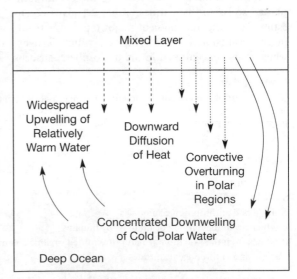

Fig. 2.2 Global mean energy flows between the ocean surface layer and the deep ocean.

Box 2.1 The greenhouse effect and real greenhouses

The term 'greenhouse effect' is used to refer to the tendency of the atmosphere to create a warmer climate than would otherwise be the case. However, the physical mechanisms by which the presence of the atmosphere warms the climate and the primary mechanism that causes a greenhouse to be warm are in fact quite different. A greenhouse heats up by day as the air within the greenhouse is heated by the sun. Outside the greenhouse, near-surface air that is heated through absorption of solar radiation by the ground surface is free to rise and be replaced with colder air from above. This cannot happen in a greenhouse – the heated air is physically prevented from rising and being replaced with colder air. The so-called greenhouse effect does not involve preventing the physical movement of air parcels. Rather, it involves the net trapping of infrared radiation, which occurs independently of the movement of individual air parcels or the lack thereof. There is, nevertheless, a weak similarity between the greenhouse effect and what happens in a greenhouse: the net trapping of infrared radiation occurs because the atmosphere absorbs part of the radiation emitted from the Earth's surface, and then re-emits a smaller stream of radiation owing to the fact that the atmosphere is colder than the Earth's surface. The glass enclosing a greenhouse will absorb close to 100% of the radiation emitted from within the greenhouse, but will re-emit a smaller amount of radiation to the sky if the outer surface of the glass is colder than the interior of the greenhouse.

80% of the cooling that would otherwise occur. This is referred to as *latent heat* transfer. The remaining 20% is accounted for by rising warm air and sinking cold air – the *sensible heat* flux.

Heat is also transferred horizontally – as sensible and latent heat in the atmosphere, and predominantly as sensible heat in the oceans. These transfers cause the equatorial regions to be cooler than they would otherwise be (by removing excess radiative energy) and cause the polar regions to be warmer than they would otherwise be. Changes in the rate of horizontal heat transfer can be an important source of regional climatic change.

Important heat transfers also occur between the surface layer and deeper layers of the ocean. There are three heat transfer processes: advection, diffusion, and convection. They are illustrated schematically in Fig. 2.2. Advection refers to the movement of a water (or air) parcel from one place to another. In nature, cold water sinks in polar regions and spreads throughout the world ocean, causing the deep ocean to be cold (0–2°C). This sinking is balanced by the upwelling of relatively warm water in low and middle latitudes. As explained in Box 2.2, this large-scale overturning is driven by horizontal variations in the density of water. The density of seawater in turn depends on its temperature and salinity, so the density-driven overturning is referred to as the *thermohaline overturning*. Although the water that upwells as part of the thermohaline overturning is colder than the surface water above it (and therefore cools the surface in the region where it upwells), it is warmer than the water that sinks, so the net effect is an upward heat transfer. However, because the ocean temperature decreases with increasing depth at most places, diffusive mixing – which always transfers heat from warm to cold – results in a downward heat transfer. Convection occurs when surface water becomes cold enough (or, in some cases, salty enough) for its density to become greater than that of the underlying water. When this happens, the surface water sinks and mixes with the underlying subsurface water. Since convection usually occurs in response to a cooling of the surface water, the water that sinks is generally colder than the water that rises, so that convection results in a net upward heat flux.

For an unchanging climate (in which deep ocean temperatures are also constant), the net heat flow to or from the deep ocean must be zero. However, if the surface temperature warms, a net heat flow into the deep ocean will arise which will tend to slow down the subsequent warming until the deep ocean has warmed up – a process that takes thousands of years to complete. As the surface warms, the downward diffusion of heat would increase, while the intensity of convection (which transports heat upward) would decrease, since convection is driven mainly by the occurrence of cold surface water. Conversely, if the intensity of ocean sinking and upwelling were to change, there would be a temporary net flow of heat between the deep ocean and surface which would drive a change in the surface climate. In particular, a reduction in overturning would lead to a temporary surface cooling because the upward heat transfer caused by overturning would be reduced.

Box 2.2 Explaining the thermohaline overturning

To understand how a horizontal variation in the density of sea water can drive a vertical overturning circulation cell, consider a basin with a removable separation in the centre, as illustrated in Fig. 2.2.1(a). Assume that high-density water occurs on one side of the barrier and low-density water on the other side. Pressure will increase faster with increasing depth in the high-density water than in the low-density water, since the pressure at any depth is equal to the weight of overlying water per unit area plus the atmospheric pressure. The contours for various constant-pressure surfaces will be as illustrated in Fig. 2.2.1(a), and result in a pressure gradient force that is directed from the high-density to the low-density water and increases with increasing depth. When the barrier is removed, water will flow from the high-density to the low-density side, thereby raising the water level on the low-density side and creating the pressure contour configuration shown in Fig. 2.2.1(b). This in turn would drive the circulation cell that is shown. As long as the density contrast can be maintained in spite of the homogenizing effect of the circulation, the circulation will continue.

In nature, the equator-to-pole decrease in temperature causes denser water to occur in polar regions. This drives a circulation cell with sinking at high latitudes and upwelling at low latitudes. Salinity tends to decrease towards polar regions, which would tend to make polar waters less dense. The salinity gradient therefore tends to oppose the temperature gradient. Abrupt changes in the intensity of the thermohaline circulation can occur as the balance between the driving force from the horizontal temperature gradient and the breaking force from the horizontal salinity gradient changes.

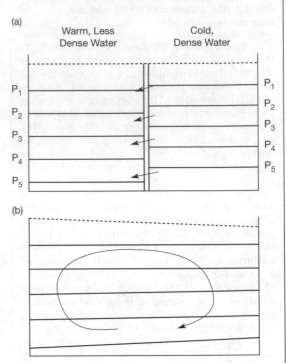

Fig. 2.2.1 (a) Distribution of constant-pressure surfaces in a tank with warm, low-density water on the left side and cold, high-density water on the right side. P₁, P₂, and so on refer to successively higher pressures. Pressure decreases more rapidly with depth in the cold water, creating a pressure gradient force that is directed from the dense to less dense water and increases in magnitude with increasing depth, as indicated by the arrows. (b) Orientation of the pressure surfaces and the resultant circulation after the barrier shown in (a) is removed.

2.3   The hydrological cycle

The hydrological cycle involves the evaporation and precipitation of water, transport of water vapour by the atmosphere, runoff from the continents to the oceans, and the net transfer of freshwater by ocean currents. As noted in Section 2.2, evaporation and precipitation account for some 80% of the required non-radiative energy transfer from the Earth's surface to the atmosphere. To the extent that condensation and the associated release of latent heat occur in different regions from where evaporation occurred, the hydrological cycle also effects important horizon-

tal energy transfers. Geographical variations in the difference between precipitation + runoff onto and evaporation from the oceans are responsible for creating regional variations in surface salinity which, as noted in Section 2.2, contribute to driving (or opposing) the ocean's thermohaline circulation. Finally, water vapour is the single most important GHG, and both its amount and distribution within the atmosphere are of critical importance to climate.

Figure 2.3 illustrates the hydrological cycle. Evaporation from the oceans and land surface is driven by the difference in vapour pressure between the surface and the immediately overlying air, but because evaporation tends to cool the surface and

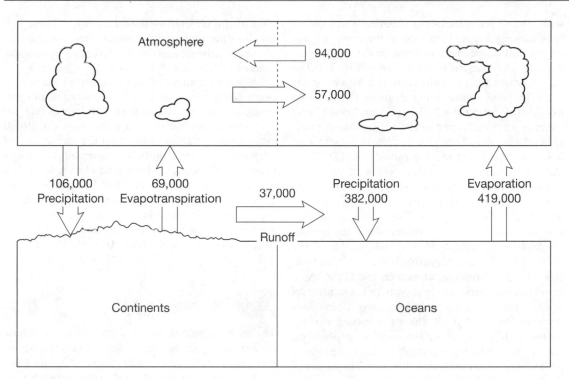

Fig. 2.3 The global hydrological cycle, showing fluxes in Gt (billions of tonnes) of H_2O. Flux estimates were derived from Christopherson (1992, his Fig. 7–10).

thereby reduce the surface vapour pressure, a steady source of heat energy is required to power the evaporation process. Water vapour is transported upward and horizontally by atmospheric winds and through widespread turbulent mixing. Concentrated upward transport also occurs as condensed liquid water in convective clouds; the condensate may fall back to the surface as rain, might partially re-evaporate as falling rain, or might re-evaporate near the cloud top as the air flow detrains and mixes with drier surrounding air. In the latter case, the cloud serves as a conduit for pumping water vapour from near the surface to the middle or upper troposphere, where it is more effective in reducing the emission of infrared radiation to space (for reasons explained in Section 3.1). Subsidence of relatively dry, upper tropospheric air between cumulus clouds tends to push water vapour in the boundary layer (the lowest 1–2 km of the atmosphere) closer to the surface.

Once it reaches the surface, some rainfall is intercepted by vegetation and can easily re-evaporate, and the remainder is partitioned between infiltration into the soil and runoff. The nature of the vegetative cover, the intensity of the rainfall, the soil type and antecedent soil moisture conditions, and the local land surface slope all influence the partitioning of rainfall into these three components. When precipitation falls as snow, a large amount of moisture can accumulate, to be released within a relatively short period of time during the spring snowmelt period. The timing of this release affects the availability of soil moisture near the end of summer; earlier release can result in soil moisture becoming depleted before the end of the summer.

Distinctly different processes govern the vertical distribution and amount of water vapour in the atmosphere at middle-to-high and at low latitudes. At middle and high latitudes, water vapour is transferred vertically by large-scale winds, associated primarily with travelling storm systems. The humidity in the free troposphere is closely coupled to that of the atmospheric layer next to the Earth's surface (the so-called boundary layer). At low latitudes (30°S–30°N), the atmospheric circulation is dominated by the Hadley cells. The Hadley cells consist of a flow of surface air toward the equator (or thereabouts) in both hemispheres, rising motion with intense precipitation in the zone where the air flow converges, poleward flow in the upper troposphere, and descending motion in the subtropical regions. This pattern is illustrated in Fig. 2.4. The region where the two cells converge, known as the Inter-Tropical Convergence Zone or ITCZ, shifts northward during the NH (northern hemisphere) summer and southward in the SH (southern hemisphere) summer. This can be clearly seen in Fig. 2.4. The descending motion outside the ITCZ creates an inversion (a layer where temperature increases with increasing height) just above the boundary layer that suppresses mixing with the underlying boundary layer. The water vapour content of much of the atmosphere above the boundary layer in the subtropics is related to that of the very cold and hence very dry air that detrains from the tops of convective clouds in the ITCZ. As a result, atmospheric water vapour in the subtropics is decoupled from the underlying, moist boundary layer. In the ITCZ itself, the water vapour content of the middle and upper troposphere depends on the extent to which ice crystals from the detraining air are able to sublimate as they fall. The extent of detrainment varies with the intensity of convection and hence with the moisture content and temperature of the boundary layer. Some descending motion occurs in the ITCZ in the region between convective clouds (which cover only a few percent of the area), and the water vapour content of the descending air is tied to the moisture content of the air that detrains from the nearby convective columns. These in turn are coupled to the local atmospheric boundary layer. This vertical zonation, with a turbulent boundary layer, an upper troposphere dominated by detrainment from convective columns, and a middle troposphere dominated by subsiding air, is illustrated in Fig. 2.5.

To summarize, the moisture content of the middle and upper troposphere in the ITCZ is indirectly coupled to the moisture content of the local (nearby) boundary layer, since the boundary layer provides the moisture source for the convective columns. The convective columns cover only a few percent of the ITCZ but feed the air that descends between the columns. Outside the ITCZ but still in the Hadley cell region, the middle and upper tropospheric humidity is not directly related to the local boundary layer. Finally, the middle and upper tropospheric humidity in middle and high latitudes is directly coupled to the underlying boundary layer. These distinctions are potentially of great importance to the change in climate caused by increasing concentrations of GHGs, as will be discussed in Chapter 9.

The entire amount of water vapour in the atmosphere is completely replaced by evaporation and precipitation once every eight days. This implies that, for averaging times longer than a few days, the global rate of precipitation must equal the global rate of evaporation since, otherwise, the amount of water vapour in the atmosphere would change in such a way as to bring the two rates into balance. Since the evaporation rate is directly related to surface temperature, this short turnover time also implies that the amount of water vapour in the atmosphere responds almost instantaneously to changes in the climate.

2.4   The carbon cycle

The carbon cycle is critically important to climate because it regulates the amount of two important greenhouse gases in the atmosphere: CO_2 and CH_4. Carbon, like water, continuously cycles between various "reservoirs" or temporary holding areas, but this cycling is considerably more complicated for CO_2 than for water. First, carbon occurs in quite different chemical forms in the different reservoirs. Second, the transfer processes from one reservoir to another are considerably more complicated and varied than for water. Third, whereas the amount of water vapour in the atmosphere is directly and instantly driven by changes in the climate, the CO_2 concentration in the atmosphere responds to changes in the climate on time scales ranging from a few months to thousands of years.

Carbon reservoirs and fluxes

Figure 2.6 illustrates the major carbon reservoirs as a series of interconnected boxes. The number in each box is the estimated amount of carbon in the reservoir in units of gigatonnes (Gt, or billions of tonnes) prior to human disturbance. Carbon occurs in the atmosphere primarily as CO_2 and to a much smaller extent as CH_4; in the living biota and soils as organic matter; and in the oceans primarily as dissolved CO_2, HCO_3^- (bicarbonate ion), and CO_3^{2-} (carbonate ion). The CO_2, HCO_3^-, and CO_3^{2-} are collectively

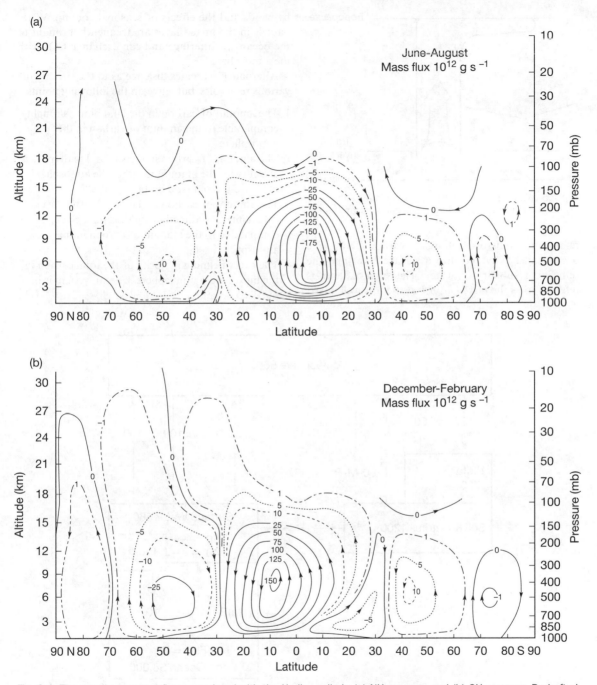

Fig. 2.4 The zonally averaged flow associated with the Hadley cells in (a) NH summer, and (b) SH summer. Redrafted from Newell (1978).

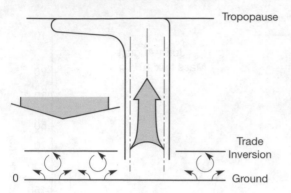

Fig. 2.5 Water transfer processes associated with cumulus convection. Reproduced from Sun and Lindzen (1993).

referred to as dissolved inorganic carbon (DIC). The ocean can be divided into a surface layer and the remaining subsurface water. The surface layer is referred to as the *mixed layer* because it is well mixed by winds and the effects of seasonal cooling. Water parcels in the mixed layer are frequently brought to the ocean–air interface and can exchange CO_2 with the atmosphere.

Alternative estimates disagree as to the sizes of the various reservoirs, but agree on the following points:

1 The amount of carbon in the land biota is roughly comparable to the amount of carbon in the atmosphere.
2 The amount of carbon in the soil and in detritus is about twice the amount in either the atmosphere or above-ground land biota.
3 The amount of carbon in the ocean mixed layer, which interacts directly with the atmosphere, is comparable to the amount of carbon in the atmosphere itself.
4 The overwhelming majority of the total carbon in the biosphere + atmosphere + ocean system is in the deep ocean.

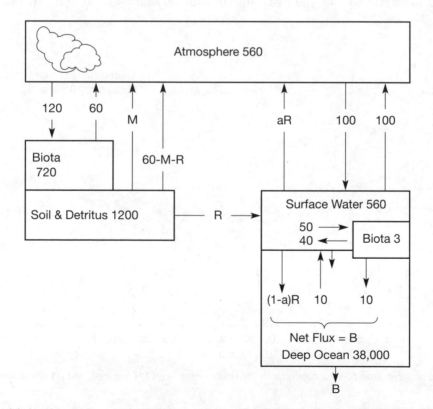

Fig. 2.6 The global carbon cycle, as compiled from a wide range of sources. Numbers in boxes are the amounts of carbon in each reservoir as Gt C, while numbers next to arrows between boxes are the annual rates of carbon transfer in Gt C. M = methane flux, R = riverine flux of particulate carbon, and B = burial flux.

The ocean can be thought of as a giant flywheel in the carbon cycle, while the atmosphere contains a comparatively small amount of carbon. Thus, a small percentage loss of carbon from the ocean to the atmosphere would translate into a large increase in the amount of carbon in the atmosphere.

Also shown in Fig. 2.6, as numbers next to the arrows between the boxes, are the rates of carbon transfer (fluxes) in units of Gt per year. Carbon dioxide is incorporated into living plant material through photosynthesis and released to the atmosphere through respiration (the process by which organic matter is "burned" to provide energy, either by the plant itself, or by organisms that consume and thereby decompose organic litter). Some of the biospheric carbon is returned to the atmosphere as CH_4 rather than CO_2. This is represented by the flux "M" in Fig. 2.6, which is probably on the order of 0.2 Gt C per year. This flux is broken down into various biospheric sources in Table 2.2, where it is seen that the major sources are emissions from wetlands and termites. The CH_4 is largely oxidized to CO_2 within the atmosphere after about 10 years on average. Some of the biospheric carbon is also delivered to the oceans by rivers (the flux "R" in Fig. 2.6), where it either is oxidized and returned to the atmosphere as CO_2, or sinks into the deep ocean. This flux is probably on the order of 0.5 Gt C per year. Carbon is transferred between the atmosphere and the ocean mixed layer through diffusion of gaseous CO_2 across the air–sea interface. When the carbon cycle is in a balanced state, the total flux into each reservoir will equal the total flux out of the reservoir. However, because of the return flow of riverine carbon, the net CO_2 flux between the atmosphere and ocean was not zero even prior to human perturbations. Photosynthesis, respiration, and related processes also occur in the ocean, as explained more fully

Table 2.2 Estimated emissions of methane from the terrestrial biosphere and oceans (natural emissions), taken from Lelieveld *et al.* (1998). $1Tg = 10^{12}$ gm.

Source	Emission (Tg per year)
Wetlands	145 ± 30
Termites	20 ± 20
Oceans	10 ± 5
Wild ruminants	5 ± 5
Freshwater	5 ± 5
CH_4 from sediments	5 ± 5
Total natural	190 ± 70

below. There is a small net flux into the deep ocean that, for a balanced system, will be equal to the rate of burial of carbon. Over geological time scales, the burial flux will be balanced by a variety of processes, including volcanic outgassing of CO_2. These processes are not shown in Fig. 2.6.

Atmosphere–ocean mixed layer–deep ocean exchange processes

When CO_2 enters sea water, the following chemical reactions take place:

$$CO_2 \text{ (gas)} + H_2O \text{ (liquid)} \rightarrow H_2CO_{3(aq)} \quad \text{(carbonic acid)} \quad (2.1)$$

$$H_2CO_3 \rightarrow H^+ + HCO_3^- \quad (2.2)$$

$$CO_3^{2-} + H^+ \rightarrow HCO_3^- \quad (2.3)$$

giving the net reaction:

$$H_2O + CO_2 + CO_3^{2-} \rightarrow 2HCO_3^- \quad (2.4)$$

About 90% of the inorganic carbon in the oceans is in the form of HCO_3^-, about 10% is in the form of CO_3^{2-}, and less than 1% is in the form of CO_2. The transfer of CO_2 between the atmosphere and ocean is driven by the difference in CO_2 partial pressures (pCO_2). If, for example, the pCO_2 in the atmosphere is greater than that of the ocean surface, there will be a net flow of CO_2 into the ocean. The fact that only 1% of oceanic carbon is in the form of CO_2 means that the pCO_2 in the surface water is much smaller than it would be otherwise. This in turn allows the ocean to hold a large amount of carbon without creating a large back-pressure that would drive CO_2 back into the atmosphere.

In the surface layer of the ocean, two biologically driven processes of importance occur: the construction of soft organic tissues by photosynthesis, which can be represented by the reaction

$$CO_2 + H_2O \rightarrow CH_2O + O_2 + \quad (2.5)$$

and the construction of calcareous skeletons, through the net reaction

$$Ca^{2+} + 2HCO_3^- \rightarrow CaCO_3 + H_2O + CO_2 \quad (2.6)$$

Photosynthesis in the oceans is estimated to take up about 50 Gt C per year. Most of this goes into short-lived micro-organisms, and is rapidly returned to ocean surface water when these organisms die or are eaten and "burned" by larger organisms. However, some of the soft tissue and skeletal material produced in the mixed layer ends up in the deep ocean through sinking dead micro-organisms. The combined flux is crudely estimated to be about $10\,Gt\,C$ per year, and is referred to as the *biological pump*. About 60–80% of this flux is due to falling soft tissue, and the remainder is due to falling $CaCO_3$ particles. These and other processes by which carbon is transferred between the surface layer and deep ocean are illustrated in Fig. 2.7. Most of the organic tissue decomposes in the upper 1km of the ocean, while most of the $CaCO_3$ dissolves below a depth of 3–4km. The net result is to deplete the surface water of nutrients (primarily phosphate and nitrate) and DIC, to create maximum nutrient con-

centrations around a depth of 1 km, and to create a maximum DIC concentration at a somewhat greater depth in the ocean. This is illustrated in Figure 2.8, which shows the globally averaged profiles of phosphate (PO_4^{3-}) and DIC in the world ocean. Biological activity transfers DIC from areas of low DIC (the surface water) to areas of large DIC (the deep water), which is against the concentration gradient. This transfer is therefore referred to as the "biological pump", since pumps are devices that transfer air or liquid against a pressure gradient.

Because of the sharp increase in the DIC concentration with increasing depth in the upper ocean, turbulent mixing (diffusion) tends to transfer DIC (and

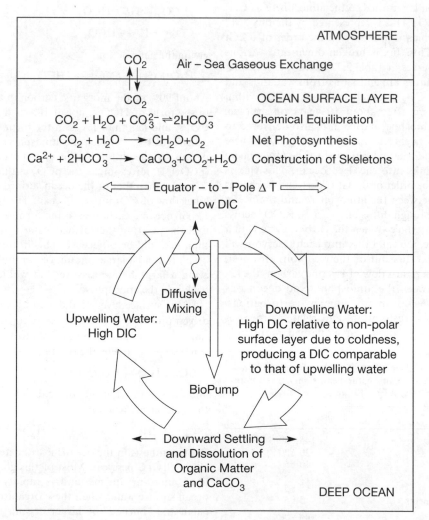

Fig. 2.7 Processes that transfer carbon between the atmosphere, the ocean mixed layer, and the deep ocean. Reproduced from Harvey (1996a).

nutrients) upward. This is the main process that balances the downward transfer of DIC by the biological pump. Convection also causes an upward flow of DIC, since the water that sinks tends to have a lower DIC concentration than the water that rises. One might expect that advection would also transfer DIC upward, since the concentration of DIC in the surface layer – which feeds the downwelling water in polar regions – is much less than in the water that upwells from below. However, the solubility of DIC in seawater increases with decreasing temperature, and as warm surface waters flow poleward they are able to absorb more CO_2 from the atmosphere as they cool. As a result, the DIC concentration in sinking water is about 8% greater than the global average surface concentration and, it turns out, very close to that in the water that upwells at lower latitudes. Consequently, the net vertical transfer of DIC due to advective overturning is rather small. However, the effect of the solubility variation alone is to cause a downward transfer of DIC, and this downward transfer is referred to as the "solubility pump".

We can now see that there are two reasons why the atmospheric CO_2 concentration is so low, and why most of the carbon is in the oceans. The first reason is related to the chemistry of DIC and the fact that most of the oceanic carbon does not occur as dissolved CO_2, as discussed above. The second reason is that the DIC concentration is comparatively low in the surface mixed layer but high in the deeper ocean, combined with the fact that the atmosphere is in direct contact only with the mixed layer but is largely isolated from the deep ocean. The mixing processes within the mixed layer are quite intense, but the mixing or exchange of DIC between the mixed layer and the deeper water is much slower. This is why the biological pump is able to maintain a low DIC concentration in the mixed layer in spite of much greater concentrations immediately below. The low DIC concentration in the surface water combined with the chemistry of DIC results in a low pCO_2 of surface water, and since the atmospheric pCO_2 tends to adjust to the pCO_2 of the water with which it is in contact, the atmospheric pCO_2 is also low.

Weathering and formation of carbonate sediments

The weathering of rocks on land removes carbon dioxide from the atmosphere. Rocks can be divided into two broad categories based on chemical composition: *silicates,* which contain minerals built around the elements silicon (Si) and oxygen (O),

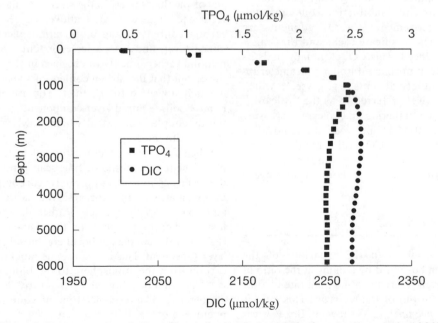

Fig. 2.8 The globally averaged vertical profile of dissolved phosphate (from Levitus *et al.*, 1993) and dissolved inorganic carbon (from Takahashi *et al.*, 1981) in the ocean.

and *carbonates,* which contain minerals which are built around CO_3^{2-}. Most chemical weathering at the Earth's surface occurs through reaction with carbonic acid (H_2CO_3), which forms when CO_2 dissolves in H_2O. The net weathering reactions of carbonate and silicate rocks can be represented schematically as

$$CaCO_3 + H_2O + CO_2 \rightarrow Ca^{2+} + 2HCO_3^- \quad (2.7)$$

for carbonate rocks, and

$$CaSiO_3 + 2CO_2 + H_2O \rightarrow Ca_2^+ + 2HCO_3^- + SiO_2 \quad (2.8)$$

for silicate rocks. Two CO_2 molecules are removed from the atmosphere for each Ca^{2+} released by the weathering of silicates, and one CO_2 is removed for each Ca^{2+} released by the weathering of carbonates. All of the Ca^{2+} delivered to the oceans, whether from the weathering of carbonates or silicates, is available to form $CaCO_3$ through the reverse of reaction (2.7). This reaction adds only one CO_2 to the atmosphere + ocean system for each Ca^{2+} that is precipitated. If the Ca^{2+} came from the weathering of a carbonate rock, then the permanent burial of $CaCO_3$ has no net effect on the amount of carbon in the atmosphere + ocean system. However, when weathering of silicate rocks is followed by the burial of $CaCO_3$, there has been a net conversion from silicate to carbonate rocks and a net removal of CO_2 from the atmosphere + ocean system. However, as ocean sediments enter subduction zones through plate tectonic movements, $CaCO_3$ sediments are transformed back to silicate rocks and the "missing" CO_2 is produced. This CO_2 eventually finds its way back to the atmosphere through volcanic eruptions and hydrothermal vents (e.g. geysers), which are often collectively referred to as the "volcanic" source. Since the weathering and volcanic degassing fluxes are about 0.2 Gt C per year, the removal of CO_2 by weathering + $CaCO_3$ sedimentation, and its restoration from volcanic degassing, completely replaces the atmospheric CO_2 content once every 3000 years or so.

Response time scales

Considerable insight into the behaviour of the carbon cycle can be gained by looking at the ratio of the size of a given reservoir to the total rate of flow of carbon into or out of the reservoir. This ratio is called the *residence time*, τ. As shown in Box 2.3, τ is the average length of time that a C atom spends in a reservoir before being transferred to another reservoir; it is also the length of time required, on average, to completely replace all the C atoms in a given reservoir. From Fig. 2.6 it is seen that, based on the fluxes between the atmosphere, land biota, and soil carbon and detritus, the atmospheric reservoir of carbon is completely replaced about once every 5 years on average (τ_{atm} = 5 years). The fluxes between the atmosphere and ocean mixed layer are also sufficient to completely replace the atmospheric CO_2 once every 5 years. Photosynthesis is sufficient to completely replace the land biota once every 6 years on average, giving τ_{biota} = 6 years. About 60 Gt carbon enter and leave the soil and detritus pool every year, and since its size is about 1200 Gt, the mean residence time for this reservoir, τ_{soil}, is about 20 years. The oceanic mixed layer exchanges carbon with both the atmosphere and deep ocean; exchange with the atmosphere is sufficient to completely replace the surface layer carbon about once every 6 years, whereas exchange with the deep ocean is sufficient to completely replace surface layer carbon about once every 50 years. On the other hand, the deep ocean carbon is completely replaced only once every 3000 years or so.

As shown in Box 2.3, the residence time gives the characteristic time scale with which the carbon content of a given reservoir will adjust to a change in any of the fluxes between the reservoirs. Based on the residence times estimated above, we see that the carbon content of the biota, atmosphere, ocean, mixed layer, and – to a lesser extent – soils, will respond rather quickly to changes in the inter-box fluxes, but that the carbon content of the deep ocean is much slower to fully adjust. The atmosphere + biota + soils + mixed layer components thus form a tightly coupled subsystem which slowly exchanges carbon with the deep ocean.

When fossil fuel carbon is added to the atmosphere, the relevant response time scale is not given by the residence time of atmospheric carbon based on the exchange with the other reservoirs with which the atmosphere rapidly interacts. Rather, the response to fossil fuel carbon is given by the residence time of carbon in the coupled atmosphere–biosphere–mixed layer subsystem. This is because the rapid transfer of carbon from the atmosphere to the biota or mixed layer is quickly followed by a return flow to the atmosphere. The residence time of coupled atmosphere + biota + soils + mixed layer subsystem is given by the total mass of carbon in the subsystem

Box 2.3 Residence time and adjustment time

The ratio of the size of a carbon reservoir to the total carbon flux in or out of the reservoir gives the average turnover time of carbon in the reservoir. Here, we show that, if the reservoir is thoroughly mixed and the outflow varies in direct proportion to the size of the reservoir, then this ratio also represents both the average lifetime of molecules in a reservoir and the time required for the number of molecules originally in the reservoir at time $t = 0$ to decrease to e^{-1} of the number originally present.

If the flux of CO_2 out of a reservoir is linearly proportional to the number of molecules in the reservoir, we can write

$$\frac{dN}{dt} = -kN \qquad (2.3.1)$$

where N is the size of the reservoir. For the present atmosphere, $N = 770$ Gt and dN/dt due to photosynthesis is 110 Gt yr^{-1} for present conditions, so $k = 0.14$ yr^{-1}. It is convenient to introduce a variable τ, given by $1/k$. Then τ (or $1/k$) is none other than the turnover time (flux N divided by throughput dN/dt). Integrating Eq. (2.3.1) gives

$$N(t) = N_0 e^{-kt} = N_0 e^{-t/\tau} \qquad (2.3.2)$$

which gives the decrease in the number of molecules originally present in the atmosphere at time $t = 0$, considering only removal due to photosynthesis. When $t = \tau$, N equals $1/e$ times N_0.

The average time $\bar{t}$ spent by a molecule in the reservoir is given by the number of molecules, $N(t)$, in each small range of lifespan dt, summed over all the possible lifespans and divided by the total number of molecules originally present. That is,

$$\bar{t} = \frac{1}{N_0} \int_0^\infty N_0 e^{-t/\tau} dt = \tau \qquad (2.3.3)$$

In sum, for well-mixed reservoirs where the rate of removal depends on the amount of material present, dividing the steady-state mass of the reservoir by the steady-state outflow gives the amount of time required to completely replace the mass originally present. This is known as the turnover time. The turnover time is equal to the time constant for the exponential decrease in the number of atoms originally present, and also equals the average length of time spent by an atom in the reservoir.

The preceding analysis assumes that the average lifespan τ is constant. However, in the case of CH_4, emissions of the gas into the atmosphere increase the atmospheric lifespan of all the CH_4 already present. As a result, a *perturbation* in the atmospheric concentration of CH_4 decreases more slowly than the exponential decrease given by Eq. (2.3.2) using the unperturbed value of τ. This slower decrease is referred to as the *adjustment time*, and is about 20–60% longer than the residence time (Section 7.5).

(about 3100 Gt) divided by the rate of exchange with the deep ocean (about 10 Gt C but highly uncertain). The result is a residence time of about 300 years. This is a very crude representation of highly complex processes, which are discussed in more detail in Chapter 8, but serves to illustrate in an intuitively simple manner how two very different response time scales for atmospheric CO_2 (5–6 years and 300 years) can arise and how we can get a feeling for what the magnitude of the response time scales should be.

2.5   The sulphur cycle

The sulphur cycle involves the emission of sulphur(s) from land plants, the oceans, and volcanoes in many different chemical forms (Bates *et al.*, 1992). The main climatic effect of S is through the formation of

sulphate (SO_4^{2-}) aerosols. These aerosols directly affect climate by reflecting a portion of the incoming solar radiation back to space. They have other, much less certain, effects through their role as cloud condensation nuclei (CCN). The greater the availability of CCN, the easier it is to form cloud droplets. For a given amount of liquid water, a greater concentration of cloud droplets results in smaller cloud droplets, which in turn increases the cloud reflectivity. The linkage between cloud droplet size and cloud reflectivity is explained in Box 2.4. In addition, the reduced size of the cloud droplets may very well increase the cloud lifespan and reduce precipitation rates.

Figure 2.9 shows the components of the sulphur cycle and the associated fluxes that directly involve sulphate. Biological processes on land result in the emission of S primarily as H_2S (hydrogen sulphide), while marine phytoplankton emit S as $S(CH_3)_2$ (dimethylsulphide or DMS). The vast majority of

Box 2.4 Effect of cloud droplet size on cloud reflectivity

When light passes from one medium to another, such as from air to water or from water back to air, it bends or changes direction as it crosses the interface between the two media. This process is called *refraction*. As light travels through a water droplet, it will be refracted twice – once as it enters the droplet, and once as it exits from the droplet. This is the process by which light is scattered by cloud droplets; repeated encounters with cloud droplets can completely turn the light around. In between the two refraction events is an opportunity for partial absorption of light by the droplet. The larger the droplets, the greater the absorption relative to scattering. Thus, if an increase in the concentration of cloud concentration nuclei leads to an increase in the number of cloud droplets with no change in the cloud liquid water content, then the average droplet size will be smaller. This will make the cloud more reflective and less absorptive, which has a cooling effect on climate.

the biogenic S emissions to the atmosphere occur as DMS. Hydrogen sulphide is oxidized by reaction with the hydroxyl radical (OH) to SO_2, while DMS can be oxidized to either methanosulphuric acid (MSA, CH_3SO_3H) or SO_2. About 80% of the SO_2 is oxidized to SO_4^{2-} (sulphate) under present-day conditions, while the balance is deposited at the Earth's surface (Roelofs *et al.*, 1998). The sulphate forms aerosol particles. The transformation of SO_2 to SO_4^{2-} can occur in the gas phase or inside cloud droplets or sea salt water aerosols, and all three routes are indicated in Fig. 2.9. Interaction with sea salt aerosols can account for roughly 20–80% of the oxidation of SO_2 to SO_4^{2-} (Suhre *et al.*, 1995). About 60% of the oxidation of SO_2 to SO_4^{2-} is carried out by H_2O_2 (hydrogen peroxide), with the balance split roughly equally between OH and O_3 (Roelofs *et al.*, 1998). Sulphur is delivered to the ocean in rainwater and in river water as SO_4^{2-}, and occurs in ocean water overwhelmingly as SO_4^{2-}. Oceanic SO_4^{2-} can be directly released to the atmosphere as a component of sea salt aerosols that are formed when air bubbles at the ocean surface burst. The bursting air

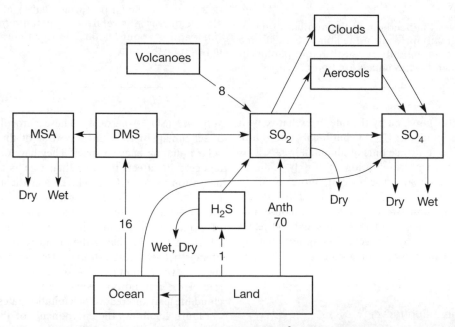

Fig. 2.9 Components of the global sulphur cycle involving sulphate (SO_4^{2-}). 'Dry' and 'Wet' refer to dry and wet deposition, respectively; MSA = methanosulphuric acid; DMS = dimethylsulphide; H_2S = hydrogen sulphide. The numbers next to arrows between boxes are the approximate annual rates of sulphur transfer in Tg S. Based on Schlesinger (1997), Charlson *et al.* (1992), and other sources.

bubbles produce film drops from the disintegration of the bubble film, and produce jet drops from the collapse of the bubble cavity, as illustrated in Fig. 2.10. As can be seen from Table 2.1, sea salt aerosols are thought to be the second most abundant aerosol type in the atmosphere, after mineral dust.

Anthropogenic emissions of S occur largely as SO_2. The annual anthropogenic emission in the late 1980s is estimated to have been around 70 Tg S per year (Tg = teragram, or 10^{12} g), about nine times the volcanic SO_2 emission of 8 Tg S per year and about three times the total natural S emission of 25 Tg S per year. There are sizeable uncertainties associated with all of these emission estimates. A significant fraction of the sulphate produced from oxidation of SO_2 (whether directly emitted by human activities or a product of natural DMS emissions) becomes incorporated into sea salt aerosols (O'Dowd et al., 1997). In this way, sulphate–sea salt interaction might significantly modulate the effect of anthropogenic (and natural) SO_2 emissions on cloud properties. The production of sea salt aerosols depends on the wind strength at the ocean surface, and could therefore change as the climate changes.

The major removal process for sulphate aerosols is through rainout. Since this removal process is very effective, and water vapour itself cycles through the atmosphere once every 8 days on average, the average lifespan of sulphate aerosol in the atmosphere is also only a few days (4.7 according to Roelofs et al., 1998). This implies, first, that aerosol concentrations (and their climatic effects) will respond essentially instantaneously to changes in emissions, and second,

that considerable regional variations in the sulphate aerosol concentration are possible, since the atmospheric lifespan is short compared to the length of time required for long-distance transport.

2.6 The nitrogen cycle

Nitrogen is an essential element for life, and it is important to climate through its effect on the rate of photosynthesis on land and in the ocean, through the greenhouse effect of N_2O (nitrous oxide), through its role in the chemistry of ozone in both the stratosphere and troposphere, and as a likely source of CCN. Nitrogen occurs in the atmosphere overwhelmingly as molecular nitrogen (N_2, 3.9×10^9 Tg), but also as N_2O (1.3×10^3 Tg N), as NO + NO_2 (nitric oxide and nitrite, on the order of 1 Tg N but highly uncertain), as NH_3 (ammonia, about 1 Tg N), and as NH_4^+ (ammonium) and NO_3^- (nitrate) aerosols (about 2 Tg N) (Jaffe, 1992). There are several abiotic subcycles involving nitrogen, but here we shall focus on the biological cycling of nitrogen. This cycle is illustrated in Fig. 2.11 and discussed below.

The atmosphere is 78% N_2, but nitrogen in this form cannot be used by plants. Rather, nitrogen must first be converted to NH_4^+ by a process called *nitrogen fixation*, which is carried out by several types of bacteria and by blue-green algae. Some of these nitrogen fixers exist freely in soils (asymbiotic), while others form symbiotic relationships with the roots of higher plants. Once converted to NH_4^+, nitrogen can be taken up by plants through one of two routes: by direct absorption of ammonium, or through *nitrification* – in which bacteria convert NH_4^+ to nitrite (NO_2) and nitrate (NO_3^-) – followed by absorption of nitrate and its assimilation into organic matter.

A number of processes occur during the decomposition of organic matter. *Mineralization* refers to the processes that release nutrients in inorganic forms. In the case of N, the mineralization process is *ammonification,* which converts organic N to NH_4^+. Most mineralization reactions are the result of enzymes excreted by soil microbes. Microbes are very efficient in absorbing the nutrients that are released by mineralization, resulting in *immobilization* of the nutrients. Microbes not only sequester nutrients from decomposing organic matter, but can

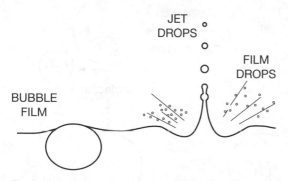

JET
DROPS

FILM
DROPS

BUBBLE
FILM

Fig. 2.10 Processes by which a bursting air bubble at the ocean surface injects sea salt aerosols into the atmosphere. Reproduced from Blanchard (1983).

also take nutrients from the soil solution that would otherwise be available for plant growth. Much of the mineralized N found in the soil is released from dead microbes, not directly from soil organic matter. Some of the NH_4^+ that is released by mineralization and not taken up by plants or microbes is converted to NO_3^- as a result of nitrification, and some diffuses to the atmosphere as NH_3 (ammonia), where it is oxidized to NO and N_2O (nitrous oxide). The NO_3^- produced from mineralized NH_4^+ can be taken up by plants or microbes, or converted back to N_2 in anaerobic environments through a process called *denitrification*. During both nitrification and denitrification, a small amount of N_2O is formed. N_2O can diffuse to the stratosphere, where it catalyses the destruction of O_3 **and is destroyed by ultraviolet radiation**. Like ozone, N_2O is a greenhouse gas, so its release to the atmosphere directly affects climate, as well as indirectly through its effect on ozone.

Nitrogen fixation in the sea is less intense than on land, amounting to about 40 Tg N/yr globally (compared to about 150 Tg N/yr on land). Other important sources of nitrogen to the ocean are river inputs and washout (estimated by Schlesinger (1997) at 36 Tg N/yr and 30 Tg N/yr, respectively). The main process removing nitrogen from the ocean is denitrification. N_2O is produced in the ocean and diffuses to the atmosphere in association with both nitrification and denitrification. In most regions of the ocean, nitrate is not measurable in the surface water, suggesting that nitrogen is a limiting nutrient for biological productivity in the oceans (Schlesinger, 1997). However, the oceanic component of the N cycle is poorly understood. There has been considerable uncertainty concerning the role of the oceans as a source or sink of N_2O (Jaffe, 1992), although recent work suggests that the oceans are a source of about 4 Tg N/yr (Nevison *et al.,* 1995).

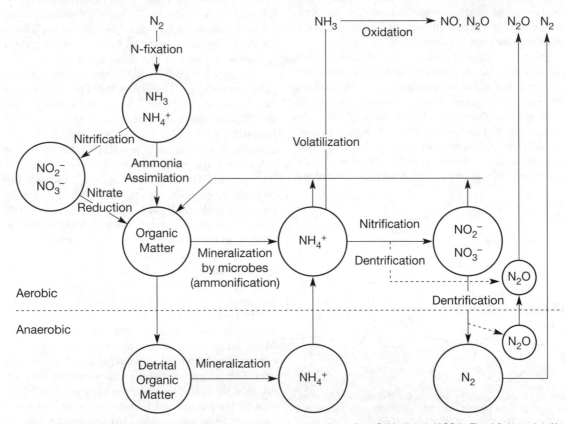

Fig. 2.11 Biological processes involved in the global nitrogen cycle, based on Schlesinger (1991, Fig. 12.1) and Jaffe (1992, Fig. 12.1).

The nitrogen oxides NO and NO_2 are produced naturally by lightning and from natural fires. The oxidation of NO_2 produces HNO_3 (nitric acid), which can then react with NH_3 to form the ammonium nitrate (NH_4NO_3) aerosol. Ammonium nitrate is likely to be an important source of cloud condensation nuclei (Jaffe, 1992).

Humans have perturbed the N cycle through direct emissions of NO_x (the sum of NO and NO_2) and N_2O, and through the production of nitrogen fertilizers. The components and fluxes in the nitrogen cycle involving or affecting N_2O are illustrated in Fig. 2.12. Natural nitrogen fixation on land is estimated to be about 150 Tg N/yr, with another 40 Tg N/yr due to the production of fertilizers. Some of the fixed nitrogen is returned to the atmosphere as N_2O through the process of denitrification, and the production of fertilizers may have increased the rate of production of N_2O through denitrification by 2–5 Tg N/yr. Industrial processes and biomass burning are estimated to add an additional 1–3 Tg N/yr as N_2O (a more detailed breakdown of anthropogenic emissions of N_2O is given in Chapter 4). As noted above, N_2O is destroyed when it reaches the strato-sphere.

There might be additional, oceanic and soil, sinks for N_2O, but these are not shown in Fig. 2.12.

The combustion of fossil fuels also adds about 40 Tg N/yr to the atmosphere as NO_x, much of which is redeposited on land as a component of acid rain. This anthropogenic N input could have stimulated terrestrial photosynthesis enough to remove an extra 0.5–1.0 Gt C per year from the atmosphere, as discussed later in Section 8.5. However, excessive inputs of N can lead to the loss of mycorrhizal fungi, thereby exacerbating P deficiency and reducing long-term productivity.

The uncertainty concerning the role of the ocean in the N_2O cycle is important to the projection of future N_2O trends. If there is a natural source of N_2O substantially larger than the currently estimated natural source of about 10 Tg N/yr, then the direct + indirect anthropogenic contribution of about 5 Tg N/yr is small and there is a large unidentified sink. In this case, future increases in N_2O concentration will be comparatively small. However, if the natural sources and sinks are around 10 Tg N/yr (as appears to be the case), then the anthropogenic input is important (Jaffe, 1992).

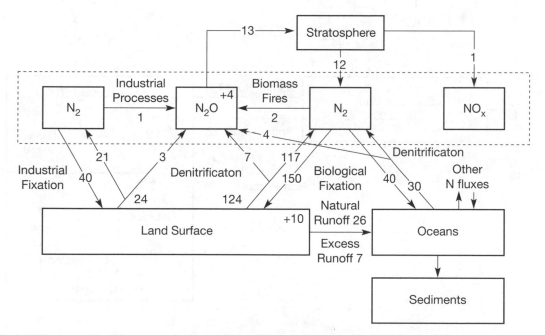

Fig. 2.12 Components of the global nitrogen cycle involving N_2O. The atmospheric N_2 reservoir is shown as two boxes so that the natural and anthropogenic fixation (and induced denitrification) fluxes can both be shown without cluttering the diagram. The numbers next to arrows between boxes are the annual rates of nitrogen transfer in Tg N, while numbers in boxes (where present) indicate the rate of change (Tg N per year) in the size of the reservoir. Based on Jaffe (1992) and Schlesinger (1997), with some fluxes adjusted so that the natural cycle is balanced.

2.7   The phosphorus cycle

The global phosphorus cycle is illustrated in Fig. 2.13. Unlike carbon, nitrogen, or sulphur, the exchanges between the atmosphere and either the terrestrial biosphere or the oceans are very small compared to other phosphorus fluxes. Phosphorus is released from rocks through chemical weathering on land and dissolved into the soil water. It occurs as either H_3PO_4, $H_2PO_4^-$, HPO_4^{2-}, or PO_4^{3-} (phosphate), with the proportions in each form depending on the pH of the soil water. These forms of phosphorus are collectively referred to as dissolved inorganic phosphorus (DIP), and are analogous to the multiple forms of dissolved inorganic carbon (DIC) discussed earlier. Some DIP is delivered to the oceans by rivers, some is taken up by plants and converted to organic phosphorus (that is, it becomes part of the plant organic matter), and the balance reacts with other minerals to produce

forms of phosphorus that are not available to plants. When plant material decays, the organic phosphorus can be converted back to DIP and reused by plants. Some of the dead organic matter is carried by rivers to the oceans as small particles – forming particulate organic phosphorus (POP). Marine photosynthesis and decay also produce POP. As terrestrial and marine POP settle into the deep ocean, some will be consumed along the way, while the rest will be buried and eventually become part of new rocks, thereby completing the cycle.

Both nitrogen and phosphorus appear to be limiting nutrients for marine biological productivity. Phosphorus is also likely to be a limiting nutrient for biological productivity in tropical terrestrial ecosystems. The marine phosphorus cycle stands in contrast to the marine nitrogen cycle in that atmospheric inputs are less important for phosphorus and there are no gaseous losses, so that river inputs are largely balanced (over long time scales) by permanent burial.

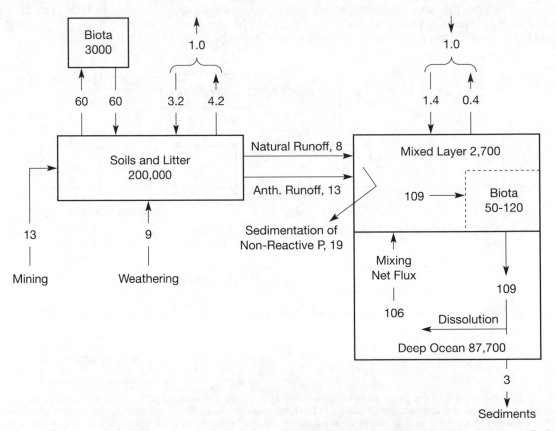

Fig. 2.13 The global phosphorus cycle. Numbers in boxes are the amounts of phosphorus in each reservoir as Tg P, while numbers next to arrows between boxes are the annual rates of phosphorus transfer in Tg P. Based on Schlesinger (1991, 1997), with some fluxes adjusted so that the natural cycle is in steady state.

In the case of nitrogen, river inputs are largely balanced by net air–sea exchange, as the rate of permanent burial of nitrogen is small. The residence time of nitrogen in the ocean is on the order of years, whereas that of phosphorus is on the order of 30,000 years. However, rapid changes in the surface concentration of either nutrient can occur through changes in the rate of upwelling. This in turn could lead to important changes in the strength of the biological carbon pump and hence in the amount of CO_2 in the atmosphere.

2.8   Linkages between the biogeochemical cycles through atmospheric chemistry

Chemical reactions in the atmosphere provide a number of linkages between the carbon, nitrogen, and sulphur cycles. A central player in tropospheric chemistry is the hydroxyl radical (OH), which has a concentration of only 3×10^{-14} but is the main oxidizing agent in the atmosphere. It oxidizes and thereby destroys most trace gases in the atmosphere, with the notable exception of CO_2 and N_2O. It is produced largely by reaction with O_3, water vapour, and ultraviolet radiation. Some OH is also produced by reaction of HO_2 (hydrogen dioxide) with NO (Eisele *et al.*, 1997). Increases in the concentration of O_3 or in emissions of NO will therefore tend to increase the concentration of OH, thereby decreasing the concentration of most other trace gases in the atmosphere. A warmer climate would be associated with more water vapour in the troposphere, which would also tend to increase the OH concentration. Figure 2.14 shows the latitude–

height distribution of the zonal (east–west) mean OH concentration in January and July, as computed by a 3-D atmospheric chemistry model (the OH distribution cannot be directly measured easily). The concentration is highest at low latitudes, where there is more water vapour for dissociation by UV radiation, and is greater in summer than in winter, owing to the greater availability of sunlight and water vapour. Table 2.3 lists the main atmospheric gases that are removed by reaction with OH.

About 50–60% of the hydroxyl radicals react with and thereby remove CO (carbon monoxide) and most of the rest react with and remove CH_4. Reaction of OH with CO results in the following:

- no net consumption of OH in NO-rich environments, because consumption of an OH when it combines with CO is followed by further reactions that lead to the production of an OH;
- production of O_3 if the NO concentration exceeds an amount that varies from 4 pptv (parts per trillion, or 10^{12}, by volume) at the ground to 2 pptv at the tropopause (the boundary between the troposphere and the stratosphere);
- destruction of OH and O_3 in NO-poor environments, such as occurred over large parts of the troposphere prior to the Industrial Revolution.

Reaction of an OH radical with CH_4 results in the following:

- a net gain of 0.5 OH radicals and 3.7 O_3 molecules in NO-rich environments;
- a net loss of 2–3.5 OH radicals and 0–1.7 O_3 molecules in NO-poor environments.

Further details can be found in Crutzen (1988).

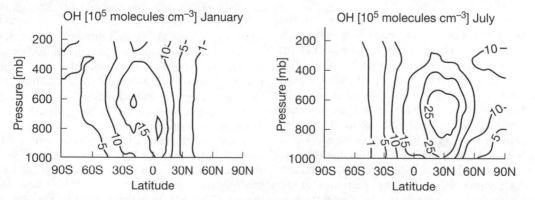

Fig. 2.14 Zonal mean concentration of OH under present-day conditions, as simulated by Wang *et al.* (1998b) for January and July.

Table 2.3 Selected atmospheric trace gases which are removed in part or entirely through reaction with OH, from Schlesinger (1991) except the lifespans of CH_4 and CO, which are taken from Table 2.4, and SO_2, which is taken from Roelofs *et al.* (1998).

Trace gas	Concentration in NH (ppbv)	Tropospheric lifespan	Contribution of OH-sink reaction to removal (%)
CH_4	1780	7.9 years	90
CO	250	20–50 days	100
NMHCs[a]	2–10	1–100 days	50–100
SO_2	0.2	2.4 days	20[b]
COS	0.5	5 years	30
H_2S	—	4 days	100
$(CH_3)_2S$	—	1 day	50
NO, NO_2	0.1	1 day	100[c]
NH_3	1	14 days	10

[a] NMHCs = Non-methane hyrocarbons.
[b] Taken from Roelofs *et al.* (1998).
[c] According to Wang *et al.* (1998b), half of NO_x is removed by hydrolysis of N_2O_5 in aerosols and half by reaction with OH.

It can be seen from the above that the nitrogen cycle, through its control of NO, regulates the amount of O_3 and OH in the troposphere and, through its effect on OH, influences the concentration of almost all the greenhouse gases in the atmosphere. It does this by influencing the reaction pathways involved in the oxidation of CO and CH_4, as well as by directly contributing to the production of OH through reaction with HO_2.

The nitrogen cycle also influences the production of O_3 (and hence OH) independently of reactions involving CO and CH_4, through the following reactions:

$$NO_2 + h\nu \rightarrow NO + O \qquad (2.9)$$

$$O + O_2 \rightarrow O_3 \qquad (2.10)$$

giving the net reaction

$$NO_2 + O_2 \rightleftharpoons NO + O_3 \qquad (2.11)$$

This is an equilibrium reaction, meaning that the concentrations of NO_2, NO, and O_3 will adjust such that the reaction occurs at an equal rate in both directions. A high concentration of NO_2 will drive the reaction to the right, creating O_3, while a high concentration of NO will drive the reaction to the left, destroying O_3. Both NO and NO_2 are produced as a byproduct of the combustion of fossil fuels and as a result of burning of biomass, but the two are so tightly coupled by chemical reactions that they are referred to as NO_x. Most NO_x is emitted as NO but is quickly converted to NO_2 by reaction with O_3, thereby tending to reduce the O_3 concentration. However, the NO_2 then serves as a catalyst in reaction cycles with hydrocarbons or CO that create O_3 (that is, the NO_2 is regenerated during each cycle, so that one NO_2 molecule can ultimately create several O_3 molecules). For this reason, NO_x emissions tend to create O_3, although the production rate depends on a number of other conditions in the atmosphere, conditions that vary with time and location.

NO_x has a lifespan of about 1 day, which means that essentially none can reach the stratosphere. However, N_2O can reach the stratosphere (its lifespan is 120 years), where it is dissociated by ultraviolet radiation into NO through the reactions

$$N_2O + h\nu \rightarrow N_2 + O \qquad (2.12)$$

and

$$N_2O + O \rightarrow 2NO \qquad (2.13)$$

NO then reacts with and destroys O_3 through the reverse of reaction (2.11). This converts NO to NO_2, which then forms HNO_3 and diffuses to the troposphere, where it is washed out in rain. Photochemical dissociation of N_2O in the stratosphere is the only removal process for N_2O (except for a possible ocean sink), which explains its long atmospheric lifespan.

In summary, N emissions (as NO_x) directly create O_3 in the troposphere, while N emissions (as N_2O) lead to the destruction of O_3 in the stratosphere. NO_x emissions also indirectly lead to the creation of O_3 in the atmosphere by shifting the pathways involved in the oxidation of CO and CH_4 from those that destroy O_3 toward those that create O_3. The *presence* of O_3 in turn leads to the production of OH in the presence of sunlight. The concentration of NO also determines whether OH is created, destroyed, or left unchanged during the *production* of ozone, and NO is directly involved in the production of OH through reaction with HO_2. By influencing the concentration of OH in the atmosphere (both directly and indirectly), N emissions also influence the concentrations of all the gases listed in Table 2.3, two of which – CH_4 and SO_2 – have important direct and/or indirect effects on climate.

The above discussion has highlighted a number of linkages between emissions of CO, CH_4, and NO_x through their effect on atmospheric O_3. The atmospheric distribution of O_3, however, is highly variable owing to the fact that both O_3 and its precursors have very short lifespans within the atmosphere. Table 2.4 presents the atmospheric lifespans of O_3 and the main gases involved in O_3 chemistry in the troposphere.

Other linkages between the biogeochemical cycles through atmospheric chemistry arise through chemical reactions within or on the surface of aerosols. For example, sulphate aerosol contributes to the removal of NO_x, a process that is accounted for in the atmospheric chemistry of Krol *et al.* (1998). Sea salt-water aerosols contribute to the conversion of SO_2 into SO_4^{2-} (Section 2.5) and hence to sulphate aerosols. Mineral aerosols (dust) react with sulphate in several ways: the rate of conversion from SO_2 to SO_4^{2-} on the surfaces of mineral aerosols might be sufficient to double the amount of sulphate in the atmosphere in regions of high dust loading, but the effectiveness of this sulphate in reflecting solar radiation would be reduced owing to the fact that the sulphate would now be occurring as part of larger particles (Dentener *et al.*, 1996). The adsorption of

sulphate by both dust and soot would increase their chemical reactivity, thereby increasing the role of dust as cloud condensation nuclei (CCN) but reducing the number of sulphate CCN (Tegen *et al.*, 1997). The net effect on the number and effectiveness of CCN is unknown. Finally, interactions of N_2O_5, O_3, and HO_2^- radicals with dust can decrease the tropospheric O_3 concentration by up to 10% in and near the dust source areas (Dentener *et al.*, 1996). The presence of sulphate aerosols also slightly reduces O_3 concentrations (Dentener and Crutzen, 1993).

2.9　Linkages between the biogeochemical cycles and climate

The preceding sections outlined the flows of energy and mass (as H_2O, C, S, N, and P) that link the various components of the climate system. The various energy and mass flows are closely interlinked, and in this section the major linkages are outlined. These linkages lead to a series of *feedbacks* between climate and the various biogeochemical cycles.

A feedback occurs when a change in some quantity, ΔA, leads to a change in a second quantity, ΔB, and the change ΔB tends to provoke a further change in A. If ΔB provokes a change in A that is in the same direction as the initial change, then a *positive feedback* occurs, whereas if it provokes a change in A that is opposite to the initial change, a *negative feedback* occurs. Positive feedbacks amplify the initial change and tend to destabilize a system, whereas negative feedbacks dampen the initial change, thereby stabilizing a system.

Table 2.4 Atmospheric lifespan of O_3 and of gases involved in O_3 chemistry in the troposphere.

Gas	Lifespan	Reference
OH	1–10 seconds	Krol *et al.* (1998)
NO_x	1 day	Schlesinger (1991)
NMHCs[a]	1–100 days	Schlesinger (1991)
CO	20 days, lower troposphere 50 days, upper troposphere	Mauzerall *et al.* (1998)
O_3	10 days, surface tropics 100–200 days, upper tropical troposphere and high-latitude surface	Wang *et al.* (1998c)
CH_4	6.2 years, pre-industrial (1850) 7.9 years, present (1992)	Lelieveld *et al.* (1998)

[a] NMHCs = Non-methane hydrocarbons.

The tightest linkage is between the energy and water vapour flows, in that evaporation and condensation of water are the major processes that offset the radiative heating of the surface and the radiative cooling of the atmosphere. As the ocean surface temperature increases, relatively more of the net radiation goes into evaporating water, so that the overall intensity of the hydrological cycle and the atmospheric water vapour content tend to increase. Since water vapour is a greenhouse gas, this leads to further warming, a further increase in evaporation, and so on, as part of a positive feedback loop.

The rates of photosynthesis and respiration on land are directly affected by climate, through their dependence on both temperature and soil moisture, and through the dependence of photosynthesis on cloud cover. An increase in photosynthesis will tend to reduce the atmospheric CO_2 concentration, while an increase in the rate of respiration will tend to increase atmospheric CO_2, thereby provoking further changes in climate. Climate is likely to affect the strength of the oceanic biological pump through its effect on the rate of vertical overturning circulation and turbulent mixing within the oceans, which supply the nutrients to the surface layer that are needed for photosynthesis. Alterations in the strength of the biological pump will, in turn, affect the atmospheric CO_2 concentration and feed back onto climate.

The atmospheric methane concentration depends on the magnitude of the emission sources and on the rate of removal. The primary natural emission sources are wetlands, where anaerobic decomposition of organic matter produces CH_4. A warmer climate could increase CH_4 emissions in two ways: by increasing the rate of photosynthesis and hence the supply of organic matter for decomposition, and through the direct effect of warmer temperatures on the rate of decomposition. Conversely, drier conditions in wetlands would tend to decrease CH_4 emissions. A warmer climate would also increase the rate of removal of CH_4 from the atmosphere, by increasing the OH concentration (in turn a result of a higher water vapour content in the atmosphere).

It is not known at present how climate, the availability of sunlight, or nutrient conditions affect the production of DMS by marine bacteria, but the potential exists for a strong negative or positive feedback. If a warmer climate increases the production of DMS and this leads to increased CCN, this will serve as a negative feedback since cloud reflectivity of solar radiation would increase. The transformation of DMS to sulphate is a chemical reaction that depends on the OH concentration, so that processes that alter the atmospheric OH concentration could influence climate through changes in cloud reflectivity and lifespan resulting from changes in sulphate concentrations. These processes include changes in the water vapour content of the atmosphere, in the amount of tropospheric O_3, in emissions of CO and CH_4, and in changes in the amount of stratospheric O_3 (which would alter the amount of UV radiation reaching the troposphere). There is also a direct linkage between the carbon and sulphur cycles in that a change in marine biological productivity will probably alter the production of DMS.

The nutrient cycles are tied to the hydrologic and carbon cycles in a number of ways. First, the rate of photosynthesis depends on the availability of nutrients. Second, in plants with symbiotic nitrogen fixation, the rate of fixation is often tied to the rate of NPP (Net Primary Production, equal to gross photosynthesis minus respiration losses by the plant). This is because nitrogen fixers consume a non-negligible fraction of NPP for their own energy needs. For the same reason, asymbiotic nitrogen fixation is often limited to soils with a high organic matter content (Schlesinger, 1991, page 149). Increases in NPP may therefore stimulate greater rates of nitrogen fixation. The nutrient cycles are directly tied to the hydrological cycle in that river runoff is the source of almost all of the phosphate to the ocean surface layer and much of the nitrogen that arrives in a form usable by marine plants (with the balance of the usable nitrogen delivered by precipitation). Also, the balance between evaporation, precipitation, and runoff strongly influences the surface salinity distribution in the ocean, which in turn drives (in part) the overturning circulation that brings nutrients from the deep ocean to the mixed layer.

As previously noted, sea salts are an important aerosol, and their production is directly tied to the wind strength over the ocean. Winds are also responsible for lifting mineral dust aerosols into the atmosphere. As climate changes, the atmospheric loading of these aerosols would probably change, producing both direct and indirect (cloud-related) effects on the absorption and scattering of solar radiation.

2.10 Natural causes of climatic change

Climate can change naturally for a variety of reasons. Some of the driving factors of climatic change operate at time scales of hundreds of millions of years, whereas others fluctuate over a time period of only a few years. In this section the main causes of natural climatic change are discussed in terms of the key processes involved, the time scales of change, and the magnitude of change.

Changes in the composition of the Earth's atmosphere

The atmospheric CO_2 content can change as a result of driving forces that are external to the climatic system, or as a result of internal feedbacks between climate and the carbon cycle. Examples of externally driven changes in atmospheric CO_2 include (i) a change in the rate of volcanic degassing of CO_2 due to a change in the rate of sea-floor spreading and the associated subduction and metamorphism of $CaCO_3$ sediments; (ii) an increase in the rate of removal of CO_2 from the atmosphere due to an increase in the rate of weathering as a result of tectonically driven periods of mountain building; and (iii) a change in the strength of the oceanic biological pump due to a change in the latitudinal and seasonal distribution of solar radiation as the Earth's orbit changes. The first two driving factors operate at periods of many tens to hundreds of millions of years, while the third factor would operate at the time scale of the important orbital frequencies, namely 20,000–40,000 years.

An extensively studied, potential internally derived cause of changing atmospheric CO_2 concentration is a shift in carbon between the deep ocean and the mixed layer + atmosphere system due to a change in the large-scale overturning circulation within the ocean. Other potential internal causes include feedbacks between climate and the size of the terrestrial biosphere, and climatically induced changes in the rate of delivery of nutrients to the ocean. Sudden changes in the ocean circulation could alter the atmospheric CO_2 concentration within 100 years, whereas a change in the rate of delivery of P to the ocean would require about 30,000 years to have a noticeable effect, since the turnover time of oceanic P is about 30,000 years (see Fig. 2.13).

In most of the examples cited above, the change in atmospheric CO_2 concentration could be the initial driver of climatic change, or it could act as a positive or negative feedback (depending on the case) to amplify or dampen a prior climatic change. Thus, changes in the rate of weathering could be the initial cause of a change in atmospheric CO_2, but if atmospheric CO_2 increases (or decreases) for other reasons, the weathering rates will increase (or decrease) in response, thereby tending to stabilize the atmospheric CO_2 concentration at some new value. Similarly, if ocean circulation changes as part of some internal, self-sustained oscillation, the resultant CO_2 changes would drive changes in the climate. However, if the ocean circulation change is driven by some prior change in climate, the resultant change in atmospheric CO_2 concentration could either amplify or diminish the initial climatic change (depending on the direction of the induced CO_2 changes).

Changes in topography, land–sea geography, and bathymetry

The topography of the land surface affects wind and rainfall patterns, thereby exerting an immediate effect on weather patterns and climate, and – at high latitudes – determines the ease or difficulty of initiating ice ages. The uplift of major mountain ranges, such as the Western Cordillera in the Western Hemisphere, or the Himalayan–Tibetan plateau, exerted a profound influence on the regional climate. Similarly, the development of high-latitude plateaux just below the present-day snowline allowed widespread permanent snowcover to arise in response to a modest lowering of the snowline, thereby permitting rapid extension in the area of permanent snow cover and glacierization (Williams, 1979).

On a longer time scale, continental drift leads to changes in the proportion of ocean and land at different latitudes. Since continents tend to absorb less solar radiation than oceans (owing to their higher albedo), shifting a continent from low latitudes to higher latitudes (where there is less radiation to be reflected) results in a net increase in the global average absorption of solar radiation, thereby tending to warm the global mean climate. Changes in the rate of sea-floor spreading from mid-ocean ridges altered the volume of the ocean basins and hence the portion of the Earth's surface covered in water, thereby changing the average surface albedo of the Earth and hence the climate.

Finally, changes in the shape of ocean basins (bathymetry) and, in particular, such things as the closing of the former gap between North and South America or the opening of the gap between Antarctica and South America, altered ocean circulation patterns worldwide. The altered circulation patterns in turn led to large changes in regional climate by redistributing heat from one region to another. In cases where large new areas of snowcover developed (such as with cooling of Antarctica), this would have also led to changes in the globally averaged climate.

Changes in solar luminosity

The energy output from the sun varies on at least two different time scales. At one extreme, the luminosity has systematically increased from about 70–75% of its present value 3.5 billion years ago, to its present value. This is a robust result of models of stellar evolution (Newman and Rood, 1977; Gough, 1981). At the other extreme, the solar luminosity has been observed to vary by about ±0.07% over the 11-year sunspot cycle (Hoffert *et al.,* 1988). Observations of other stars suggest that variations of a few tenths of a percent could occur in the sun over an 80-year period (Baliunas and Jastrow, 1990). There have been some claims that larger changes in solar luminosity have occurred during the past century, but these claims are best addressed in the context of trying to explain observed temperature changes (Section 11.4). Changes in the flux of ultraviolet radiation, in association with overall changes in solar luminosity, might induce changes in the amount of stratospheric O_3 that significantly reduce the net heating or cooling effect of changes in solar luminosity (Haigh, 1994).

Changes in the Earth's orbit

Owing to the perturbing influence of the gravitational force from other planets in the solar system, the Earth's orbit changes in a systematic and calculable way. The three parameters that describe the Earth's orbit are its eccentricity, the tilt of the rotational axis from a line perpendicular to the orbital plane (the obliquity of the axis), and the time of year when the Earth is closest to the sun (the time of perihelion). Table 2.5 gives the present value, the range of variation, and the dominant frequencies of variation for each of these orbital parameters. Changes in the obliquity and the time of perihelion have no effect on the annual and global mean availability of solar energy at the top of the atmosphere, but lead to rather large changes (up to ±15%) in the amount of sunlight in a given season at high latitudes. Changes in the Earth's orbit are regarded as the driving mechanism behind glacial–interglacial oscillations in climate, although several feedback mechanisms – including changes in the atmospheric concentrations

of CO_2 and CH_4 – are required to fully explain ice age cooling. The Earth's orbit changes over periods of many thousands of years, so it is not relevant to climatic change during the next few centuries.

Volcanic activity

Volcanic activity influences climate through the emission of SO_2 during eruptions. The SO_2 is converted to sulphate aerosols once in the atmosphere, but because much of the SO_2 is injected into the stratosphere (rather than the troposphere), the aerosols persist for a few years rather than a few days and are therefore able to spread globally. These aerosols cool the climate by reflecting sunlight, but the cooling effect – like the aerosols themselves – lasts only a few years. However, if many volcanic eruptions occur at closely spaced intervals, then a longer-term effect on climate is possible. Major volcanic eruptions during the past 140 years caused a global mean cooling of 0.1–0.2°C during the first two years after the eruption, although North America and Eurasia warmed by several degrees in the first or second winter after major eruptions owing to shifts in wind patterns (Robock and Mao, 1995).

Internal variability of the atmosphere–ocean system

The various natural causes of climatic change discussed above can, with one exception, be regarded as *external* causes of climatic change. That is, the factor causing the climatic change (such as continental drift, volcanic activity, or a changing orbit or solar luminosity) is not influenced by the climatic system itself. The exception to this generalization is potential changes in the atmospheric concentration of CO_2 that arise as a result of internally generated changes in ocean circulation. However, whether or not such internally generated changes lead to changes in the

Table 2.5 The Earth's orbital parameters, their present values, the range of variation over time, and the major periods of variation.

Orbital parameter	Present value	Range of values	Major period (years)
Eccentricity	0.0167	0.01–0.04	100,000
Obliquity	23.45°	22–25°	41,000
Timing of perihelion	3 January	365 days	23,000

atmospheric CO_2 concentration, such changes are an important cause of regional (and perhaps global) climatic change in their own right.

A number of computer models of the atmosphere–ocean system have been built in which self-sustained oscillations in climate occur without any external driving force. Oscillations at a time scale of several hundred to several thousand years involve alternating heat flows into or out of the deep ocean, combined with feedbacks between the vertical or latitudinal temperature and salinity structure in the oceans and the oceanic circulation. Oscillations at longer time scales can be generated through interactions between the oceans and continental-scale ice sheets, while decadal-scale oscillations can be generated through changes in ocean mixing and currents in the top few hundred metres of the ocean (Weaver and Sarachik, 1991; Delworth et al., 1993). Decadal time-scale fluctuations in regional climate can also arise through interactions between the atmosphere, the ocean surface temperature, and the extent of sea ice (Mysak and Manak, 1989). Temperature and moisture anomalies on land that persist through an entire summer can arise through feedbacks between precipitation, soil moisture, and surface temperature (Atlas et al., 1993). In particular, initially dry conditions can cause reduced rainfall (as less moisture is recycled through evaporation), which tends to perpetuate the dry conditions until an influx of moisture from elsewhere breaks the positive feedback loop.

Year-to-year or even decade-to-decade oscillations do not constitute a change in climate, but rather are part of the variability of weather that serves to define the climate of a given region. Climate is conventionally defined as the average and variability of weather averaged over a 30-year sampling period. Thus, a change in climate occurs when there is a change in the averages and/or variabilities from one 30-year period to the next.

Internally generated climatic variability is largest at the subcontinental scale, but decreases in magnitude as the scale under consideration increases. Thus, year-to-year fluctuations in regional temperature of a few degrees are common, but year-to-year fluctuations in global mean temperature are generally less than 0.1 or 0.2°C. This is because much of the internally generated variability involves redistributing heat horizontally from one region to another, effects which cancel out when averaged over the entire planet.

Summary

Table 2.6 summarizes the various natural causes of climatic change in terms of the time scales and magnitude of change.

2.11 Human causes of climatic change

Humans are altering climate at a global scale by altering the atmospheric concentrations of GHGs, both by direct emissions of GHGs into the atmosphere, and indirectly by inducing changes in the chemistry of the atmosphere which lead to changes in the concentrations of some GHGs. Humans also affect the climate on a continental scale through the emission of a variety of aerosols. Changes in land cover (e.g. deforestation) also influence climate, largely at the local scale but possibly with some global-scale effects through large-scale waves in the atmosphere.

Water vapour is the strongest contributor to the natural greenhouse effect, but, among the GHGs, it is most directly linked to climate and therefore least directly controlled by human activity. This is because evaporation is strongly dependent on surface temperature, and because water vapour cycles through the atmosphere quite rapidly, once every 8 days on average. Concentrations of the other natural GHGs, in contrast, are strongly and directly influenced by emissions associated with the combustion of fossil fuels, by some agricultural activities, and by the production and use of various chemicals. There are also a number of entirely artificial GHGs that humans are adding to the atmosphere.

Increasing concentrations of well-mixed GHGs

The extent to which the concentration of a gas in the atmosphere varies with location depends on the lifespan of the gas in the atmosphere compared to the length of time required for atmospheric winds to transport the gas from one place to another, and how geographically concentrated the sources of the gas and its removal processes (sinks) are. If the lifespan of the gas is much longer than the transport time, then there will be an opportunity for mixing to produce a fairly uniform concentration of the gas before it is removed, even with concentrated sources and

Table 2.6 Summary of the major natural causes of climatic change in terms of time scale and magnitude.

Cause	Time scale (years)	Global mean magnitude, Lower and upper limits
Volcanic activity	1–4	0.4°C[a]
Internal variability	10^1–10^3	0.2–0.4°C[b]
Transitions to new climate states[c]	10^1–10^2	2–3°C[d]
Changes in solar luminosity	10^1–10^9	0.1°C[e]
Changes in GHG concentrations	10^2–10^9	2–3°C[f]
Changes in the Earth's orbit	10^4–10^5	4–6°C[g]
Changes in land–sea geography	10^7–10^8	up to 5°C[h]

[a] This is the largest inferred effect during the past 130 years.

[b] Based on millennial-scale simulations with computer climate models and multi-century reconstructions of NH mean surface-air temperature using proxy paleoclimatic data.

[c] The time scale here is the fastest time seen for the transition from one climatic state to another in the paleoclimatic record. The timing of the transitions is highly variable, and such transitions might be triggered by slower and more gradual changes.

[d] Based on the finding of transitions in polar ice cores corresponding to roughly half of the difference between glacial and interglacial climates in these regions. If the cause of the polar climatic change is local (e.g., from a change in nearby ocean currents), then the global mean temperature change as a fraction of the global mean difference between glacial and interglacial climates would have been smaller.

[e] This is a reasonable estimate of the effect of solar variability of no more than a few tenths of a percent during the past 100 years (see Section 11.4). Since solar luminosity is estimated to have increased from 70% of its present value to its present value over the past 3 billion years, the associated change in temperature would have been extremely large were it not for largely counteracting changes in other variables (such as atmospheric GHG concentrations).

[f] Based on estimates that decreases in GHG concentrations during the peak of the last ice age can explain about half of the inferred global mean cooling of 4–6°C. Much larger changes in CO_2 concentration (up to 4–10 times the present concentration) occurred over the past 500 million years, as illustrated in Fig. 1.2.

[g] This is the estimated difference in globally averaged surface temperature between glacial and interglacial periods. The glacial–interglacial transitions were triggered by changes in the Earth's orbit, but a number of slow positive feedback mechanisms – including changes in GHG concentrations and expansion of ice sheets — are required to explain the full amplitude of temperature change.

[h] Based on the maximum simulated effect of changes in the distribution of continents between now and the Cretaceous Period, when the globally averaged surface temperature was at least 6°C warmer than today.

sinks. The important mixing times in the atmosphere are as follows: east–west mixing in the troposphere, a few weeks; mixing between the hemispheres, one year; mixing from the surface layer to the upper troposphere, one month; mixing between the troposphere and stratosphere, about three years. All of the greenhouse gases except O_3 and a few minor gases have atmospheric lifespans well in excess of one year, and so have almost the same concentration everywhere in the atmosphere. Such gases are said to be *well mixed*. A few greenhouse gases (such as chloroform, HCFC-123, and HFC-152a) have a lifespan of around one year; these gases are fairly well mixed within the troposphere, but have a sharply lower concentration within the stratosphere.

The lifespan of a gas in the atmosphere not only plays a role in determining how uniform its concentration will be, but also plays a role in determining its

effect on climate. This is because the greater the lifespan, the greater the buildup of the gas in the atmosphere that will occur for a given rate of emission. The relationship between emission rate, lifespan, and concentration is explored in Box 2.5. The other factor determining the climatic effect of a given gas is its ability to trap heat on a molecule-by-molecule basis. Table 2.7 lists the important well-mixed GHGs, the average atmospheric lifespan for each gas, and the ability of each gas to trap heat on a molecule-per-molecule basis compared to the heat-trapping ability of CO_2.[1] Many are several hundred to several thou-

---

[1] As previously noted, the removal processes for CO_2 are considerably more complicated than for other gases, and it is not correct to speak of a single average lifetime in the case of CO_2. This issue and the specific removal processes for other gases are discussed in Chapter 8.

Table 2.7 Greenhouse gases directly emitted through human activities, relative heat trapping on a molecule-per-molecule basis compared to CO_2, average lifetime in the atmosphere, concentration in 1995, and global mean radiative heating in 1995 as computed using the formulae given in Table 2.9. For a more complete listing of greenhouse gases, see Table 7.1.

Gas	Relative heat trapping ability[a]	Atmospheric lifespan (years)	Concentration (ppbv)		Heating perturbation in 1995 ($W m^{-2}$)
			Pre-industrial	1995	
CO_2	1	Variable	278,000	360,000	1.40
CH_4	26	7.9[b]	700	1725	0.47
N_2O	206	120	275	311	0.14
CFC-11	12,400	50	0.000	0.272	0.082
CFC-12	15,800	102	0.000	0.532	0.205
HFC134a	9570	14.6	0.000	0.0016	0.0003
SF_6	≤36,000	3200	0.000	0.0032	0.0020
CF_4	5600	50,000	0.000	0.075	0.0071

[a] From Shine et al. (1990), except for CH_4 (Lelieveld and Crutzen, 1992), SF_6 (derived from Ko et al., 1993), and CF_4 (derived from Schimel et al., 1996).
[b] Based on Lelieveld et al. (1998). The value of 12.2 years given in Schimel et al. (1996) is the adjustment time. The difference between the atmospheric lifespan and the adjustment time is explained in Box 2.3.

sand times more effective in trapping heat than is CO_2, and some of the GHGs will remain for hundreds to thousands of years in the atmosphere. The last three columns of Table 2.7 give the pre-industrial concentration, the concentration in 1995, and the estimated globally averaged rate of heat trapping in 1995. The 1995 heat trapping depends on the effectiveness of the gas in trapping heat and the increase in concentration that occurred up to 1995; the latter depends on the cumulative emission, the timing of past emissions, and the average lifespan of the gas in the atmosphere. Although CO_2 is the least effective in trapping heat on a molecule-per-molecule basis, the amount of CO_2 emitted is so much larger than that of any other GHGs that the CO_2 increase accounts for about 55% of the total heat trapping that has occurred so far. The total heat trapping in 1995 due to well-mixed GHGs was about 2.4 $W m^{-2}$.

Changes in the concentration of ozone

Ozone differs from the other greenhouse gases in that its atmospheric lifespan is relatively short – from 10 to 200 days (Table 2.4). As a result, O_3 cannot spread very far before it is removed, so large differences in its concentration occur from one region to another. Ozone also differs from the other GHGs in that it is not directly emitted into the atmosphere; rather, it is produced through photochemical reactions involving other substances – referred to as *precursors* – that are directly emitted. There are also

Box 2.5 Relationship between lifespan and steady-state concentration

The atmospheric lifespan of most greenhouse gases is long enough that the atmosphere can be treated as a well-mixed reservoir, and the rate of removal of the gas varies directly with the concentration. Changes in the concentration, C, of the gas in the atmosphere are given by the imbalance between the flux input, F, and the rate of removal, C/τ. That is,

$$\frac{dC}{dt} = F - \frac{C}{\tau} \qquad (2.5.1)$$

where τ is the average residence time (or turnover time). The steady-state concentration, C_0, is such that $dC/dt = 0$, from which it follows that

$$C_0 = \tau F_0 \qquad (2.5.2)$$

where F_0 is the initial flux input. If there is a perturbation in the flux, such that $F = F_0 + \Delta F$, then the new steady-state concentration will be

$$C' = (C + \Delta C) = \tau(F + \Delta F) \qquad (2.5.3)$$

Subtracting Eq. (2.5.2) from Eq. (2.5.3) gives the result that $\Delta C = \tau \Delta F$.

Thus, for a given additional flux into the atmosphere, the steady-state change in the concentration of the gas varies directly with the average lifespan of the gas. This is because the longer the lifespan, the greater the buildup in gas concentration that is required before the rate of removal (which is proportional to the amount of gas present) equals the rate of input.

several chemical reactions involved in the destruction of O_3, and the O_3 concentration at any given place and time depends on the balance between production, destruction, and transport to or from other regions. The effectiveness of the gases involved in producing or destroying ozone depends on where they occur and on the concentrations of other gases involved in the chemistry of O_3. Since 1979, the total amount of O_3 in a vertical column has decreased at an average rate of 3–4% per decade at middle and high latitudes in the NH, and by up to 6% per decade at high latitudes in the SH. As discussed in Section 5.6, this is the net result of a decrease in the amount of O_3 in the stratospheric O_3 and an increase in the amount of O_3 in the troposphere. Various pieces of observational evidence – also summarized in Section 5.6 – indicate that the amount of ozone in the troposphere has roughly doubled since the Industrial Revolution.

It is not possible to give a single number for the effectiveness of O_3 in trapping heat on a molecule-per-molecule basis. This is because the effectiveness of any greenhouse gas in trapping heat depends on its temperature compared to that of the underlying surface (as explained in Section 3.1), and hence depends on its height within the atmosphere. Since well-mixed gases have a uniform concentration – both vertically as well as horizontally – it is possible to compute the heat trapping of an "average" molecule (as given in Table 2.7). However, there is no fixed vertical distribution for changes in the O_3 concentration, so it is not possible to work out a generally applicable, average heat trapping for O_3 molecules.

The climatic impact of changes in the amount of tropospheric O_3 is estimated using global-scale chemical models that simulate the past and present distribution of ozone in the troposphere. These models require estimates of past and present-day emissions of the key O_3 precursors. Results from recent calculations, which are reviewed in Section 7.3, indicate a global mean radiative heating of 0.4–0.7 W m^{-2} (compared to 2.4 W m^{-2} for all well-mixed gases).

Losses in stratospheric ozone are more recent, so we have an observational baseline against which to compare recent concentrations. However, computation of the climatic effects of losses in stratospheric ozone is quite uncertain because of the possibility of very large indirect effects of losses in stratospheric O_3 on the planetary energy balance, as discussed in Section 7.5.

Increasing concentrations of aerosols

Aerosols, like GHGs, are produced both naturally and through human activity; natural aerosols include sea salt, dust, and volcanic aerosols, while anthropogenic aerosols are produced from the burning of biomass and fossil fuels, among other sources. Some aerosols, such as dust, are directly emitted into the atmosphere. The majority of aerosols, however, are not directly emitted but – like tropospheric O_3 – are produced through chemical transformation of precursor gases. The most important anthropogenic aerosol is thought to be sulphate, which is produced from SO_2 that is released with the combustion of sulphur-containing coal, from the refining of oil, and from the smelting of certain metals. Other anthropogenic aerosols include non-absorbing organic compounds produced from the burning of biomass and the oxidation of hydrocarbons, light-absorbing aerosols (soot or black carbon) produced from the incomplete combustion of fossil fuels and biomass, nitrate aerosols, and anthropogenically induced increases in the production of dust. All tropospheric aerosols have a short lifespan (days) in the atmosphere due to the fact that they are rapidly washed out with rain. For this reason and because the strength of the emission sources varies strongly from one region to another, the amount of aerosols in the atmosphere varies considerably from one region to another.

The effect of aerosols on the radiative balance of the Earth is highly uncertain. With the exception of soot, aerosols have a cooling effect, but the magnitude is uncertain. This uncertainty arises in part because aerosols affect the radiative balance directly by scattering sunlight in cloud-free areas, and indirectly by altering the radiative properties of clouds (as noted in Section 2.5 in our discussion of the sulphur cycle). The calculated global mean direct effect of pure sulphate aerosols ranges from –0.3 W m^{-2} to –0.7 W m^{-2}, while the calculated indirect effect ranges from –0.4 to –4.8 W m^{-2}. The way in which these estimates were made and the associated uncertainties are discussed in Section 7.4, while indirect observational constraints on the aerosol cooling effect are discussed in Section 11.4.

Changes in the land surface

It has long been suggested that human-induced changes in the land surface albedo could have had a noticeable effect on regional- and global-scale temperature. Among the first estimates of such effects are those of Sagan *et al.* (1979), who suggested that changes in land cover (primarily desertification, deforestation, and salinization) caused a global mean cooling tendency of 0.2°C during the preceding 25 years, and up to 1°C cooling during the preceding several thousand years. A more recent analysis by Bonan (1997), focusing on the USA, indicates that conversion of forest to cropland caused a summer cooling of up to 2°C over a wide region of the central United States. This result was obtained using a 3-D atmospheric model coupled to a land surface module that accounts for differences in albedo, leaf area, and stomatal conductance among different vegetation types. More recently, Chase *et al.* (1999) compared the global distribution of surface air temperature as simulated using a 3-D atmospheric model with the current global distribution of vegetation types and with estimated natural vegetation types (i.e. prior to human disturbance). The largest temperature changes occurred in NH mid-latitudes over land, with current vegetation causing temperatures to be up to 1.5°C warmer in winter (the season with the largest effects) over eastern North America, western Europe, and Siberia. These changes are largely the result of changes in *tropical* land cover that have already occurred, and not a result of changes in the areas experiencing the largest temperature changes. A reduction of tropical forest cover reduces the intensity of tropical convection over land, thereby reducing upper-air outflow from the tropics to the mid-latitudes. This in turn causes changes in wind patterns which explain the temperature change patterns. Over the ocean, there was a tendency for air temperature to decrease, but this was suppressed in the experiments performed by Chase *et al.* (1999) because sea surface temperatures (SSTs) were held fixed (owing to the absence of an ocean model that could predict SST changes). Allowing SSTs to fall would have moderated the warming over land to some extent.

2.12 Relative magnitudes of natural and future human causes of climatic change

Having identified the various natural and anthropogenic causes of climatic change in Sections 2.10 and 2.11, the most pertinent question one can ask is, 'Which of these factors, if any, is likely to dominate climatic change during the 21st century?' Much of the rest of this book is, in fact, devoted to answering this question. As a preview of what is to come, it is helpful at this point to outline some of the key steps that are required in order to answer this question.

Changes in the surface climate, whatever the cause, will be driven by an imbalance between the absorption of solar radiation by the surface–atmosphere system and the emission of infrared radiation to space, or by an imbalance between the net radiative budget (solar + infrared) and the net heat transfer to or from the deep ocean. The change of surface temperature in response to either of these imbalances will be determined by a series of feedbacks that will be introduced in Chapter 3 and discussed more extensively in later chapters. Although there is uncertainty concerning the absolute climatic response to a given energy imbalance, the climatic responses to different driving factors will be roughly proportional to the associated energy imbalances. Thus, by ranking the magnitude and time scale of the energy imbalances created by different natural and anthropogenic causes of climatic change, we can determine their relative importance for future climatic change. As will be discussed in Chapter 13, the heat trapping due to increases in CO_2 and other GHGs could easily reach 4–6 W m^{-2} by the end of the 21st century. This is much larger than the perturbation that can be reasonably expected due to natural fluctuations over a time span as short as 100 years. Furthermore, the aerosol cooling effect, which at present must be substantially less than the GHG heating effect (Section 11.4), is not likely to grow as fast as the GHG heating effect. Thus, irrespective of how responsive the climate is to natural and anthropogenic driving forces, the natural driving forces will almost certainly be overwhelmed by the heating effect of increasing GHG concentrations during the 21st century.

CHAPTER 3

The physics of the greenhouse effect, radiative forcing, and climate sensitivity

As noted in Section 2.2, the term "greenhouse effect" refers to the reduction in outgoing infrared radiation to space due to the presence of the atmosphere. It can be and has been directly measured, based on the surface emission (which depends only on temperature) and the emission at the top of the atmosphere as observed by satellites. In this chapter the underlying physics of this effect is explained in very simple terms. This is followed by a discussion of the heating perturbation that occurs when GHG or aerosol concentrations change – the so-called "radiative forcing". The concept of fast and slow feedbacks and of climate sensitivity is then introduced, followed by a discussion of the relationship between radiative forcing and climate sensitivity for a variety of important radiative forcing mechanisms.

3.1   Physics of the greenhouse effect

All objects above absolute zero (0 K or –273°C) emit electromagnetic radiation. For objects at typical earth-atmosphere temperatures, the emitted radiation falls almost entirely in the infrared (IR) part of the spectrum (wavelengths of 4–50 μm). The maximum amount of radiation that can be emitted is given by

$$F_{max} = \sigma T^4 \tag{3.1}$$

where $\sigma = 5.664 \times 10^{-8} \, W \, m^{-2} \, K^{-4}$ and is called the Stefan–Boltzmann constant, and T is the absolute temperature in kelvins. Equation (3.1) is known as the *Stefan–Boltzmann Law*. Objects that emit the maximum amount of radiation are called *blackbodies*. The ratio of actual emission to blackbody emission is called the *emissivity*, ε. Actual emission is thus given by

$$F = \varepsilon \sigma T^4 \tag{3.2}$$

The land and ocean surface and clouds thicker than cirrus clouds (thin, wispy clouds) emit almost as blackbodies ($\varepsilon \approx 1.0$), the clear-sky atmosphere has an emissivity of 0.4–0.8, and cirrus clouds have a typical emissivity of 0.2. The Earth, atmosphere, and clouds also absorb IR radiation. The ratio of absorbed radiation to incident radiation is called the absorptivity. Kirchhoff's Law states that absorptivity is equal to emissivity, which implies that a blackbody will absorb all of the incident infrared radiation, while the atmosphere will absorb only part of the incident radiation, the balance being transmitted. Reflection or scattering of IR radiation is generally negligible and so will not be considered here.

With this background, we can analyse the greenhouse effect with the aid of the simple two-box model shown in Fig. 3.1. The lower box represents the Earth's surface, which has temperature T_s and emits σT_s^4 (since $\varepsilon_s = 1.0$). The upper box represents the atmosphere, which has temperature T_a and emissivity ε_a. It absorbs $\varepsilon_a \sigma T_s^4$ from below, allows $(1 - \varepsilon_a)\sigma T_s^4$ to be transmitted to space, and re-emits an amount $\varepsilon_a \sigma T_a^4$ to space. The total emission to space is given by

$$F_{space} = (1 - \varepsilon_a)\, \sigma T_s^4 + \varepsilon_a \sigma T_a^4 \tag{3.3}$$

which can be rewritten as

$$F_{space} = \sigma T_s^4 - \varepsilon_a(\sigma T_s^4 - \sigma T_a^4) \tag{3.4}$$

Equation (3.4) represents the emission to space as the sum of the surface emission and a term that depends on the atmosphere. Since, in general, $T_a < T_s$, the second term (with the negative sign) is negative, that is, it subtracts from the surface emission, thereby reducing the emission to space. This term is properly called the "atmospheric effect", but corresponds to what is popularly known as the "greenhouse effect".

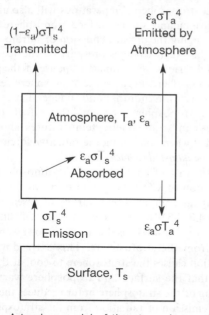

$(1-\varepsilon_a)\sigma T_s^4$
Transmitted

$\varepsilon_a \sigma T_a^4$
Emitted by
Atmosphere

Atmosphere, T_a, ε_a

$\varepsilon_a \sigma T_s^4$
Absorbed

σT_s^4
Emisson

$\varepsilon_a \sigma T_a^4$

Surface, T_s

Fig. 3.1 A two-box model of the atmosphere–surface system, showing the transfer of IR radiation. T_a and T_s are the atmospheric and surface temperatures, respectively, and ε_a is the atmospheric emissivity.

In physical terms, what is happening is that the atmosphere absorbs a portion of the radiation stream from the surface ($\varepsilon_a \sigma T_s^4$) and replaces this stream with its own emission ($\varepsilon_a \sigma T_a^4$). As long as the atmosphere is colder than the surface, the replacement stream is less than the original stream, so that the net emission of infrared radiation to space is reduced. Equation (3.4) makes it clear that the larger the atmospheric emissivity, the greater the reduction in outgoing radiation to space (even though the contribution from the atmosphere itself increases). Thus, the greenhouse effect (the second term in Eq. (3.4)) depends on two factors: the difference between surface and atmospheric temperatures, and the atmospheric emissivity. The greenhouse effect increases as either of these terms increases.

The real atmosphere does not consist of a single layer with a single temperature. Rather, there is a vertical variation of temperature, with temperature decreasing with increasing height within the troposphere. Thus, T_a in Eq. (3.3) or (3.4) can be regarded as the *effective* emitting temperature of the atmosphere. It depends on the variation of temperature with height, weighted by the fraction of total emissivity in each layer of the atmosphere, and summed over all layers. The effective radiating temperature

also depends on the extent to which radiation from lower atmospheric layers is absorbed by upper layers (in the case of upward radiation), and vice versa in the case of downward radiation. Since the surface is adjacent to the warmest part of the atmosphere, the effective radiating temperature for downward emission is greater than for upward emission. For this reason, the downward IR emission from the atmosphere is greater than the upward emission in nature (as illustrated in Fig. 2.1), but it is the same ($\varepsilon_a \sigma T_a^4$) in our two-box model (Fig. 3.1) because we have only a single atmospheric temperature.

The primary gases that contribute to the atmospheric emissivity are H_2O, CO_2, O_3, and CH_4. Carbon dioxide and CH_4 are fairly evenly mixed throughout the atmosphere. The ozone concentration varies primarily with height, but this variation is relatively fixed in time. The water vapour concentration, however, varies strongly in space and in time. If a given amount of water vapour were to be distributed higher in the troposphere, this would increase the emissivity of the cold layers and decrease the emissivity of the warm layers, thereby reducing the effective radiating temperature. From Eq. (3.4), this would increase the greenhouse effect (the emission to space would be reduced). Another way to view this result is as follows: increasing the emissivity of a warm layer or a cold layer by a given amount would increase the absorption of surface radiation by the same amount, but the increase in emission from the atmosphere would be smaller if the extra emissivity is added to the colder layer. Thus, water vapour (or any other GHG) is more effective as a greenhouse gas in the upper troposphere (where temperatures are colder) than in the lower troposphere.

There is another reason why a given change in the amount of water vapour in the upper troposphere will have a greater impact on the greenhouse effect than the same change in the lower troposphere: the increase in emissivity resulting from a given increase in the amount of water vapour is greater the smaller the initial amount of water vapour (Spencer and Braswell, 1997). Thus, a given increase in water vapour is more effective in trapping heat if the increase occurs in the upper troposphere than in the lower troposphere because the change in emissivity will be larger in the upper troposphere, and (as discussed above) because the effect of a given change in emissivity is greater in the upper troposphere.

Although the atmosphere emits radiation to the surface, this downward emission does not constitute the greenhouse effect. The trapping of outgoing radiation is far more important to the surface temperature than is the downward IR emission. This is because the atmosphere and surface are tightly coupled by sensible and latent heat fluxes. Downward emission redistributes heat between the surface and atmosphere without affecting the total amount of heat available, whereas the partial trapping of radiation emitted to space alters the heat balance of the combined atmosphere + surface system. This point can be demonstrated quantitatively (albeit approximately) by modifying the two-box model to include latent and sensible heat fluxes between the surface and atmosphere, as shown in Box 3.1.

3.2   Radiative forcing

In the preceding section it was stressed that the effect of IR-active gases on surface climate arises largely from the reduction of infrared radiation to space, rather than from the downward emission from the atmosphere. This is because the surface and atmosphere are tightly coupled by non-radiative heat exchanges. We can refine this picture, however, by distinguishing between the stratosphere and troposphere and by noting that, whereas the troposphere and surface are tightly coupled, the coupling between the stratosphere and troposphere is rather weak. Thus, if the solar luminosity were to change, or the concentration of GHGs or aerosols were to change, the surface–troposphere system and stratosphere can respond more or less independently of each other. For this reason, changes in the surface temperature are driven by changes in the net radiation *at the tropopause*, not at the surface or at the top of the atmosphere. This change in net radiation is called the *radiative forcing*, because it is what forces or drives the change in surface climate. The radiative forcing involves the change in the net downward flux of solar energy at the tropopause, the change in the upward emission of infrared radiation at the tropopause, and the change in the downward emission of infrared radiation from the stratosphere. The change in net radiation at the tropopause before any temperatures are allowed to change is called the *instantaneous radiative forcing*. As stratospheric temperatures change in response to the perturbation in the stratospheric radiative energy balance, the downward emission of infrared radiation will change, which will alter the subsequent change in the tropospheric and surface temperatures. Of course, any change in surface–troposphere temperatures will also alter the radiative balance of the surface–troposphere system (by changing the upward emission of infrared radiation), but such changes are part of the feedbacks which determine the ultimate response of the surface and troposphere. Because the stratosphere responds quickly (within months) and independently of the surface–troposphere system, the effect of any changes in stratospheric temperatures should be included when computing the radiative forcing. This gives the *adjusted radiative forcing*.

In the case of a doubling of the atmospheric concentration of CO_2, the global mean instantaneous forcing, as computed by a variety of researchers, is about 4.0–4.5 W m^{-2} (Cess *et al.*, 1993). Of this, about 1.5 W m^{-2} – or 40% – is due to extra downward emission from the stratosphere. This extra downward radiation causes the stratosphere to cool at the same time that the surface and troposphere warm. The cooling of the stratosphere in turn reduces the downward emission of radiation from the stratosphere by about 0.5 W m^{-2} (for a doubling of CO_2), thereby offsetting part of the initial increase in downward emission caused by the increase in CO_2. Hence, the global mean adjusted radiative forcing for a CO_2 doubling is 3.5–4.0 W m^{-2}. On the other hand, the increase in downward radiation at the surface is only about 1.4 W m^{-2} in the global mean (Section 7.1), but – as shown in Box 3.1 – the surface temperature change does not depend on the direct surface forcing. These flux changes and the associated changes at the top of the atmosphere are shown in Fig. 3.2. Note that, before adjustment of stratospheric temperatures, the reduction in upward radiation at the top of the atmosphere is considerably less than the forcing at the tropopause, whereas after adjustment, the two are the same.

To summarize, then, the surface and troposphere respond as a tightly coupled system to any of the factors that alter the Earth's radiative energy balance. To a good approximation, this coupled response is governed by the change in net radiation at the tropopause, after accounting for any adjustments in stratospheric temperatures that might occur.

Comparison of instantaneous and adjusted forcings

Table 3.1 lists the instantaneous and adjusted direct radiative forcings for changes in the concentration of selected GHGs. Details on how these and other

---

Box 3.1 Further analysis of the natural greenhouse effect

In this box it will be demonstrated that the effect on surface temperature due to the downward emission of infrared radiation by greenhouse gases is an order of magnitude smaller than the effect on surface temperature arising from the reduction in outgoing emission to space.

Consider a two-box model, where Q_a^* and Q_s^* are the rate of absorption of solar energy by the atmosphere and surface, respectively, $L\uparrow$ and $L\downarrow$ are the upward and downward infrared fluxes at the surface, and H is the sum of the sensible and latent heat transfer between the surface and atmosphere. The emission to space, L_{sp}, is equal to $L\uparrow - G$, where G is the greenhouse effect. For a steady-state climate, the net energy flow into each box is zero. Thus, we can write

$$Q_a^* + G + H - L\downarrow = 0 \tag{3.1.1}$$

and

$$Q_a^* + L\downarrow - L\uparrow - H = 0 \tag{3.1.2}$$

If a heating perturbation is applied, the temperatures of both the surface and atmosphere, and the associated energy flows, will adjust in such a way as to restore zero net energy input for both boxes. If $F(T_0)$ is the value of the energy flux F for the initial steady-state climate, characterized by temperature T_0, then the value of F for some new temperature T can be approximated by the first-order Taylor Series expansion, $F(T) = F(T_0) + (dF/dT)\Delta T$. Now suppose that we are going to change $L\downarrow$ by an amount $\Delta L\downarrow$. For simplicity, we shall neglect changes in the absorption of solar energy $(Q_a^* + Q_s^*)$, as such changes do not significantly change the results. The equations expressing the new balance that will eventually be achieved by the atmosphere and surface are

$$-\Delta L\downarrow + \left(\frac{\partial G}{\partial T_a} + \frac{\partial H}{\partial T_a} - \frac{\partial L\downarrow}{\partial T_a}\right)\Delta T_a + \left(\frac{\partial G}{\partial T_s} + \frac{\partial H}{\partial T_s}\right)\Delta T_s = 0 \tag{3.1.3}$$

and

$$\Delta L\downarrow + \left(\frac{\partial L\downarrow}{\partial T_a} - \frac{\partial H}{\partial T_a}\right)\Delta T_a - \left(\frac{\partial L\uparrow}{\partial T_s} + \frac{\partial H}{\partial T_s}\right)\Delta T_s = 0 \tag{3.1.4}$$

respectively. Equations (3.1.3) and (3.1.4) can be added to yield

$$\left(\frac{\partial G}{\partial T_a}\right)\Delta T_a + \left(\frac{\partial G}{\partial T_s} - \frac{\partial L\uparrow}{\partial T_s}\right)\Delta T_s = 0 \tag{3.1.5}$$

As discussed in Section 3.1, G is given by $\varepsilon_a(\sigma T_s^4 - \sigma T_a^4)$, where T_a is the effective radiating temperature of the atmosphere. The atmospheric emissivity, ε_a, depends on the concentration of gases such as CO_2 and CH_4, which we can regard as being independent of temperature, and on the concentration of water vapour. As temperature increases, the saturation vapour pressure and hence the amount of water vapour in the atmosphere increase rapidly, but as water vapour amount increases the emissivity saturates at a maximum value of 1.0. For temperature deviations around some reference temperature T_0, it is convenient to approximate ε_a as

$$\varepsilon_a = a + b\,(T_a - T_0)^n \tag{3.1.6}$$

Thus, the greenhouse effect G is given by

$$G = (a + b\,(T_a - T_0)^n)(\sigma T_s^4 - \sigma T_a^4)$$

The total turbulent flux H can be written as $H = c\,(T_s - T_a)$, while $L\uparrow = \sigma T_s^4$ and $L\downarrow = \varepsilon_a(T_a + \delta T)^4$, where δT accounts for the fact that the effective temperature for downward radiation is greater than for upward radiation (as discussed in Section 3.1). Given a present-day atmospheric emissivity, ε_0, of about 0.73 and choosing $n = 1$ and $T_0 = 200\,K$, we require that $a = 0.4086$ and $b = 0.00824$. Appropriate values of the other constants appearing in the above expressions are as follows: $\sigma = 5.67 \times 10^{-8}\,W\,m^{-2}\,K^{-4}$, $T_s = 289\,K$, $T_a = 239\,K$, $\delta T = 30\,K$, and $c = 20\,W\,m^{-2}\,K^{-4}$. We can then evaluate the partial derivatives appearing in Eqs (3.1.3)–(3.1.5) as follows:

$$\frac{\partial G}{\partial T_a} = \frac{d\varepsilon_a}{dT_a}\,(\sigma T_s^4 - \sigma T_a^4) - \varepsilon_0 4\sigma T_a^3 = 1.734 - 2.260 = -0.526\,W\,m^{-2}\,K^{-1} \tag{3.1.7}$$

$$\frac{\partial G}{\partial T_s} = \varepsilon_0 4\sigma T_s^3 = 4.00\,\mathrm{W\,m^{-2}\,K^{-1}} \tag{3.1.8}$$

$$\frac{\partial L\!\uparrow}{\partial T_s} = 4\sigma T_s^3 = 5.474\,\mathrm{W\,m^{-2}\,K^{-1}} \tag{3.1.9}$$

$$\frac{dL\!\downarrow}{dT_a} = 4\varepsilon_0\sigma(T_a + \delta T)^3 = 3.22\,\mathrm{W\,m^{-2}\,K^{-1}} \tag{3.1.10}$$

$$\frac{\partial H}{\partial T_s} = -\frac{\partial H}{\partial T_a} = c = 20\,\mathrm{W\,m^{-2}\,K^{-1}} \tag{3.1.11}$$

The first term on the right-hand side of Eq. (3.1.7) is the effect of increasing atmospheric emissivity as atmospheric temperature increases, and it is positive. The second term is the effect of increasing atmospheric temperature with fixed emissivity and surface temperature, and it is negative. That is, the greenhouse effect decreases as atmospheric temperature increases with fixed surface temperature and emissivity.

Inserting the appropriate derivative values into Eq. (3.1.5) leads to the result that $\Delta T_a = -2.80\Delta T_s$. From Eq. (3.1.4) we obtain

$$\Delta T_s = \frac{\Delta L\!\downarrow}{2.80\left(\dfrac{\partial L\!\downarrow}{\partial T_a} - \dfrac{\partial H}{\partial T_a}\right) + \left(\dfrac{\partial L\!\uparrow}{\partial T_s} + \dfrac{\partial H}{\partial T_s}\right)} = \frac{\Delta L\!\downarrow}{90.5} \tag{3.1.12}$$

and from the above, it follows that

$$\Delta T_a = -\frac{\Delta L\!\downarrow}{32.3} \tag{3.1.13}$$

Thus, an increase in $L\!\downarrow$ by $10\,\mathrm{W\,m^{-2}}$ would cool the atmosphere by about 0.3 K and warm the surface by about 0.1 K. The reason the temperature responses are so low is the strong coupling between the atmosphere and surface, and the most important contributor to this is the turbulent heat flux. This is reflected in the fact that the largest terms in the denominator to Eq. (3.1.12) are the $\partial H/\partial T_a$ and $\partial H/\partial T_s$ terms.

Now, assume that instead of perturbing $L\!\downarrow$, we perturb G by an amount ΔG. As before, T_a and T_s will adjust so as to restore zero net energy balance for both the atmosphere and surface. The equations expressing the new balance are

$$\Delta G + \left(\frac{\partial G}{\partial T_a} + \frac{\partial H}{\partial T_a} - \frac{\partial L\!\downarrow}{\partial T_a}\right)\Delta T_a + \left(\frac{\partial G}{\partial T_s} + \frac{\partial H}{\partial T_s}\right)\Delta T_s = 0 \tag{3.1.14}$$

and

$$\left(\frac{\partial L\!\downarrow}{\partial T_a} - \frac{\partial H}{\partial T_a}\right)\Delta T_a - \left(\frac{\partial L\!\uparrow}{\partial T_s} + \frac{\partial H}{\partial T_s}\right)\Delta T_s = 0 \tag{3.1.15}$$

From Eq. (3.1.15) it follows that $\Delta T_a = 1.1\Delta T_s$. Substituting into Eq. (3.1.14) yields

$$\Delta T_s = \frac{\Delta G}{1.1\left(\dfrac{\partial G}{\partial T_a} + \dfrac{\partial H}{\partial T_a} - \dfrac{\partial L\!\downarrow}{\partial T_a}\right) + \left(\dfrac{\partial G}{\partial T_s} + \dfrac{\partial H}{\partial T_s}\right)} = \frac{\Delta G}{2.12} \tag{3.1.16}$$

and, from the above

$$\Delta T_a = -\frac{\Delta G}{1.93} \tag{3.1.17}$$

Thus, an externally imposed $10\,W\,m^{-2}$ increase in G requires a warming of about $5\,K$. Careful inspection of Equation (3.1.16) indicates that the two turbulent heat flux terms, $\partial H/\partial T_a$ and $\partial H/\partial T_s$, largely cancel, so that the temperature response is governed by the radiative damping (which is much weaker). Furthermore, the total radiative damping appearing in Eq. (3.1.16) is related, and very similar in magnitude, to the total radiative damping to space, $\partial L_{sp}/\partial T_a$, as the reader should be able to show.

To summarize, when the downward infrared flux is altered, there are opposing radiative perturbations for the atmosphere and surface, and the response of both is governed by turbulent heat exchange. When a net heating perturbation is applied to the atmosphere + surface system, the response is governed by the radiative damping to space. Since radiative damping to space is much weaker than turbulent heat flux damping, the surface (or air) temperature response is much larger in the second case than in the first case.

anthropogenic forcings were computed are given in Chapter 7, after the mathematical tools used for their computation have been introduced. The instantaneous and adjusted radiative forcings are almost identical in all cases except for CO_2, where the adjusted forcing is about 7% smaller than the instantaneous forcing, and for depletion of stratospheric O_3, where the two forcings differ in sign.

The size of the difference between the instantaneous and adjusted radiative forcings is a direct reflection of the extent of adjustment in stratospheric temperature that occurs. The main energy balance in the stratosphere (in the global mean) is between absorption of ultraviolet radiation by O_3

and emission of infrared radiation by CO_2. Thus, an increase in CO_2 concentration cools the stratosphere, causing the adjusted forcing to be smaller than the instantaneous forcing. Because O_3 is the prime gas responsible for heating the stratosphere, a decrease in stratospheric O_3 causes a large cooling of the stratosphere and a reduction in downward radiation that more than offsets the extra penetration of solar radiation. This in turn changes the sign of the surface–troposphere forcing from positive to negative. A more extensive analysis, explaining the reasons for the differences between CO_2 and other well-mixed gases, is presented in Section 10.1 (Box 10.1).

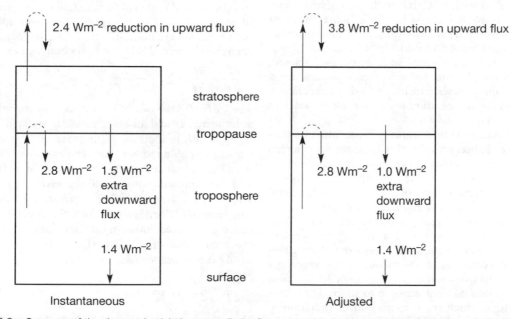

Fig. 3.2   Summary of the changes in global mean radiative fluxes associated with a doubling of the atmospheric CO_2 concentration, both before ("Instantaneous") and after ("Adjusted") statospheric temperatures have fully responded to the change in CO_2 concentration. After stratospheric adjustment, the change in net radiation at the top of the atmospheric equals the change (net forcing) at the tropopause, since otherwise further changes in stratospheric temperature would be required.

Table 3.1 Instantaneous and adjusted direct radiative forcings for changes in the concentration of various greenhouse gases, as given by Shine *et al.* (1995), except for SF_6, which is the average of the clear and cloudy sky forcings as computed by Myhre and Stordal (1997). Instantaneous and adjusted radiative forcings for a wide range of other gases can be found in Myhre and Stordal (1997) and Christidis *et al.* (1997). See also Table 7.1. The difference between the instantaneous and adjusted radiative forcing is due to the change in stratospheric temperature (ΔT_{st}) that is induced by the change in gas concentration, which is given in the last column.

Forcing mechanism	Instantaneous forcing ($W\,m^{-2}$)	Adjusted forcing ($W\,m^{-2}$)	ΔT_{st}
CO_2: 300 → 600 ppmv	4.63	4.35	Cooling
CH_4: 0.28 → 0.56 ppmv	0.53	0.52	Slight cooling
N_2O: 0.16 → 0.32 ppmv	0.96	0.93	Slight cooling
CFC-11: 0 → 1 ppbv	0.21	0.22	Warming
CFC-12: 0 → 1 ppbv	0.26	0.28	Warming
SF_6: 0 → 1 ppbv	0.56	0.62	Warming
Stratospheric O_3: 50% reduction	0.26	−1.23	Strong cooling
Solar luminosity: 2% increase	4.48	4.58	Warming

3.3   Concept of radiative damping

The temperatures of the atmosphere and surface tend to adjust themselves such that there is a balance between the absorption of energy from the sun and the emission of infrared radiation to space. If, for example, there were to be an excess of absorbed solar energy over emitted infrared radiation (as occurs with the addition of GHGs to the atmosphere), temperatures would increase but, in so doing, the emission of infrared radiation to space would increase. This would reduce the initial imbalance, and eventually a new balance would be achieved, but at a new, warmer temperature. The more rapidly infrared emission to space increases with increasing temperature, the less the temperature must increase in order to restore balance.

However, as temperatures increase, other quantities also change, which further alters the net emission of radiation to space. For example, an increase in the amount of water vapour as temperature increases tends to counteract the effect of warmer temperatures in increasing the emission of infrared radiation to space. As a result, infrared emission would increase more slowly and a greater temperature increase would be required to restore radiative balance. A reduction in the amount of solar energy reflected to space would also tend to offset the effect of increasing infrared emission to space. The resulting rate at which the net emission of radiation to space increases is called the *radiative damping*, λ. The term "damping" arises because it is the increase in net emission to space as temperatures warm that limits or dampens the increase in temperature. A

larger radiative damping results in a smaller temperature increase. The radiative damping is given by

$$\lambda = \frac{dF}{dT} - \frac{dQ}{dT} \tag{3.5}$$

where F is the global mean emission of infrared radiation to space, Q is the global mean absorption of solar radiation, and T is the global mean temperature (strictly speaking, F should be evaluated at the tropopause and Q should be based on the absorption of solar radiation by the surface and troposphere only, since we are interested in surface temperature change). The term dF/dT can be decomposed as

$$\frac{dF}{dT} = \frac{\partial F}{\partial T} + \sum_i \frac{\partial F}{\partial I_i}\frac{dI_i}{dT} \tag{3.6}$$

where $\partial F/\partial T$ is the rate of increase in F when temperature increases and all other variables are held constant, $\partial F/\partial I_i$ is the rate of increase in F as some climatic variable I_i other than temperature changes, dI_i/dT is the rate of change in I_i with temperature, and the summation is over all the variables I_i that affect the emission of infrared radiation to space. The term $\partial F/\partial T$ arises solely from the direct dependence of infrared emission on temperature through the Stefan–Boltzmann Law (Eq. (3.1)). The term dQ/dT can be decomposed as

$$\frac{dQ}{dT} = \sum_i \frac{\partial Q}{\partial I_i}\frac{dI_i}{dT} \tag{3.7}$$

Here, there is no $\partial Q/\partial T$ term because the absorption of solar radiation does not depend directly on temperature in any way, but only through the effect of temperature on other variables (such as the amount of snow and ice, the amount of water vapour, or the amount and properties of clouds). Given λ, the final or steady-state temperature change is given by

$$\Delta T = \frac{\Delta R}{\lambda} \tag{3.8}$$

where ΔR is the adjusted radiative forcing.

Linkage to climatic feedbacks

In Section 2.9 we defined a feedback as a situation where a change in some quantity, ΔA, leads to a change in a second quantity, ΔB, and the change ΔB tends to provoke a further change in A. Here, ΔA is the initial heating perturbation due to the radiative forcing (ΔR), and ΔB is the change in temperature. Temperature changes lead to further changes in radiative heating through the dependence of various internal variables on temperature, dI_i/dT, and the dependence of net emission to space on these internal variables, $\partial F/\partial I_i - \partial Q/\partial I_i$. Since an increase in F has a cooling effect and an increase in Q has a heating effect, the change in radiative heating due to feedback involving variable i, ΔJ_i, is given by

$$\Delta J_i = -\left(\frac{\partial F}{\partial I_i} - \frac{\partial Q}{\partial I_i}\right)\frac{dI_i}{dT}\Delta T = \frac{\partial N_i}{\partial I_i}\frac{dI_i}{dT}\Delta T \tag{3.9}$$

where $N = Q - F$ is the net radiation at, strictly speaking, the tropopause. If ΔR and ΔJ_i are both positive or both negative, then the feedback involving variable i is a positive feedback since it reinforces the initial heating perturbation. If $\Delta R > 0$, then $\Delta T > 0$. To have $\Delta J_i > 0$, we must have $(\partial N/\partial I_i)dI_i/dT > 0$, which means that $(\partial F/\partial I_i)dI_i/dT < 0$ or $(\partial Q/\partial I_i)dI_i/dT > 0$. Both of these conditions reduce the radiative damping (see Eq. (3.5)). From Eq. (3.8), this leads to a larger temperature change, which is what one expects for a positive feedback. The reverse applies to a negative feedback. Further analysis is presented in Box 3.2.

3.4 Temperature response without feedbacks

In the absence of any feedbacks except the increase in temperature itself, the only way that the emission of radiation to space increases is through the direct dependence of emission on temperature through the Stefan–Boltzmann Law (Eq. (3.1)). This is equivalent to the statement $\lambda = \partial F/\partial T$, but $\partial F/\partial T$ must be evalu-

ated using the effective radiating temperature T_e of the combined earth–atmosphere system. T_e is the temperature of a blackbody with the same emission to space as the earth–atmosphere system, so it is given by

$$T_e = \left(\frac{F}{\sigma}\right)^{0.25} \tag{3.10}$$

Given that $F = 240\,\mathrm{W\,m^{-2}}$, $T_e = 255\,\mathrm{K}$. $\partial F/\partial T$ is given by the derivative of Eq. (3.1) evaluated at T_e, that is,

$$\frac{\partial F}{\partial T} = 4\,\sigma T_e^3 \tag{3.11}$$

Thus, $\lambda = \partial F/\partial T = 3.76\,\mathrm{W\,m^{-2}\,K^{-1}}$. As discussed in Section 3.2, a doubling of atmospheric CO_2 causes a global mean radiative surplus ΔR of 3.5–$4.0\,\mathrm{W\,m^{-2}}$. The temperature increase required to restore radiative balance in this case would be $\Delta R/\lambda = 1.0\,\mathrm{K}$ (a warming of $1.0°C$). However, as noted above, the very change in temperature would cause other atmospheric and surface properties to change, which would lead to further alterations in the energy balance and require further temperature changes through a series of feedback processes. These feedback processes are identified below and discussed in more detail in Chapter 9.

3.5 Climate feedbacks

Feedbacks play a major role in translating a radiative forcing into a change in climate. They do this by altering the radiative damping. Among the feedbacks that have to be considered are the following.

Water vapour amount

In a warmer climate the atmospheric concentration of water vapour will increase. Because water vapour is a greenhouse gas, this is a positive feedback – that is, it amplifies the initial change.

Water vapour distribution

If, as the total water vapour amount increases, the percentage increase is greater at higher altitudes than at lower altitudes, this will result in a greater climate warming than if the water vapour amount increases by the same percentage at all altitudes. This is because the higher the water vapour, the more effective it is as a greenhouse gas, as explained in Section 3.1. The case where the percentage increase in water vapour is larger at higher altitude is equivalent to increasing the water vapour amount by the same percentage at all

Box 3.2 Linear feedback analysis

In this box we introduce and use concepts taken from engineering linear systems analysis to gain further insights into the interaction between multiple positive and negative feedbacks in the climate system. Schlesinger (1985) was among the first to apply these concepts to climatic feedbacks.

Consider first the case where there are no feedbacks except through the direct dependence of infrared emission on temperature. If ΔR is the radiative forcing, the temperature response ΔT can be written as

$$\Delta T = G_0 \Delta R \tag{3.2.1}$$

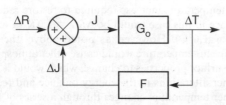

Fig. 3.2.1 Block diagram illustrating a climate feedback loop. ΔR is the radiative forcing, G_0 is the system gain, and ΔJ is the subsequent change in net radiation due to internal feedbacks F. Redrafted from Schlesinger (1985).

where G_0 is referred to as the *system gain*. By comparison of Eq. (3.2.1) with Eq. (3.8), we see that $G_0 = (\partial F / \partial T)^{-1}$. When indirect feedbacks are allowed, the change in temperature leads to a further change, ΔJ, in the net radiation, which then feeds back into the system, as shown in Fig. 3.2.1. The temperature response is now given by

$$\Delta T = G_0(\Delta R + \Delta J) \tag{3.2.2}$$

On the assumption that the individual feedbacks are independent of one another, then $\Delta J = \Sigma \Delta J_i$. The ΔJ_i depend on ΔT according to the relationship given in Eq. (3.9), which can be written as $\Delta J_i = F_i \Delta T$. Letting $F = \Sigma F_i$, we can rewrite Eq. (3.2.2) as

$$\Delta T = G_0(\Delta R + F \Delta T) \tag{3.2.3}$$

Solving for ΔT, we obtain

$$\Delta T = \left(\frac{G_0}{1 - f}\right)\Delta R \tag{3.2.4}$$

where $f = F G_0 = \Sigma f_i$, and

$$f_i = G_0 \frac{\partial N}{\partial I_i}\frac{dI_i}{dT} = G_0 \frac{\Delta J_i}{\Delta T} \tag{3.2.5}$$

The variables f and f_i are referred to as the *feedback factor*. The magnitude of f_i is a measure of the strength of the feedback involving variable i; it is positive for a positive feedback and negative for a

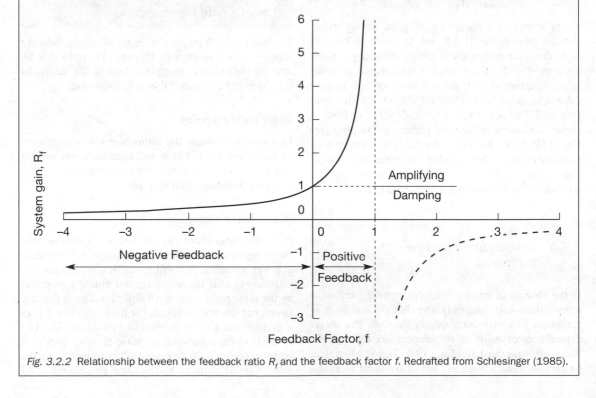

Fig. 3.2.2 Relationship between the feedback ratio R_f and the feedback factor f. Redrafted from Schlesinger (1985).

negative feedback. The ratio of ΔT with all feedbacks to ΔT without indirect feedbacks, R_f, is equal to $1/(1-f)$. Figure 3.2.2 shows R_f as a function of f. In deriving the results leading up to Fig. 3.2.2, we have assumed that (i) the strength of each feedback is independent of all the other feedbacks, and (ii) the strength of the feedback does not change as the magnitude or sign of ΔR and ΔT changes. In other words, we have assumed that the system is *linear*. Nevertheless, the increase in ΔT with increasing f (as more positive feedbacks are added) is decidedly non-linear: a given increase in f has a greater absolute effect on ΔT the greater the initial value of f. Thus, the effect of a given feedback on ΔT depends on what other feedbacks are already present. Another important lesson that can be drawn from Fig. 3.2.2 is that an infinitely strong negative feedback cannot reverse the sign of the temperature response, but rather drives the response to zero.[a]

The assumption of constant and independent feedbacks has been evaluated by Schlesinger (1985) and is a reasonably valid approximation except when combining multiple feedbacks involving clouds (such as cloud amount and cloud optical depth), in which there are important interactions between the feedbacks. However, this does not invalidate the important insights that we have derived from Fig. 3.2.2.

---

[a] Hansen *et al.* (1997a) found that, when O_3 is added to the surface layer only in their atmospheric model, the resultant negative cloud feedback can reverse the sign of the surface temperature response, but this reversal is dependent on the direct forcing being very small, and on opposing temperature changes occurring at the surface and in the lowest atmospheric layer – something that does not happen with less idealized forcings.

altitudes, then shifting the distribution upward. This serves as a positive feedback. Conversely, a downward shift in the water vapour distribution as the climate warms would serve as a negative feedback.

Clouds

Changes in clouds can serve as a positive or negative feedback on climate. Clouds have a cooling effect by reflecting sunlight; this effect depends on the difference between the cloud and surface albedos, and on the amount of incident solar radiation. The smaller the underlying albedo and the greater the incident solar radiation, the greater the increase in the reflection of solar radiation to space due to clouds. The cooling effect thus depends on where and when the clouds occur. Clouds exert a warming effect by absorbing the infrared radiation emitted by the surface and re-emitting a smaller amount of radiation owing to the fact that the cloud top is colder than the surface. Since higher clouds are colder and thus emit less radiation, the warming effect of clouds depends on how high they are. High clouds tend to have a net warming effect on climate, while low clouds have a net cooling effect. The net feedback of changes in clouds depends on a large number of variables, so even the sign of the feedback is uncertain.

Atmospheric temperature structure

Changes in the rate of decrease of tropospheric temperature with height (or *lapse rate*) will alter the relationship between surface temperature and infrared emission to space, thereby exerting a feedback effect on climate. If the lapse rate decreases as climate warms, then the upper troposphere warms faster than the lower troposphere, and by more than if the lapse rate were constant. This in turn means that the infrared emission to space increases faster as the surface temperature increases than for a constant lapse rate, so that less surface temperature warming is required in order to restore radiative balance. Thus, a decrease in lapse rate with warming serves as a negative feedback. Conversely, an increase in lapse rate with warming serves as a positive feedback.

Areal extent of ice and snow

A reduction in the area of sea ice and seasonal snow-cover on land as climate warms will reduce the surface reflectivity, thereby tending to produce greater warming (a positive feedback). However, concurrent changes in cloud cover that could be induced by the change in ice or snowcover could significantly alter the net feedback (Randall *et al.*, 1994).

Vegetation

Changes in the distribution of different biomes or in the nature of vegetation within a given biome can also lead to changes in the surface reflectivity, thereby exerting a feedback effect on climatic change.

The carbon cycle

The effect of climate on the terrestrial biosphere and the oceans is likely to alter the sources and sinks of CO_2 and CH_4, leading to changes in their atmospheric concentrations and hence causing a radiative feedback.

Of these feedbacks, those involving water vapour, the lapse rate, and clouds respond essentially instantaneously to climatic change, while those involving sea ice and snow respond within a few years. These feedbacks are therefore referred to as "fast" feedbacks. Some vegetation and carbon cycle processes are relevant on a time scale of decades. Other feedbacks, such as dissolution of carbonate sediments in the ocean and enhanced chemical weathering on land (both of which tend to reduce the atmospheric CO_2 concentration), require hundreds to thousands of years to unfold. Feedbacks involving continental ice sheets occur over periods of thousands of years.

3.6   Climate sensitivity: definition

The term "climate sensitivity" refers to the ratio of the steady-state increase in the global and annual mean surface air temperature to the global and annual mean radiative forcing. It has units of $K(Wm^{-2})^{-1}$. It is standard practice to include only the fast feedback processes, including changes in water vapour, in the calculation of climate sensitivity, but to exclude possible induced changes in the concentrations of other GHGs (as well as other slow feedback processes). The various fast feedbacks alter the climate sensitivity by altering the radiative damping.

It is common practice to use CO_2 doubling as a benchmark for comparing the climate sensitivities as computed by different computer climate models that simulate the fast feedbacks. The climate sensitivity is often quoted as the globally averaged warming once the climate has fully adjusted to an imposed doubling of atmospheric CO_2. This is generally expected to be between 1.5°C and 4.5°C although, as discussed in Chapter 9, the lower half of this range seems more likely. However, it is more appropriate to restrict the term "climate sensitivity" to the ratio of the global mean temperature response to the global mean radiative forcing, and to refer to the 1.5–4.5°C range as the equilibrium temperature response for a CO_2 doubling rather than calling it the climate sensitivity. Nevertheless, it is often convenient to refer to the CO_2 doubling response as the "sensitivity", and this will be done here when it is clear from the context what is meant.

The climate sensitivity computed for a CO_2 doubling, which entails a global mean radiative forcing of 3.5–4.0 Wm^{-2} with a particular latitudinal and vertical structure, is often used to estimate the climatic response for smaller and larger global mean forcings and when several different forcings are operative at the same time. The validity of this procedure depends on the following assumptions:

- that the climate sensitivity is constant and therefore independent of the magnitude (and sign) of the forcing, so that the temperature response varies linearly with the forcing; and
- that the climate sensitivity does not depend on the spatial (latitudinal) pattern of the forcing, so that it is the same for different forcing mechanisms.

To the extent that these conditions are true, one can simply add up the individual radiative forcings to get the total global mean radiative forcing, and then apply the climate sensitivity obtained for a CO_2 doubling to determine the global mean temperature response. However, there are important circumstances in which these assumptions are not valid, as will be discussed.

Apart from the question of whether or not the global mean temperature response depends on the latitudinal variation in the forcing (and not just its global mean value), there is the additional question of the extent to which the *local* temperature response depends on the *local* forcing rather than on the global mean forcing. This issue and the assumptions underlying the use of the climate sensitivity concept will now be discussed.

3.7   Dependence of climate sensitivity on the magnitude, nature, and spatial pattern of the radiative forcing

The concept of climate sensitivity is useful only to the extent that the climate sensitivity is largely independent of the magnitude, nature, and spatial pattern of the radiative forcing. Otherwise, the climate sensitivity would have to be recomputed for each and every case of interest. The extent to which these conditions are satisfied is discussed below.

Dependence of climate sensitivity on the magnitude and sign of the forcing

Experiments using three-dimensional atmospheric models – the characteristics of which are outlined in Chapter 6 – indicate that the climate sensitivity is indeed close to constant. Marshall *et al.* (1994) obtained a close to linear variation of global mean surface air temperature as the solar luminosity varied from $1300\,W\,m^{-2}$ to $1450\,W\,m^{-2}$ (a forcing change of $26\,W\,m^{-2}$), while Oglesby and Saltzman (1992) found the global mean surface temperature to vary with the natural logarithm of the CO_2 concentration as the concentration varied from 150 ppmv to 1000 ppmv. Since the radiative forcing also varies with the natural logarithm of CO_2 concentration (as discussed in Section 7.7), the temperature changed in direct proportion to the radiative forcing in their simulations. Intuitively, one might expect the climate sensitivity to be larger for a decrease in solar luminosity or CO_2 concentration than for an increase, and to decrease the warmer the climate becomes. This is because, when the limits of ice and snow advance equatorward by a given distance, the area of the latitudinal strip involved is greater than for a retreat, and the incoming solar radiation is larger. Consequently, the snow- and ice-albedo feedback should be stronger for a decrease in solar luminosity or CO_2 than for an increase. This is what Hansen *et al.* (1997a) found. However, there are a number of other non-linearities in climate models, involving the amount of water vapour in the atmosphere and what is known as the "continuum absorption" of infrared radiation by water vapour (Lal and Ramanathan, 1984), that apparently offset the non-linearity in the surface albedo feedback in some models.

Dependence of climate sensitivity on the nature of the forcing

The extent to which climate sensitivity varies with the nature of the forcing depends on the extent to which different forcings induce different feedbacks involving clouds. In the absence of cloud feedbacks, both the vertical and the latitudinal distribution of the forcing have, in most cases, very little effect on the global mean temperature response.

Consider the case without cloud feedbacks. The global mean changes in infrared fluxes that would occur if CO_2 were instantly doubled, after allowing for changes in stratospheric temperatures only, have already been shown in Fig. 3.2. Of the total surface–troposphere heating of about $4.0\,W\,m^{-2}$, 75% is a direct heating of the troposphere and only 25% is a direct heating of the surface. If the solar luminosity were to increase by 1.7%, the total surface–troposphere heating perturbation would also be about $4.0\,W\,m^{-2}$, with the surface absorbing on average an extra $3\,W\,m^{-2}$ and the troposphere an extra $1\,W\,m^{-2}$. In this case, the surface–troposphere partitioning of the heating perturbation is the opposite of that for a CO_2 increase. However, it can be shown through the simple analysis given in Box 3.3 that the final temperature changes for the troposphere and surface should be almost identical to each other in both cases, with the troposphere warming up slightly more than the surface in both cases. The reason for this is that the troposphere and surface are sufficiently tightly coupled that the partitioning of the initial heating perturbation between the two is not important. Thus, as soon as the troposphere begins to warm (in the case of a CO_2 increase), the downward emission of IR radiation to the surface increases and the upward flux of sensible heat decreases. Both of these changes inhibit further tropospheric warming and promote surface warming, but as the surface warms the heat fluxes to the troposphere increase, so that the troposphere does continue to warm up, but closely in step with the surface warming. This raises another important point: although the direct increase in downward emission at the surface due to a CO_2 increase is small ($<1\,W\,m^{-2}$), the increase in downward emission due to the subsequent increase in tropospheric temperature is very large (about $10\,W\,m^{-2}$ in equilibrium). This is another way of explaining why the direct effect of CO_2 on the downward flux at the surface is not important to the surface temperature response.

The preceding analysis assumes that the radiative damping to space is unaffected by the vertical distribution of the forcing. As noted above, this is not valid when cloud feedbacks are allowed. The most important differences in the sensitivity are between well-mixed GHGs, changes in both stratospheric and tropospheric O_3, and additions of absorbing aerosols (primarily soot and sulphate mixed with soot); the differences in climate sensitivity among different well-mixed GHGs are less than 20% according to simulations with 3-D atmospheric models (Wang *et al.*, 1991; Hansen *et al.*, 1997a). The underlying principle is that cloud cover tends to decrease in the atmospheric layers having the greatest direct heating (owing to absorption of solar radiation in the case of absorbing aerosols, or to the net absorption of infrared radiation in the case of O_3). Since clouds at different heights have different net effects on the radiative energy balance, the cloud feedback depends on the vertical distribution of the radiative forcing. This effect is most important for changes in O_3 and in

---

Box 3.3 Role of vertical partitioning of the radiative heating perturbation

In Box 3.1 we showed, as a byproduct of our primary analysis, that the atmospheric and surface temperature responses to a reduction in the outgoing infrared emission to space are almost identical. In this box we explore in more detail the dependence of the atmosphere/surface temperature response ratio on the partitioning of a heating perturbation between the atmosphere and surface. To do this, we recast Eqs (3.1.1) and (3.1.2) in the form

$$Q_a^* + L\!\uparrow + H - L\!\downarrow - L_{sp} = 0 \tag{3.3.1}$$

and

$$Q_a^* + L\!\downarrow - L\!\uparrow - H = 0 \tag{3.3.2}$$

Assume that heating perturbations ΔF_a and ΔF_s are applied to the atmosphere and surface, respectively. The equations expressing the new balance that will eventually be achieved by the atmosphere and surface are then

$$\Delta F_a - \left(\frac{\partial L\!\downarrow}{\partial T_a} - \frac{\partial H}{\partial T_a} + \frac{\partial L_{sp}}{\partial T_a}\right)\Delta T_a + \left(\frac{\partial L\!\uparrow}{\partial T_s} + \frac{\partial H}{\partial T_s} - \frac{\partial L_{sp}}{\partial T_s}\right)\Delta T_s = 0 \tag{3.3.3}$$

and

$$\Delta F_s + \left(\frac{\partial L\!\downarrow}{\partial T_a} - \frac{\partial H}{\partial T_a}\right)\Delta T_a - \left(\frac{\partial L\!\uparrow}{\partial T_s} + \frac{\partial H}{\partial T_s}\right)\Delta T_s = 0 \tag{3.3.4}$$

which can be rewritten as

$$\Delta F_a - \left(\alpha_1 + \frac{\partial L_{sp}}{\partial T_a}\right)\Delta T_a + \left(\alpha_2 - \frac{\partial L_{sp}}{\partial T_s}\right)\Delta T_s = 0 \tag{3.3.5}$$

and

$$\Delta F_s + \alpha_1 \Delta T_a - \alpha_2 \Delta T_s = 0 \tag{3.3.6}$$

where

$$\alpha_1 = \frac{\partial L\!\downarrow}{\partial T_a} - \frac{\partial H}{\partial T_a} = 23.22 \ \mathrm{W\,m^{-2}\,K^{-1}} \tag{3.3.7}$$

and

$$\alpha_2 = \frac{\partial L\!\uparrow}{\partial T_s} + \frac{\partial H}{\partial T_s} = 25.474 \ \mathrm{W\,m^{-2}\,K^{-1}} \tag{3.3.8}$$

In Eqs (3.3.5) and (3.3.6), the radiative damping to space terms ($\partial L_{sp}/\partial T_a$ and $\partial L_{sp}/\partial T_s$) have been separated from the other heat flux damping terms. The latter have been lumped together as α_1 and α_2, where α_1 gives the rate at which the heat flux from the atmosphere to the surface increases as T_a increases, and α_2 gives the rate at which the heat flux from the surface to the atmosphere increases as T_s increases.

From Eq. (3.3.6) we obtain

$$\Delta T_s = \frac{\Delta F_s + \alpha_1 \Delta T_a}{\alpha_2} \tag{3.3.9}$$

The first term in the numerator of Eq. (3.3.9) is the direct heating of the surface and the second term is the additional heating due to the warming of the atmosphere. The surface temperature response is thus given by the total heat input to the surface divided by the ability of the surface to dispose of extra heat (the surface damping). Inserting Eq. (3.3.9) into Eq. (3.3.5), we obtain the following:

$$\Delta T_a = \frac{\Delta F_a + \left(1 - \frac{1}{\alpha_2}\frac{\partial L_{sp}}{\partial T_s}\right)\Delta F_s}{\dfrac{\partial L_{sp}}{\partial T_a} + \dfrac{\alpha_1}{\alpha_2}\dfrac{\partial L_{sp}}{\partial T_s}} \tag{3.3.10}$$

$$\Delta T_s = \frac{\Delta F_a + \left(1 + \frac{1}{\alpha_1}\frac{\partial L_{sp}}{\partial T_a}\right)\Delta F_s}{\frac{\alpha_2}{\alpha_1}\frac{\partial L_{sp}}{\partial T_a} + \frac{\partial L_{sp}}{\partial T_s}}$$

(3.3.11)

With reference to Box 3.1, the $\partial L_{sp}/\partial T_a$ and $\partial L_{sp}/\partial T_s$ terms can be evaluated as

$$\frac{\partial L_{sp}}{\partial T_s} = \frac{\partial L\uparrow}{\partial T_s} - \frac{\partial G}{\partial T_s} = 1.474\ \mathrm{W\,m^{-2}\,K^{-1}}$$

(3.3.12)

$$\frac{\partial L_{sp}}{\partial T_a} = -\frac{\partial G}{\partial T_a} = 0.526\ \mathrm{W\,m^{-2}\,K^{-1}}$$

(3.3.13)

Thus, $(1/\alpha_1)\partial L_{sp}/\partial T_a$ and $(1/\alpha_2)\partial L_{sp}/\partial T_s$ are both much less than 1. As a result, both ΔT_a and ΔT_s depend on the total radiative forcing ($\Delta F_a + \Delta F_s$), with almost no dependence on the breakdown of the total between ΔF_a and ΔF_s. The stronger the surface–atmosphere coupling (i.e., the larger α_1 and α_2) the less dependent ΔT_a and ΔT_s are on the breakdown of the total forcing, while the smaller the climate sensitivity (i.e., the larger $\partial L_{sp}/\Delta T_a$ and $\partial L_{sp}/\Delta T_s$), the more important the breakdown is.

It should be noted that the foregoing results are valid only as long as the various partial derivatives used in the analysis do not themselves depend on the nature of the forcing and its distribution. This is expected to be a good assumption except for cases in which a particular forcing is able to induce different cloud feedbacks from what would otherwise occur. This in turn is expected to occur in the case of moderately to strongly absorbing aerosols and for O_3.

absorbing aerosols, both of which involve strong vertical variations. The difference in the cloud feedback for changes in CO_2 and solar luminosity is also important, with increases in solar luminosity tending to have a greater climate sensitivity than increases in CO_2 because of greater decreases in the amount of low level cloud in the former case (low clouds have a net cooling effect, so their decrease enhances the warming). However, this effect is more than offset – in the simulations of Hansen et al. (1997a) – by differences in surface albedo feedback, so that the final climate sensitivity is modestly smaller (by 20%) for an increase in solar luminosity than for an increase in CO_2.

Different heating perturbations also have a different latitudinal structure. Since both cloud and surface albedo feedbacks vary in strength with latitude, differences in the latitudinal variation in the forcing can also lead to differences in the global mean climate sensitivity. However, as noted above, simulations with atmospheric models show that the climate sensitivities for changes in CO_2, for other well-mixed GHGs, and in solar luminosity are the same to within 20%. There is a tendency for a greater sensitivity for a CO_2 increase than for an increase in solar luminosity because more of the forcing is in

high latitudes in the case of CO_2, where positive snow and ice feedbacks operate. The two sensitivities are within 20% of each other in the simulations of Hansen et al. (1997a), once allowance is made for the fact that they think that the cloud-surface albedo interaction is too strong in their model. Concentrated regional forcing by essentially non-absorbing aerosols (such as pure sulphate) also does not alter the climate sensitivity. This is implied by the simulation results of Cox et al. (1995), in which the global mean temperature response to aerosols scales with the global mean forcing in the same way as for well-mixed GHGs. These results can be understood with the aid of the two-box model discussed in Box 3.4.

In summary, then, neither the different vertical structure nor the different latitudinal structure of the heating perturbation associated with increases in different GHGs or with changes in solar luminosity is particularly important to the global mean climate sensitivity. This is true even when cloud feedbacks are allowed when calculating the sensitivity, and also applies to regionally concentrated non-absorbing aerosols. The climate sensitivities for changes in CO_2, other GHGs, solar luminosity, or non-absorbing aerosols are within ±20% of each other.

Dependence of the local temperature response on the spatial forcing pattern

When the spatial differences in the forcings are not too large (such as between different well-mixed gases or between well-mixed gases and changes in solar luminosity), the variation in the temperature response with latitude is largely independent of the latitudinal structure of the radiative forcing, and depends mainly on the global mean forcing. That is, there is a roughly constant temperature response pattern, and the absolute responses get multiplied everywhere by roughly the same factor when the global mean forcing changes.

This behaviour is demonstrated by simulations (using 3-D atmospheric models) of the temperature response to changes in CO_2 and solar luminosity. Figure 3.3 compares the latitudinal variation in surface–troposphere forcing for a doubling of atmospheric CO_2 and a 2% increase in solar luminosity, while Fig. 3.4 gives the latitudinal variation in the zonally (east–west) averaged surface-air warming as simulated by Hansen et al. (1997a). In both cases the heating perturbation is strongest at low latitudes, with a particularly sharp dropoff with increasing latitude for an increase in solar luminosity. Nevertheless, in both cases the simulated temperature response is smallest at low latitudes and largest at high latitudes, with roughly the same shape to the

variation of the temperature response with latitude. This is because the latitudinal structure of the temperature response is largely determined by feedbacks that are driven by overall warming and not by the spatial structure of the heating perturbation. Thus, although the direct heating effect of an increase in solar luminosity is smallest in polar regions, the warming at lower latitudes induces an increase in the north–south heat flux that overwhelms the direct effect of the heating perturbation at high latitudes and is further amplified by feedbacks involving the melting of ice and snow. The response patterns are most similar when cloud properties and amounts are held constant (Fig. 3.4(a)), but even when clouds are allowed to change, the difference between the two perturbations is still small (Fig. 3.4(b)). A very similar response pattern occurs when the atmospheric loading of natural aerosols increases, even though the sign of the forcing in this case is different between low latitudes and polar regions (Harvey, 1988a). Furthermore, similar response patterns are seen for a variety of past climates, both warmer and colder than the present climate (Section 10.1 and Fig. 10.3). Thus, for many cases, both the global mean temperature response and the regional temperature response are expected to vary in rough proportion to the global mean radiative forcing, and are largely independent of the geographical pattern of the forcing.

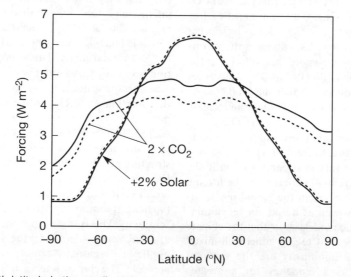

Fig. 3.3   Variation with latitude in the zonally and annually averaged instantaneous (solid lines) and adjusted (dashed lines) radiative forcing for a doubling of atmospheric CO_2 and for a 2% increase in solar luminosity, as computed by Hansen *et al.* (1997a). These results were kindly provided in electronic form to facilitate regraphing.

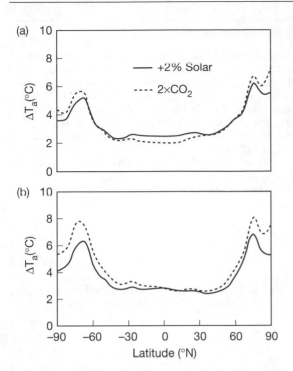

Fig. 3.4 Variation with latitude in the zonally and annually averaged surface air warming in response to a doubling of atmospheric CO_2 and a 2% increase in solar luminosity, as computed by Hansen *et al.* (1997a) for cases with (a) fixed cloud amounts and properties, and (b) variable cloud amounts and properties. These results were kindly provided in electronic form to facilitate regraphing.

3.8 Concept of an equivalent CO_2 increase

Because the global mean temperature response to radiative forcing perturbations involving a mixture of GHG increases depends only on the global mean forcing (to within ±20%), it is useful to determine the increase in CO_2 concentration alone that would give rise to the same global mean forcing as the particular combination of GHG increases that actually occurs. This is referred to as the *equivalent CO_2 increase*. It is not appropriate to include forcings for which the climate sensitivity is different from that for a CO_2 increase, nor is it recommended to include forcings with a strong enough regional variation to produce a markedly different response pattern than for increases in CO_2 or other well-mixed GHGs. Thus, the global mean radiative forcing due to increases in tropospheric aerosols and O_3 should not be lumped in with other radiative forcings before computing the equivalent CO_2 increase.

As an example, consider the radiative forcings shown in Table 2.7 for the increases in GHG concentrations that had occurred by 1995. The CO_2 concentration had increased by 28% and caused an adjusted radiative forcing of 1.40 W m^{-2}. However, the total radiative forcing from well-mixed gases by 1995 was 2.37 W m^{-2}, which is what would be produced by CO_2 alone if it were to increase in concentration by 55%. The equivalent CO_2 increase is thus 55%, and the equivalent CO_2 concentration in 1995 was 430 ppmv.

An exception to the above may occur if the total forcing is large enough that some critical threshold is crossed and the climate system abruptly jumps to some completely different state (having, for example, a completely different ocean circulation pattern). In this case, the regional climate patterns and probably even the global mean temperature response would no longer vary in proportion to the global mean forcing. Furthermore, when the regional variation in the forcing is sufficiently large (and different for different forcings), the regional response patterns will depend on the specific forcings involved and not just on the global mean forcing. This is the case for anthropogenic aerosols, which are highly concentrated spatially. This result is also implied by the analysis in Box 3.4.

3.9 Summary

To summarize this chapter, the response of surface temperature to a perturbation in the radiative fluxes is governed by the change in net radiation at the tropopause, after allowing for adjustment of stratospheric temperatures. This change in net radiation is referred to as the adjusted radiative forcing. The climate sensitivity is the ratio of the global mean temperature response to the adjusted global mean radiative forcing. Three questions related to climate sensitivity and radiative forcing were addressed in this chapter:

1 Is the climate sensitivity constant, such that the temperature response varies linearly with the magnitude of the radiative forcing?

Box 3.4 Role of horizontal partitioning of the radiative heating perturbation

In order to determine the dependence of the low- and high-latitude temperature response on the break-down of radiative forcing between low and high latitudes, we shall consider a two-box climate model. The low-latitude box represents latitudes 0–60° and covers a fraction $A_L = 0.87$ of the Earth, while the high-latitude box represents latitudes 60–90° and covers a fraction $1 - A_L = 0.13$ of the Earth. We do not need to distinguish between the surface and atmosphere for this exercise. The energy budget of each box involves the absorption of solar energy (Q^*), the emission of infrared radiation to space (F), and a transfer of heat (F_M) between the two boxes that we shall assume depends linearly on the difference in temperature between the two boxes. That is, $F_M = \alpha(T_L - T_H)$, where T_L and T_H are the temperatures of the low and high latitude boxes, respectively. Given an observed meridional heat flux at 60°N of about 3×10^{15} W, an interbox temperature difference of 20 K, and a hemispheric area of 1.3×10^{14} m², $\alpha = 1.15$ W m⁻² K⁻¹. The energy balance equations for the two boxes are

$$Q_L^* - F_L - \frac{\alpha}{A_L}(T_L - T_H) = 0 \tag{3.4.1}$$

$$Q_H^* - F_H + \frac{\alpha}{1 - A_L}(T_L - T_H) = 0 \tag{3.4.2}$$

When low and high latitude radiative heating perturbations, ΔF_L and ΔF_H, are applied, the low and high latitude surface temperatures will adjust themselves so as so re-establish zero net energy input for each box. The equations expressing the new balance are

$$\Delta F_L + \left(\frac{\partial Q_L^*}{\partial T_L} - \frac{\partial F_L}{\partial T_L}\right)\Delta T_L - \frac{\alpha}{A_L}(\Delta T_L - \Delta T_H) = 0 \tag{3.4.3}$$

$$\Delta F_H + \left(\frac{\partial Q_H^*}{\partial T_H} - \frac{\partial F_H}{\partial T_H}\right)\Delta T_H + \frac{\alpha}{1 - A_L}(\Delta T_L - \Delta T_H) = 0 \tag{3.4.4}$$

We shall assume that $\partial Q_L^*/\partial T_L = 0$ (where any changes in ice and snow cover can be neglected) and that $\partial F_L/\partial T_L = \partial F_H/\partial T_H = B$. Equations (3.4.3) and (3.4.4) can be solved to yield

$$\Delta T_H = \frac{\Delta F_H + \left[\dfrac{\alpha A_L}{(\alpha + A_L B)(1 - A_L)}\right]\Delta F_L}{\left(\dfrac{\alpha}{1 - A_L}\right) - \left(\dfrac{\alpha^2}{(\alpha + A_L B)(1 - A_L)}\right) + B - \dfrac{\partial Q_H^*}{\partial T_H}} \tag{3.4.5}$$

$$\Delta T_L = \frac{\Delta F_L + \left[\dfrac{\alpha A_L}{\left(\dfrac{\alpha}{1 - A_L}\right) + B - \dfrac{\partial Q_H^*}{\partial T_H}}\right]\Delta F_H}{\dfrac{\alpha}{A_L} - \left(\dfrac{\alpha^2/A_L}{\alpha + (B - \partial Q_H^*/\partial T_H)(1 - A_L)}\right) + B} \tag{3.4.6}$$

For the extreme case where $\alpha = 0$ (no meridional heat transport), we obtain

$$\Delta T_H = \frac{\Delta F_H}{B - \partial Q_H^*/\partial T_H} \tag{3.4.7}$$

$$\Delta T_L = \frac{\Delta F_L}{B} \tag{3.4.8}$$

Thus, the local temperature response depends on only the local forcing and radiative damping when there is no horizontal heat transport, as expected.

In the more general case, it can be seen that

- the lower the climate sensitivity (larger B), the smaller the coefficients in front of the second term in the numerator of Eqs (3.4.5) and (3.4.6), and the more the local response is tied to the local radiative forcing;
- the larger the climate sensitivity, the greater the tendency for the local response to be tied to the global mean radiative forcing.

Substituting $B = 2\,\mathrm{W\,m^{-2}\,K^{-1}}$, $\partial Q_H^* / \partial T_H = 1.0\,\mathrm{W\,m^{-2}}$ (a reasonable value when averaged over the high-latitude box), and the values for α and A_L given above, we obtain

$$\Delta T_H = \frac{\Delta F_H + 2.663\Delta F_L}{6.326} \tag{3.4.9}$$

$$\Delta T_L = \frac{\Delta F_L + 0.172\Delta F_H}{2.244} \tag{3.4.10}$$

These equations display the following interesting properties:

- For a globally uniform forcing of $4\,\mathrm{W\,m^{-2}}$, the low- and high-latitude warmings are $2.09\,\mathrm{K}$ and $2.32\,\mathrm{K}$, respectively, compared with a warming of $2\,\mathrm{K}$ that would occur in both cases without high-latitude albedo feedback.
- The low-latitude radiative forcing has a disproportionately greater effect on the high-latitude response than the high-latitude radiative forcing, owing to heat transport between the boxes and the much greater area of the low-latitude box.
- Because of the above, it is possible for high-latitude warming to occur when the direct forcing at high latitudes is one of cooling, if $|\Delta F_H| < 2.66\,|\Delta F_L|$.
- For a sufficiently large difference in the magnitude of the regional forcings, the sign of the local response will be equal to that of the local forcing, in spite of heat transport between the boxes.

Although these results have been obtained using a highly simplified two-box model, they capture the essence of what occurs in much more complex models, and thus provide considerable insight into the behaviour of these models and also of the real world.

2 Under what conditions does the global mean temperature response depend only on the total global mean radiative forcing, and not on the spatial structure of the forcing and hence on the particular causes of the radiative forcing?

3 To what extent is the spatial pattern of the temperature response independent of the spatial pattern of the radiative forcing, and dominated instead by regionally varying feedbacks that are driven by overall climatic warming?

With regard to the first question, the climate sensitivity is not constant, but the change in its strength over a modest range of forcing magnitudes is likely to be quite small compared with the underlying uncertainty (a factor of three) in the magnitude of the climate sensitivity. With regard with the second question, the global mean temperature response and its spatial structure do not depend to any great extent on the spatial structure of the heating perturbation for changes in the concentration of well-mixed gases, in solar luminosity, or in natural aerosols. Rather, the global mean temperature response and its spatial variation are governed by the global mean radiative forcing and the feedbacks that come into play. For forcings involving large increases in non-absorbing aerosols, the geographical pattern of the temperature response (but not the global mean temperature response) depends on the geographical pattern of the forcing. For increases in absorbing aerosols and for changes in stratospheric and tropospheric O_3, the climate sensitivity appears to be different from those of other forcings, owing to the fact that different cloud feedbacks come into play. However, in this case the possible differences in the climate sensitivity between aerosol increases and other forcing factors are small compared with the enormous uncertainty in the aerosol forcing. These conclusions are summarized in Table 3.2.

Table 3.2 Characterization of different forcing mechanisms based on present understanding. It is assumed that no abrupt changes in the climate system occur.

Forcing change	Climate sensitivity same as for a change in CO_2?	Regional temperature response varies with global mean forcing?
Well-mixed GHGs	Yes, within ±20%	Yes
Solar luminosity	Yes, within ±20%	Yes
Stratospheric O_3	No[a]	No
Tropospheric O_3	No[b]	No
Stratospheric aerosols[c]	Yes, within ±20%	Yes
Non-absorbing tropospheric aerosols	Yes	Yes or No[d]
Absorbing tropospheric aerosols	No	Yes or No[d]

[a] According to Hansen *et al.* (1977a), the climate sensitivity is comparatively small for changes in O_3 within about 10 km of the tropopause, and comparatively large for changes higher in the stratosphere. However, as discussed in Section 7.2, the apparent sensitivity to changes in stratospheric O_3 depends on how the tropopause is defined – a factor not taken into account in the Hansen *et al.* (1997) analysis. Nevertheless, the sensitivity of climate to changes in stratospheric O_3 probably does differ from that for changes in CO_2 owing to differing changes in cloud cover in the upper troposphere and lower stratosphere.

[b] According to Hansen *et al.* (1997a), climate sensitivity is comparatively large for changes in O_3 in the lowest 5 km of the troposphere.

[c] Assumed to be non-absorbing.

[d] Depends on how regionally concentrated the aerosol forcing is. For increases in natural aerosols with a latitudinal structure such as might have plausibly occurred during the last ice age, the response pattern is largely independent of the forcing pattern (see Harvey, 1988a). For increases in anthropogenic aerosols, the local response is strongly influenced by the local forcing because of the much stronger regional variations in such forcing.

Given that the climatic response to changes in well-mixed GHGs varies in proportion to the global mean radiative forcing, and does not depend on the detailed spatial structure of the forcing, it is valid to represent the net effect of increases in a variety of well-mixed GHGs in terms of the equivalent CO_2 increase – that is, the increase in CO_2 concentration that would produce the same net global mean radiative forcing. It is not valid to include forcings that have a different associated climate sensitivity when computing the equivalent CO_2 increase.

CHAPTER 4

Factors driving anthropogenic emissions to the atmosphere

Emissions of greenhouse gases arise from the whole spectrum of human activities, involving energy use, deforestation and other land surface modifications, agricultural activities, and the chemical industry. In addition, emission of sulphur from the burning of biomass and sulphur-containing fossil fuels leads to the formation of sulphate aerosols, while biomass burning and some uses of fossil fuels also produce carbon-containing aerosols. In this chapter we identify the major human or human-influenced emission sources for the important greenhouse gases, precursors for tropospheric O_3, and aerosols and aerosol precursors. Quantitative emission factors and/or estimated global inventories for present-day emissions of GHGs, ozone precursors, and aerosol precursors are then presented. This is followed by data on the historical variation of CO_2 emissions and CO_2 concentrations, which, combined with other evidence to be discussed here, establishes beyond doubt that the observed buildup of CO_2 is indeed due to human emissions of CO_2. Lastly, the variation in the atmospheric CH_4 and N_2O concentrations since the Industrial Revolution and estimates of the variation in the emissions of aerosol precursors are presented.

4.1 Emission sources and emission inventories for GHGs, ozone precursors, and aerosols or aerosol precursors

In this section we identify the major human activities that contribute to emissions of GHGs, ozone precursors, and sulphate aerosol precursors; and we present emission factors for selected energy sources or industrial activities.

Carbon dioxide

When fossil fuels are burned to produce energy, the carbon in the fuel combines with oxygen in the air to produce carbon dioxide. The production of CO_2 is thus inextricably tied with the very use of fossil fuels, and the only way to produce less CO_2 from a given fossil fuel is to use less of it. The three major types of fossil fuels are coal, oil, and natural gas. Fossil fuels currently provide about 85% of the world's total energy supply, and future CO_2 emissions will be given by the product of the following factors:

- population (P),
- economic output (dollars) per person ($\$/P$),
- average energy consumption (joules) per dollar of economic output ($J/\$$),
- average CO_2 emission per joule of energy consumption (E/J).

Thus,

$$\text{Emission} = P \times (\$/P) \times (J/\$) \times (E/J)$$

This equation makes it clear that future emissions depend on the future human population, the growth of the global economy, the kinds of economic activities undertaken and the efficiency with which energy is used (which together determine $J/\$$), and the mix of energy sources used and the efficiency with which they are transformed into intermediate energy forms such as electricity and refined petroleum products, which together determine E/J.

Table 4.1 gives the CO_2 emission per unit of energy for various fossil fuels (and in the production of certain chemical products, which are discussed later). Among the fossil fuels, coal produces the most CO_2 per unit of energy provided, while natural gas produces the least. Renewable energy sources such as

Table 4.1 Carbon dioxide emission factors for combustion of various fossil fuels (given as kg C per GJ of fuel energy) and the manufacture of various chemical products (given as kg C per kg of manufactured product). The emission factors in the latter case reflect CO_2 released during the chemical reactions involved in the manufacture of the product, but do not include emissions associated with the energy needed to make the product. Combustion emission factors are based on the higher heating value of the fuel.
Source: see references below.

Source	CO_2 emission factor
Combustion Sources (kg C/GJ energy)	
Anthracite coal	23.5–26.6[a,b]
Lignite coal	22.2–25.9[c,d]
Bituminous coal	23.9–24.5[d]
Sub-bituminous coal	24.8–25.7[c,d]
Oil	17–20[a,b]
Natural gas	13.5–14.0[a,b]
Process sources (tonne C/tonne product)	
Cement	0.08–0.26[a,e]
Lime	0.215[a]
Ammonia	0.431[a]

[a] Jaques (1992).
[b] Nakicenovic *et al.* (1996, p.75).
[c] Calculations based on information given by DOE (1992, p. 124).
[d] Calculations based on information given by IEA (1991a, p.24).
[e] Chen (1998).

solar and wind energy entail no direct emissions of CO_2. Thus, CO_2 emissions can be reduced by shifting from coal and oil to natural gas within the fossil fuel part of the energy supply mix, or by shifting from fossil fuels to renewable forms of energy. Conversely, the production and use of synthetic oil and gas made from coal entails substantially greater CO_2 emissions than the use of conventional oil and gas or the direct use of coal.

A second major source of CO_2 emissions is deforestation. The emissions arise in this case either through the decay of plant debris left on the ground after the extraction of timber, or as a result of forest fires that are set in order to clear land for agricultural purposes (a common practice in tropical countries). The accounting is complicated by the fact that deforested land is often abandoned after a few years, and regrowth of forest on abandoned land re-absorbs CO_2 from the atmosphere, although the forest rarely regrows to its original carbon content. A further complicating factor is that difficult-to-measure changes in the amount of carbon in soils also occur.

For these reasons, estimates of the CO_2 released from deforestation are highly uncertain. This is in contrast to fossil fuel CO_2 emissions, where the emission is directly tied to the amount of fuel used, which in turn is well known because it involves commercial transactions. Houghton *et al.* (1983) presented an early analysis of land use emissions that is still of interest because of its detailed discussion of methodological issues, while Hall and Uhlig (1991) provide a more recent discussion of methodological problems.

A third major source of CO_2 emissions is the chemical release of CO_2 in the production of cement, lime, and ammonia. This CO_2 is in addition to the CO_2 released from the combustion of the fossil fuels that are used to provide the energy for the manufacture of these products. Note the particularly high emission factors for the production of lime and ammonia.

Table 4.2 presents an inventory of CO_2 emissions from fossil use, chemical products, and land use changes for the 20 countries with the largest total emissions in 1990. The emission estimates were obtained from World Resources Institute (1998), except where indicated. Emission inventories prepared as part of national commitments under the United Nations Framework Convention on Climate Change can be obtained from the web site http://www.unfccc.de/, but inventories for all the nations shown in Table 4.2 were not available at the time of this writing. Another source of emission estimates is the Global Change web site of the Pacific Institute for Studies in Development, Environment, and Security (http://www.globalchange.org/). Annual per capita CO_2 emissions range from a low of 1.2 tonnes CO_2 per year (India) to a high of 23.5 tonnes CO_2 per year (Australia). Table 4.3 gives the total global emissions of CO_2 (by mass of carbon)[1] due to each type of fossil fuel, due to the manufacture of cement, and due to land use changes. The global fossil fuel emission of CO_2 is currently about 6.0 Gt C per year, while tropical deforestation is estimated to result in emissions of 1.6 ± 1.0 Gt C per year. Given a current human population of about 6 billion, the world average fossil fuel emission is about one tonne of carbon per person per year (or 3.67 tonnes of CO_2 per person per year).

---

[1] To convert from tonnes of C to tonnes of CO_2, multiply by 3.67.

Table 4.2 National inventories of CO_2 emission (Mt CO_2) associated with use of fossil fuels; manufacture of cement, ammonia, and lime; and deforestation. Inventories are given for the 20 countries with the largest total emissions. Also given are estimated total CH_4 emissions (Mt CH_4). Source: WRI (1998), except where indicated.

| Country | CO_2 emission (Mt CO_2) | | | | Per capita CO_2 (t CO_2) | Methane emission (Mt CH_4) |
	Fossil fuels	Chemical products	Land use changes	Total		
Australia	286.8	3.0	130	420	23.5	6.24
Brazil	236.5	12.7	1200[a]	1450	9.3	—
Canada	430.4	5.3	—	436	14.8	3.51
China	2970.4	222.0	9	3200	2.7	33.83
France	329.6	10.5	−37	303	5.3	2.83
Germany	815.2	19.9	−20	815	10.0	5.20
India	873.9	34.9	150[a]	1060	1.2	—
Indonesia	286.4	9.7	455[a,b]	750	3.8	3.75
Iran	255.7	8.1	—	264	3.8	—
Italy	392.5	17.4	−37	373	6.6	3.91
Japan	1081.7	45.1	−90	1037	8.3	1.32
Mexico	345.9	11.9	89	447	4.9	2.98
North Korea	255.7	1.3	—	257	11.1	—
Poland	331.1	6.9	—	338	8.8	2.47
Russia	1799.9	18.1	−587	1230	8.3	27.36
South Korea	346.1	1.3	—	345	8.3	—
South Africa	301.3	4.5	—	306	7.4	—
Ukraine	432.7	5.5	52	490	7.5	9.46
United Kingdom	535.9	6.2	−6	536	9.2	3.88
United States	5430.2	38.3	−532	4936	18.5	28.17

[a] Taken from Myers (1989).
[b] WRI has Indonesia an implausibly large sink of 822 Mt CO_2, when the country is almost certainly a large source.

Table 4.3 Total global CO_2 emissions (Gt C) in 1995 as a result of the combustion of solid, liquid, and gaseous fossil fuels, as a result of gas flaring and cement manufacture, and due to land use changes. Taken from Marland *et al.* (1998), except for the land use emission, which is taken from Schimel *et al.* (1996).

Source	Emission (Gt C)
Solids	2.45
Liquids	2.57
Gases	1.14
Gas flaring	0.06
Total fossil fuel	6.22
Cement manufacture	0.19
Land use changes	0.6–2.6
Total	7.0–9.0

Table 4.4 Methane emission factors for various fossil fuels. *Source*: see references below.

Fuel	CH_4 emission factor (kg CH_4/GJ)
Coal mining	0.13–0.53[a]
Underground	0.46–0.49[b]
Surface	0.12–0.13[b]
Oil	≤0.03[c,d]
Natural gas	0.18–0.19[c,e,f]

[a] Based on CH_4 emission factors given in Beck *et al.* (1993) and references therein.
[b] Based on 1989 coal production with CH_4 emission estimates from Kirchgessner *et al.* (1993). The emission factor depends on the average annual coal heating value assumed.
[c] Barnes and Edmonds (1990).
[d] Emissions occur when gas is vented but not flared, during oil extraction.
[e] Gas Research Institute (1997).
[f] This is the emission factor for each 1% leakage of natural gas, from the point of extraction to the point of use.

Methane

Methane is emitted to the atmosphere as a byproduct of fossil fuel use and through a variety of other human activities. Methane is produced whenever organic matter decomposes in the absence of oxygen (anaerobic conditions). Coal consists of former swamp deposits that were buried and underwent partial decomposition, so that methane is trapped in most coal seams. The process of mining coal allows this methane to seep into the atmosphere. Methane is also found in association with oil, and is often flared during the extraction of oil. Finally, natural gas is 90–95% methane, and some methane leaks into the atmosphere during the extraction and distribution of natural gas. Unlike CO_2, there is considerable scope for reducing methane emissions while continuing to use a given fossil fuel. For example, some of the methane in coal seams and oil fields can be collected and used rather than allowing it to escape into the atmosphere, while measures can be taken to reduce the leakage of methane in association with the extraction and distribution of natural gas. Table 4.4 lists methane emission factors per unit of energy extracted from the ground for typical present-day conditions and practices. Coal mines can be divided into two categories: open surface mines and underground mines, with much lower emissions from the former. Natural gas pipelines can extend over thousands of kilometres, and leaks in these pipelines can be responsible for up to half of natural gas loss, with loss rate estimates ranging from 0.2% to over 5% (Beck *et al.,* 1993). The remaining natural gas loss occurs during extraction, flaring, gas processing, and distribution. The emission factor given in Table 4.4 represents methane emissions for a net natural gas loss of 1% of total production, which includes emissions from each of these stages. Methane is also released from the slow decomposition of organic matter in land that is flooded for hydro-electric power production. Conservative estimates indicate that the global warming effect of hydro-electric dams in the Brazilian Amazon can be *several times* the global warming effect of fossil fuel power plants producing the same amount of electricity (Fearnside, 1995; Rosa and Schaeffer, 1995).

Methane is released to the atmosphere from a wide range of human activities other than the production of energy. These include the burning of biomass and a number of activities where partial decomposition of organic matter under anaerobic conditions occurs. The latter includes cultivation of rice, landfills, sewage treatment plants, and the raising of ruminant

Table 4.5 Estimates of global anthropogenic methane emissions from various sources, as summarized by Prather *et al.* (1995).

Methane source	Emission (Tg CH_4/year)
Coal mining	15–45
Coal combustion	1–30
Extraction of oil	5–30
Extraction and use of natural gas	25–50
Total fossil	46–155
Sewage treatment plants	15–80
Sanitary landfills	20–70
Domestic animals	65–100
Animal waste	20–30
Rice paddies	20–100
Biomass burning	20–80
Total biospheric	160–460
Total	206–615

animals such as cattle. Agricultural sources of methane, and emission reduction options, are reviewed by Mosier *et al.* (1998a). Estimates of the global inventories of methane emission due to various human activities, as summarized by Prather *et al.* (1995), are given in Table 4.5. In most cases the uncertainties are a factor of 3–4 or more, with an overall uncertainty of a factor of 3. The last column of Table 4.2 gives estimated methane emissions by 20 of the countries with the largest emissions.

Nitrous oxide

The primary sources of nitrous oxide (N_2O) emissions are industrial processes and nitrogen fertilizers, with the latter estimated to be responsible for 50–75% of total emissions (Table 4.6). This inventory contains an overall uncertainty of about a factor of 3. Nitrous oxide emissions from the combustion of fossil fuels depend on the temperature of combustion, the extent to which controls to reduce NO_x emissions are implemented (NO_x contributes to the buildup of ground-level ozone or "smog"), and the technology used to control NO_x. Catalytic

Table 4.6 Estimates of global anthropogenic N_2O emissions, as summarized by Prather *et al.* (1995).

Nitrous oxide source	Emission (Tg N/year)
Cultivated soils	1.8–5.3
Biomass burning	0.2–1.0
Industrial sources	0.7–1.8
Cattle and feed lots	0.2–0.5
Total	2.9–8.6

converters are increasingly used in automobiles in order to reduce NO_x emissions, but they increase N_2O emissions. Similarly, N_2O emissions from fertilizer use (on cultivated soils) vary widely depending on the nature of the fertilizer and especially depending on the soil water content and the timing of fertilization. Emission factors for N_2O are easily as fluid as they are for CH_4. Mosier et al. (1998b) provide a recent review of agricultural N_2O emissions and reduction options.

Halocarbons

The term "halocarbon" refers to compounds containing either chlorine, bromine, or fluorine, together with carbon, many of which act as powerful GHGs. These include the chlorofluorocarbons (CFCs) and their replacements, the hydrochlorofluorocarbons (HCFCs) and the hydrofluorocarbons (HFCs). The chlorine- and bromine-containing halocarbons are involved in the depletion of the stratospheric ozone layer. Halocarbons are produced by the chemical industry and are used in a variety of applications, mostly involving refrigeration and cooling. The emission of halocarbons to the atmosphere depends on, first, the amount that is used in a given application (such as air conditioning), and second, the extent to which leakage to the atmosphere can be prevented. In the case of air conditioning, the amount used depends on the design and efficiency of the cooling equipment and on the overall cooling requirements of the building, while the leakage depends on how well maintained the equipment is and how servicing and disposal of cooling equipment are carried out. Thus, unlike CO_2, there are no hard and fast emission factors for halocarbons.

Aerosols and aerosol precursors

Aerosols can be classified as "primary" and "secondary". Primary aerosols are those that are directly emitted, while secondary aerosols are produced through the chemical transformation of precursor gases. Table 4.7 summarizes estimates of the total anthropogenic emissions of primary aerosols or aerosol precursors, or of the rate of production of secondary aerosols, during the 1980s. The main aerosols of interest are: (i) sulphate (SO_4^{2-}) aerosols, (ii) soot or "black carbon" aerosols, which are partially absorbing, and (iii) non-absorbing organic carbon aerosols. The uncertainty in sulphur emissions, at ±25% of the central estimate, is comparatively small; for some other aerosol types, the uncertainty is a factor of 5.

Table 4.7 Estimates of the global anthropogenic emissions of major aerosol types during the 1980s, as summarized by Jonas et al. (1995), except where indicated.

Aerosol type	Estimated flux (Tg/year)	
	Range	Preferred
Primary		
Industrial dust	40–130	100
Soot, total	5–25	10
from fossil fuels[a]	7–8	
from biomass burning[a]	5–6	
Secondary		
SO_x emission (Tg S/year)	66–94[b]	67[c]
SO_4 production from SO_x (Tg S/year)[d]	40–60	47
NO_x emission (Tg N/year)	21–30[e]	25[f]
Nitrates from NO_x (Tg N/year)[g]	6–15	12
Organic matter from biomass burning	50–140	80
Organic matter from non-methane hydrocarbons (NMHCs)	5–25	10

[a] Based on Cooke and Wilson (1996) and Liousse et al. (1996).
[b] This is the range for estimates of fossil fuel SO_x emissions for the inventory years 1985 and 1986, as given by Benkovitz et al. (1996), plus 2 Tg S/yr to account for biomass-related emissions. On average, about 80% of emitted SO_x is converted to sulphate before being removed from the atmosphere.
[c] This is the preferred estimate of Benkovitz et al. (1996), including 2 Tg S/yr from biomass burning.
[d] The estimates by Jonas et al. (1995) are in mass of SO_4, but have been converted here to mass of S to permit comparison with the SO_x emission, which is also given in mass of S. In the S chemistry model of Roelofs et al. (1998) about 80% of the emitted SO_2 is converted to sulphate, the rest being directly deposited at the Earth's surface.
[e] This is the range for estimates of total anthropogenic NO_x emissions for the inventory years 1985 and 1986, as given by Benkovitz et al. (1996) and IEA (1991b), the latter being shown in disaggregated from in Table 4.12.
[f] This is preferred estimate of 21 Tg N/yr of Benkovitz et al. (1996), which largely excludes biomass emissions, plus a middle estimate (from Table 4.12) for emissions due to biomass burning of 4 Tg N/yr.
[g] The estimates by Jonas et al. (1995) are in mass of NO_2, but have been converted here to mass of N to permit comparison with the NO_x emission, which is also given in mass of N.

As noted in Section 2.5, sulphate aerosols are produced from the chemical transformation of SO_2 that is emitted during the combustion of S-containing coal, the refining of S-containing oil, and the smelting of minerals such as zinc, copper, and lead, which are mixed with sulphur. Sulphur dioxide is thus the precursor of sulphate aerosols. The sulphur content of crude petroleum ranges from 0.1% to 2.6% (Spiro *et al.*, 1992). For coal, the sulphur content can be as low as 0.5% for anthracite and over 7.5% for some types of lignite (Hargreaves *et al,* 1994; Spiro *et al.,* 1992; EIA, 1998). Depending on the energy content of the coal, corresponding uncontrolled sulphur emission factors can range from less than 0.2 to over 2.0 kg S per gigajoule of energy.

Table 4.8 summarizes sulphur emission factors for various processes with uncontrolled emissions.

Sulphur emissions are currently being reduced in many developed countries, due to their impact on regional air pollution and their role in acid rain formation. Canadian emissions had fallen by 43% from the 1980 level by 1995 while, under the Clean Air Act, emissions in the United States will have fallen by 40% from the 1980 level by 2010 (Acidifying Emissions Task Group, 1997). Under the 1994 *Protocol to the 1979 Convention on Long-Range Transboundary Air Pollution on Further Reduction of Sulphur Emissions* (henceforth, the Second Sulphur Protocol), total emissions from the European Community are to be reduced by 62% from the 1980 level by 2000, with some countries accepting reduction requirements of 80% or more (UN, 1994). Under the Second Sulphur Protocol, many eastern European countries are required to reduce emissions by 40–70% by 2010. Combustion-related emissions can be reduced by switching form high-sulphur to low-sulphur fuels, through boiler additives that reduce sulphur emissions by 50–70%, and with scrubbers, that can remove over 95% of SO_2 from flue gas

(Princiotta, 1991). It is also possible to remove 97–99% of the sulphur from coal prior to its combustion if the coal is gasified first (NAPAP, 1987). Under prevailing practice, 50–90% of the sulphur in nickel ores is prevented from entering the atmosphere during smelting, while 90–95% is contained for copper, lead, and zinc smelting (Charles Ferguson, Inco Limited, personal communication, 1999). Under best practice, 99% of sulphur is contained during copper smelting.

The Second Sulphur Protocol is based on a "critical loads" approach; that is, rates of deposition at which harmful effects supposedly begin are estimated based on ecological knowledge, then the emission reductions required to avoid exceeding these rates of deposition are calculated using computer models of atmospheric chemistry and transport. A similar approach is gaining credence in North America. The Second Sulphur Protocol requires the parties to encourage research aimed at achieving even further emission reductions, something that appears to be required if sensitive ecosystems in the UK and perhaps elsewhere in Europe are indeed going to be protected (Jenkins *et al.*, 1998). To protect almost all lakes in North America from the effects of acid rain, total sulphur emissions would need to be reduced by 75% from existing caps. This would give total reductions comparable to the most stringent reductions already required in Europe. Inasmuch as discussions among the various stakeholders concerning further significant reductions in emissions are underway (e.g. Acidifying Emissions Task Group, 1997), a reduction of this magnitude may very well be achieved.

Carbon (or carbonaceous) aerosols occur in two forms: as soot, which is directly emitted during incomplete combustion of fossil fuels or organic matter, and as reflective organic compounds that are produced photochemically from volatile (off-gassed) organic compounds. The latter can be subdivided into organic gases produced during the burning of biomass, and as non-methane hydrocarbons (NMHCs) that are released to the atmosphere largely through the use of refined petroleum products (such as gasoline). Organic carbon aerosols are also produced naturally through the photochemical production of gases (such as terpenes) that are released by vegetation.

Global inventories of the emissions of the major aerosols or aerosol precursors are available on a latitude–longitude grid. Benkovitz *et al.* (1996) have published gridded inventories of SO_x and NO_x emissions, and they provide an Internet address through which their inventories can be obtained.

Table 4.8 Sulphur emission factors as (kg S/kg C) for the use of coal and oil, or as (kg S/tonne metal) for the smelting of metals. *Source*: see references below.

Process	Uncontrolled emissions
Coal combustion	0.2–2.2 kg S/GJ[a,b]
Oil refining	0.1–0.4 kg S/GJ[c,d,e]
Primary copper smelting	1060 kg/tonne[f]
Secondary copper smelting	225 kg/tonne[f]
Primary lead smelting	149 kg/tonne[f]
Secondary lead smelting	42.6 kg/tonne[f]
Primary zinc smelting	490 kg/tonne[f]

[a] EIA (1998).
[b] Range of sulphur emission factors is for coal with a sulphur content ranging from 0.5% to 5%.
[c] Naturalgas.com: *The Environment: Acid Rain*, at http://naturalgas.com/environment/acid/html.
[d] Syncrude Canada (1998).
[e] This range of emission factors is for fuel oil with a sulphur content of 0.5% to 2%.
[f] Spiro *et al.* (1992).

Table 4.9 Estimated inventory of direct anthropogenic emissions of CO in 1986 (teragrams (10^{12} g) of C per year) by geopolitical region, as given in Tables A-7 and A-8 of IEA (1991b).

Source	OECD[a]	CPE[b]	LDC[c]	World
Industrial	0.4	0.6	0.2	1.2
Residential/commercial	5.3	13.7	36.9	55.7
Transportation	54.7	8.0	25.1	87.8
Electricity generation	0.6	0.5	0.1	1.2
Other transformation	0.2	0.2	0.1	0.4
Total energy-related	61.2	22.8	62.4	146.4

[a] OECD = Organization for Economic Cooperation and Development.
[b] CPE = Centrally Planned Economies.
[c] LDC = Less Developed Countries.

Table 4.10 Estimated total global anthropogenic emissions of CO in 1992, including both direct emissions and emissions arising from the anthropogenic production of CH_4 and NMHCs (non-methane hydrocarbons).

Source	Emission (Tg C/yr)
Direct emissions	
Energy use	230[a], 350[b]
Biomass burning	170[a,b]
Indirect emissions	
Oxidation of CH_4	243[a,c]
Oxidation of NMHCs	50[a]

[a] Lelieveld et al. (1998).
[b] Berntsen et al. (1997).
[c] Derived from an assumed anthropogenic CH_4 emission of 304 Tg C/yr and CO production rate of 0.8 moles per mole of emitted CH_4.

Precursors of tropospheric ozone

As discussed in Section 2.8, tropospheric ozone is produced as a byproduct of the oxidation of CO and CH_4 if the concentration of NO is sufficiently high. It is also produced as a byproduct of the oxidation of NMHCs in the presence of adequate NO. Emissions of CO, CH_4, NMHCs, and NO_x (= NO + NO_2) therefore all lead to the production of O_3 in the troposphere, so these gases are ozone precursors. Tables 4.9–4.12 present estimates of the global emissions of CO, NMHCs, and NO_x (CH_4 emission estimates were given in Table 4.5). There are uncertainties in the energy-related emissions of CO and NO_x of probably at least ±50%, while the uncertainty associated with biomass emissions is even larger. Table 4.9 shows energy-related CO emissions by region, as given by the International Energy Agency (IEA, 1991b), while Table 4.10 gives the global CO emissions due to various direct and indirect anthropogenic sources as adopted by Lelieveld et al. (1998) for chemical modelling purposes. The total global energy-related emissions adopted by Lelieveld et al. (1998) is within 60% of the total given by IEA (1991b), which is a comparatively small disagreement. Biomass burning is a significant source of NO_x and especially CO emissions, while transportation is the largest fossil fuel-related source of CO and NO_x emissions on a global basis.

It follows from our discussion in Section 2.8 that the efficiency of NO_x in creating O_3 (or even whether or not O_3 is created) depends strongly on where the NO_x is emitted. In general, NO_x emitted from the surface in the SH is twice as efficient in producing O_3 as NO_x emitted in the NH (Wang et al., 1998b), while NO_x emitted in the upper troposphere (from aircraft) is about 20 times more efficient at producing O_3 than NO_x emitted from the surface (Johnson et al., 1992). Furthermore, an O_3 molecule in the upper troposphere is about 1.5 times more efficient in trapping heat than an O_3 molecule near the Earth's surface. As a result, NO_x emissions from aircraft are about 30 times more effective in trapping heat than NO_x emissions from surface sources at mid-latitudes in the NH. Thus, changes in total global NO_x emissions will not be a good indicator of the changing climatic effect of NO_x emissions if the relative importance of different emission sources changes.

Table 4.11 Estimated inventory of global emissions (Tg C/year) of various non-methane hydrocarbons, as given by Wang et al. (1998a).

Species	Anthropogenic				Natural vegetation	Total
	Industrial	Biomass	Oxidation	Total		
Ethane	6.3	2.5	0	8.8	0	8.8
Propane	6.8	1.0	0	7.8	0	7.8
$\geq C_4$ alkanes	30	0	0	30	0	30
$\geq C_3$ alkenes	10.4	12.6	0	23	0	23
Isoprene	0	0	0	0	597	597
Acetone	1.0	8.9	12.4	22.3	15	37.3
Total	54.5	25.0	12.4	91.9	612	703.9

Table 4.12 Estimated inventory of global anthropogenic emissions of NO_x in 1986 (teragrams N per year), as given in Tables A-9 and A-10 of IEA (1991b) for all sources except biomass burning, which is based on Crutzen and Andreae (1990) and Andreae (1995). N emissions given by IEA (1991) are reported in terms of the mass of NO_x, but have been converted here to mass of N by assuming that all of the NO_x is NO_2.

Source	OECD	CPE	LDC	World
Industrial	1.2	1.4	0.5	3.2
Residential/commercial	0.6	0.9	0.6	2.1
Transportation	6.0	1.9	2.4	10.3
Electricity generation	3.2	3.4	1.1	7.7
Other transformation	0.4	0.5	0.3	1.2
Total energy-related	11.5	8.1	4.9	24.5
Biomass burning	—	—	—	2–8

See Table 4.9 for definitions of the acronyms used here.

As previously summarized in Table 2.4, the lifespan of CO in the atmosphere ranges from less than 20 days in the lower 4 km of the atmosphere, to close to 50 days in the upper troposphere. For this reason and because emission sources are concentrated regionally, the CO concentration also varies regionally. The lifespan of NO_x is even shorter, on the order of 1 day. As a result, and because the O_3 lifespan is also rather short (10–200 days, Table 2.4), the highest O_3 concentrations are found over and downwind from the major sources of NO_x, CO, and NMHCs.

Summary of emissions of trace gases from changes in land use

In the preceding discussion, reference has been made to emissions of trace gases (other than CO_2) in association with the burning of biomass. Biomass burning is, however, only one of three or four steps in the transformation of tropical forests to other land uses. The first step is the clearing (cutting) of standing vegetation. This is followed by the burning of felled vegetation, planting of grain and root crops, and finally, conversion to pasture with periodic burning to control regrowth. Emissions of trace gases, or changes in the natural emissions or sinks of trace gases, are associated with each of these steps.

Table 4.13 presents a qualitative summary of the effects of clearing, burning, and agricultural or pastoral use of former tropical forests on emissions of CO, CH_4, NMHCs, NO_x, N_2O, and aerosol particles or aerosol precursors, based on the work of Keller *et al.* (1991) and Scholes *et al.* (1996). Table 4.14 presents a range of quantitative emission factors from biomass burning for these substances. Biomass

Table 4.13 Effects of deforestation on trace gas and aerosol emissions, based on Keller *et al.* (1991) and Scholes *et al.* (1996).

	CO	CH_4	NMHC	NO_x	N_2O	Aerosols
Clearing	?	–	–	+	+	?
Burning	++	++	++	++	+	++
Use	?	+	–	++	++	?

Key:
– decreased emissions
+ increased emissions
++ greatly increased emissions

Table 4.14 Emission factors from biomass burning, based on summaries given in Keller *et al.* (1991) and Scholes *et al.* (1996). All emission factors are in units of moles of gas or aerosol per mole of CO_2.

CH_4	0.8–1.6%
CO	8–15%
NMHC	1.1–1.7%
N_2O	$1.5–3.0 \times 10^{-4}$
NO_x	2×10^{-3}
Aerosols:	
Grasslands	$5–7 \times 10^{-3}$
Savannas	$5–6 \times 10^{-3}$
Forests	$5–50 \times 10^{-3}$

burning causes large emissions of these substances, most of which serve as ozone precursors. A large fraction (about 15%) of the carbon that is released to the atmosphere is released as CO. This CO is eventually oxidized to CO_2 (on a time scale of about 20 days), but tends to produce O_3 in the process owing to high NO concentrations (see Section 2.8). Up to 2% of the carbon is released as CH_4. This CH_4 is also eventually oxidized to CO_2 (on a time scale of about 10 years), but in the interim, each CH_4 molecule is 26 times as effective as a CO_2 molecule in contributing to the greenhouse effect (see Table 2.7). Substantial quantities of sooty (absorbing) and non-absorbing carbon aerosols are also released, particularly from smouldering woody material. These aerosols have a variety of climatic effects, as discussed in Section 7.4. Estimates of the total global emissions of CH_4, N_2O, carbon aerosols, CO, NO_x, and NMHCs due to biomass burning were given in Tables 4.5–4.12.

Agricultural use of land (after it has been cleared and burned) has its largest effects on N_2O and NO_x emissions, which tend to increase sharply. These effects arise through changes in soil temperature, moisture, and biological productivity; and through a reduction in the extent of re-absorption of NO_2 by plants (in dense forest canopies, NO is released from the soil and oxidized to NO_2, much of which is then absorbed onto leaf surfaces). The main cause of increased N_2O emissions is the application of N fertilizers although, as noted earlier, the magnitude of the emissions depends strongly on soil water content and the timing of fertilization.

Comparison of natural and anthropogenic emissions

In preceding sections of this chapter we have presented estimates of the anthropogenic emissions of the major GHGs and of the precursors of O_3 and aerosols. It can be instructive, but also misleading, to directly compare the magnitudes of the natural and anthropogenic emissions. The potential for being misled involves CO_2. Total anthropogenic emissions are currently around 6–8 Gt C yr^{-1}. This is very small compared to natural emissions, which amount to about 60 Gt C yr^{-1} due to respiration of plant detritus and soil carbon, and about 100 Gt C yr^{-1} due to gaseous diffusion from the oceans. One might therefore conclude that anthropogenic emissions cannot be important. The reason that this conclusion is wrong is that CO_2 is able to continuously cycle between the atmosphere and other reservoirs.

All GHGs or aerosol precursors except CO_2 are removed from the atmosphere entirely, or almost entirely, through chemical destruction (usually by reacting with OH or through photodissociation). The rate of removal is proportional to the concentration within the atmosphere (although not linearly so in the case of CH_4, as discussed in Section 8.10). As a result, a 10% increase in the total emission would eventually lead to a roughly 10% increase in concentration, as this brings the removal rate back into balance with emissions. In the case of anthropogenic CO_2, the main removal processes are photosynthesis and inflow into the ocean, not irreversible chemical destruction. As the removal processes increase in strength (due to an initial increase in concentration), the return flows or natural emissions also increase. Thus, as photosynthesis increases, the supply of organic matter increases, and this allows the respiration flux to increase. Similarly, as CO_2 flows into the ocean, the back-pressure of CO_2 in the water increases (and quite rapidly, as discussed in Chapter 8), so the outflow also increases. The removal process becomes permanent only as CO_2 works its way into the deep ocean, a process that takes on the order of 1000 years. This makes the atmospheric CO_2 concentration particularly sensitive to additional emissions. To illustrate this point, note that total natural emissions are about 160 Gt C yr^{-1}. Total anthropogenic emissions are about 5% of this (6–8 Gt C yr^{-1}) but, if sustained at this level, would eventually lead to CO_2 concentrations *many times* the initial concentration.

It is sometimes said that the reason that anthropogenic emissions of CO_2 are important is that the emissions are large compared to the natural residual or imbalance between natural emissions and removal processes. This imbalance is very close to zero. This is not the correct explanation, since the natural imbalance for the other gases is also close to zero. Rather, the difference is in the reversibility of the removal processes for CO_2 and other gases.

With this explanatory preamble behind us, we are now in a position to directly compare natural and anthropogenic emissions. This is done in Table 4.15 for all the gases of interest except CO_2, because such a comparison is not meaningful in the case of CO_2. Also shown in Table 4.15 are the estimated natural and anthropogenic rates of nitrogen fixation, as this is an important perturbation to the nitrogen cycle, and the estimated natural and anthropogenic emissions of aerosols or aerosol precursors. Anthropogenic emissions are a significant fraction of natural emissions for CH_4, SO_2, N_2O, CO, NO_x, and many aerosols.

Table 4.15 Comparison of total natural and anthropogenic emissions for gases that are removed from the atmosphere through irreversible chemical destruction. Also given are nitrogen fixation rates on land and the estimated natural and anthropogenic emissions of aerosols or aerosol precursors. As explained in the text, it is not meaningful to directly compare natural and anthropogenic emissions of CO_2. Except where indicated, anthropogenic emissions are taken from other tables in this chapter, while natural emissions are taken from Chapter 2.

Gas	Units for emission	Natural emissions	Anthropogenic emissions
CH_4	Tg CH_4 per year	190 ± 70[a]	340 ± 160[a]
N_2O	Tg N per year	6–12[b]	3.9–7.6
SO_x	Tg S per year	8	60–100[c]
All S species	Tg S per year	16–31[d]	60–100
CO	Tg C per year	360[e]	700
NO_x	Tg N per year	0.5–10[f]	21–30
N fixation	Tg N per year	150	40
C aerosols:			
Soot	Tg C per year	0	5–25
Organic C from fires	Tg C per year	?	50–140
Organic C from NMHCs	Tg C per year	55	5–25

[a] From Lelieveld *et al.* (1998). In combining natural and anthropogenic emissions, the total source must equal 600 ± 80 Tg/yr. The uncertainty range for anthropogenic emisions given here largely overlaps the range given in Table 4.5 from the summation of individual anthropogenic sources.
[b] From Prather *et al.* (1995).
[c] From references discussed in Section 7.4.
[d] Range used by authors cited in Table 7.4.
[e] From Lelieveld *et al.* (1998).
[f] From Table 12.3 of Jaffe (1992).

4.2 Variation in GHG concentrations and in GHG and aerosol emissions up to the present

Systematic atmospheric measurements of the concentration of CO_2, CH_4, and N_2O began in 1958, 1978, and 1976, respectively. Concentrations prior to these dates have been determined by measurements of the concentrations of these gases in air bubbles from ice cores extracted from Antarctica and Greenland. Prior to presenting composite data that combine these two sources of information, there are a number of issues concerning the interpretation of the ice core data that need to be discussed.

Interpretation of recent ice core concentration data

The air bubbles in ice cores form from air between the grains of snow as the snow is converted to firn (and eventually to ice). The age of the firn at the time that the enclosed air bubbles become sealed from the overlying air will be smaller the faster the rate of snow accumulation. Thus, at high snow-accumulation sites in the Antarctic (near the coast), the firn is about 40 years old when it becomes sealed, while at the South Pole (where the accumulation rate is very low), the firn is about 1000 years old when it becomes sealed (Etheridge *et al.*, 1992). If there were unrestricted movement of air between the atmosphere and the depth where the air bubbles become sealed, the air would be zero years old at the time of sealing. However, owing to limited movement within the firn, the air has already aged to some extent at the time of sealing. At the coastal DE08 site in Antarctica the firn is sealed at an age of 40 years but the air on average is 10 years old at the time the firn is sealed, which means that the air in the ice is 30 years younger than the surrounding ice. However, the firn at this site is sealed over a period of about 8 years, so the air at this site is a sample of air that existed at a time 26–34 years younger than the ice that encloses the air bubbles (Etheridge *et al.*, 1996). This is relevant information, since the age of the ice can be exactly determined (for the last few hundred years) by counting annual layers.

Other issues pertaining to the ice core data involve understanding the circumstances under which the gas

Table 4.16 Characteristics of the observational networks for the measurement of atmospheric CO_2, CH_4, and N_2O concentrations, and summary results, based largely on Boden *et al.* (1994).

Gas	Concentration		Growth rate (%/year)	Year in which direct observations began	Number of stations in current network
	Pre-industrial	1992			
CO_2	280 ± 5	356	0.3–0.4	1958	49
CH_4	0.82 ± 0.03	1.714	0.5–0.7	1978	39
N_2O	0.276 ± 0.003	0.313	0.2–0.3	1976	12

concentration can change after the air bubbles have become sealed, changes that occur during the measurement process, and the calibration of different instruments used by different labs or at different times. These issues have all been carefully addressed, and what turn out to be only minor corrections to the raw data are routinely made (Etheridge *et al.,* 1992, 1996).

Concentrations of CO_2, CH_4, and N_2O

The first systematic observations of the atmospheric concentration of CO_2 began in 1958 at Mauna Loa (Hawaii). This was followed by the establishment of further stations, leading to the current global network of 49 observing stations for CO_2. There is also a reasonable network for CH_4, but relatively few sampling sites for N_2O (Table 4.16). The importance of having a global network, rather than one measurement site, arises from the fact that there are subtle spatial variations in the concentration. These differences have been used in the case of CO_2 to deduce rough estimates of the spatial distribution and magnitude of the various sinks (as discussed in Section 8.4), and have been used in the case of CH_4 to deduce the magnitude and spatial distribution of various sources (Hein *et al.,* 1997). Figure 4.1 shows the variation in the atmospheric concentration of CO_2, CH_4, and N_2O since 1600 based on a composite of ice core data from Antarctica and directly measured concentrations at the South Pole. The CO_2 and CH_4 data are from the DE08 core (mentioned above), which was drilled in 1987. The youngest air sample is from ice dating to 1948, which contains air centred at 1978 – the year that direct measurements of CH_4 began! As seen from Fig. 4.1, there is no discontinuity between the ice core data and the directly measured concentration data for either CO_2 or CH_4, which serves to confirm the validity of the ice core

data. The N_2O ice core data are also from a high-accumulation core, collected by the Japanese Antarctic Research Expedition in 1991 (Machida *et al.,* 1995). The most recent air sample is centred around 1964 in this case.

Figure 4.2 provides a more detailed look at the change in CO_2 concentration from 1960 to 1998, as measured at the Mauna Loa observatory in Hawaii. A seasonal variation in the CO_2 concentration is seen, superimposed on an overall upward trend. The seasonal cycle is due to the drawdown of atmospheric CO_2 in NH spring and summer, when photosynthesis exceeds respiration, and a net release in fall and winter, when the opposite occurs. Although barely discernible in Fig. 4.2, the seasonal variation has increased by about 15% (from 6.52 ppmv averaged over 1960–1963, to 7.65 ppmv averaged over the last three years of the record). The increase in the seasonal variation can be partly explained as a result of greater seasonal growth (and hence decay) of plants due to the stimulatory effect of higher atmospheric CO_2 (Houghton, 1987), which will be discussed in Section 8.1.

Figure 4.3 provides a detailed look at the change in CH_4 concentration from 1984 to 1997. A pronounced decrease in the rate of growth of atmospheric CH_4 concentration is evident, from about 15 ppbv per year to less than 5 ppbv per year (Fig. 4.3(b)). Several factors contributed to these changes. The gradual decrease in growth rate from 1984 to 1990 is probably due to an increase in OH concentration which, as discussed in Section 5.7, is believed to have occurred during this time period. The 1984–1990 decrease in CH_4 growth rate occurred in spite of increasing emissions (as shown below). The abrupt drop in growth rate by 1992 is probably due to the decrease in stratospheric O_3 following the eruption of Mt Pinatubo in June 1991, which would have temporarily allowed greater penetration of UV radiation into the troposphere and even greater production of OH

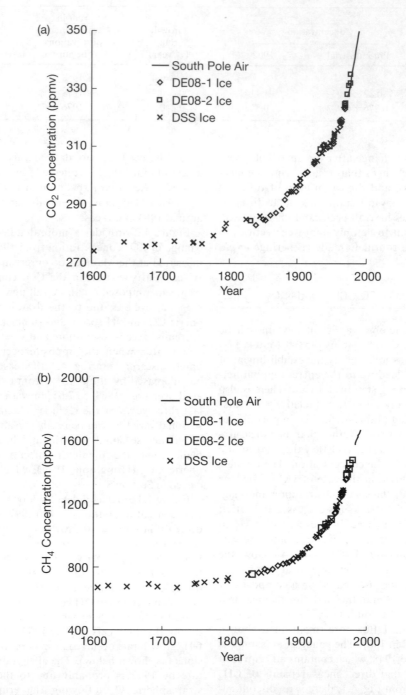

Fig. 4.1 Variation in the observed atmospheric concentration of (a) CO_2, (b) CH_4, and (c) N_2O from 1600 to 1995. The data sources are as follows: ice core CO_2, Etheridge *et al.* (1996); ice core CH_4, Etheridge *et al.* (1992); ice core and atmospheric N_2O, Machida *et al.* (1995); atmospheric CO_2 and N_2O, *Trends '97* (web site: http://cdiac.esd.ornl.gov). Dave Etheridge and Toshinobu Machida kindly provided their data in electronic form.

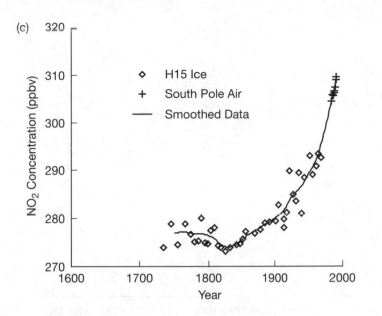

Fig. 4.1 Continued

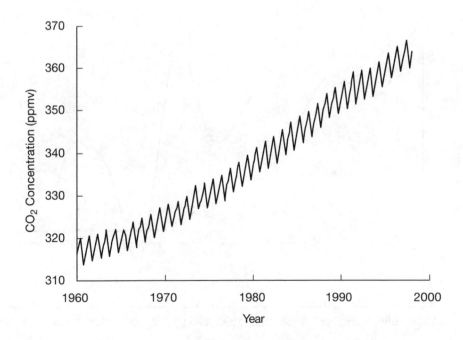

Fig. 4.2 Variation in atmospheric CO_2 concentration from January 1960 to December 1997, as measured at Mauna Loa observatory and given by Keeling and Whorf (1998) and available at the web site http://cdiac.esd.ornl.gov.

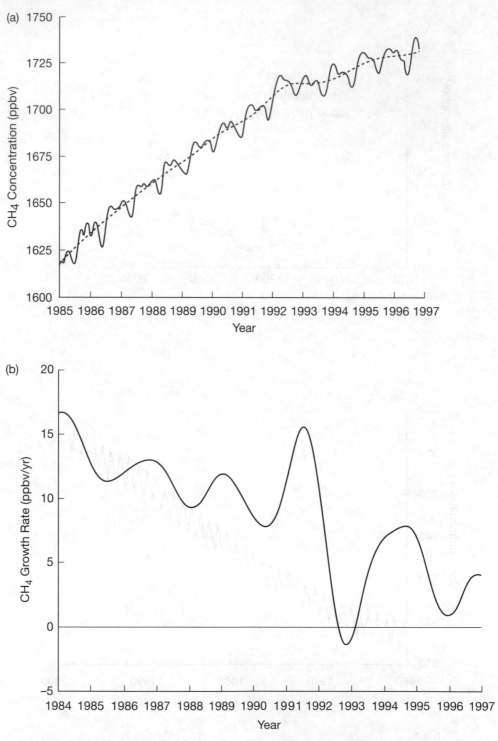

Fig. 4.3 (a) Globally averaged variation in atmospheric CH_4 concentration from 1984 to 1997. (b) Rate of growth in average atmospheric CH_4 concentration from 1984 to 1997. Reproduced from Dlugokencky *et al.* (1998).

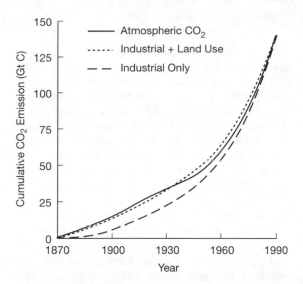

Fig. 4.4 Comparison of the buildup of atmospheric CO_2 concentration with cumulative emissions of CO_2 since 1850 due to fossil fuels alone or due to fossil fuels and changes in land use. Cumulative fossil fuel emissions were multiplied by a factor of 0.427, while cumulative fossil fuel plus land use emissions were multiplied by a factor of 0.647.

(Bekki *et al.,* 1994). The continued but less marked reduction in the growth rate after 1992 could be due to a reduction in the rate of leakage of CH_4 from the natural gas pipeline system in the former Soviet Union (Law and Nisbet, 1996). Finally, year-to-year fluctuations in the rate of growth of atmospheric CH_4 can be partly explained by the dependence of both the natural emission of CH_4 (from wetlands) and its removal (by reaction with OH) on temperature (Bekki and Law, 1997). It should also be noted that, with constant emissions, the atmospheric CH_4 concentration would level off after a time comparable to the atmospheric lifespan of CH_4, which is on the order of 10 years. With only a small decrease in global CH_4 emissions, the atmospheric CH_4 concentration would begin to decrease. As noted above, there is considerable scope for reducing methane emissions.

CO_2 emissions

Plate 1(a) shows the variation in total global emissions of CO_2 due to the burning of fossil fuels and broken down by fuel type, while Plate 1(b) shows an estimate of net emissions due to all land use changes and broken down by ecosystem. Plate 2 directly compares total fossil fuel and total net land use emis-

sions. The land use emission estimates are from Houghton (1998), and reach 2.2 Gt C yr-1 by 1990, which is near the high end of the 1.6 ± 1.0 Gt C yr^{-1} range given earlier. According to Plate 2, land use emissions exceeded fossil fuel emissions up to about 1900; the crossover time would have occurred sooner if land use emissions were smaller than those given in Plate 1. According to Plate 1(b), land use emissions increased slowly from 1870 to 1940, then rose steadily and sharply after 1940. Emissions from non-tropical forest ecosystems (i.e., from developed countries) exceeded tropical emissions until about 1940, at which point tropical emissions began to increase sharply while emissions from other ecosystems decreased. By 1990, essentially all of the land use emissions were from the tropics.

The land use CO_2 emissions given in Plate 1(b) were derived from estimates of historical rates of land cover change and of the carbon content of different ecosystems before and after disturbance by humans. Holdsworth *et al.* (1996; their figure 5) present alternative estimates of the CO_2 emission from land use changes based on the difference between CO_2 concentrations in NH and SH ice cores, and based on carbon isotope variations in tree rings. These and other data from ice cores indicate that there was a significant pulse of CO_2 to the atmosphere associated with deforestation in North America, with peak emissions around 1889–1910 as high as 3 Gt C per year.

Figure 4.4 compares the change in atmospheric CO_2 content since 1870 with the cumulative emissions since 1870 from (a) fossil fuels alone, and (b) fossil fuels and land use changes. The cumulative emissions are multiplied by a constant scaling factor so as to make the three curves match up in 1990. The scaling factor is thus the fraction of the cumulative emissions between 1870 and 1990 that still remains in the atmosphere. This fraction is commonly referred to as the "airborne fraction" and is assumed, for the sake of argument, to have been constant. Using fossil fuel emissions alone, the inferred airborne fraction is 0.647, while using fossil fuel plus Houghton's (1998) land use emissions, the inferred airborne fraction is 0.427. As seen from Fig. 4.4, allowing for fossil fuel emissions only with a relatively high airborne fraction does not do particularly well in matching the observed buildup of atmospheric CO_2. However, a much better fit to the shape of the observed atmospheric CO_2 buildup is obtained when the estimated land use emissions shown here (or alternative estimates) are included. That is, the rate of increase of atmospheric CO_2 during the past 120 years has increased more or less in proportion to the increase in the annual anthropogenic emissions. This match is one of several lines of evidence that allow us to state with complete certainty that the increase in atmospheric CO_2 since the Industrial Revolution was caused by human emissions. These lines of evidence are summarized in Box 4.1.

Box 4.1 How can we be certain that human CO_2 emissions have caused the recent increase in atmospheric CO_2?

One of the few things that is known with absolute certainty is that the increase in the atmospheric CO_2 concentration during the past two centuries is a direct result of human emissions of CO_2, and not the result of some natural fluctuations that occurred coincidentally during and after the Industrial Revolution. The evidence supporting this statement is as follows.

- The recent increase in atmospheric CO_2 far exceeds the bounds of natural variability experienced during the preceding 250,000 years.
- The rate of increase during the past century has increased in step with increasing anthropogenic emissions (as shown in Fig. 4.4).
- The atmospheric CO_2 concentration is slightly higher in the northern hemisphere, where the bulk of human emissions occur, and the inter-hemispheric concentration difference has been increasing over time.
- The atmospheric ratios of ^{14}C to ^{12}C and of ^{13}C to ^{12}C have declined during the past two centuries in the manner expected if human emissions are responsible for the buildup of atmospheric CO_2.
- Various lines of evidence indicate that the undisturbed terrestrial biosphere and the oceans – the only possible alternative candidates to explain the CO_2 buildup – have both been sinks rather than sources of atmospheric CO_2.

In spite of this overwhelming evidence, there are still sometimes suggestions in the fringe or pseudo-scientific literature (e.g. Seitz, 1994) that the current CO_2 increase might not be due to human emissions. Such suggestions are based on a gross misunderstanding of how the oceanic part of the carbon cycle works – a topic that is discussed in great detail in Chapter 8.

Methane emissions

Plate 3 shows the historical variation in anthropogenic emissions of methane, as estimated by Stern and Kaufmann (1996a). The 1990 emission of about 370 Tg CH_4 falls in the middle of the uncertainty range of 206–615 Tg CH_4 given in Table 4.5. According to Plate 3, fossil fuels contributed 20% of the total anthropogenic emission, which is in accord with the ^{14}C-based estimates discussed below. There is considerable uncertainty in CH_4 emission estimates. Nevertheless, some qualitative features of Plate 3 are most probably correct: that there were already substantial anthropogenic emissions by 1870; that most of the growth since then has been due to emissions from fossil fuels, livestock, and landfills; that emissions from biomass burning and rice paddies have also grown since 1870, but by a smaller factor; and that the rate of emission growth was larger after 1940 than before 1940.

Trends in the average $^{14}C/^{12}C$ ratio of atmospheric CH_4 serve to set a constraint on the relative importance of fossil fuel and geological sources of CH_4 (which are ^{14}C-free due to their great age) and CH_4 emissions from modern biomass (i.e., all of those given in Table 2.2 except emissions from sediment). As reviewed by Lelieveld et al. (1998), various estimates of the amount of "old" CH_4 emitted as a fraction of total (natural + anthropogenic) emissions range from 16% to 30%. Lelieveld et al. (1998) adopt a value of 20 ± 4%. The total source must equal the sum of all sinks plus the observed rate of increase, and this sum is estimated to amount to 600 ± 80 Tg/yr. Hence, they deduce an emission of old CH_4 of 120 ± 40 Tg/yr, of which 10 ± 5 is estimated to be from sediments and recent thawing of permafrost. Thus, the total fossil fuel emission is estimated to be 110 ± 45 Tg/yr. This serves to constrain the sum of the fossil fuel emissions given in Table 4.5, which otherwise implies that fossil fuel emissions could be as high as 50% of the total (natural + anthropogenic) emissions.

Sulphur emissions

Past concentrations of aerosols cannot be measured. However, since the lifespan of aerosols in the troposphere is so short (a few days), the concentration responds essentially instantaneously to changes in emissions. Although the global mean concentration

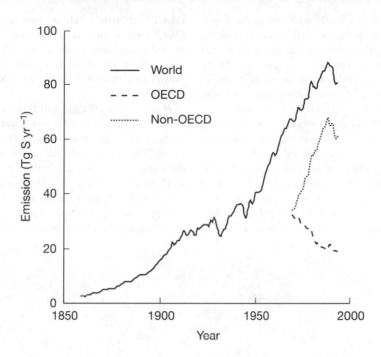

Figure 4.5 Historical variation in global anthropogenic S emission (Tg S yr^{-1}), as estimated by Stern and Kaufmann (1996b) and given at the web site http://cres.anu.edu.au/~dstern/anzsee/datasite.html. Also shown, since 1970, are separate emissions from OECD and non-OECD countries. This breakdown is not available for times prior to 1970.

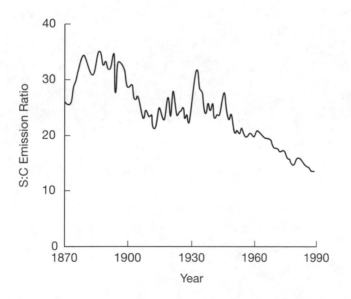

Fig. 4.6 Variation in the S:C mass emission ratio (Tg S per Gt C) due to the combustion of fossil fuels, based on the CO_2 emission data given in Plate 1(a) and the S emission estimates of Fig. 4.5.

does not vary exactly linearly with the global emission, variations in the estimated emission can be used as a rough proxy for the variation in global mean atmospheric concentration. Figure 4.5 shows the past variation in anthropogenic sulphur emissions as estimated by Stern and Kaufmann (1996b), based on inventories of fossil fuel use and metal smelting. The 1985 emission was 83 Tg S according to Fig. 4.5; alternative estimates of the total emission for 1985 range from 65.0 Tg S (Chin *et al.,* 1996) to 92.2 Tg S (Pham *et al.,* 1996). According to Fig. 4.5, global sulphur emissions rose rapidly until 1910, experienced less rapid growth from 1910 to 1940, rose sharply between 1940 and the late 1980s, and then decreased. From 1970 to the mid-1990s,

OECD emissions fell by about 40%, while non-OECD emissions rose sharply. The decrease in global emissions after 1990 is largely due to a decrease in non-OECD emissions (in particular, falling emissions in eastern Europe and the former Soviet Union).

A critical factor in determining anthropogenic effects on future climate is the ratio of sulphur to carbon emissions due to the burning of fossil fuels alone. This ratio is shown in Fig. 4.6, based on the data shown or used in computing the curves shown in Plate 1(a) and Fig. 4.5. As can be seen from Fig. 4.6 there has been a steady downward trend in the S:C emission ratio, and there is no reason not to expect this trend to continue for a very long time.

CHAPTER 5

Observed changes in the climate system and sea level during the recent past

A number of lines of evidence indicate that the Earth's climate has warmed during the 20th century. Foremost among these are compilations of the variation in global mean sea surface temperature and in surface air temperature over land and sea. Supplementing these indicators of surface temperature change is a global network of balloon-based observations of atmospheric temperature since 1958. There are also several indirect or *proxy* indications of temperature change, including satellite observations of microwave emissions from the atmosphere and records of the width and density of tree rings. The combination of surface-, balloon-, and satellite-based indicators provides a more complete picture than can be obtained from any indicator alone, while proxy records from tree rings and other indicators allow the temperature record to be extended back several centuries at selected locations. Apart from temperature, changes in the extent of alpine glaciers, sea ice, seasonal snow cover, and the length of the growing season have been documented that are consistent with the evidence that the climate is warming. Less certain changes appear to have occurred in precipitation, cloudiness, and inter-annual temperature and rainfall variability. Finally, information from tide gauges scattered along the world's coastlines indicates that sea level has been rising more rapidly during the past few decades than during the preceding several centuries, and this is also consistent with a recent warming of the climate.

In this chapter we critically examine the various atmospheric and sea surface temperature data-sets and other indicators of a changing climate. Particular attention is paid to the wide array of adjustments that must be made to the raw data in almost every case before trends that are thought to truly reflect changes in climate (rather than changing instruments or procedures) can be extracted. Where multiple sources of data exist for the same variable (such as mid-tropospheric temperature), the strengths and weaknesses of the different records are outlined and, where possible, apparent conflicts are reconciled. In addition to discussing sea level and the usual variables of climatic interest – temperature, ice and snow, and rainfall – this chapter also critically examines records concerning (i) the present-day distribution of water vapour in the atmosphere and of trends in the amount and distribution of water vapour, (ii) trends in the amount of ozone in both the stratosphere and troposphere, and (iii) trends in the amount of OH (hydroxyl) and CO (carbon monoxide) in the atmosphere. We have already alluded to the importance of water vapour and its vertical distribution for climate sensitivity in Section 3.5. Determining the trends in ozone (and particularly the vertical distribution of the trends) is important for determining the present-day radiative forcing due to changes in ozone – a forcing that could be modestly important compared to the forcing due to changes in the concentrations of the well-mixed GHGs, as will be discussed in Chapter 7. Trends in OH and CO are good indicators of changes in the overall chemistry of the atmosphere, with OH having important implications for the atmospheric lifespan of many GHGs and the production of aerosols (Section 2.8).

5.1 Indicators of global mean temperature change

Indicators of changing global mean temperature can be classified as thermometer-based and proxy-based. The thermometer-based indicators consist of records of sea surface temperature (SST), surface-air temperature, and temperatures in the free atmosphere as measured by sensors suspended from balloons. Borehole temperature measurements are also thermometer-based, but a mathematical inversion procedure is required to translate the measured variation of temperature with depth in the ground to the variation in surface temperature through time, so this method will be classified as a proxy indicator. Other proxy-based indicators include satellite measurements of microwave emission and records of tree-ring width and density.

Surface and surface-air temperature records

Thermometer-based measurements of air temperature were systematically recorded at a number of sites in Europe and North America as far back as 1760, but the set of observing sites did not attain sufficient geographic coverage to permit a rough computation of the global average land temperature until the mid-19th century. Sea surface temperatures (SSTs) and marine air temperatures (MATs) have been systematically collected by merchant ships since the mid-19th century, but these temperature records are restricted to the routes followed by commercial ships, so that there are relatively few data from large regions in the southern hemisphere even today. Land-based, marine air, and sea-surface temperature datasets all require various corrections to account for changing conditions and measurement techniques. The main global datasets and the correction techniques used are described below.

There are three global-scale datasets of what are largely land surface-air temperature variations: a dataset produced at the University of East Anglia (UK) by Phil Jones and colleagues (Jones *et al.,* 1986a,b; Jones, 1988; Jones *et al.,* 1994; hereafter J); a dataset produced by Jim Hansen and colleagues at the Goddard Institute for Space Studies (GISS) in the USA (Hansen and Lebedeff, 1987; Hansen *et al.,* 1996a; hereafter HL); and a dataset produced by Konstantin Vinnikov and colleagues at the State Hydrological Institute in St Petersburg, Russia (Vinnikov *et al.,* 1987, 1990; hereafter V). In all three datasets, ocean areas are represented by scattered island stations, although HL and J supplemented their marine air temperature dataset with observations from fixed-position meteorological ships. J have more stations in their dataset, especially in the early years, which permitted them to begin their time series in 1851 rather than 1880 (as in the other two datasets). They also carried out detailed station-by-

station assessments of consistency through time in the measurement techniques and conditions, and have corrected many of the individual station time series. Inconsistencies can arise from changes in the measurement site, instruments, instrument shelters, or the way in which monthly averages were computed, or due to the growth of cities around the sites (which would lead to a spurious warming trend due to a growing urban heat island). An attempt has been made to remove the effects of urbanization in the East Anglia dataset by comparison of trends at individual urban stations with nearby rural stations, but not in the HL dataset. Jones *et al.* (1990) believe that any residual urban heat island effect in their dataset is a warming of less than 0.05°C during the 20th century, which is an order of magnitude smaller than the inferred warming of 0.5°C. HL screened their data for gross errors, but assumed that remaining errors other than the urban heat island cancelled out for hemispheric and global averages. They estimated the bias due to urbanization by repeating their analysis without sites from cities with populations over 100,000. This reduced their estimated 1880–1980 warming from 0.7°C to 0.6°C, and HL speculated that any remaining urban heat-island effect would not reduce the global mean warming to less than 0.5°C. V compiled their NH data independently of J. They used the J dataset for the SH but averaged the data in a different way.

The important information in the global temperature analysis is the year-by-year deviation of the temperature from the average during some reference period. These temperature deviations are referred to as "anomalies", although this is quite an inappropriate term. All three datasets must use exactly the same reference period if anomalies from the three datasets are to be properly compared. Figure 5.1 compares the global mean temperature anomalies from all three datasets using slightly different reference periods. Although the effects of urbanization and gross errors have been accounted for in these datasets, there are likely to be systematic biases in the early part of the record owing to the fact that the Stevensen Screen (the aerated box that houses meteorological instruments on land) did not become the universal standard until well into the 20th century. The most likely error is that temperature anomalies in the tropics are about 0.2°C too warm from 1900 to 1930 (Parker, 1994). This implies that the true global mean warming since 1900 is somewhat larger than indicated in Fig. 5.1.

There are two global marine datasets: the United Kingdom Meteorological Office (UKMO) marine data bank (Bottomley *et al.,* 1990), and the Comprehensive Ocean-Atmosphere Data Set (COADS) (Woodruff *et al.,* 1987). The COADS contains 40% more individual SST observations than UKMO, but does not distinguish between nighttime and daytime marine air temperatures (Jones *et al.,* 1991). The analysed part of the UKMO set contains only nighttime marine air temperatures (NMATs) to avoid potentially variable effects of solar heating of the ship decks (Parker and Folland, 1991). Figure 5.2 compares the marine datasets with each other or with land data in four different ways and, as discussed below, these comparisons indicate that all of the marine datasets require extensive corrections.

Figure 5.2(a) shows the variation over time in the difference between the UKMO NMAT anomalies and the COADS MAT anomalies. Although absolute MAT and NMAT temperatures are expected to differ, the deviations of each time series from its reference period should vary more or less in unison, yet the difference in the anomalies has undergone marked fluctuations. This could have arisen if the MAT series contains a varying mix of daytime and nighttime observations, or if the UKMO NMAT data are contaminated with varying amounts of daytime data, or if changing ship characteristics or observation techniques affected daytime and nighttime observations differently. Figure 5.2(b) shows the variation through time in the difference between hemispheric mean SST anomalies as given by the two datasets. The variation in the anomaly difference is smaller than for air temperatures, but the reasons for the variation are not clear (Jones *et al.,* 1991). Figure 5.2(c) shows the variation in the difference between hemispheric mean SSTs and MATs as given by the COADS. If instrumentation had remained constant, the difference between SST and MAT should have remained roughly constant. Finally, Figure 5.2(d) shows the difference in the hemispheric mean anomalies of the UKMO NMAT dataset and the anomalies in the coastal and island data from the Jones *et al.* (1986a,b) dataset.

There are a number of reasons why SST and MAT data need to be corrected. Up to 1941, most SST measurements were made in buckets of seawater that were hoisted onto the ship deck, while after 1941, most measurements were made at the engine water intakes. Figure 5.3 shows photographs of some of the different kinds of buckets that were in use in the

19th century. Between about 1856 and 1910, there was a transition from wooden buckets to canvas buckets, but the timing of the transition is uncertain. Evaporative cooling of the bucket after having been placed on the ship deck would have occurred in both cases, but to a much greater extent with canvas buckets. There was a gradual transition from sailing ships to steamships between 1876 and 1910, which altered the height of the ship deck and the speed of the ship. A change in the height of ship decks would have affected MATs, while the change in ship speed would have affected the evaporative cooling of the buckets.

The simplest way to correct the marine data is to adjust the datasets to achieve consistency with other. This is the approach that was followed by Jones *et al.* (1986c). The first step is to adjust the MAT (or NMAT) anomalies so that the marine and adjacent land temperature anomalies move in parallel at time scales of a decade and longer. This requires that the land data have already been corrected to ensure internal consistency. In the second step, the SST anomalies are adjusted so that they move in parallel with the adjusted MAT (or NMAT) anomalies. Jones *et al.* (1991) refer to this as the a *posteriori* approach. An alternative approach is to compute the required correction based on first principles and one's understanding of the causes of errors – what Jones *et al.* (1991) call the a *priori* approach. This has been used to correct both NMAT and SST anomalies.

Folland *et al.* (1984) corrected NMAT data based on the historical increase in the average height of ship decks (from 6 m before 1890 to 17 m by the 1980s) combined with a knowledge of the vertical temperature gradient near the ocean surface under various conditions. This correction removed a spurious cooling between the late 19th and 20th centuries of 0.2°C, but does not agree at all with the required correction implied by Fig. 5.2(d). Bottomley *et al.* (1990) and Parker *et al.* (1995) added further adjustments, including replacing NMAT anomalies with SST anomalies in places where the NMAT anomalies did not seem to be reasonable. The final result in these cases is a mixture of a *priori* and a *posteriori* adjustments. Plate 4 compares the original NMAT anomalies and successive corrections by Folland *et al.* (1984), Bottomley *et al.* (1990), and Parker *et al.* (1995). Figure 5.4 compares the most recently corrected NMAT anomalies with uncorrected SST anomalies, and clearly illustrates the discordance between the two time series prior to 1941.

To correct SST anomalies due to the shift from wooden to canvas buckets, and then to engine-intake measurements, requires using a mathematical model that takes into account the temperature, air–sea temperature difference, relative humidity, and windspeed. The two major unknowns are the length of time between taking the sample and the temperature measurement, and the proportions of

wooden and canvas buckets prior to 1910. Folland and Parker (1989) assumed that all buckets were already canvas by 1856, while Jones *et al.* (1991) assumed that all buckets in use at this time were wooden and perfectly insulated. Folland and Parker (1995) assumed that 80% of the buckets were wooden in 1856, with the proportion decreasing linearly to zero by 1920.

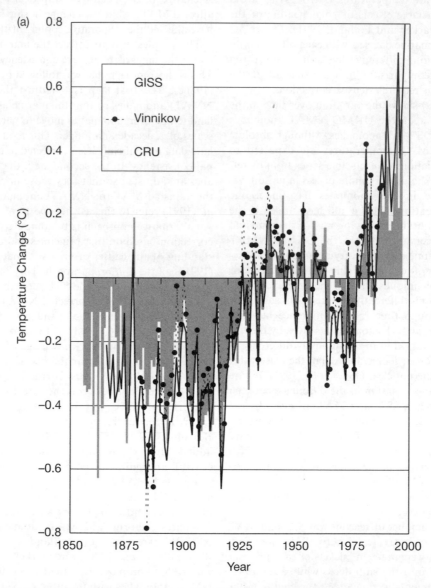

Fig. 5.1 Derivations (a) NH, and (b) SH average temperature from the 1961–1990 mean as given by Hansen and Lebedeff (1987) and updated from http://www.giss.nasa.gov (GISS), as given by Vinnikov *et al.* (1987, 1990), and as given by Jones *et al.* (1986a, b) and updated from http://www.cru.uea.ac.uk (CRU).

Folland and Parker's (1995) estimate of the changing proportion of wooden buckets is linked to their estimate of the length of time that either wooden or canvas buckets sat on the ship decks before the temperature was measured. They first estimated the time interval during which canvas buckets sat on the deck by initially assuming that only canvas buckets were used. Using a mathematical model for the cooling of a canvas bucket, they determined the time interval that minimizes the error in the amplitude of the annual cycle, with the correct annual cycle determined from the seasonal variation of the measured SST obtained when intake measurements were solely used. This is based on the fact that cooling effects will be greatest in winter, so that the greater the time available for cooling, the greater the distortion in the seasonal cycle. The

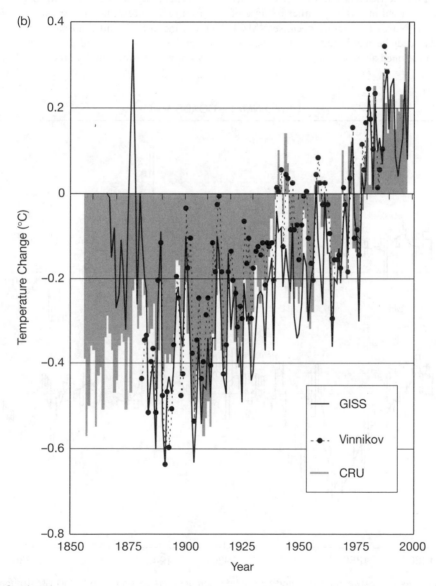

Fig. 5.1   (b) Continued

implied exposure time is 4–6 minutes. Next, they computed the exposure times for wooden buckets using the same procedure, but the computed exposure times were unrealistically long, so they used the exposure time (4 minutes) given in published instructions to observers. Having adopted exposure times for canvas and wooden buckets, the proportion of wooden buckets was chosen so as to give the closest agreement between SST and NMAT anomalies in the 19th century in the tropical Indian Ocean and the tropical Pacific Ocean. Other areas could not be used in this way, either because of insufficient data (North Pacific) or because NMAT anomalies had already been adjusted based on the SST anomalies (as mentioned above).

Plate 5 shows the uncorrected global mean SST anomalies, and as corrected by Folland *et al.* (1984), Bottomley *et al.* (1990), and Parker *et al.* (1995). Successive correction attempts have resulted in markedly different reconstructed SST anomalies for the period prior to 1900. The most recently calculated correction increases steadily from about 0.1°C in 1856 to 0.4°C in 1940, as the proportion of canvas buckets steadily increased, then abruptly drops to zero in 1941 when engine-intake measurements began.

Figure 5.5 presents a composite of the corrected global mean SST and land surface air temperature anomalies for the period 1861–1998. Temperature deviations for individual years, as well as a running

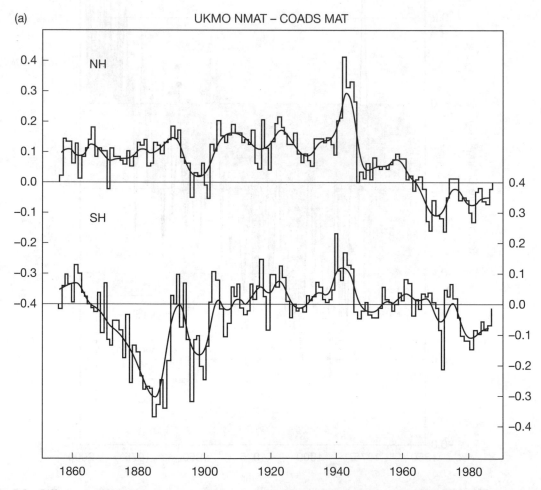

Fig. 5.2   Differences (°C) between uncorrected hemispheric temperature datasets: (a) UKMO NMAT anomalies minus COADS MAT anomalies; (b) UKMO SST minus COADS SST; (c) COADS MAT minus COADS SST; (d) coastal and island temperature data of Jones *et al.* (1986,a,b) minus UKMO MAT. If no corrections to any of the data were required, all of these differences would be very close to zero. Redrafted from Jones *et al.* (1991).

average (using 5 years centred on the current year), are shown. An overall warming of about 0.6–0.7°C is seen, with the warming occurring in two steps: between 1910–1940 and after 1970. 1998 was by far the warmest year on record, with a temperature about 0.6°C warmer than the average for the period 1961–1990.

The number of observation sites and their geographical range varied substantially over the time period represented by the preceding figures. Figure 5.6 shows the geographical distribution of meteorological stations used by HL for 1870, 1900, 1930, and 1960. A circle of 1200-km radius is drawn around

each station, as HL found that this is the distance over which interannual temperature anomalies are reasonably well correlated. Figure 5.7 shows the variation in the number of stations in the HL analysis and in the fraction of the NH and SH within 1200 km of at least one station. Figure 5.8 shows the variation in the annual number of observations of NMAT and SST. Also shown in Fig. 5.8 is the number of annual observations that could be available if a number of datasets that have not yet been digitized and checked for quality were included. These datasets would fill a number of important geographic gaps during the early part of the record.

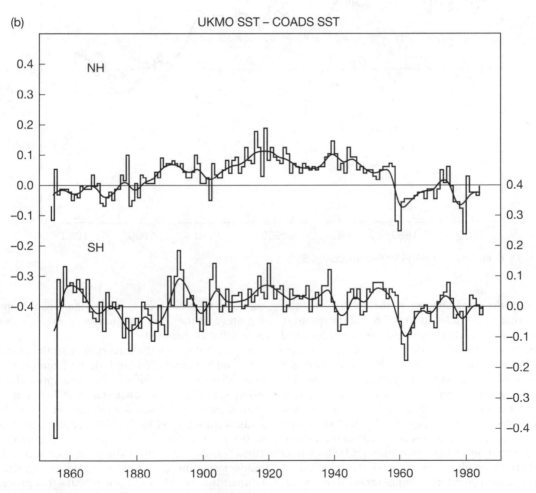

Fig. 5.2   *Continued:* (b) UKMO SST minus COADS SST.

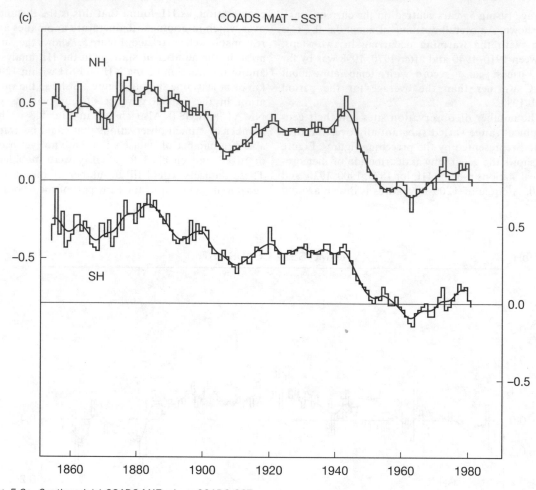

Fig. 5.2   *Continued:* (c) COADS MAT minus COADS SST.

Given the large changes in the observation network that are evident in Figs 5.6–5.8 the question arises as to what effect these changes might have had on the inferred hemispheric and global mean temperature trends. This question has been addressed in a variety of ways, and a consistent result is that changes in the observation network had, at most, a small effect on the overall trend during the past century. However, the ranking of individual years with similar mean temperatures does depend on the spatial distribution of the observation grid. Short-term oscillations involve changes in winds that produce limited spatial correlations, while longer-term oscillations are more related – it appears – to external factors that give better spatial correlations. Thus, the longer the time scale under consideration, the greater the correlation between spatially separated monitoring sites and the fewer the number of evenly spaced points that are required.

Figure 5.9 shows the uncertainty in the global, NH, and SH combined land air–SST temperature anomalies due to changes in the observational network, as computed by Jones *et al.* (1997). Using the central estimate ± 1 standard deviation, the global mean warming from 1851 to 1995 could be as small as 0.5°C or as large as 0.8°C (i.e. 0.65 ± 0.15°C). However, if one applies the estimated errors by random amounts each year, estimates the resultant temperature trend, and then repeats this process 1000 times to determine the distribution of temperature trends so computed, one obtains an estimated trend and uncertainty of about 0.65 ± 0.05°C

(d)   COASTAL LAND – UKMO NMAT

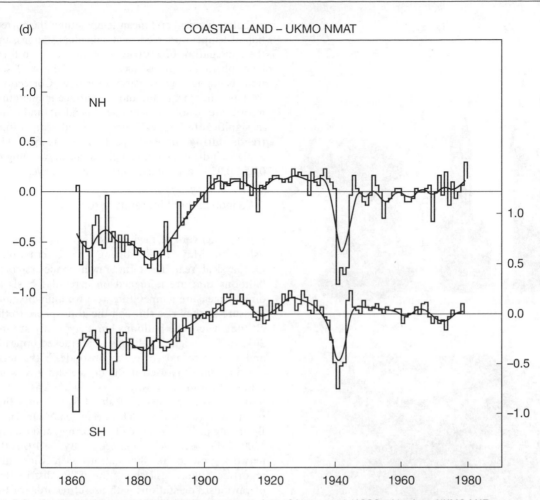

Fig. 5.2   Continued: (d) coastal and island temperature data of Jones *et al.* (1986,a,b) minus UKMO MAT.

(Phil Jones, personal communication, November 1998).[1] The exact uncertainty depends on how correlated one assumes the errors to be from one year to the next.

The latitudinal variation in the zonally (east–west) averaged temperature, and the difference in warming between the NH and SH, are of great importance to climatic change and the understanding of its causes, as will be seen in subsequent chapters. Figure 5.10 shows the latitudinal variation in the zonally averaged change in the combined SST/land surface–air dataset that is available from the University of East

Anglia (http://www.cru.uea.ac.uk). Changes are shown for the two periods when most of the warming of the last 130 years occurred, 1910–1940 and 1970–1998 (see Fig. 5.5), as well as for the period 1910–1998. During the first warming step, the warming was strongly suppressed in NH mid-latitudes, while during the second warming step the warming was most pronounced in NH mid-latitudes. For the period 1910–1998, there is very little variation of the warming with latitude, and very little difference in the warming between the two hemispheres.

Since 1982, satellite-based estimates of SST have also been available (Reynolds, 1988). The satellites measure SST at the interface between the ocean and atmosphere – what is referred to as the "skin" temperature – whereas the other data pertain to a depth

[1]   This procedure is referred to as a Monte Carlo estimation technique.

Fig. 5.3 Various kinds of buckets used in the 19th and early 20th centuries for measuring sea surface temperature. Reproduced from Folland and Parker (1995).

of at least 1 m below the sea surface. The satellite data thus have to be adjusted for comparability, and corrected for variations in the amount of water vapour and aerosols between the surface and the satellite, and for drift in the satellite sensors. Thus, inclusion of satellite data into the existing data network will not be straightforward.

To summarize this subsection, there are two critical issues with regard to the construction of time-varying global or hemispheric mean temperature: the correction of temperature records at individual stations, and the aggregation of individual station data to form hemispheric and global means when the mix of stations being aggregated varies over time. Corrections for land air, marine air, and sea surface temperature records are to some extent interdependent, and result in significant changes in the inferred temperature trends during the early part of the record. The analyses indicate that a real global mean warming of $0.65 \pm 0.05°C$ occurred during the past 130 years.

Balloon-based temperature records

A global network of rawindsonde stations was established in May 1958, during the International Geophysical Year. The term "rawindsonde" refers to balloons that are launched on a regular basis to directly measure temperature, relative humidity, and pressure as they rise through the atmosphere to the altitude where the balloons burst (typically around 20 km). The original balloons were tracked optically and were referred to as "radiosondes"; the term "rawindsonde" is used if the device can give wind information in some way (as is now the case). The initial network consisted of about 540 stations that reported once daily at 0000 Greenwich Mean Time. Beginning in 1963, twice-daily reporting, at 0000 and 1200 GMT, began to be important. By the 1970s the network grew to 700–800 stations with twice-daily reporting. Figure 5.11 shows the global distribution of rawindsonde stations with useful records for the period 1973–1994.

Gaffen (1994) and Parker and Cox (1995) discuss potential discontinuities and spurious trends in the rawindsonde temperature data due to changing instruments and operational procedures. The major causes of discontinuities and spurious trends are: (i) changes in the temperature sensor used; (ii) changes in the way that the effects of solar heating of the instrument package are accounted for, or the initial introduction of a computed correction; (iii) changes in the shield around the instrument package; (iv) changes in the length of the line between the instrument package and the balloon (which is heated by the sun in the case of daytime measurements and produces a warm wake that impinges upon the instrument package); and (v) changes in the heights and temperatures at which the balloons burst. The adjustments to the measured temperature to account

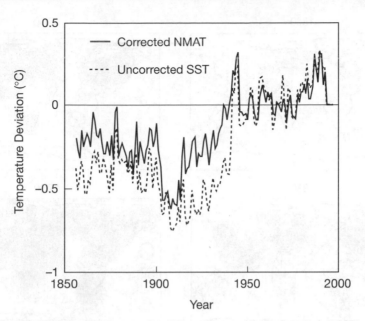
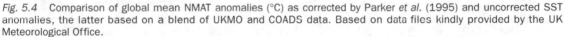

Fig. 5.4 Comparison of global mean NMAT anomalies (°C) as corrected by Parker *et al.* (1995) and uncorrected SST anomalies, the latter based on a blend of UKMO and COADS data. Based on data files kindly provided by the UK Meteorological Office.

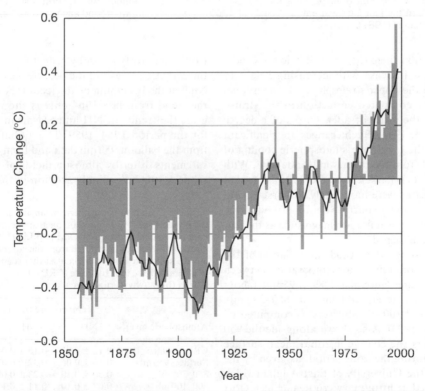

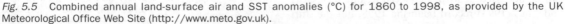

Fig. 5.5 Combined annual land-surface air and SST anomalies (°C) for 1860 to 1998, as provided by the UK Meteorological Office Web Site (http://www.meto.gov.uk).

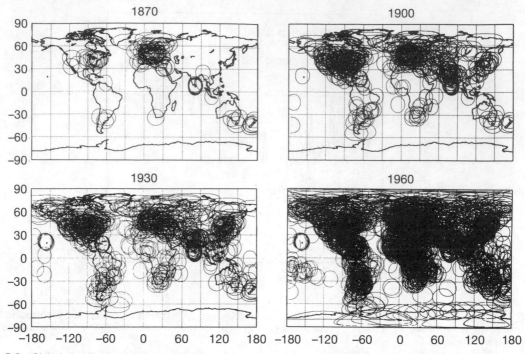

Fig. 5.6 Global distribution of meteorological stations with surface-air temperature records used by Hansen and Lebedeff (1987) at the four indicated dates. A circle of 1200 km radius is drawn around each station. Reproduced from Hansen and Lebedeff (1987).

for solar radiation (item (ii)) are well below 1°C near the surface but increase with increasing height to about 20°C in the upper stratosphere; the occasional changes in the computed correction in the stratosphere are on the order of a few tenths of a degree Celsius (Gaffen, 1994). Such changes are significant compared to the long-term stratospheric cooling of 1.5°C inferred from the raw rawindsonde data. With regard to item (v), Parker and Cox (1995) state that the early balloons were more likely to burst in cold conditions, creating a spurious cooling trend in the lower stratosphere of unknown magnitude as the balloons were strengthened.

Figure 5.12 shows the latitudinal variation of the change in the zonal mean air temperature between December 1963 and November 1989 at 950 mb (about 500 m above sea level) and for the 850–300 mb, 300–100 mb, and 100–50 mb layers as computed by Oort and Liu (1993). Also shown alongside the 950 mb variation is the change in zonal mean surface temperature during the same period, as given by the dataset from the University of East Anglia (which can be accessed at http://www.cru.uea.ac.uk). Oort and Liu's (1993) record begins in 1958, but the shift

from once- to twice-daily reporting in 1963 biased the early trends, as did the occurrence of a strong El Niño at the beginning of the record. For this reason, the trend from late 1964 only is shown. Table 5.1 gives the trends in NH and SH mean temperature for the period 1964–1989. The global mean trend from the balloon 950 mb data and from surface measurements differ by almost a factor of two, and the shape of the latitudinal variation in the trend mag-

Table 5.1 Trends in hemispheric mean air temperature for the period 1964–1989, based on measurements taken from balloons and as analysed by Oort and Liu (1993). Also given are the trends computed from the combined surface air–sea surface temperature data set that is available from the Climate Research Institute (CRU), University of East Anglia (http://www.cru.uea.ac.uk).

Atmospheric layer	Temperature trend (°C/10 yr)		
	NH	SH	Global
CRU–surface	0.12 ± 0.03	0.16 ± 0.02	0.13 ± 0.02
Surface–950 mb	0.13 ± 0.06	0.32 ± 0.07	0.24 ± 0.06
850–300 mb	0.20 ± 0.10	0.23 ± 0.09	0.22 ± 0.09
300–100 mb	−0.03 ± 0.10	−0.11 ± 0.11	−0.07 ± 0.09
100–50 mb	−0.38 ± 0.14	−0.43 ± 0.16	−0.40 ± 0.12

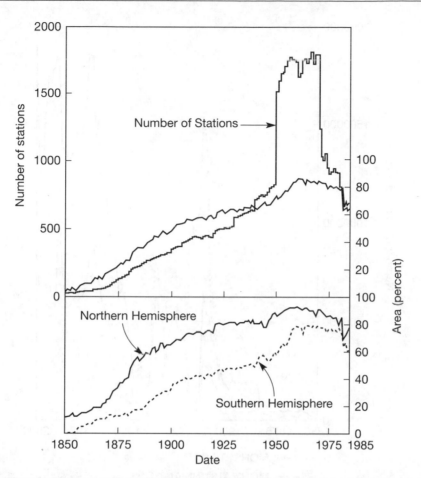

Fig. 5.7 (a) Number of stations used in the Hansen and Lebedeff analysis, and percentage of globe within 1200 km of at least one station. (b) Percentage of hemispheric area location within 1200 km of at least one station. Reproduced from Hansen and Lebedeff (1987).

nitude is opposite for the two datasets. The balloon but not the surface-based data indicate a significantly greater near-surface warming in the SH than in the NH. Nearly identical warming is seen for the 850–300 mb layer. A weak cooling is seen in the upper troposphere (300–100 mb) and a strong cooling in the stratosphere. In light of the above discussion of spurious cooling trends, the stratospheric results should be regarded with suspicion but, as discussed next, can be compared with independently derived trend estimates.

Satellite-based proxy records

Microwave sensors, known as microwave sounding units (MSUs), have been placed aboard satellites and are able to measure microwave radiation (at a wavelength of 4 mm) emitted by O_2. Since the atmospheric concentration of O_2 is constant in space and in time, variations in the emitted radiation should be directly related to variations in the temperature of emission. Two different channels sample the radiation at slightly different frequencies, which correspond to emissions in different regions of the atmosphere: MSU channel 2R samples the middle and lower troposphere (mainly the 850–300 mb layer), while channel 4 largely samples the stratosphere (mainly the 50–100 mb layer). The channel 2R data are in fact derived by subtracting measurements made with the satellite looking toward the horizon (near-limb observations) from measurements made with the satellite looking more or less straight down (near-nadir observations) (Gaffen, 1998).

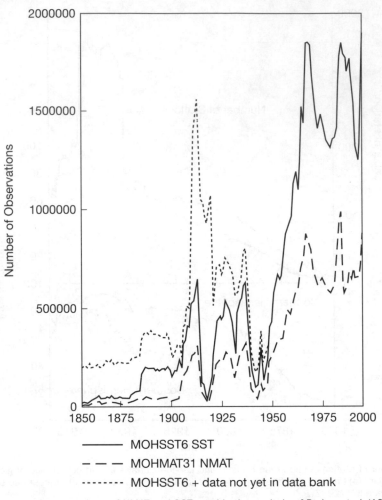

Fig. 5.8   Annual number of observations of NMAT and SST used in the analysis of Parker *et al.* (1995). Also given, as a dotted line, is the number of SST observations potentially available if known additional data sources were to be used. Reproduced from Parker *et al.* (1995).

The MSU data have the distinct advantage of providing global coverage, unlike the rawindsonde and surface-based measurements, both of which are concentrated over land and in the northern hemisphere. There are, however, a number of potential problems with the MSU data. First, the 1979–1994 record is based on eight different satellites. Although there was always some temporal overlap between the outgoing satellite and its replacement, to allow intercalibration, the additional noise introduced by these changes makes the analysis of trends somewhat problematic (Hurrel and Trenberth, 1997, 1998; Christy *et al.,* 1997, 1998; Trenberth and Hurrel, 1997). Second, there was a spurious warm-

ing of 0.03–0.04°C/yr after 1990 due to a drift in the orbit times which, however, can be estimated and corrected for (Christy and Spencer, 1995). Third, the distribution of heights sampled by the MSU can change as the amount of ice crystals and raindrops changes. If these water substances extend to greater altitude as the climate warms, the MSU will sample higher and thus colder parts of the troposphere, thereby giving a smaller temperature increase than occurred in reality (Hansen *et al.,* 1995b). Most significantly of all, the height of each satellite has gradually decreased over time owing to the frictional effect of the outer atmosphere. This altered the near-limb observations more than the

(a) Global mean decadal temperature anomaly

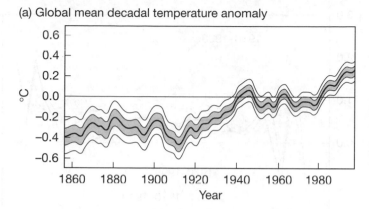

(b) Northern hemisphere decadal temperature anomaly

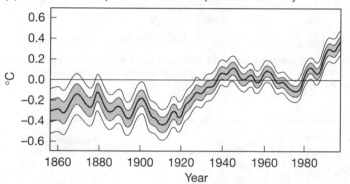

(c) Southern hemisphere decadal temperature anomaly

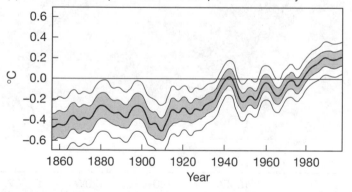

Fig. 5.9   Combined land–air and SST temperature anomalies (°C) for (a) the globe, (b) northern hemisphere, and (c) southern hemisphere, with the central estimate (heavy lines), ±1 standard deviation (shaded), and ±2 standard deviations (thin lines). Reproduced from Jones *et al.* (1997).

near-nadir observations, and since the deduced lower tropospheric temperature depends on the difference between the two, this introduced a spurious trend – a cooling – of 0.12°C/decade (Wentz and Schabel, 1998).

Figure 5.13 shows the variation in global mean temperature deduced for MSU-2R and MSU-4, along with the variation – since 1964 – in the rawindsonde temperatures from comparable layers in the atmosphere, and in surface air temperature. The MSU-2R

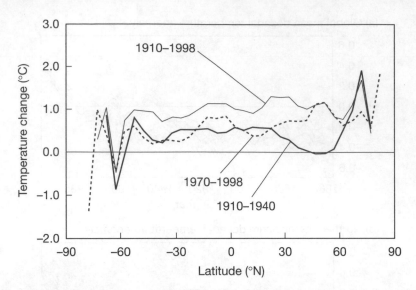

Fig. 5.10 Change in the zonally average composite SST/land surface–air temperature over the periods 1910–1940, 1970–1997, and 1910–1997. Based on gridded data obtained from the University of East Anglia web site (http://www.cru.uea.ac.uk).

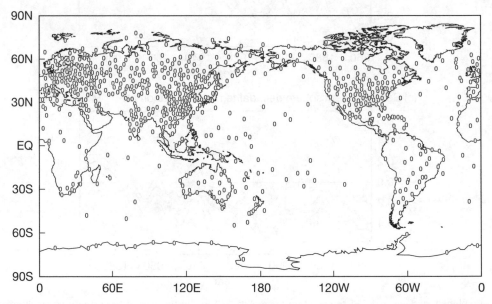

Fig. 5.11 Distribution of land-based rawindsonde stations that have relatively long records during the period 1973–1994. Also included are some Ocean Weather Stations, most of which are no longer active. Reproduced from Elliott (1995).

temperatures shown in Fig. 5.13(b) are those of Wentz and Schabel (1998), in which the effects of changing satellite altitude have been corrected for. Two rawindsonde time series are shown – one by Oort and Liu (1993), based on the global rawindsonde network of 700–800 stations, and the other by Angell (1997), based on a carefully chosen subset of 63 stations. In order to facilitate intercomparison, all of the temperature series shown in Fig. 5.13(b) have been shifted to give zero deviation in 1979. There is good

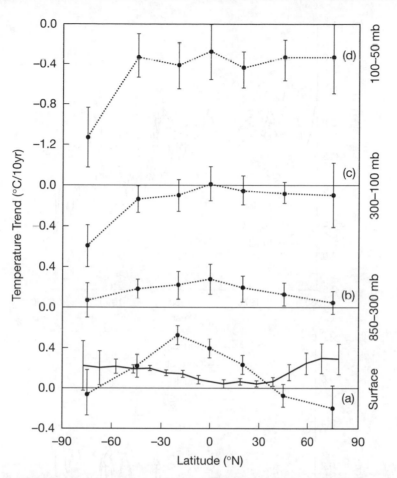

Fig. 5.12 Variation in global mean temperature for 1964–1989 as deduced from balloon-based measurements at (a) the 950 mb level (surface), and for (b) the 850–300 mb layer, (c) the 300–100 mb layer, and (d) the 100–50 mb layer. Redrafted from figure 4 of Oort and Liu (1993). Shown alongside the 950 mb balloon data is the trend in zonally averaged surface temperature (solid line) as obtained from the Climate Research Unit at the University of East Anglia.

agreement between the balloon-based and corrected satellite-based trends. The global mean tropospheric temperatures experienced a much greater decrease in response to the eruption of Mt Pinatubo (in mid-1991) than did the surface temperature, but both rebounded to the same deviation (relative to 1979) by the end of 1998. This is quite reasonable, given the moderating effect of the large thermal mass of the ocean surface layer compared to the atmosphere.

Given the continual discovery of new sources of error in the satellite data, it is not surprising that the temperature trends deduced from the satellite data for a given time interval have changed many times. This is in contrast to the temperature trends from the land-based record over the same period, which

are quite stable. Because of the shortness of the record (still less than 20 years), the deduced temperature trend depends strongly on how data at the beginning and end of the data period are treated. Christy *et al.* (1995) deduced a global mean cooling of 0.05°C/decade using all the original data for the period 1979–1994, but obtained a warming of 0.09°C/decade over this period after removing the effect of an El Niño at the beginning of this time period and the effect of a large volcanic eruption (Mt Pinatubo) at the end. Later, Christy *et al.* (1998) deduced a global mean cooling of 0.046°C/decade over the period 1979–1997. However, correction of the aforementioned effects of decreasing satellite altitude turns this cooling into a warming of

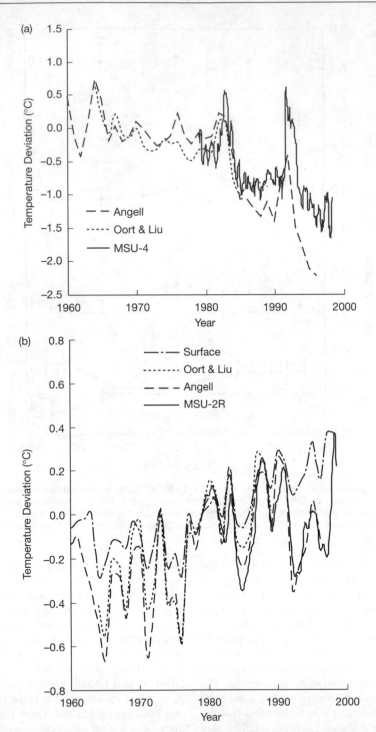

Fig. 5.13 (a) Comparison of temperature records for the stratosphere (50–100 mb). Shown are the MSU-4 record (solid line) and the balloon-based records of Oort and Liu (1993) and Angell (1997). (b) Comparison of tropospheric and surface temperature records. Shown are the MSU-2R record for the troposphere (mainly the 850–300 mb layer) from Wentz and Schabel (1998), the balloon-based records of Oort and Liu (1993) and Angell (1997) for the 800–350 mb layer, and the surface record prepared at the University of East Anglia (http://www.cru.uea.ac.uk). The Angell (1997) data were obtained from the Carbon Dioxide Information and Analysis Center (http://cdiac.esd.ornl.gov/ftp/ndp008).

0.07°C/decade for the period 1979–1995 (Wentz and Schabel, 1998). Prabhakara *et al.* (1998) obtained an even larger warming trend, 0.11°C/decade for the period 1980–1996, after accounting for changes in the satellite altitude.

By comparison, Jones (1994) computed temperature trends for the period 1979–1993 of 0.10°C/decade based on rawindsonde data for the 850–300 mb layer, and 0.17°C/decade based on combined land-surface air and SST data, respectively [2] Part of the difference between MSU/rawindsonde trends and surface-based temperature trends could be related to a real difference in the surface and mid-tropospheric temperature changes. If this is true, the difference is opposite to that which would be expected if GHGs alone were responsible for the warming (see Section 10.1). Part of the difference could also be due to uneven sampling of surface air temperature. Karl *et al.* (1994) find that the spatial variability of temperature trends tends to decrease as the time scale increases. Thus, incomplete spatial sampling can introduce significant errors in decadal time-scale trends (comparable to the length of the MSU record), but errors in century time-scale trends are one to two orders of magnitude smaller than the inferred trend itself. However, the rawindsonde data should be subject to the same sampling problems, yet the rawindsonde and MSU data give essentially the same trends during the period of overlap.

Turning to the stratosphere, the trend for MSU-4 from 1979 through to April 1997 is a cooling of 0.49°C per decade (Christy *et al.*, 1998). The rawindsonde data of Angell (1997) show a much stronger cooling over this period (Fig. 5.13a). The Oort and Lia (1993) dataset agrees with the MSU–4 data during the period of overlap (1979–1989). This agreement does not validate the rawindsonde trend prior to 1979, because most of the known procedural or instrument changes that would have caused spurious trends occurred prior to 1979. The real trend prior to 1979 may have been smaller than shown in Fig. 5.13(a).

---

[2] These trends were computed after removing the warming effect of an El Niño at the beginning of the record, and the cooling effect of Mt Pinatubo just after 1991. Prior to adjustment, the surface trend was 0.1°C/decade. The effect of Mt Pinatubo would be smaller for the time period (1979–1995) considered by Wentz and Schabel (1998), so removal of the effect of Mt Pinatubo would have had a smaller effect on the trend.

Ground temperature profiles

Variations in surface temperature over time will propagate into the Earth's subsurface with some attenuation, producing a vertical variation in temperature at any given time that reflects the temporal variation in surface temperature. It is therefore possible to reconstruct surface temperature variations from borehole records of the temperature variation with depth. For such reconstructions to have any climatic relevance, it must first be determined that the vegetation at the site has not been altered over the time interval of interest, as changes in vegetation can cause changes in the surface temperature. Analysis of borehole records from widespread sites that pass this test indicates that surface warming has occurred during the 20th century at almost all sites examined (Nicholls *et al.*, 1996, Section 3.2.5.2; Harris and Chapman, 1997).

Changes in alpine glaciers, seasonal snow cover, and sea ice

In almost all parts of the world where mountain glaciers occur, retreat of the glaciers has occurred during the 20th century. Although the mass budget of glaciers depends on the amount of snowfall, as well as on temperature, the observed retreat is consistent with a warming in alpine regions of 0.6–1.0°C (Oerlemans, 1994).

The annual mean extent of snowcover on land in the NH decreased by about 10% during the period 1972–1992, with the largest decreases in spring and autumn (Groisman *et al.*, 1994). Figure 5.14 shows the variation in the annual and seasonal extent of snowcover in the NH (excluding Greenland) and the variation in surface air temperature in regions where snowcover has decreased. The snowcover extent was determined from satellite images in the visible part of the solar spectrum. The temperature and snowcover variations are closely related.

Estimation of the area of sea ice from satellite data is more difficult, since it is necessary to be able to determine the fraction of open water within an image pixel, as well as the extent of sea ice. The area of sea ice depends on both the extent or limits of sea ice, and the fractional coverage or concentration of sea ice within the area bounded by the sea ice limits. Early observations used visible and infrared images. The most consistent and continuous records began in 1978 with microwave images taken at several different

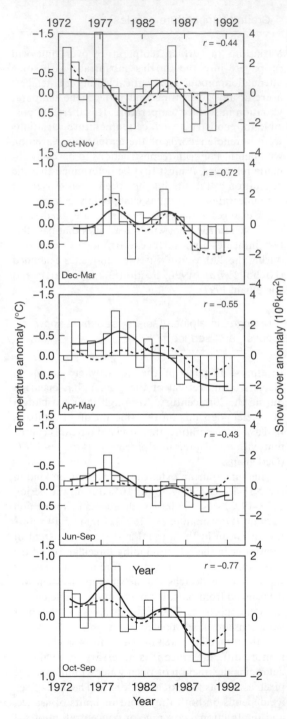

Fig. 5.14 Variation in the mean annual and seasonal extent of snowcover on land in the NH (bars, yearly values; solid lines, smoothed values), and of smoothed surface temperature (dashed lines) over regions of transient snow cover. Reproduced from Nichols *et al.* (1995).

wavelengths. Three satellites span the period from 25 October 1978 to 20 September 1995: the Scanning Multichannel Microwave Radiometer (SMMR), on board Nimbus 7, from 25 October 1978 to 20 August 1987; and two US Defense Meteorological Satellite Program (DMSP) Special Sensor Microwave Imagers (SSM/I F8 from 9 July 1987 to 18 December 1991, and SSM/I F11 from 3 December 1991 to 30 September 1995). These sensors provide data for both polar regions. The usable overlap between SSMR and SSM/I F8 is only 3 weeks, while that between the SSM/I satellites is only 2 weeks. This is unfortunate, since a large adjustment in the SSM/I F8 data is required based on the overlapping measurements. Furthermore, the original sea ice areas computed from the raw SSMR data and provided to the research community were recently revised, based on an improved algorithm, and this altered the adjustment to the SSM/I sea ice areas that were derived from the period of overlap.

Figure 5.15 shows the variation in the extent of SH sea ice based on the original and revised analysis of the SSMR data, as computed by Stammerjohn and Smith (1997). The two trends diverge strongly from around 1986, just before the switch from the SSMR to the first SSM/I satellite. The original analysis showed a strong upward trend in the area of sea ice in all seasons, while the revised analysis shows no upward trend (in winter and spring) or a much weaker trend (summer and autumn) between 1978 and 1996. The weak trend in total SH sea ice area is the result of an increase in the area of sea ice in some sectors and decreases in others. However, evidence from whaling ships indicates that the total area of SH sea ice decreased in extent by about 25% between the mid-1950s and the early 1970s (de la Mare, 1997) – prior to the period of satellite observations.

Figure 5.16 shows the inferred variation in the extent of sea ice in the northern hemisphere as computed by Maslanik *et al.* (1996), using the original analysis of the SSMR satellite data. In this case, a clear downward trend in the area of sea ice is evident. The nature of the Arctic data is such that the re-analysis of the satellite SSMR data, mentioned above, is not likely to significantly alter the deduced trends. Consistent with the trend of decreasing extent of sea ice, Smith (1998) finds that the annual number of days in which melting of sea ice occurs increased at a rate of 5.3 days (8%) per decade over the period 1979–1996.

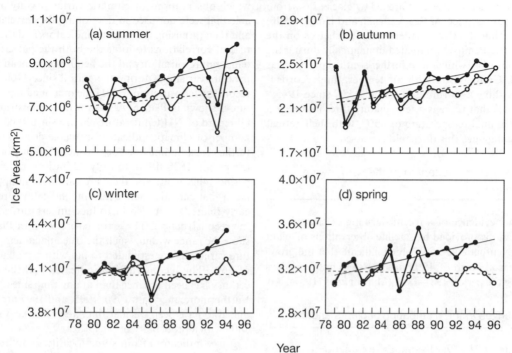

Fig. 5.15 Variation in seasonal sea ice area in the southern hemisphere during the period 1979–1996 as inferred from satellite data, using the original (solid circles) and updated (open circles) analysis of the raw data. This is an update of Stammerjohn and Smith (1997), using a datafile kindly provided by Sharon Stammerjohn.

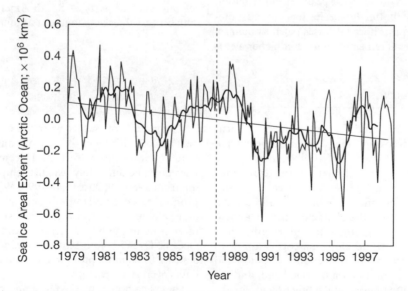

Fig. 5.16 Variation in the extent of sea ice during the period 1979 to September 1998 as measured by satellites for the northern hemisphere. This is an update of Maslanik *et al.* (1996), kindly provided by James Maslanik.

In addition to sea ice, the Antarctic continent is fringed by ice shelves that are fed by glacier flow from the land-based ice masses. Vaughan and Doake (1996) report that the five northernmost ice shelves on the Antarctic Peninsula retreated dramatically during the past 50 years, while those further south show no clear trend. Comparison with air temperature records – which show a warming of about 2.5°C since 1945 – indicates that there is a temperature threshold near a mean annual temperature of –5°C, with shelf retreat beginning once this threshold is crossed.

Changes in the length of the NH growing season

Further evidence of a recent warming of climate over NH land is provided by satellite observations of plant photosynthetic activity, which indicate that the growing season north of 45°N has increased by 12 ± 4 days during the period 1981–1991 (Myneni *et al.,* 1997).

Changes in the distribution of plants and animals

Epstein *et al.* (1998) summarize evidence indicating that the distributions of a number of plant and animal species have changed in a manner that is consistent with a warming of the climate. For example, upward displacements of plant distributions have been documented on 30 alpine peaks, the distribution of a number of butterfly species has shifted poleward, and mosquito-borne diseases (such as malaria and dengue fever) are being reported at higher elevations than before.

Long proxy records

Proxy temperature indicators at a number of locations allow us to extend temperature records back several centuries. Mann *et al.* (1998) estimated the variation in NH mean annual temperature since AD 1400 based on such indicators as tree-ring width and density, the chemical composition and annual growth rate in corals, the characteristics of annual layers in ice cores, and long historical records. Their approach involved (i) developing a correlation with the large-scale patterns of temperature variation and the NH average variation, on the one hand, and the individual climatic indices on the other hand, for the period 1902–1980, when both pieces of information are available; (ii) testing the predictions made using

the climate indicators over the period 1854–1901, for which observations of climatic variability are available that were not used in developing the correlation; and then (iii) using the climatic indicators and the statistical correlations to infer the both the patterns of temperature variation and the hemispheric mean temperature variations during the period since 1400. The reconstructed variation in NH mean annual temperature is shown in Plate 6. Also shown is the instrumental record of NH temperature change since 1900. The number of climatic indicators available decreases the further back in time one goes (from 112 indicators for the period 1820–1980, to only 22 indicators for the period 1400–1450), and there are various assumptions (such that climatic change and the indicators are linearly related) that also introduce uncertainties into the reconstruction. The error bars shown in Plate 6 show the range within which the true annual temperature variations are estimated to fall with a probability of 96%. The reconstruction indicates that the 20th century has been warmer than at any time since 1400. Furthermore, the years 1990, 1995, and 1997 are estimated to have been warmer than any other year since 1400 with a probability of 99.7%.

Proxy indicators from some localities can extend the available temperature records back even further. Tree-ring records from the northern Urals indicate that summer temperatures in that region were warmer in the 20th century than at any time since AD 914 (Briffa *et al.,* 1995), although there are regions (such as parts of South America) where the 20th century does not stand out as unusual.

5.2   Geographic, seasonal, and diurnal patterns of warming

The magnitude of surface warming during the recent past varied with location, season, and time of day. Plate 7 shows the pattern of December–January–February and June–July–August temperature warming from 1950–1959 to 1990–1998. Warming is most pronounced over NH continents at high latitudes, especially in winter, but there are also broad regions of cooling in the North Atlantic and North Pacific Oceans. Over SH mid-latitude oceans, the warming appears to be more pronouned in summer (although data coverage is sparse).

There has been a marked asymmetry in the warming between night and day. Analysis of non-urban land stations over the period 1950–1993 and representing about 50% of the global land area indicates that nighttime minimum temperatures have warmed more than twice as fast as daytime maximum temperatures – at a rate of 1.79°C per 100 years compared to 0.84°C per 100 years. Consequently, the diurnal temperature range has decreased at a rate of 0.79°C per 100 years (Easterling *et al.*, 1997).[3] Regions and seasons showing the strongest reduction of diurnal temperature range are associated with statistically significant increases in cloudiness.

5.3   Stratospheric temperature

Changes in stratospheric temperature could, in principle, help in determining the causes of observed changes in surface climate, because different causal factors have different effects on stratospheric temperatures. As discussed in Section 5.1, the only long-term data for stratospheric temperature (extending back to 1958) are the rawindsonde data, but these are not considered to be reliable for stratospheric temperature trends during the early part of the record. The raw data indicate a cooling of 1.5°C from 1958 to 1995. For the period since 1979, satellite-based proxy data are available, and these indicate a similar cooling to that given by the rawindsonde data (Fig. 5.13). This is also the time when changes in stratospheric temperature were overwhelmingly influenced by the loss of ozone (Section 11.4). Thus, for the period for which there are reliable data for stratospheric temperature trends, the observed signal is dominated by a factor that is already well established, so that these trends are not helpful in deciphering the relative importance of other causes of climatic change. Conversely, for the time period when other factors could possibly have been identified based on trends in stratospheric temperature, the

---

[3] The trend in DTR does not equal the trend in maximum temperature minus the trend in the minimum temperature for two reasons. (1) Discontinuities (due to changes in instrumentation) in all three temperature series at each station were corrected only if the discontinuity exceeded a specified threshold. Sometimes discontinuities in DTR exceeded the threshold while those for the maximum and minimum temperature did not. (2) Some stations have only maximum or minimum temperatures, so different sets of stations go into the computation of the mean trends for the three temperature series.

data are not reliable. We shall return to this issue in Section 11.3, in which the issue of detecting the effects of increasing GHG concentrations on the Earth's climate is discussed.

5.4   Trends in precipitation, storminess, and El Niño

A climatic parameter that is every bit as important as temperature is the amount (and distribution) of precipitation, but precipitation – whether as rain or snow – is much harder to measure than temperature. Difficulties arise not only from the tendency of precipitation gauges to underestimate the amount of rainfall owing to the effects of turbulence around the gauge created by the gauge itself, but also because precipitation varies strongly from one point to the next. It is therefore difficult to derive meaningful areal-mean precipitation amounts unless a very dense network of precipitation gauges is used – a condition that is rarely satisfied. However, as long as changes in the precipitation gauges are properly accounted for, it should be possible to correctly deduce at least the direction in which precipitation is changing, even if the absolute areal-mean precipitation amounts still contain significant errors.

Mean annual precipitation

There are two global precipitation datasets: "Hulme" (Hulme, 1991; Hulme *et al.*, 1994), and the "Global Historical Climate Network" (GHCN), described by Vose *et al.* (1992) and Eischeid *et al.* (1995). Precipitation over land tends to be underestimated by up to 10–15% owing to the effects of airflow around the collecting dishes, and a gradual reduction in the underestimation has resulted in a spurious upward trend in precipitation. After correcting for this spurious trend, there is evidence for an overall increase in mean annual precipitation over land by about 1% – a difference that is hard to distinguish from zero. Clearer trends are evident on a regional basis, as illustrated in Fig. 5.17 for the periods 1955–1974 to 1975–1994 and 1900–1994. Precipitation appears to have increased in most regions except southern Europe, the Sahel region of Africa, Indochina, Japan, and Chile.

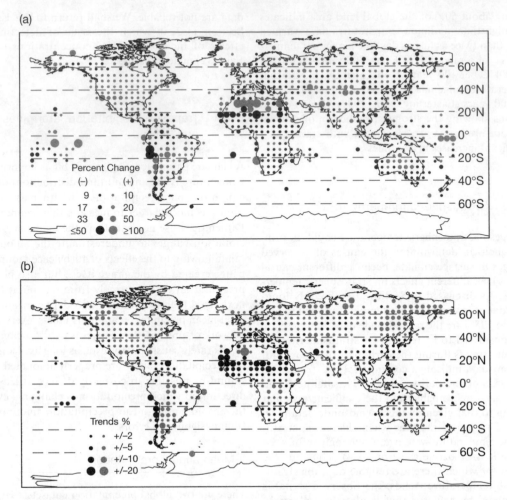

Fig. 5.17 (a) Changes in precipitation over land from 1955–1974 to 1975–1994, expressed as a percentage of the 1955–1974 mean. (b) Changes in precipitation over land from 1900 to 1994, expressed as a percentage of the 1961–1990 or 1951–1980 mean (depending on the region). Reproduced from Nicholls *et al.* (1996).

Heavy precipitation

Karl *et al.* (1995) found a trend towards a greater proportion of rainfall occurring as heavy-rainfall events in the USA (1911–1992), the former Soviet Union (1935–1989), and China (1952–1989). In an update of their analysis for the US, Karl and Knight (1998) found that the proportion of total annual rainfall accounted for by the rainiest 10% of days increased from 36% at the beginning of the record to 40% at the end of the record. During this period, total annual rainfall in the United States increased by about 10%. Suppiah and Hennessy (1998) report trends of both increasing and decreasing total rainfall and heavy rainfall in Australia, depending on the region.

Snowfall in Antarctica and Greenland

Walsh (1995) summarizes evidence that indicates that snowfall over Antarctica increased by 5–20% in recent decades. This evidence consists of the changing thickness of annual layers in Antarctic ice cores, and estimates of the convergence of atmospheric water vapour fluxes over Antarctica. In contrast, Fischer *et al.* (1998) report evidence from an ice core in central Greenland that the rate of snow accumulation has decreased by 20% during the past 50 years. This is consistent with data from other ice cores, that they cite, and with evidence – discussed in Section 12.1 – of an inverse relationship between regional temperature and snowfall in Greenland during the Holocene Epoch (the last 10,000 years).

Tropical cyclones

There appears to be a weak downward trend in the intensity and frequency of the most intense hurricanes in the Atlantic Ocean (Landsea *et al.*, 1996). Figure 5.18 shows the average peak intensity reached by all hurricanes in each year during 1944–1995, and the maximum intensity reached by the most intense hurricane of the year. Both peak and average intensities show a downward trend. Figure 5.19 shows the variation in the frequency of intense Atlantic hurricanes (having sustained surface winds of at least $50 \, \text{m s}^{-1}$ at some point) and in the remaining hurricanes; a downward trend is evident for intense hurricanes, with the 1995 hurricane season standing out as an unusually active season. In the case of the western North Pacific Ocean, the frequency of hurricanes decreased from the early 1960s to about 1979, and has been increasing since then but has not yet reached previous activity levels (Chan and Shi, 1996).

Extratropical cyclones

Figure 5.20 shows the 11-year running mean of the number of intense winter extratropical cyclones in the NH for the period 1899–1991, as deduced by Lambert (1996) based on the analysis of sea level pressure maps. The Pacific data prior to 1930 are not considered to be reliable. There was little or no trend in the frequency of winter cyclones between 1930 and

1970, followed by a sharp increase after 1970 in the Pacific Ocean sector.

El Niño

El Niño-like conditions existed in the tropical Pacific from 1990 to June 1995, the longest such El Niño on record. There has also been a trend for more frequent El Niño events since 1976. Trenberth and Hoar (1996, 1997) performed a statistical analysis of an El Niño index for the period 1882–1997, and found that the likelihood of the recent changes occurring due to natural variability alone, given the previous record, is only once in 2000 years. The long El Niño of 1990–1995 was followed by one of the strongest El Niños on record during 1997–1998, which reinforces the viewpoint that the statistical characteristics of El Niño have changed during the last 20 years.

5.5   Changes in the amount of water vapour in the atmosphere

Determining the way in which the amount and vertical distribution of water vapour in the atmosphere have varied as the surface climate varied over the last few decades would be of great value in assessing the long-term sensitivity of climate to radiative forcings. This is because changes in the amount of water

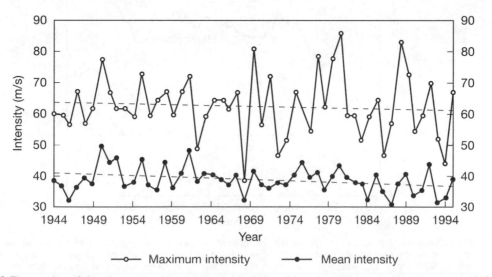

Fig. 5.18 Time series of the mean annual peak sustained windspeed attained in Atlantic hurricanes and of the windspeed attained by the strongest hurricane. Reproduced from Landsea *et al.* (1996).

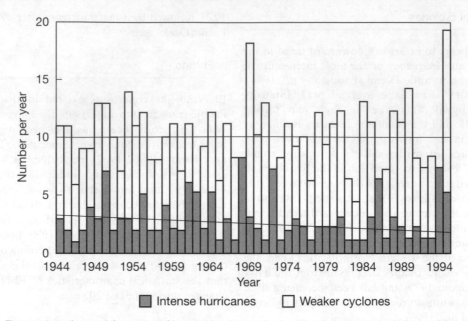

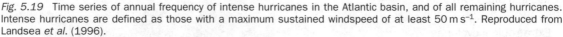

Fig. 5.19 Time series of annual frequency of intense hurricanes in the Atlantic basin, and of all remaining hurricanes. Intense hurricanes are defined as those with a maximum sustained windspeed of at least 50 m s^{-1}. Reproduced from Landsea *et al.* (1996).

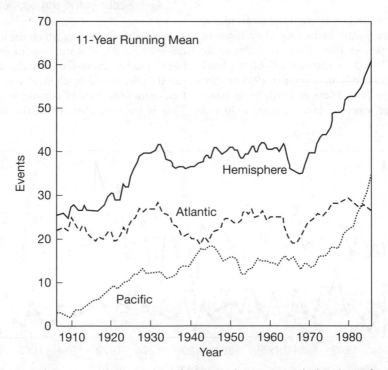

Fig. 5.20 Eleven-year running means of the number of intense winter extratropical cyclones for the whole NH and for the Atlantic and Pacific sectors. An intense cyclone is defined as the occurrence of a grid point with a mean sea level pressure of 970 mb or less which is lower than each of the four surrounding grid points. Reproduced from Lambert (1996).

vapour in the atmosphere as the climate changes are thought to be the single most important positive feedback in the climate system (Sections 3.5 and 9.1). Water vapour in the upper troposphere and lower stratosphere is particularly important in this respect. In addition, increases in the amount of water vapour in the stratosphere resulting from an increase in the concentration of CH_4 (and oxidation of the excess CH_4) constitute an additional radiative forcing that should be taken into account when evaluating the radiative effect of CH_4 emissions. Empirical determination of the trends in the amount of water vapour throughout the atmosphere is therefore of great importance.

The total amount of water vapour in an overhead column is referred to as the *precipitable water*. It is given as the depth of liquid water that would form if all of the water vapour were to be condensed out. Current datasets can be used to determine how the amount of precipitable water has changed with surface temperature over the last few decades. However, current datasets are inadequate for assessing long-term trends in the amount of water vapour in the upper troposphere (above the 500 mb pressure level). Water vapour in this region does not make a large contribution to precipitable water, but has a disproportionately large effect on the natural greenhouse effect.

There are three separate issues pertaining to observations of the vertical distribution of water vapour in the atmosphere: determining the *climatology* or long-term average of the water vapour distribution; determining the vertical structure of *interannual variations* in the amount of water vapour in the atmosphere; and determining the vertical structure of *long-term trends* in the amount of water vapour in the atmosphere. The climatology of the water vapour distribution provides insight into the relative importance of different processes that operate in the real atmosphere and that would therefore be involved in climatic feedbacks; the observed climatology also serves as one test of the water vapour processes in computer simulation models that are used to forecast long-term changes in climate. The computed climatology of the water vapour distribution is less sensitive to changes in instrumentation than are computed trends, which depend on small differences obtained at different times. The nature of the interannual variability in atmospheric water vapour as given by observed datasets is also more reliable than the long-term trends, since it is less affected by instrumentation problems, but might not be fully applicable to long-term climatic change. Thus, the trends that are most relevant to long-term climatic change are the least reliable, while the most reliable information is of questionable utility for projecting future climatic change.

In this section we first describe and assess the various methods of measuring the vertical distribution of water vapour in the atmosphere. This is followed by the resultant climatology, short-term variability, and long-term trends as deduced from the available datasets. A discussion of what these observations may or may not tell us about long-term climate sensitivity is deferred to Section 9.2.

Methods for observing atmospheric water vapour

The rawindsonde network described in Section 5.1 provides concurrent observations of temperature and relative humidity (which is directly measured). Absolute atmospheric humidity is obtained from the relative humidity and temperature data, and so will be affected by changes in either the relative humidity or temperature sensors. Thus, all of the factors discussed in Section 5.1 that affect the temperature measurements will also affect the calculated absolute humidity. In addition, there have been numerous changes in the humidity sensors themselves, as well as occasional changes in the algorithm used to convert the electrical signal from the sensor to a relative humidity value (Elliott, 1995). Finally, at US-controlled stations, it was the practice from 1973 to 1993 to record relative humidity values below 20% as 20%, and to ignore data for temperatures below –40°C (Elliott and Gaffen, 1991; Spencer and Braswell, 1997). This resulted in a bias towards high average moisture in the upper troposphere.

Satellite observations provide a second source of data on atmospheric humidity. The longest record, running from February 1979 to the present, is the Tiros Operational Vertical Sounder (TOVS) system (Elliott, 1995; Soden and Lanzante, 1996). It measures precipitable water vapour in three layers, 1000–700 mb, 700–500 mb, and 500–300 mb, based on the characteristics of emitted infrared radiation at wavelengths of 8.3 μm, 7.3 μm, and 6.7 μm, respectively. TOVS provides near-global coverage with measurements 2–4 times per day. Temperature must also be estimated from the satellite observations and used to convert the measured radiation into water vapour amounts. Water vapour can be determined only for areas where clouds cover less than 75% of the region.

Where partial cloud cover occurs, the results are applicable only to the cloud-free regions. The Stratospheric Aerosol and Gas Experiment (SAGE II) instrument on board the Earth Radiation Budget Satellite has measured water vapour in the stratosphere and upper troposphere since late 1984. SAGE II uses a *solar occultation* technique, with near-global coverage from the mid-troposphere to a height of about 45 km. Satellite microwave observations that can be used to infer water vapour amounts have been available since mid-1987. These observations have been made using the Special Sensor Microwave Imager (SSM/I) on board the Defense Meteorological Satellite Program (DMSP) series that has also provided information on sea ice (discussed above). Three channels are available that are sensitive to water vapour amounts over broad regions of the atmosphere, centred at about 340 mb, 480 mb, and 580 mb in tropical regions (Spencer and Braswell, 1997).

Water vapour climatology

According to Elliott (1995), rawindsonde moisture data are not reliable above the 500 mb pressure level (a height of about 5 km), except in the tropics, where the estimates are said to be "reasonable" up to the 300 mb level (a height of about 8 km). Figure 5.21 compares the areally averaged humidity profile in the 30°S–30°N region as obtained from SAGE II and

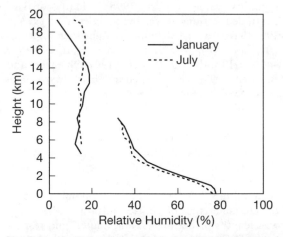

Fig. 5.21 Vertical variation in relative humidity, averaged between latitudes 30°S and 30°N. The upper two curves are based on SAGE II data, averaged over cloud-free regions only, while the lower two curves are based on rawindsonde measurements and are averaged over both clear and cloudy skies. Redrawn from Sun and Lindzen (1993).

from the average of the rawindsonde stations in this region. There is significant disagreement for the height interval where the two methods overlap. Part of this disagreement is due to the fact that the SAGE data are averaged over cloud-free regions only, whereas the rawindsonde data include cloudy conditions and so will tend to show greater average humidity. Part of the disagreement also appears to be due to the fact that the rawindsonde network is very sparse over broad regions of the tropics where the relative humidity is very low (10–20%). This is demonstrated by comparison of Fig. 5.22, which shows the relative humidity in the tropics centred at the 340 mb level as deduced from satellite microwave observations, with Fig. 5.11, which shows the locations of the rawindsonde stations. There are very large regions where the relative humidity is less than 10% but almost no rawindsonde stations sampling these regions. However, the main cause of the disagreement is due to deficiencies in the radiosonde instruments themselves. This is clearly demonstrated by Larsen *et al.* (1993), who compared radiosonde data with nearby satellite data for the year 1987. Of 10,200 SAGE II profiles and 559,000 radiosonde profiles, they found 814 radiosonde profiles that could be reliably compared with a SAGE profile that was within 250 km and 6 hours of the radiosonde profile. When the radiosonde profiles were sorted by instrument type, the SAGE/radiosonde discrepancy persisted for Goldbeaters and carbon hygristor instruments, but was largely eliminated for the thin-film capacitive-type instrument. Since the latter is thought to be most accurate, Larsen *et al.* (1993) conclude that the difference when comparing average profiles is largely due to errors in the radiosonde data, the vast majority of which are based on the first two types of sensors. As a gradual transition to the thin-film capacitive sensors occurs, a spurious downward trend in upper tropospheric relative humidity will appear in the radiosonde dataset.

Soden and Lanzante (1996) examined the geographical distribution of the differences between mean tropospheric humidity as computed from radiosonde data over the period 1979–1991 with TOVS data for the same time period. Relative to this satellite dataset, the radiosonde data were systematically too moist over the former Soviet Union and China (where Goldbeaters sensors are used) and too dry elsewhere (where carbon hygristor and thin film capacitive sensors are used). The total difference in relative humidity is 18–20% of saturation (i.e., comparable to the disagreement seen in Fig. 5.21 for the

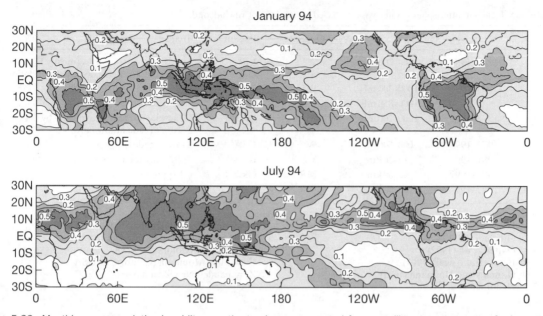

Fig. 5.22 Monthly average relative humidity over the tropics as computed from satellite measurements of microwave radiation, for January and July 1994, from Spencer and Braswell (1997). See Fig. 5.11 for the locations of the tropical rawindsonde stations.

30°–30°N region). This difference is consistent with the slower response time of the Goldbeaters sensor, such that it still "remembers" the higher humidity of the lower atmosphere once the balloon has reached the upper troposphere, therefore giving too high a reading. This implies that radiosonde-based upper tropospheric humidities averaged over the NH extra-tropics will also contain significant errors.

The dry regions shown in Fig. 5.22 correspond to regions of large-scale subsidence (recall the discussion of water vapour in Section 2.3). Smetz and van de Berg (1994), Salathé *et al.* (1995), and Chen *et al.* (1996) compared water vapour amounts as inferred from a variety of satellite observations and as simulated by three-dimensional atmospheric models (or AGCMs, whose characteristics are described in Section 6.1). The model relative humidities in the upper troposphere tend to be too moist in the subsidence regions (by 10–20% of saturation) and too dry in the ITCZ. An exception is in the model analysed by Soden and Bretherton (1994), where the moisture errors show no systematic relationship with the large-scale circulation. Since the dryness of the subsidence regions depends in large part on what happens to water vapour that is detrained from the tops of cumulus convective columns, as discussed in Section 2.3, the inability of AGCMs to simulate the observed

dryness could be due to deficiencies in the treatment of these processes in AGCMs. Conversely, these errors might imply that the Hadley circulation in the models is too weak, or that there is too much mixing between the moist boundary layer and the overlying subsiding layer of the atmosphere. The possible implications of these errors for the simulation of future climatic change will be discussed in Chapter 9.

Long-term trends

Table 5.2 summarizes a number of recent analyses of the decadal and longer trends in atmospheric water vapour. Efforts have been made to remove the effects of most known changes in instruments and data processing procedures, so the trends in the total precipitable water and in the amount of water vapour in the lower troposphere should be reliable. The results summarized in Table 5.2 consistently show that both total precipitable water and the amount of water vapour in the lower troposphere have increased during the past 1–2 decades, at the same time that the global mean temperature has increased. Most of the instances where the amount of precipitable water has decreased coincide with areas where temperatures have also decreased.

Table 5.2 Trends in atmospheric water vapour, based on analysis of rawindsonde data.

Source	Years	Layer	Quantity	Region	Trend
1	1965–1984	500–700 mb	W	EIP	Increasing at 10 stations (mostly by 3–5%/decade), slight decrease at 4 stations
2	1973–1990	1000–300 mb	W	TWP	6%/decade
		1000–850 mb	RH	TWP	4%/decade
3	1973–1990	1000–200 mb	W	Global	Generally increasing, typically by 5–10%/decade at low latitudes
4	1973–1993	1000–500 mb	W	USA	3–4%/decade
5	1981–1994	10–26 km	W	B	3–10%/decade in 2 km layers
6	1991–1997	40–60 km	MR	Global	0.13 ppmv/year
7	1992–1997	40–60 km	MR	C, NZ	0.15 ppmv/year

Sources:
1 = Hense *et al.* (1988)
2 = Gutzler (1992)
3 = Gaffen *et al.* (1992)
4 = Elliott and Angell (1997)
5 = Oltmans and Hofmann (1995)
6,7 = Nedoluha *et al.* (1998)

Regions:
EIP = equatorial Indian and Pacific Oceans
TWP = tropical western Pacific Ocean
B = Boulder (Colorado), USA
C = Table Mountain (California), USA
NZ = Lauder, New Zealand

Quantities:
W = precipitable water
RH = relative humidity
MR = mixing ratio

Also shown in Table 5.2 (as items 6–7) are results from balloon-based, satellite-based, and ground-based instruments designed specially for measuring the amount of water vapour in the stratosphere. Although these records are shorter than the others, they also show an upward trend in water vapour amount, albeit with considerable scatter. The mixing ratio trend of 0.13–0.15 ppmv/yr is on an initial water vapour amount of about 5.5 ppmv. About half of the upward trend can be explained by the observed increase in the stratospheric CH_4 (which is oxidized to H_2O), while the other half can be explained by an increase in temperature at the tropical tropopause by 0.1 K/yr (this would permit more water vapour to enter the stratosphere by increasing the saturation vapour pressure, which governs how much moisture rising tropospheric air can hold).

Interannual variability

Considerable interannual variability in the amount of water vapour in the atmosphere occurs in association with the El Niño oscillation. Measurement of the vertical structure of such variations, including the amplitude of water vapour changes in the upper troposphere, is not subject to the problems of changing instruments and analysis procedures that complicate

the determination of long-term trends, because the observations need span a much shorter period of time. We shall discuss in considerable detail the observed interannual variability and what it might be able to tell us about long-term climate feedback processes in Chapter 9, where the question of climate sensitivity is addressed.

5.6   Changes in the amount of ozone in the atmosphere

The total amount of ozone in the atmospheric column at a given location can be measured using three different approaches: (i) using ground-based instrumentation; (ii) through instruments launched on balloons; and (iii) using satellite-based remote sensors. In some cases, information on the vertical distribution (or profile) of ozone can also be obtained. Surface concentrations can also be directly measured from ground-based instruments. In this section the major ozone datasets and their limitations are briefly described, followed by a summary of the most recent information concerning trends in total ozone and in the amounts in the stratosphere and troposphere.

Surface-based ozone data

The earliest ozone data consist of measurements of the concentration in surface air, some going as far back (with breaks) as 1870. The Dobson spectrophotometer for measuring total overhead ozone was designed in the 1920s, at which time regular observations began (Bojkov and Fioletov, 1995). There were only about 20 observing stations until the International Geophysical Year (1957–1958), when a wider observing network using standardized methods and calibrations was established. Today there are more than 90 continuously operating stations worldwide.

The Dobson spectrophotometer can also be used to infer the vertical variation in the ozone concentration by making measurements at different solar zenith angles. The calculated ozone amounts are referred to as Umkehr observations. Umkehr ozone amounts are considered to be reliable only between heights of about 19 and 43 km (WMO, 1994). However, this height interval nicely complements that which can be reliably inferred by other techniques, and so provides useful information.

Balloon-based ozone data

The second source of information on ozone trends comes from measurements of ozone concentration made from instruments on balloons. The vertical profile of ozone concentration that is collected as the balloon rises through the atmosphere is referred to as an "ozonesonde", and by comparing seasonally averaged ozonesondes over a period of years, the vertical variation of changes in ozone concentration can be determined. The balloons are generally launched once per week and reach a height in excess of 30 km before bursting, but the results are usable only up to 27 km due to instrument limitations (WMO, 1998). Balloon-based observations began between 1966 and 1970 at most sites, but unfortunately the number of sites is rather limited: six sites in North America, three in Europe, three in Japan, and one each in the Pacific Ocean (at Mauna Loa), in Australia, in Brazil, and in Antarctica. These data therefore cannot provide global mean trends.

A number of adjustments and corrections are required before the raw ozonesonde data can be used to deduce trends in ozone concentration (Logan, 1994). First, the vertical sum of the ozone amounts measured from balloons for each small atmospheric layer does not usually agree with the total amount of ozone in the atmospheric column as independently measured by surface instruments. To make the two totals agree, the ozone profile amounts are multiplied by a scaling factor that usually falls between 0.8 and 1.2 (i.e., the required correction is ±20%). This is based on the assumption that the percentage errors are the same at all heights, although this is unlikely to be true in reality. Before the scaling factor can be computed, the amount of ozone above the balloon burst height has to estimated and added to the balloon total, and then compared with the independently estimated total ozone amount. There is a puzzling trend in the required scaling factor at most stations of a few tenths of a percent per year since 1970. The ground-based column ozone amounts – which are used to correct the ozonesonde data – have themselves been adjusted based on comparison with TOMS satellite data (see below), based on correlations between 100 mb temperature and column ozone amounts, and based on comparisons between neighbouring stations. The trends in the scaling factors and the adjustments to the ground-based data each alter the calculated trends in tropospheric ozone by less than 20–25%, and in stratospheric ozone by less than 10–20% in all except two cases.

Other factors that alter the calculated ozone trends are (i) changes in the type of instruments used at some stations, which introduced discontinuities in the measured amount of ozone of 10–40% in the troposphere and of 2–5% in the stratosphere; (ii) changes in the time of day when the balloons are launched at some stations, which caused differences of about 20%; and (iii) trends in the amount of SO_x (SO + SO_2) and NO_x (NO + NO_2) pollution in the atmosphere (the presence of these pollutants interferes with the detection of ozone, so a decrease in pollutant loading appears as an increase in the measured amount of O_3). It should also be noted that absolute values of the surface ozonesonde concentrations (20–30 ppbv) differ significantly from ground-based measurements of the surface ozone concentration (30–45 ppbv) at the Canadian sites (Tarasick *et al.,* 1995) and possibly elsewhere. The reasons for this discrepancy are not known. Finally, there is significant natural variability from one year to the next in the amount of ozone in the stratosphere and upper troposphere, which makes the detection of statistically significant trends difficult.

In summary, then, it is clear that there are major difficulties in what little balloon-based ozone data are available. These difficulties preclude using the data to determine quantitatively accurate trends, although they can probably provide reliable qualitative information (as discussed below).

Satellite data

There are four major satellite-based ozone datasets. Satellite observations of ozone began with the Total Ozone Mapping Spectrometer (TOMS), on board the Nimbus 7 satellite. TOMS collected daily, vertically integrated ozone amounts between October 1978 and May 1993. It could not see O_3 below the tops of clouds, so cloud height had to be estimated, and climatological (long-term average) ozone amounts were added to the air not seen. The raw data have been re-analysed several times as computational procedures and supplementary data have improved. Comparison of TOMS O_3 variations with ground-based data indicated a long-term drift in the TOMS sensors, which has been accounted for; the drift from 1978 to 1987 amounted to 2.8% (Fishman *et al.*, 1990). A second TOMS instrument was launched on board the Russian Meteor 3 satellite in August 1991 and remained in operation until December 1994. Unfortunately its orbit has drifted, and periodically all of the data are collected at very large zenith angles, which makes the use of these data difficult. A third TOMS instrument was launched on board the Earth Probe (EP) satellite in August 1996, and remained in operation as of this writing.

The second satellite-based ozone dataset is from the Stratospheric Aerosol and Gas Experiment (SAGE) instruments. This dataset was collected in two phases; the first, SAGE I, extended from 18 February 1979 to 18 November 1981, and the second, SAGE II, began on 24 October 1984 and was still being collected as of this writing. SAGE gives the vertical profile of O_3, with 1 km resolution, from cloud tops to a height of 55 km (SAGE I) or 65 km (SAGE II). Interference with aerosols makes it difficult to accurately determine trends below a height of 20 km, and significant problems still remain (WMO, 1998). Because there is no overlap in the SAGE I and SAGE II data, the combination of these two datasets introduces uncertainties when it comes to the analysis of trends. According to Fishman *et al.* (1990), the systematic offset between the two datasets is 4% or less. A re-examination of the pre-launch calibration data for the SAGE I instrument package indicated that the original profile data were offset by about 300 m from the correct altitudes. Adjusting the heights assigned to the inferred ozone concentrations by this amount improves the agreement with other estimates of the ozone profiles (WMO, 1994).

The third satellite-based ozone dataset is from the Solar Backscatter Ultraviolet (SBUV) instrument. SBUV was launched on Nimbus 7 (along with TOMS) and collected data from November 1978 to June 1990. It can determine the vertical profile of ozone in the stratosphere and above. A second SBUV instrument (SBUV-2) was launched on board the NOAA-11 satellite, and collected data from January 1989 to October 1994. This satellite's orbit also drifted, from an equatorial crossing time of 1:30 pm to 4:30 pm. Effects related to the changing zenith angle may have altered the measured ozone trends (WMO, 1994). The difference between SBUV-1 and SBUV-2 ozone amounts during the 18 months of overlap is generally less than 1.5%. The SBUV is best for detecting ozone trends in the 25–45 km height interval (WMO, 1998).

The fourth satellite-based ozone dataset is from the TIROS-N operational vertical sounders (TOVS) on NOAA[4] polar orbiting satellites. Global coverage is available, from February 1979 to the present. Although the satellite measures total ozone in principle, the deduced trends are more representative of the low stratosphere (WMO, 1998).

Trends in total ozone

The easiest ozone trends to decipher are the trends in the total amount of ozone in the atmospheric column at a given location. Reported trends in the amount of ozone are based on statistical regression models that attempt to remove the effects of variations in solar activity (as indicated by sunspot number) and in stratospheric winds, which cause short-term variations in the amount and distribution of ozone. Figure 5.23 shows the variation with latitude in the trend in total mean annual overhead ozone, as measured by ground-based Dobson spectrophotometers and as measured by the three TOMS satellites. The trends are given as percent per decade, but were computed over the period January 1979 to December 1997. Trends near the equator are not sta-

---

[4]   National Oceanographic and Atmospheric Administration, US government.

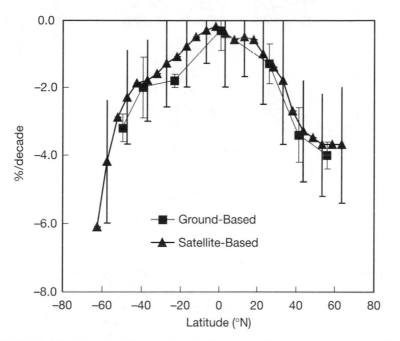

Fig. 5.23 Variation with latitude in the trend in annual mean total overhead ozone as measured by Dobson spectrophotometers and by TOMS satellite instruments. Also given are estimated uncertainties, but to avoid clutter, uncertainties are given for only every other latitude for the satellite-based data. Based on data in Tables 4.5 and 4.6 of WMO (1998).

tistically significant, while the losses reach 6%/decade at high latitudes in the SH and 4%/decade at high latitudes in the NH. The typical uncertainty is ±25%. There has been a noticeable slowing in the rate of O_3 loss in the 1990s that parallels the slowdown in the rate of increase of the chlorine loading in the stratosphere (WMO, 1998).

Trends in stratospheric ozone

Trends in the amount of ozone in individual atmospheric layers, such as the upper and lower stratosphere and the troposphere, are much less reliable than the trends in total ozone. This is unfortunate, since the climatic impact of changes in the amount of ozone depends critically on the vertical variation in the changes, as discussed in Section 7.2. Of particular importance are changes in the amount of ozone near the tropopause, in both the lower stratosphere and upper troposphere. Figure 5.24 shows the vertical distribution of the trend in ozone at latitudes 30°N–50°N as inferred from SAGE, SBUV, and Umkehr data. Particularly strong disagreements are evident in the critical 15–25 km layer, next to the tropopause.

Trends in tropospheric ozone: changes since the Industrial Revolution

Surface ozone observations were made at various sites in Europe as far back as 1839 (when ozone was first discovered). Ozone concentrations in the middle and late 19th century were measured using the Schönbein method, which consisted of soaking a strip of blotting paper in a solution containing starch and potassium iodide, then exposing the paper to the air while shielding it from the sun and rain. After exposure the strips were moistened, and they developed a bluish colour, the intensity of which served as an indicator of the relative ozone concentrations. Standardized procedures, paper, and a precise chromatic scale graduated from 0 to 21 had been developed by 1858 in France. In 1876, a more reliable method of measuring ozone, using arsenite ions, was developed at the Montsouris Observatory in Paris and compared with the Schönbein method over a period of 7 years. The resulting correlation between the Schönbein and arsenite methods, combined with modern refinements to account for differences in relative humidity, forms the basis for interpreting late 19th century measurements that

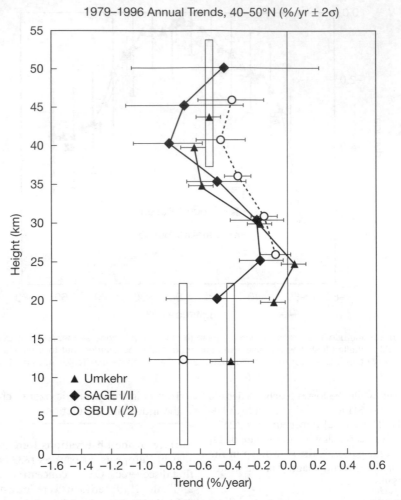

1979–1996 Annual Trends, 40–50°N (%/yr ± 2σ)

Fig. 5.24 Vertical variation of the trend in the annual mean amount of ozone as measured by SBUV, SAGE, and Umkehr instruments. Reproduced from WMO (1998).

were made throughout the world using the Schönbein method. Marenco *et al.* (1994) review these early measurements and successive attempts to improve the Schönbein-arsenite calibration. They caution that the Montsouris calibration may not be valid if paper other than that used at Montsouris was used. Furthermore, most of the early measurements were made next to the surface, at low elevation, where ozone concentrations tend to be lower than in the free troposphere (above the boundary layer). Other measurements were made under polluted conditions, which also tend to reduce the measured ozone concentration.

However, there are early measurements in Europe that were made at high-altitude sites under pollution-free conditions. The earliest such observations are

from the Pic du Midi observatory for the period 1874–1909. The Pic du Midi observatory is near the Atlantic coast of France at an elevation of 3000 m, so the measurements made there should be representative of a broad region of the free troposphere in the NH mid-latitudes. These measurements were made using the same paper and procedures as at Montsouris, so the arsenite calibration can be applied once differences in pressure, temperature, and relative humidity between Paris and Pic du Midi are accounted for. Accounting for the first two variables is straightforward. To account for differences in humidity, Marenco *et al.* (1994) used only measurements taken at relative humidities between 35 and 85% (corresponding to the range at Paris), then used

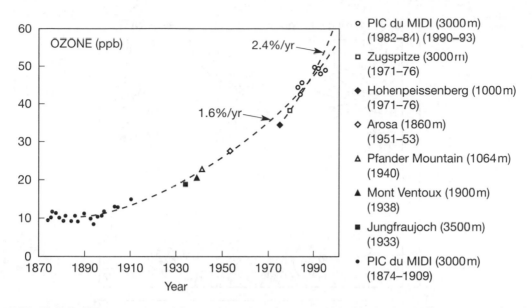

Fig. 5.25 Historical measurements of the concentration of ozone at high-altitude sites in Europe, as analysed and compiled by Marenco *et al.* (1994). Reproduced from Marenco *et al.* (1994).

modern measurements to determine what effect this filtering of the observations would have on the average ozone concentration (the estimated bias is only 3%). The resultant ozone concentration fluctuates between 8 and 14 ppbv, with a clear upward trend developing by about 1890.

Figure 5.25 shows the original Pic du Midi observations along with more recent observations from the Pic du Midi and other high-altitude sites in Europe. There is a very clear increase in the ozone concentration in the free troposphere over Europe, by a factor of 5, during the past 120 years. The present-day concentration is around 50 ppbv. Surface ozone was also measured at Montevideo (Uruguay) during 1883–1885 and at Cordoba (Argentina) during 1886–1892 using the Schönbein method. Sandroni *et al.* (1992) have analysed these data using the Montsouris calibration and have attempted to account for differences in relative humidity. Comparison of these measurements with ozone concentrations today at remote sites at the same latitude in the southern hemisphere indicates that the surface ozone concentration in clean air has roughly doubled (from annual mean values of about 10 ppbv to about 20 ppbv) during the past century. This increase is attributed to the transport of ozone and NO_x from the northern hemisphere, although NO_x pro-

duced from biomass burning in the southern hemisphere would also have contributed.

A second indication of the increase in the amount of ozone in the SH relies on observed seasonal variations. There are large seasonal variations in the amount of tropospheric ozone in the tropics, related to seasonal variations in biomass burning which, combined with regional differences in the amount of ozone, can be used to set lower and upper limits to the increase in tropospheric ozone in the tropics due to biomass burning. Using this approach, Portmann *et al.* (1997) inferred that the vertically integrated increase in tropospheric ozone during the biomass burning season (August–September–October) is from two to almost four times that over the tropical South Atlantic region.

A final indication of the relative increase in tropospheric ozone since the late 19th century is provided by meridional profiles of the zonal mean mid tropospheric O_3, as measured from aircraft. Figure 5.26 presents results from Marenco and Said (1989) for profiles aligned along the east coast of North America/west coast of South America (southbound flights) and along the east coast of South America/west coast of Africa (starting from Dakar) and Europe (northbound flights). Assuming that the two hemispheres would have roughly comparable amounts of O_3 in the absence of anthropogenic

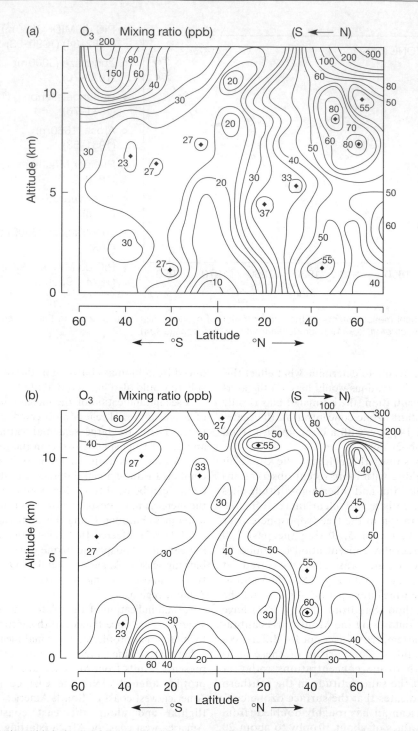

Fig. 5.26 Variation with latitude of the present-day concentration of ozone, as measured from airborne campaigns.
(a) Southbound flights, aligned roughly along the east coast of North America and the west coast of South America.
(b) Northbound flights, aligned roughly along the east coast of South America, the west coast of Africa north of Dakar,
and the west coast of Europe. Reproduced from Marenco and Said (1989).

emissions of O_3 precursors, these data indicate that the ozone concentration in the mid-troposphere in much of the NH has roughly doubled (intrusion of O_3-rich stratospheric air at the time of the measurements explains the even greater NH enhancement in the upper troposphere).

Trends in tropospheric ozone: changes in recent decades

Recent trends in the amount of ozone in the troposphere can be determined (i) from ozonesonde data, or (ii) by subtracting the amount of ozone above the tropopause, as determined from SAGE data, from the concurrent amount of total ozone as determined from TOMS data. The ozonesonde data, after taking into account the discontinuities and required corrections that were described above, indicate the following (Logan, 1994): (i) either no evidence (according to Logan, 1994) of a decrease in tropospheric ozone over northern Canada during the period 1980–1991, or a statistically significant decrease (according to Tarasick *et al.,* 1995); (ii) no change at the Boulder (USA) ozonesonde site during the 1980s; (iii) an increase in 0–10 km ozone over Europe up to 1982, no change during the period 1982–1989, and a decrease since 1989; (iv) an increase of 0.37%/yr in winter at Mauna Loa during the period 1973–1980, but not since 1980; (v) an average increase of 1%/yr from 1970 to 1991 at Wallops Island, off the east coast of the USA; and (vi) a steady increase by 2–4%/yr right through the 1970s and 1980s in Japan. The drift in the required scaling factor and the adjustments to the ground-based column amounts (discussed above) do not alter the direction of the deduced trends given here.

Because the TOMS data give only vertically integrated ozone amounts, they normally cannot be used to infer tropospheric ozone amounts. However, by comparing total ozone amounts over the high Andes (altitude about 6 km) and the adjacent Pacific Ocean, Jiang and Yung (1996) were able to deduce the amount of ozone in the lower 6 km of the atmosphere in this region. This layer accounts for most of the tropospheric ozone. They derived an upward trend in the amount of ozone during the period 1979–1992 of $1.48 \pm 0.40\%$ per year (a cumulative increase of 14–24%).

Thus, the ozonesonde data and satellite data together indicate that the amount of ozone in the troposphere has been increasing during the last 1–2 decades in most regions. The changes in the amount of ozone in the troposphere during the past two decades, as measured by ozonesondes, result in changes in the total column amount of ozone, ranging from a decrease of 1.1%/decade to an increase of up to 1.5%/decade, with most stations producing an increase of 0.0–0.8%/decade (Logan, 1994; her Table 10). By comparison, the average column loss in NH mid-latitudes (including the effect of changes in the troposphere and stratosphere) is 2–4%/decade (Fig. 5.23). The decrease in stratospheric ozone is thus 0–40% larger than the loss in total column ozone. The increase in tropospheric ozone, and the implied greater decrease in stratospheric ozone, have opposing effects on surface climate, but ozone losses in the lower stratosphere have a stronger effect than equal absolute increases in the troposphere – a matter that will be discussed further in Chapter 7.

5.7   Very recent trends in the atmospheric OH and CO concentrations

Both CO and OH have very short atmospheric lifespans, that for CO ranging from 20 to 50 days, and that for OH ranging from 1 to 10 seconds. For this reason, their concentration varies considerably spatially (and also temporally in the case of OH). This makes the determination of trends very difficult. As discussed in Section 2.8 and summarized in Table 2.3, OH is crucial to the chemistry of many climatically important gases (and to the production of sulphate aerosols). Since CO is a major sink for OH, it is also indirectly of great climatic importance. Here, evidence concerning recent trends in the OH and CO content of the atmosphere is briefly discussed.

Hydroxyl

The global distribution of OH cannot be measured in practice, so trends cannot be measured either. Rather, recent trends in the concentration of OH are inferred from 3-D atmospheric chemistry models that are constrained to match the observed spatiotemporal variation in the concentration of methylchloroform (MCF, CH_3CCl_3). MCF is chosen because it is solely of anthropogenic origin, its rate of emission is well known, and reaction with OH is the dominant removal process. In this way, Krol *et al.* (1998) deduced an OH trend of a $0.5 \pm 0.6\%$ per year increase for the period 1978–1993. A very tentative breakdown of the possible factors responsible for this increase is given in Table 5.3. The two largest factors contributing to the increase appear to be the decrease in stratospheric O_3 (which permits more ultraviolet radiation to reach the troposphere) and the increase in NO_x emissions. The recent decrease in CO emissions (see below) and the increase in the amount of water vapour in the tropical atmosphere also appear to be important. It should be stressed, however, that the inferred increase in OH is still rather tentative. Prinn *et al.* (1995) deduced an essentially zero trend for the period 1978–1994.

Three-dimensional atmospheric chemistry models can also be used to assess the total change in OH since the beginning of the Industrial Revolution. The increase in CO and CH_4 will tend to reduce OH everywhere, but this will be counteracted most strongly in the NH by anthropogenic NO emissions and by the increase in tropospheric O_3 concentration. The result, according to Crutzen and Zimmermann (1991), has been to decrease OH by 10–30% in the SH, but to increase OH by up to 20% in parts of the NH.

Carbon monoxide

The 3-D atmospheric chemistry model of Crutzen and Zimmermann (1991) also provides estimates of the distribution of CO within the atmosphere for pre-industrial and present-day conditions. Based on a comparison of these two distributions, it appears that the CO concentration has increased by a factor of 2 throughout much of the NH troposphere, and by a factor of 1.5 throughout much of the SH troposphere. As for recent trends, CO has been measured worldwide only since 1980, although the CO concentration can be inferred from measurements of the solar infrared radiation ($\lambda = 4.63\,\mu m$) made at astronomical observatories since 1950. These data indicate that the CO concentration increased by $0.85 \pm 0.15\%$ per year over central Europe from 1950 to 1987. No trend is evident at Cape Point, in South Africa, during this time period (Brunke *et al.,* 1990). A transition from increasing to decreasing global mean CO concentration occurred in the late 1980s, such that the average concentration decreased by 2.0% per year in the NH and 3.0% per year in the SH (Novelli *et al.,* 1998). Much of this decrease occurred between 1991 and 1993. Total CO

Table 5.3 Tentative breakdown of the causes of the inferred global mean increase in the tropospheric OH concentration by $7.5 \pm 9.0\%$ from 1978 to 1993, based on simulations with a 3-D atmospheric chemistry model by Krol *et al.* (1998). The difference between the last two lines is a result of non-linearities in the atmospheric chemistry.

Causal factor	% Change in OH
11% increase in the CH_4 concentration	–1.1
6.5% decrease in the CO concentration	+1.7
Loss of stratospheric O_3	+2.0
0.2°C temperature increase	+0.1
10% increase in tropical H_2O	+1.7
10% increase in NO_x emissions	+2.0
Sum of the above	+6.4
Simultaneous imposition of the above	+6.0

emissions appear to have decreased since the mid-1970s in North America and Europe (12.4% of automobile fuel was released as CO in the mid-1970s, compared to 5.6% by 1989, according to references cited in Bakwin *et al.,* 1994) but probably increased in Asia. Emissions due to the burning of biomass appear to have decreased in the 1990s. However, Novelli *et al.* (1994) conclude that a decrease in global CO emissions is not likely to be able to completely explain the recent trend of decreasing concentration, and that an increase in the OH concentration is also needed.

An indication of the increase in atmospheric CO concentration that has occurred due to human activities can be gleaned from Fig. 5.27, which shows the latitudinal variation in the CO concentration in the atmospheric boundary layer (the layer next to the surface). Sites in highly polluted regions have been excluded, so the variation seen in Fig. 5.27 is representative of overall conditions. The near-surface CO concentration over a broad region of the NH is around 140 ppbv, which is almost three times the concentration of 50–60 ppbv seen through much of the SH. Since about one-half of the total CO emission is from the oxidation of natural hydrocarbons produced from vegetation (according to Table 4.15), and since there is more vegetation in the NH than in the SH, some hemispheric asymmetry is expected even in the absence of anthropogenic emissions. However, the large asymmetry seen in Fig. 5.27 must be largely due to human emissions of CO. This is confirmed by the simulations of the pre-industrial CO distribution by Crutzen and Zimmermann (1991), in which they approximate hemispheric symmetry.

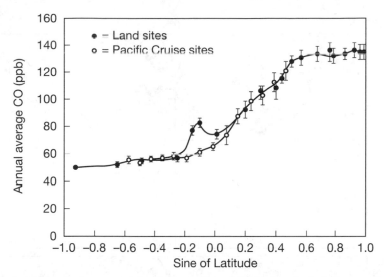

Fig. 5.27 Latitudinal variation in the present-day concentration of CO in the atmospheric boundary layer during the period 1992–1995, as given by Novelli *et al.* (1998). Land sites representing the regionally polluted atmosphere have been excluded. Reproduced from Novelli *et al.* (1998).

5.8   Sea level

Changes in sea level measured at a coastal measuring site could be due to some combination of vertical movements in the edge of the continent and of changes in the volume of the oceans. The measured change in sea level is thus the local change relative to the continental margin, not the absolute change. Vertical movements can occur due to the normal geological compaction of deltaic sediments, the withdrawal of groundwater from coastal aquifers, the uplift associated with colliding tectonic plates, or the ongoing postglacial rebound (and compensating subsidence elsewhere) associated with deglaciation at the end of the last ice age. Changes in ocean volume can occur due to human-induced withdrawals or addition of water to the ocean (a direct effect), the melting of small glaciers and ice caps, and the expansion of ocean water as it warms. Only the latter two are indicative of global-scale climatic change, so the effects of local vertical movements of the land surface and of non-climate-related changes in ocean volume need to be subtracted from the observed records, and the geographically disparate records somehow combined to produce a meaningful global average. Furthermore, the long-term trend in global mean sea level – which has been slowly rising since long before human perturbations to the heat budget of the Earth – needs to be subtracted from recent changes in order to identify the signal, if any, due to recent climatic change.

The raw data for changes in relative sea level are tide gauge observations, which are collected by the Permanent Service for Mean Sea Level (PSMSL) in England. Although there are over 1400 tide-gauge stations worldwide, there are only about 150 stations with records for longer than 50 years, mostly located in the northern hemisphere (Fig. 5.28). As in surface temperature trends (Section 5.1), the longer the time scale under consideration, the greater the correlation between spatially separated sampling sites, and the fewer the number of sites that are needed to get a globally representative trend.

Recent rates of sea level rise

Four approaches have been used to estimate recent (past 150 years) changes in global mean sea level. The first is to average all of the tide-gauge records from areas not believed to be affected by tectonic activity or glacial isostatic rebound. The problem is that glacial isostatic effects extend far beyond the areas that were covered by ice, which in any case are close to where most of the long records are located. The second approach is to estimate the long-term rate of sea level rise (SLR) during the latter Holocene based on radiocarbon-dated shorelines, to

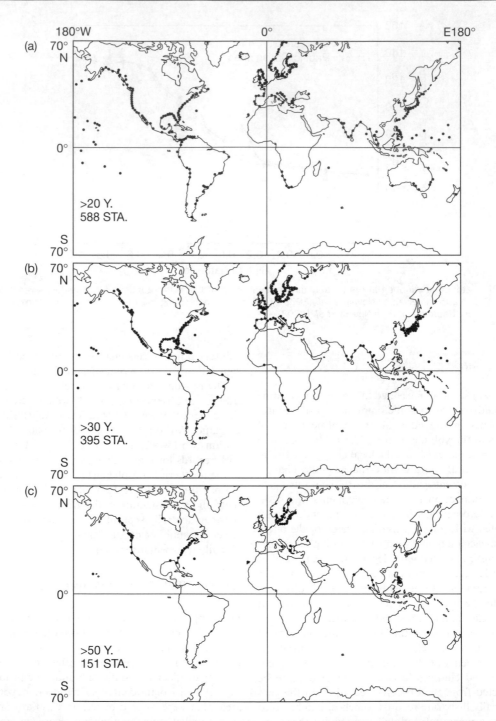

Fig. 5.28 Distribution of tide-gauge stations from the PSML having record lengths greater than (a) 20 years, (b) 30 years, and (c) 50 years. This is the map shown by Gornitz (1995), which is dated to 1991, with 10 years added to the durations to reflect close to the time that has elapsed since then.

subtract this from the closest tide-gauge rate of sea level rise to compute a residual (which will reflect the recent acceleration in sea level rise), and then to average the residuals globally. Gornitz (1995) obtained a rate of sea level rise of 1.0–1.8 mm/yr using this approach. The third approach is to use a geophysical model to determine local glacial–isostatic effects and to subtract the computed effect from the nearby tide-gauge record. The geophysical model is calibrated using ^{14}C-dated paleo-shorelines – the same data that are used directly in the second approach – but the model allows one to determine glacio-isostatic effects worldwide, not just where ^{14}C-dated shorelines exist. Using this approach, Peltier and Tushingham (1989) deduced a rate of sea level rise in the 20th century of 2.4 ± 0.9 mm/yr. Douglas (1991) used the regional isostatic rebound rates obtained by Peltier and Tushingham (1989) but a different tide-gauge dataset, in which some of the stations used by Peltier and Tushingham were rejected because of local tectonic effects. The resulting dataset consisted of 21 stations distributed throughout the NH with records of 60 years or longer. He obtained a sea level rise during 1880–1980 of 1.8 ± 0.1 mm/yr. The very small uncertainty range in this estimate (compared to that of Peltier and Tushingham) is due to the very narrow distribution in the sea level trends from individual stations in the dataset that Douglas (1991) used. Later, Douglas (1997) was able to add three more stations with records of 60 years or longer, all from the SH, and obtained the same rate of sea level rise (1.8 ± 0.1 mm/yr). The fourth approach for determining sea level rise is through satellite altimetry, a method still in its infancy, though various problems must be solved first.

Factors contributing to the recent sea level rise

An appreciation of the difficulty in accurately projecting future changes in sea level – which is the subject of Chapter 12 – can be gained by examining attempts to explain the recent 1–2 mm/yr rate of sea level rise above the long-term rate. Table 5.4 gives estimates of the various contributions to the cumulative sea level rise during the past century or so. Direct anthropogenic effects, which are discussed in some detail by Sahagian et al. (1994) and Gornitz et al. (1997), could be responsible for a fall in sea level of around 1 cm, with an uncertainty of perhaps ± 40%. Melting of small glaciers and the Greenland ice cap

may have added 4–7 cm to sea level. Thermal expansion of ocean water may have added another 2–5 cm. These effects give a sea level rise of 5–12 cm, which is less than the estimated observed sea level rise of 17–19 cm. The shortfall can be made up by changes in the mass balance of Antarctica, which is independently estimated to have caused a change in sea level of 0 ± 14 cm.

In order to move the central estimate of the directly computed increase in sea level up to the central estimate of the observed increase, we must assume that Antarctica has experienced a net loss of mass as the climate has warmed. This is contrary to observations (discussed in Section 5.4) indicating that an increase in Antarctic snowfall has occurred in recent decades, and contrary to expectations of what should happen in the 21st century. However, recent studies (summarized by Oppenheimer, 1998) indicate high rates of melting at the base of the floating ice shelves that fringe West Antarctica. Combined with estimates of a positive mass balance elsewhere, this could produce a sea level rise of 1.9 mm per year if the loss of floating ice is compensated by outflow from the land-based ice. In a simulation of the Greenland and Antarctic ice mass over the past 250,000 years, Huybrechts and de Wolde (1999) were able to close about one-third to one-half of the gap between the observed sea level rise of 18 ± 1 cm and that due to thermal expansion, melting of glaciers, direct anthropogenic effects alone (5–12 cm); for the past 100 years, they computed a contribution of Antarctica to sea level rise of 3.9 cm, without accounting for possible basal melting of the floating ice shelves.

Acceleration of the sea level rise

The warming of the past 150 years – which in many long records stands out as unusual during the past 1000 years or more – should have led to an acceleration in the rate of sea level rise for reasons explained above. Douglas (1995) reviews several lines of evidence, from sites in Europe and on the US east coast, that indicate that the rate of sea level rise prior to the mid-19th century was substantially smaller than during the past 150 years (2 mm/yr), or even indistinguishable from zero. Thus, an acceleration *has* occurred at some time in the recent past. However, analysis of a globally representative set of tidal gauge records shows that no statistically significant

Table 5.4 Estimates of the contribution of various processes to the change in sea level during the past century. Estimates of direct effects were obtained by integrating the time-dependent annual rates of change in sea level given in Figure 2 of Gornitz *et al.* (1997), except where indicated. Alternative estimates of some direct anthropogenic effects by Sahagian *et al.* (1994) are given in brackets.

Process	Contribution (cm)
Direct anthropogenic effects	
Permanent removal of water from aquifers	0.98 (1.23)[a]
Deforestation and loss of soil moisture	0.67 (0.34)
Reduction in the extent of wetlands	<0.01 (0.13)
Storage behind dams	−0.38 (−0.52)
Deep infiltration of water behind dams	−1.71
Deep infiltration of irrigation water	−1.82
Increase in atmospheric water vapour content	−0.54[b]
Effects of urbanization	1.51
Combustion of fossil fuels	0.29[c]
Total direct anthropogenic effects	−1.00
Induced by climatic change	
Melting of small glaciers	2.7 ± 0.9[d]
Melting of Greenland	3.0 ± 0.7[d]
Changes in Antarctic ice mass	0 ± 14[e]
Thermal expansion of sea water	2.2–5.1[f]
Total without Antarctica	6.3–12.4
Total with Antarctica	−8 to 26
Observed	18 ± 1

[a] The estimate by Sahagian *et al.* (1994) includes loss of water from the Caspian and Aral seas, which were not considered by Gornitz *et al.* (1997). Eliminating these contributions brings their total to 0.88 cm, which agrees better with Gornitz *et al.* (1997).

[b] This number is meant to represent the effect of enhanced evaporation from reservoirs and irrigated fields in increasing the storage of water vapour in the atmosphere, but seems to be excessive. Given that the global mean precipitable water content of the atmosphere is only 2.5 cm, this implies a 25% increase. Based on the observed trends in atmospheric water vapour reported in Table 5.2, this seems more appropriate for the increase induced by climatic warming at the global scale during the past century.

[c] Gornitz *et al.* (1997) estimate a current rate of sea level rise of 0.021 mm/yr due to the current fossil fuel use at a carbon release rate of 6 Gt per year. Given that the cumulative fossil fuel carbon release to 1990 was about 140 Gt, this corresponds to a cumulative rise of sea level of 0.29 cm (neglecting the effect of changes in the fuel mix on the $CO_2:H_2O$ emission ratio).

[d] Zuo and Oerlemans (1997).

[e] Warrick *et al.* (1996).

[f] de Wolde *et al.* (1995).

acceleration in sea level rise has occurred (Douglas, 1992). Indeed, the lack of any observed acceleration is a much more robust result than any given estimate of sea level rise, since uncertain constant contributions to the apparent sea level rise (such as post-glacial rebound) fall out of the calculation (Douglas, personal communication, 1998). There are large, interdecadal and chaotic sea level fluctuations that add a lot of noise to records that are only a few decades long, so it will be several decades before tidal gauges can be used to detect an acceleration in the rate of sea level rise, unless the interdecadal fluctuations can be understood and removed from the record (Douglas, 1992).

5.9   Summary

The trends identified in this chapter are summarized in Table 5.5. Direct measures of surface or near-surface atmospheric temperature indicate that the climate has warmed during the 20th century. This conclusion is reinforced by trends in many variables that are directly influenced by temperature: alpine glaciers have retreated worldwide, the extent of snowcover and sea ice has decreased (at least in the NH), and the growing season has increased in length. Precipitation and the proportion of total precipitation as heavy precipitation have increased in

Table 5.5 Summary of trends in observed climatic variables that are discussed in this chapter.

Variable	Analysis period	Trend or change
Surface air temperature and SST	1851–1995	0.65±0.05°C
Diurnal temperature range	1950–1993	–0.79°C/century
Stratospheric temperature	1979–1995	–0.9°C
Extent of snowcover in NH	1972–1992	10% decrease in annual mean
Extent of sea ice in NH	1973–1994	Downward since 1977
Extent of sea ice in SH	1973–1994	No change, possible decrease between mid-1950s and early 1970s
Ice shelves off Antarctic Peninsula	Last 50 years	Dramatic retreat in the north
Alpine glaciers	20th century	Widespread retreat
Length of NH growing season	1981–1991	12±4 days longer
Precipitation	1900–1994	Generally increasing outside tropics, decreasing in Sahel
Heavy precipitation	1910–1990[a]	Growing in importance
Antarctic snowfall	Recent decades	5–20% increase
Extratropical cyclones	Recent decades	Sharp increase in the Pacific sector
Tropical cylones	1945–1993	Decrease in mean annual maximum wind speeds in the Atlantic sector
	1944–1993	Decrease in frequency and intensity in the Atlantic sector
Global mean sea level	20th century	1.8±0.1 mm/year

[a] Shorter records in some regions.

many regions of the world. Changes in the frequency of tropical and extratropical cyclones, and in the intensity of tropical cyclones, have also occurred, but since projections of how these should change as the climate warms are contradictory (as will be discussed in Sections 10.5 and 10.6), these changes cannot be taken as being consistent (or inconsistent) with a warming of the climate. Sea level has been increasing during the 20th century at a greater rate than during the past 1000 years or more, although the historical record does not show an acceleration within the 20th century.

PART 2

CLIMATIC CHANGE

From emissions to climate system response

In Part One we presented an overview of the climate system, compared the major natural and anthropogenic causes of climatic change, and laid down a number of basic principles and concepts that are useful in projecting and understanding potential future climatic change. These concepts include the definition and characterization of radiative forcing, fast and slow climate feedback processes, and climate sensitivity. We also examined the changes in the composition of the Earth's atmosphere and in temperature that have occurred over the past century or so, but without, at this stage, attempting to explain past temperature changes. We were nevertheless able to identify anthropogenic emissions of GHGs and aerosols as driving factors of climatic change that will swamp natural causes of climatic change over the 21st century. We did so on the basis of the much larger radiative forcings from future anticipated anthropogenic effects than for natural forcing factors.

In Part Two we draw upon the understanding gained from Part One in order to analyse and understand the sequence of processes involved in nature in going from emissions to climatic change and sea level rise. To do this, we first introduce the tools that are used to project changes in the atmosphere composition, in climate, and in sea level: a hierarchy of mathematical models of the climate system and of the atmospheric and oceanic components of the climate system. This is the subject of Chapter 6. In Chapter 7 we explain how the present-day radiative forcing due to changes in the concentrations of well-mixed greenhouse gases, in stratospheric and tropospheric ozone, and in the concentration of aerosols is computed. These computations depend heavily on the models that are introduced in Chapter 6. We also present simple relationships between GHG concentrations and forcing, and between aerosol emissions and forcing, that can be used in projecting future changes in global average temperature. Chapters 8–12 deal with the sequence of processes involved in projecting future climatic change and sea level rise. The first set of processes, discussed in Chapter 8, involves the global carbon cycle and its response to both increasing atmospheric CO_2 concentration and the induced changes in temperature. This translates emission scenarios into concentration scenarios. We also discuss, within this chapter, the processes involved in translating emissions of other greenhouse gases and of aerosols into concentration scenarios. Next, in Chapter 9, we discuss the processes determining the global mean responsiveness (or sensitivity) of climate to a given radiative forcing, and evidence concerning how responsive the climate is. In Chapter 10 the characteristics of the climate associated with a doubling of the pre-industrial CO_2 concentration, as simulated by three-dimensional atmospheric models coupled to an ocean surface layer, are discussed. This includes an examination of simulated changes in regional temperature and soil moisture on a seasonal basis, in tropical and mid-latitude storm systems, and in such large-scale phenomena as the Asian monsoon and El Niño. In Chapter 11 we discuss the time-dependent or "transient" climatic response to changes in radiative forcing as simulated using coupled atmosphere–ocean models. Model projections of transient climatic change resulting from increases in GHG concentrations, with and without increases in aerosols, and with and without various natural forcing mechanisms, are compared with the observed climatic change over the 20th century. Chapter 12 closes Part Two of this book with a discussion of sea level rise.

CHAPTER 6

Models used in projecting future climatic change and sea level rise

In order to project the impact of human perturbations on the climate system, it is necessary to calculate the effects of all the key processes operating in the climate system. These processes can be represented in mathematical terms, but the complexity of the system means that the calculations can be performed in practice only on a computer. The mathematical formulation is therefore implemented in a computer program, which is referred to as a "model". If the model includes enough of the components of the climate system to be useful for simulating the climate (at a minimum, the atmosphere and ocean), it is commonly called a "climate model". A climate model that explicitly included all our current understanding of the climate system would be too complex to run on any existing computer. Instead, simplifications are made to a varying extent. The more that nature is simplified in constructing a model, the faster the model can be run or the less powerful the computer that is needed because fewer calculations are performed. In this chapter the major types of models and the simplifications used to create them are outlined for each of the major steps involved in simulating the climate and sea level response to anthropogenic emissions of GHGs and aerosols. The strengths and weaknesses of the major types of models are also outlined in this chapter. The discussion found in this chapter draws heavily upon Harvey *et al.* (1997).

6.1 A hierarchy of atmosphere and ocean climate models

The most detailed model of a particular process is one which is based on fundamental physical principles which are believed to be invariant. Such a model would be applicable to any climate at all. Examples of such principles include Newton's laws of motion, conservation of energy, conservation of mass, the ideal gas law, and the radiative properties of different gases. In order to represent the process in a way which can practically be used in a climate model, additional, simplifying, assumptions have to be introduced. In some cases, empirically derived relationships are included. However, because such empirical relationships are not rigorously derived from fundamental principles, the observed correlations might not be applicable if the underlying conditions change – that is, as the climate changes.

In a model, physical quantities which vary continuously in three dimensions are represented by their values at a finite number of points arranged in a three-dimensional grid (in the case of 3-D models). This is necessary because only a finite number of calculations can be performed. The spacing between the points of the grid is the "spatial resolution". The finer the resolution, the more points, and the more calculations that need to be done. Hence, the resolution is limited by the computing resources available. The typical resolution that can be used in a climate model is hundreds of kilometres in the horizontal. Many important elements of the climate system (e.g. clouds, land surface variations) have scales much smaller than this. Detailed models at high resolution are available for such processes by themselves, but these are computationally too expensive to be included in a climate model, and the climate model has to represent the effect of these sub-grid-scale processes on the climate system at its coarse grid-scale. A formulation of the effect of a small-scale process on the large scale is called a *parameterization*. All climate models use parameterization to some extent.

Another kind of simplification used in climate models is to average over a complete spatial dimension.

Instead of, for instance, a three-dimensional longitude–latitude–height grid, one might use a two-dimensional latitude–height grid, with each point being an average over all longitudes at its latitude and height. When the dimensionality is reduced, more processes have to be parameterized but less computer time is required.

Some of the main types of models for the atmospheric and oceanic components of the climate system are as follows:

One-dimensional radiative–convective atmospheric models

These models are globally (horizontally) averaged but contain many layers within the atmosphere. They treat processes related to the transfer of solar and infrared radiation within the atmosphere in considerable detail, and are particularly useful for computing the radiative forcing associated with changes in the atmosphere's composition (e.g. Lal and Ramanathan, 1984; Ko *et al.*, 1993). The change in atmospheric water vapour amount as climate changes must be prescribed (based on observations), but the impact on radiation associated with a given change in water vapour can be accurately computed. Radiative–convective models thus provide one means for determining one of the key feedbacks important to climate sensitivity through a combination of observations and well-established physical processes.

One-dimensional upwelling–diffusion ocean models

In this model, which was first applied to questions of climatic change by Hoffert *et al.* (1980), the atmosphere is treated as a single well-mixed box that exchanges heat with the underlying ocean and land surface. The absorption of solar radiation by the atmosphere and surface depends on the specified surface reflectivity and atmosphere transmissivity and reflectivity. The emission of infrared radiation to space is a linearly increasing function of atmospheric temperature in this model, with the constant of proportionality serving as the infrared radiative damping. The ocean is treated as a 1-D column that represents a horizontal average over the real ocean, excluding the limited regions where deep water forms and sinks to the ocean bottom, which are treated separately. Figure 6.1 illustrates this model. The sinking in polar regions is represented by the pipe to the side of the column. This sinking and the compensating upwelling within the

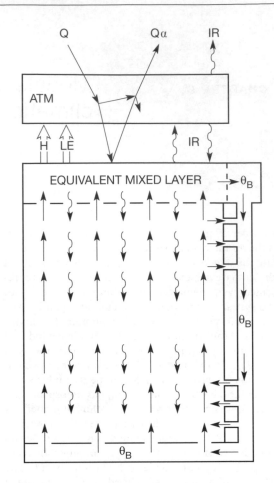

Fig. 6.1 Illustration of the one-dimensional upwelling–diffusion model. The model consists of boxes representing the global atmosphere and ocean mixed layer, underlain by a deep ocean. Bottom water forms at temperature θ_B, sinks to the bottom of the ocean, and upwells w. The temperature profile in the deep ocean is governed by the balance between the upwelling of cold water and downward diffusion of heat from the mixed layer. Latent (LE) and sensible (H) heat, and infrared radiation (IR) are transferred between the atmosphere and the surface. Q is solar radiation. Redrafted from Harvey and Schneider (1985).

column represent the global-scale advective overturning (the thermohaline circulation) (see Section 2.2).

One-dimensional energy balance models

In these models the only dimension that is represented is the variation with latitude; the atmosphere is averaged vertically and in the east–west direction, and often combined with the surface to form a single layer. The multiple processes of north–south heat transport by the atmosphere and oceans are usually represented as diffusion, while infrared emission to space is represented in the same way as in the upwelling diffusion model. Figure 6.2 illustrates a single east–west zone in the energy balance model of Harvey (1988b), which separately resolves land and ocean, surface and air, and snow- or ice-free and snow- or ice-covered regions within each land or ocean surface box. These models have provided a number of useful insights concerning the interaction of horizontal heat transport feedbacks and high-latitude feedbacks involving ice and snow (e.g. Held and Suarez, 1974).

Two-dimensional atmosphere and ocean models

Several different two-dimensional (latitude–height or latitude–depth) models of the atmosphere and oceans have been developed (e.g. Peng *et al.*, 1982, 1987, for the atmosphere; Wright and Stocker, 1991, for the ocean). The two-dimensional models permit a more physically based computation of horizontal heat transport than in one-dimensional energy balance models. In some two-dimensional ocean models (e.g. Wright and Stocker, 1991) the intensity of the thermohaline overturning is determined by the model itself, while in others (e.g. de Wolde *et al.*, 1995) it is prescribed, as in the one-dimensional upwelling–diffusion model. The one-dimensional energy balance atmosphere–surface climate model has also been coupled to a two-dimensional ocean model (Harvey, 1992; de Wolde *et al.*, 1995; Bintanja, 1995). It is relatively easy to run separate two-dimensional ocean models for each of the Atlantic, Pacific, and Indian Ocean basins, with a connection at their southern boundaries (representing the Antarctic Ocean) and interaction with a single, zonally averaged atmosphere (as in de Wolde *et al.*, 1995).

Three-dimensional atmosphere and ocean general circulation models

The most complex atmosphere and ocean models are the three-dimensional atmospheric general circulation models (AGCMs) and ocean general circulation models (OGCMs), both of which are extensively reviewed in Gates *et al.* (1996). These models divide the atmosphere or ocean into a horizontal grid with a typical resolution of 2–4° latitude by 2–4° longitude in the latest models, and typically 10–20 layers in the vertical. They directly simulate winds, ocean currents, and many other features and processes of the atmosphere and oceans. Figure 6.3 provides a schematic illustration of the major processes that occur within a single horizontal grid cell of an AOGCM. Both AGCMs and OGCMs have been used extensively in a stand-alone mode, with prescribed ocean surface temperatures in the case of AGCMs and with prescribed surface temperatures and salinities, or the corresponding heat and freshwater fluxes, in the case of OGCMs. Only when the two models are coupled together do we have what can be considered to be a climate model, in which all the temperatures are freely determined. Coupled AOGCMs automatically compute the fast feedback processes (water vapour, clouds, seasonal snow and ice) as well as the uptake of heat by the oceans, which delays and distorts the surface temperature response but contributes to sea level rise through expansion of ocean water as it warms. Table 6.1 lists the AGCM and AOGCM modelling groups that will be referred to in this book.

AOGCMs compute radiative transfer through the atmosphere (explicitly modelling clouds, water vapour and other atmospheric components), snow and sea-ice, surface fluxes, transport of heat and water by the atmosphere and ocean, and storage of heat in the ocean. Because of computational constraints, the majority of these processes are parameterized to some extent (see Dickinson *et al.,* 1996, concerning processes in atmospheric and oceanic GCMs). More detailed representations are not practical, or have not been developed, for use in a global model. Some parameterizations inevitably include constants which have been tuned to observations of the current climate.

6.2   Models of the carbon cycle

The upwelling–diffusion model that was described in Section 6.1 can be used to model the oceanic part of the carbon cycle, as in the work of Hoffert *et al.* (1981) and Jain *et al.* (1995). The global mean atmosphere–ocean exchange of CO_2, the vertical mixing of total dissolved carbon by thermohaline overturning and diffusion, and the sinking of particu-late material produced by biological activity can all be represented in this model. In the original upwelling–diffusion model, the DIC of sinking polar water needs to be prescribed, since the polar sea

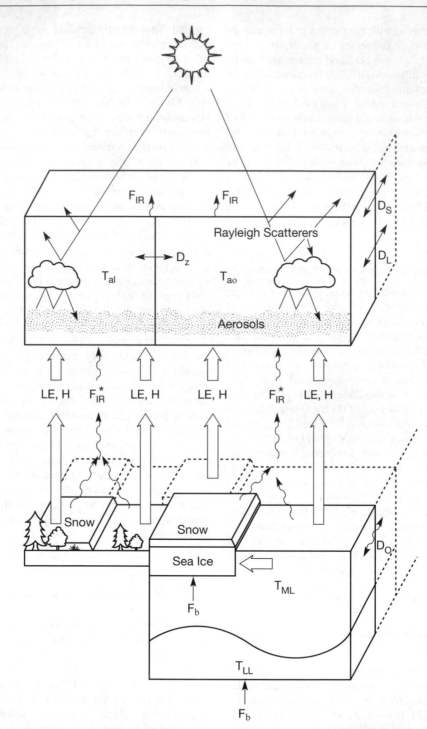

Fig. 6.2 Illustration of a single latitude zone in the energy balance climate model of Harvey (1988b). In each latitude zone, a distinction is made between land and ocean and the surface and atmosphere. Partial snowcover on land and sea ice in the ocean box can also occur. Vertical heat transfer occurs as sensible (H) and latent heat (LE), infrared radiation F_{IR}^*, and multiple scattering of solar radiation, while horizontal heat transfer is represented by diffusion.

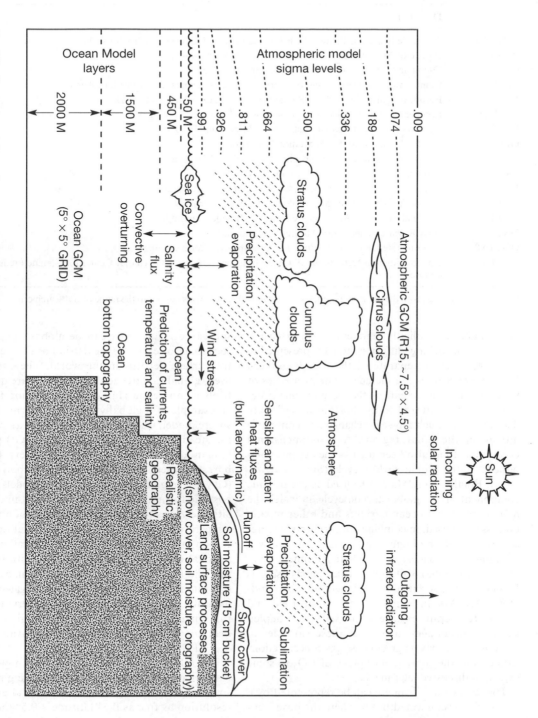

Fig. 6.3 Illustration of the major processes occurring inside a typical horizontal grid cell of a coupled atmospheric–ocean general circulation model (AOGCM). Reproduced from Washington and Meehl (1989). Many current models have twice as fine a horizontal resolution as shown here.

Table 6.1 Acronyms for the AGCM and AOGCM modelling groups referred to in this book.

Acronym	Definition
ARPEGE[a]	Action de Recherche Petite Echelle Grande Echelle, Metéo France, Toulouse, France
BMRC	Bureau of Meteorology Research Centre, Melbourne, Australia
CCC	Canadian Climate Centre, Victoria, British Columbia
CSIRO	Commonwealth Scientific and Industrial Research Organization, Aspendale, Australia
ECHAM	European Centre for Medium-Range Weather Forecasts/University of Hamburg
ECMWF	European Centre for Medium-Range Weather Forecasts, Reading, UK
GFDL	Geophysical Fluid Dynamics Laboratory, Princeton, New Jersey
GISS	Goddard Institute for Space Studies, New York
GLA	Goddard Laboratory for Atmospheres, Greenbelt, Maryland
LMD	Laboratoire de Météorologie Dynamique, Paris
LODYC	Laboratoire d'Océanographie Dynamique et de Climatologie, Paris
MPI	Max-Planck-Institut für Meteorologie, Hamburg, Germany
NCAR	National Center for Atmospheric Research, Boulder, Colorado
NRL	Navy Research Laboratory, Monterey, California
OSU/IAP	Oregon State University/Institute for Atmospheric Physics, Beijing
UKMO	United Kingdom Meteorological Office, Bracknell, UK (now the Hadley Centre for Climate Prediction and Research)

[a] This acronym pertains to the name of a model, while all the other acronyms pertain to the name of a research centre or centres.

regions are not explicitly represented. An extension of the 1-D upwelling–diffusion model is presented in Fig. 6.4, in which separate northern and southern hemisphere regions are included. The properties of water that upwells as part of the thermohaline circulation – often referred to as "bottom water" – are determined through air–sea exchange and convective mixing in the polar regions. A two-dimensional ocean model has also been used as the oceanic component of the global carbon cycle (Stocker *et al.,* 1994). Finally, OGCMs can be used as the oceanic component of the global carbon cycle, in which the model-computed ocean currents and other mixing processes are used, in combination with simple representations of biological processes and air–sea exchange (e.g. Bacastow and Maier-Reimer, 1990; Najjar *et al.,* 1992). CO_2 uptake calculations using 3-D models have so far been published only for stand-alone OGCMs, in which the circulation field and surface temperatures have been fixed. In a coupled simulation, changes in both of these variables in response to increasing greenhouse gas concentrations would alter the subsequent uptake of CO_2 to some extent, as discussed later in Chapter 8.

The terrestrial biosphere can be represented by a series of interconnected boxes, where the boxes represent components such as leafy material, woody material, roots, detritus, and one or more pools of

soil carbon. Each box can be globally aggregated such that, for example, the detrital box represents all the surface detritus in the world. The commonly used, globally aggregated box models are quantitatively compared in Harvey (1989). Figure 6.5 gives an example of a globally aggregated terrestrial biosphere model, where numbers inside boxes represent the steady-state amounts of carbon (Gt) prior to human disturbances, and the numbers between boxes represent the annual rates of carbon transfer $(Gt\,yr^{-1})$. In globally aggregated box models it is not possible to simulate separate responses in different latitude zones (e.g., net release of carbon through temperature effects at high latitudes, net uptake of carbon in the tropics due to CO_2 fertilization). Since regional responses vary non-linearly with temperature and atmospheric CO_2 concentration, extrapolation into the future using globally aggregated models undoubtedly introduces errors. An alternative is to run separate box models for major regions, rather than lumping everything together, as in van Minnen *et al.* (1996).

The role of the terrestrial biosphere in global climatic change has also been simulated using relatively simple models of vegetation on a global grid with resolution as fine as 0.5° latitude × 0.5° longitude. These models have been used to evaluate the impact on net ecosystem productivity of higher atmospheric

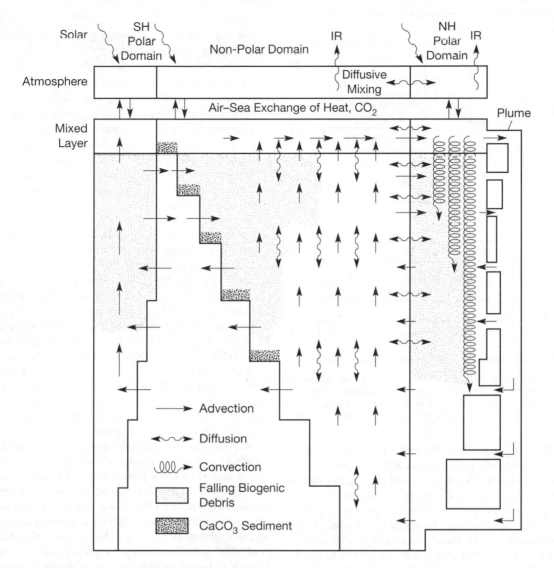

Fig. 6.4   Schematic illustration of the quasi-one-dimensional, upwelling–diffusion model of Harvey and Huang (1999), showing the processes of diffusion, advection, convective mixing, and the biological pump. Unlike the 1-D upwelling model shown in Fig. 6.1, the polar sea regions where bottom water forms or where combined upwelling and convective mixing occurs are explicitly represented. The subsurface SH polar domain is drawn separated from the non-polar domain for clarity only.

CO_2 (which tends to stimulate photosynthesis and improve the efficiency of water use by plants) and warmer temperatures (which can increase or decrease photosynthesis and increase decay processes). These models distinguish, as a minimum, standing biomass from soil organic matter. The more sophisticated varieties track the flows of both carbon and nitrogen (taken to be the limiting nutrient), and include feed-backs between nitrogen and the rates of both photosynthesis and decay of soil carbon (e.g. Rastetter *et al.*, 1991, 1992; Melillo *et al.*, 1993).

Grid point models of the terrestrial biosphere have been used to assess the effect on the net biosphere–atmosphere CO_2 flux of hypothetical (or GCM-generated) changes in temperature and/or atmospheric CO_2 concentration, but generally without

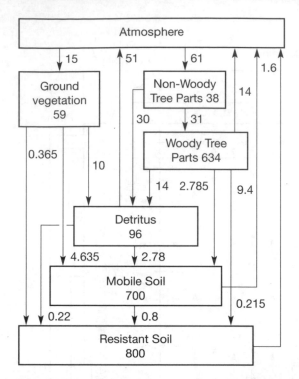

Fig. 6.5 The six-box, globally aggregated terrestrial bio-sphere model of Harvey (1989). The numbers in each box are the steady-state amounts of carbon (in Gt), while the arrows between the boxes represent the annual transfers.

allowing for shifts in the ecosystem type at a given grid point as climate changes. More advanced ecosystem models are being developed and tested that link biome models (which predict changing ecosystem types) and ecophysiological models (which predict carbon fluxes) (e.g. Plöchl and Cramer, 1995). Simulations with these and earlier models demonstrate the potential importance of feedbacks involving the nutrient cycle and indicate the potential magnitude of climate-induced terrestrial biosphere–atmosphere CO_2 fluxes. However, individual models still differ considerably in their responses (VEMAP Members, 1995). As with models of the oceanic part of the carbon cycle, such simulations have yet to be carried out interactively with coupled AOGCMs. These models also have not yet been combined with ocean carbon uptake OGCMs.

Rather detailed models of the marine biosphere, involving a number of species and interactions, have also been developed and applied to specific sites or regions (e.g. Gregg and Walsh, 1992; Sarmiento et al., 1993; Antoine and Morel, 1995).

6.3 Models of atmospheric chemistry and aerosols

As discussed in Section 2.8, atmospheric chemistry is central to the distribution and amount of ozone in the atmosphere. The dominant chemical reactions and sensitivities are significantly different for the stratosphere and troposphere. These processes can be adequately modelled only with three-dimensional atmospheric models (in the case of the troposphere) or with two-dimensional (latitude–height) models (in the case of the stratosphere). Atmospheric chemistry is also critical to the removal of CH_4 from the atmosphere and, to a lesser extent, all other greenhouse gases except H_2O and CO_2. In the case of CH_4, a change in its concentration affects its own removal rate and, hence, subsequent concentration changes. An accurate simulation of changes in the removal rate of CH_4 requires specification of the concurrent concentrations of other reactive species, in particular NO_x (nitrogen oxides), CO (carbon monoxide) and the VOCs (volatile organic compounds), and use of a model with latitudinal and vertical resolution. However, simple, globally averaged models of chemistry–climate interactions have been developed. These models treat the global CH_4–CO–OH cycle in a manner which takes into account the effects of the heterogeneity of the chemical and transport processes, and provide estimates of future global or hemispheric mean changes in the chemistry of the Earth's atmosphere. Some of the models also simulate halocarbon concentrations and the resulting atmospheric chlorine concentration, as well as radiative effects due to halocarbons (Prather et al., 1992). An even simpler approach, adopted by Osborn and Wigley (1994), is to treat the atmosphere as a single well-mixed box but to account for the effects of atmospheric chemistry by making the CH_4 lifetime depend on CH_4 concentration in a way that roughly mimics the behaviour of the above-mentioned globally averaged models or of models with explicit spatial resolution. Atmospheric O_3 and CH_4 chemistry has not yet been incorporated in AGCMs used for climate simulation purposes.

Atmospheric chemistry is also central to the distribution and radiative properties of aerosols, although chemistry is only part of what is required in order to simulate the effects of aerosols on climate. The key processes that need to be represented are the source emissions of aerosols or aerosol precursors; atmospheric transport, mixing, and chemical and physical transformation; and removal processes (primarily deposition in rainwater and direct dry deposition onto the Earth's surface). Since part of the effect of aerosols on climate arises because they serve as cloud condensation nuclei, it is also important to be able to represent the relationship between changes in the aerosol mass input to the atmosphere and, ultimately, the radiative properties of clouds. Establishing the link between aerosol emissions and cloud properties, however, involves several poorly understood steps and is highly uncertain.

Geographically distributed sulphur aerosol emissions have been used as the input to AGCMs and, in combination with representations of aerosol chemical and physical processes, have been used to compute the geographical distributions of sulphur aerosol mass using only natural emission sources and using natural + anthropogenic emission sources (e.g. Langner and Rodhe, 1991; Chin *et al.*, 1996; Pham *et al.*, 1996). Given the differences in the GCM-simulated aerosol distributions for these two sets of emission sources and a variety of assumptions concerning the aerosol optical properties, other workers have estimated the possible range of direct (cloud-free) effects on radiative forcing (e.g. Haywood *et al.*, 1997). With yet further assumptions concerning how clouds respond to sulphur aerosols, a range of indirect (cloud-induced) effects can also be computed (e.g. Boucher and Lohmann, 1995; Jones and Slingo, 1996; Kogan *et al.*, 1996; Lohmann and Feichter, 1997). The results of these and other calculations will be discussed in Section 7.4. However, simulations in which the aerosol effects on solar and infrared radiation modify the wind fields have yet to be performed.

6.4   Models of ice sheets

High resolution (20 km × 20 km horizontal grid), two- and three-dimensional models of the polar ice sheets have been developed and used to assess the impact on global mean sea level of various idealized scenarios for temperature and precipitation changes over the ice sheets (e.g. Huybrechts and Oerlemans, 1990; Huybrechts *et al.*, 1991). AGCM output has also recently been used to drive a three-dimensional model of the East Antarctic ice sheet (Verbitsky and

Saltzman, 1995), but has not yet been used to assess the possible contribution of changes in mountain glaciers to future sea level rise. Output from high-resolution ice sheet models can be used to develop simple relationships in which the contribution of ice sheet changes to future sea level is scaled with changes in global mean temperature.

6.5   Utilization of simple and complex models

As shown in the preceding discussion, models of the climate system range from relatively simple, schematic representations of the major processes, to quite complex models that attempt to represent climate processes in as much detail and with as high a resolution as computer limitations will permit. Both simple and complex models have important but different roles to play in projecting future climatic change due to human activities.

Simple models allow investigation of the basic relationships between the components of the climate system, the factors driving climatic change, and the overall response of the system. Because simple models represent only the most critical processes, they are relatively easy to understand and inexpensive to run, so that multiple diagnostic tests can be executed. They are useful mainly for exploring global-scale questions. The upwelling–diffusion model, for example, has been used to investigate the role of the oceans in delaying the climatic response to increasing greenhouse gas concentrations and the role of ocean mixing–climate feedbacks in modifying the transient response (e.g. Hoffert *et al.*, 1980; Harvey and Schneider, 1985; Morantine and Watts, 1990), in exploring the importance of natural variability in observed global mean temperature variations during the past century (Wigley and Raper, 1990; Schlesinger and Ramankutty, 1994), in setting constraints on the magnitude of the global mean aerosol cooling effect (Wigley, 1989), and in assessing the relative roles of greenhouse gases, aerosols, and solar variability in explaining global mean temperature variations during the past century (Kelly and Wigley, 1992; Schlesinger and Ramankutty, 1992).

On the other hand, three-dimensional models provide the only prospect for reliably simulating the key processes that determine climate sensitivity and the longer-term feedbacks involving the terrestrial and marine biosphere. Results from such models provide one means of determining the values of the aggregate feedback parameters that are used in simpler models. Complex models are also needed for the simulation of regional climatic change and of variability on short time scales, to identify which processes need to be included in simple models (namely, those in which the effects of small-scale variability do not average out), and to study those fundamental physical processes which can be resolved by global-scale, three-dimensional models but not simpler models (such as the role of localized oceanic convection in the large-scale ocean circulation, or the interaction between winds and large-scale heating patterns in the atmosphere). Complex models provide scenarios of time-evolving regional climatic change, as well as diurnal and seasonal patterns of climatic change and changes in interannual variability. On the other hand, complex models are computationally costly, are sometimes difficult to understand, and require high-resolution data inputs, which in some cases simply do not exist. They produce outputs which contain substantial temporal and spatial variability (sometimes referred to as "noise"); this makes analysis of their results a complicated task, as is the case for the real climate system.

The subcomponents of simple models can be constrained to replicate the overall behaviour of the more complex model subcomponents. For example, the climate sensitivity of simple models can be made to equal that of any particular AGCM by altering a single model parameter – the coefficient governing infrared emission to space – whose value implicitly accounts for the net, global mean effect of all the fast feedback processes that influence infrared radiation. Similarly, the vertical diffusion coefficient and the upwelling velocity can be readily altered such that the oceanic uptake of heat (and associated sea level rise) closely matches that of any given OGCM. Globally aggregated biosphere models can be adjusted to replicate the sensitivity to atmospheric CO_2 and temperature changes obtained by regionally distributed models. This allows the simple models to span the range of behaviour of the more detailed, regionally resolved model components, which in turn makes them ideal tools for examining interactions between the model components and for the analysis of alternative emission scenarios.

CHAPTER 7

Computation of direct and indirect radiative forcings associated with changes in the concentration of greenhouse gases and aerosols

In this chapter the procedures used to calculate the major anthropogenic forcings, and the associated uncertainties, are explained. These are the forcings due to changes in the concentration of well-mixed GHGs, the decrease in the amount of ozone in the stratosphere, the increase in the amount of ozone in the troposphere, and increases in the atmospheric loading of various aerosols. There are a number of reasons why it is useful to be able to compute present and future radiative forcings. First, to the extent that the climate sensitivity is the same for all forcings, the relative magnitudes of the different forcings indicate the relative importance of different causes of climatic change. Second, although forcings do not need to be explicitly calculated when running AGCMs (since the models respond on their own to the internally calculated spatial variation in the radiative perturbation), the ability to diagnose what the forcings are in different AGCMs is useful in understanding the reasons for different responses by different models to a given change in the atmospheric composition. Third, when simpler models – such as the 1-D upwelling–diffusion model – are used for the analysis of emission scenarios or to test the impact of alternative assumptions concerning climate sensitivity or ocean mixing, the various radiative forcings need to be prescribed for the model. Finally, a knowledge of the forcings associated with different GHGs and aerosols is one piece of information that is needed – along with information on atmospheric lifespans – in assessing the relative benefits of decreasing the emissions of different GHGs or aerosol precursors.

Apart from the direct radiative forcings associated with the measurable changes in the concentrations of the GHGs and aerosols, there are a number of indirect forcings in some cases. These indirect effects arise through the effect of a change in the concentration of one GHG on the concentration of other GHGs or aerosols. To the extent that the changes in the concentrations of all GHGs and aerosols are measured and used in computing the total radiative forcing, there is no need to explicitly consider indirect effects. However, in assessing the net effect of a change in emissions of any one GHG, the direct and indirect effects need to be taken into account.

This chapter begins by surveying estimates of the present-day radiative forcings associated with measured or estimated concentration changes, discusses the indirect radiative effects associated with changes in individual atmospheric constituents, then presents simple relationships between concentration and radiative forcing that can be used – along with estimates of future concentrations – in making projections of future radiative forcing.

7.1 Radiative forcing due to an increase in CO_2 and other well-mixed gases

The main characteristics of the radiative forcing associated with an increase in the concentration of CO_2 were identified in Section 3.2: the upward emission of infrared radiation at the tropopause decreases because more of the radiation emitted from the comparatively warm surface is absorbed, while the downward emission from the stratosphere increases. Increases in the concentration of other well-mixed GHGs provoke similar changes in the infrared radiative fluxes. The magnitude of the infrared forcing depends on three main factors: (i) the inherent ability of a given gas to absorb (and thus re-emit) infrared radiation, (ii) the extent to which other gases are already absorbing radiation at the wavelengths where a given gas is able to absorb; and (iii) the magnitude of the radiation emitted from the Earth's surface (or from the tops of clouds) at the particular wavelengths where a given gas can absorb. Some gases also absorb in the solar part of the spectrum.

These factors are illustrated in Fig. 7.1. The top panel in Fig. 7.1 shows the distribution of radiation emitted from the Earth's surface and by the Sun, while the lower panel shows the ability of the present-day atmosphere to absorb this radiation. Most of the atmosphere's absorptive ability at wavelengths less than $8.0\,\mu m$ is due to absorption by water vapour, while most of the atmosphere's absorptive ability at wavelengths greater than $15\,\mu m$ is due to absorption by CO_2. There is a distinct gap in the atmosphere's ability to absorb radiation between wavelengths of 8.0 and $12.0\,\mu m$ – a region referred to as the *atmospheric window*. Gases with strong absorption bands in this region are thus particularly important as GHGs. This includes CH_4, N_2O, O_3, and many of the halocarbons. In some cases, such as between CH_4 and N_2O, there is significant overlap in the wavelengths where the gases absorb. A detailed listing of the wavenumbers (the reciprocal of wavelength) at which a large number of gases absorb radiation is given in Table 7.1, along with other information that will be referred to later. The wavenumbers of and radiative forcings ($W\,m^{-2}$ per ppbv) of the gases listed in Table 7.1 are plotted in Fig. 7.2.

We shall now look in more detail at the latitudinal variation in the forcing due to a doubling of atmospheric CO_2, and in the components of this forcing. Figure 7.3(a) shows the annually and zonally averaged variation with latitude of the changes in these

Table 7.1 Greenhouses gases, wavenumbers and wavelengths at which they absorb infrared radiation. For the halocarbons, the concentrations in 1995 are given, as well as the instantaneous or adjusted radiative forcing per ppbv of concentration (the difference between the two is usually less than the uncertainty in their absolute values). References are listed for the absorption centres, forcing, lifespan, and 1995 concentration in that order. NA = Not applicable; NAv = Not available. Asterisks indicate the strongest absorption centre.

Gas	Formula or name	Absorption centres μ (cm^{-1})	Forcing (W m^{-2} ppbv^{-1})	Lifespan (years)	C (1995) (ppbv)	References
Major Greenhouse Gases						
H_2O	Water vapour	many	NA	8 days	NA	–,NA,1,NA
CO_2	Carbon dioxide	667	1.5×10^{-5}	NA	360,000	2,3,NA,4
O_3	Ozone	590				2
		790				2
		1041*	NA	10–200 days	NA	2,NA,5,NA
		1103				2
CH_4	Methane	1306*	0.00043	7.9	1.725	2,3,6,4
		1534				2
N_2O	Nitrous oxide	589				2
		1168				2
		1285*	0.00793	120	0.311	2,3,7,4
CFCs						
CFC-11	CFCl$_3$	846*	0.300	45	0.272	2,8,7,9
		1085				2
CFC-12	CF$_2$Cl$_2$	915*	0.386	100	0.532	2,8,7,9
		1095				2
		1152				2
CFC-13	CF$_3$Cl	783				2
		1102				2
		1210*	0.303	640	NAv	2,8,8,NAv
CFC-14	CF$_4$	632				2
		1261*	0.094	50,000	0.075	2,8,8,10
		1285*				2
CFC-113	CCl$_2$FCClF$_2$	818	0.362	85	0.084	11,8,8,9
CFC-114	CClF$_2$CClF$_2$	851	0.365	300	NAv	11,8,8,NAv
CFC-115	CF$_3$CClF$_2$	1240	0.236	1700	NAv	11,8,8,NAv
CFC-116	CF$_3$CF$_3$	714				
		1116				
		1250*	0.296	10,000	2.6	2,8,8,10

HCFCs						
HCFC-21	$CHCl_2F$	800*	0.208	1.9	NAv	12,12,13,NAv
		1080				12
		1250				12
HCFC-22	$CHClF_2$	810				2
		1110*	0.212	11.8	0.117	2,8,7,9
		1310				2
HCFC-123	CF_3CHCl_2	1280	0.191	1.4	NAv	11,14,15,NAv
HCFC-124	CF_3CHClF	1212	0.208	5.9	NAv	11,14,15,NAv
HCFC-141b	CH_3CFCl_2	755	0.149	9.2	0.0035	11,14,7,9
HCFC-142b	CH_3CF2Cl_2	1190	0.194	18.5	0.0072	11,14,7,9
HFCs						
HFC-23	CHF_3	1120				14
		1160*	0.214	243	0.0104	14,14,7,16
HFC-32	CH_2F_2	1105				14
		1115*	0.129	6	NAv	14,14,17,NAv
HFC-41	CH_3F	1065	0.032	3.7	NAv	14,14,18,NAv
HFC-125	C_2HF_5	1201	0.232	36	NAv	11,14,19,NAv
HFC-134	CHF_2CHF_2	1130	0.233	10.6	NAv	12,12,12,NAv
HFC-134a	CH_2FCF_3	1185	0.167	13.6	0.0016	11,14,7,9
HFC-152a	$C_2H_4F_2$	1140	0.124	1.5	NAv	11,14,15,NAv
Halons and other bromine-containing compounds						
Ha1211	$CBrClF_2$	820*	0.30	11	0.0031	12,7,7,20
		1100				
		1170				
Ha1301	$CBrF_3$	1085, 1209[a]	0.32	65	0.0023	11,7,7,20
CH_3Br	Methyl bromide	620				21
		950				21
		1300				21
		1450*	0.008	0.7	0.010	21,12,7,18
Other gases						
CCl_4	Carbon tetrachloride	776	0.106	35	0.103	2,8,7,9
$CHCl_3$	Chloroform	774*	0.02	0.5	NAv	2,7,7,NAv
		1220				2
CH_2Cl_2	Methylene chloride	717				2
		758*	0.03	0.46	NAv	2,7,7,NAv
		898				2
		1268				2
CH_3CCl_3	Methyl chloroform	725*	0.06	4.8	0.109	2,7,7,9
		1080				2
		1385				2
SF_6	Sulphur hexafluoride	948	0.622	3200	0.0032	22,8,8,23

References:

1 = Chapter 2.3
2 = Ramanathan *et al.* (1985)
3 = Chapter 7.7
4 = *Trends '97* (at *http://cdiac.esd.ornl.gov*)
5 = Table 2.4.
6 = Lelieveld *et al.* (1998)
7 = WMO (1998)
8 = Myhre and Stordal (1997)[b]
9 = Montzka *et al.* (1996)
10 = Harnisch *et al.* (1996)
11 = Kenny, personal communication (1999)
12 = Christidis *et al.* (1997)[b]

13 = Nimitz and Skaggs (1992)
14 = Papasavva *et al.* (1997)[c]
15 = Prather *et al.* (1995)
16 = Oram *et al.* (1998)
17 = Albritton *et al.* (1995; their Table 5.2)
18 = Wingenter (1998)
19 = Pinnock *et al.* (1995)
20 = From the Web Site: *ftp.cmdl.noaa.gov/noah*
21 = Grossman *et al.* (1997)[c]
22 = Ko *et al.* (1993)
23 = Geller *et al.* (1998)

[a] The measured difference in line strength is only 0.2%.
[b] Forcings are the average of the clear and cloudy sky adjusted forcings.
[c] Forcings are instantaneous forcings as computed for a global mean atmosphere.

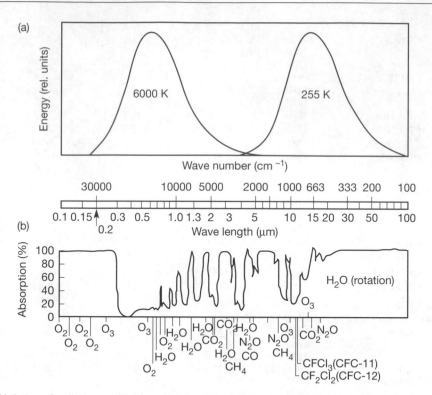

Fig. 7.1   (a) Variation of radiation emitted from the Sun (at a temperature of 6000 K) and from the Earth's surface at a surface temperature of 255 K. (b) Percent of incident radiation absorbed by the present-day atmosphere. Shown at the bottom are the gases that absorb at different wavelengths. Reproduced from Luther and Ellingson (1985).

two fluxes when the CO_2 concentration is doubled, as calculated by J. Kiehl (personal communication, 1998) using the NCAR AGCM. Also shown is the change in net downward radiation at the tropopause (the total forcing). The contribution of downward emission to the total forcing increases with increasing latitude because the stratosphere is thicker at high latitudes than at low latitudes, while the contribution of reduced upward emission decreases with increasing latitude primarily because temperatures and hence absolute emissions are smaller at high latitudes. These opposing tendencies result in a comparatively small variation in the total forcing with latitude. As noted in Section 3.2, the global mean forcing for a doubling of CO_2 is about $4.0\text{–}4.5\,\mathrm{W\,m^{-2}}$ but drops to $3.5\text{–}4.0\,\mathrm{W\,m^{-2}}$ after the cooling of the stratosphere, which results from the increased downward emission, is taken into account.

Figure 7.3(b) compares the change in net radiation at the tropopause with the reduction in outgoing infrared radiation at the top of the atmosphere (TOA) and the increase in downward radiation at the surface, when the CO_2 concentration is doubled. The increase in downward emission at the surface is slightly smaller than at the tropopause, and is substantially smaller than the net forcing at the tropopause. The TOA change of radiation is also substantially smaller than the net forcing at the tropopause. The extra downward radiation is smaller at the surface than at the tropopause in spite of the greater thickness of overlying atmosphere, because most of the atmosphere's water vapour is concentrated near the surface and hence below most of the atmosphere's CO_2 (which has a uniform concentration). Since water vapour is able to absorb IR radiation at the same wavelengths where CO_2 emits, much of the extra downward emission due to CO_2 is absorbed by water vapour (Kiehl and Ramanathan, 1982). However, since the surface temperature response to a CO_2 increase is driven by the change in net radiation at the tropopause and not at the surface, the fact that much of the extra downward emission due to a CO_2 increase is absorbed before it reaches the surface has essentially no effect on the

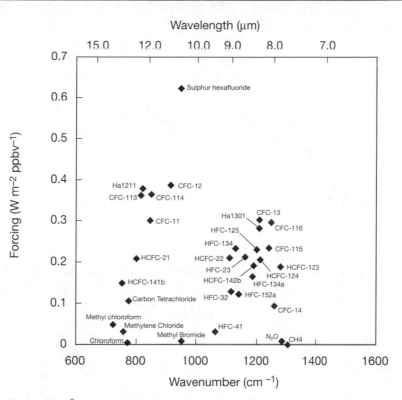

Fig. 7.2 Radiative forcing ($W m^{-2}$ per pbbv concentration) versus wavenumber for the greenhouse gases listed in Table 7.1. Where a gas absorbs radiation at several different wavenumbers, the gas is only shown next to the most important wavenumber. The corresponding wavelengths are also shown, as the top scale.

eventual surface temperature response (see Boxes 3.1 and 3.3).

The instantaneous radiative forcing for a CO_2 increase is computed based on laboratory measurements of the absorptive properties of CO_2. The uncertainty of $0.5 W m^{-2}$ arises largely from the fact that the forcing also depends on the initial amount of water vapour present in the atmosphere (since water vapour absorbs at the same wavelengths as CO_2), on the initial temperatures (which determine the absolute size of the radiative flux which is partially absorbed by extra CO_2), and on the amount and height of clouds (since there is less opportunity to absorb infrared radiation emitted from high clouds than from the surface, owing to the overlying atmosphere being thinner). However, the computations are greatly simplified by the fact that the CO_2 concentration can be assumed to be identical everywhere.

Carbon dioxide also absorbs solar radiation in the near-infrared part of the spectrum to a small extent, largely in the stratosphere. An increase in CO_2 concentration therefore leads to a reduction in the amount of solar energy reaching the surface–troposphere system. However, the heating of the stratosphere by the increased absorption of solar energy there will lead to extra downward emission of infrared radiation from the stratosphere. The net effect of these two changes is to reduce the net forcing for a CO_2 doubling by about 3–6% (Chou, 1990; Myhre *et al.,* 1998a). The absorption of solar radiation by CO_2 is of greater relevance to the temperature response of the stratosphere and to the diurnal pattern of surface temperature response, as discussed in Section 10.1.

The radiative forcing for other well-mixed gases (CH_4, N_2O, halocarbons, and SF_6) is subject to differing uncertainties. In general, the uncertainty due to overlap with H_2O is of minor importance, because most non-CO_2 GHGs absorb at wavelengths where there is relatively little absorption by water vapour. On the other hand, there is still considerable uncertainty concerning the inherent absorptive ability of some of the gases (particularly for the recently studied HCFCs and HFCs). The

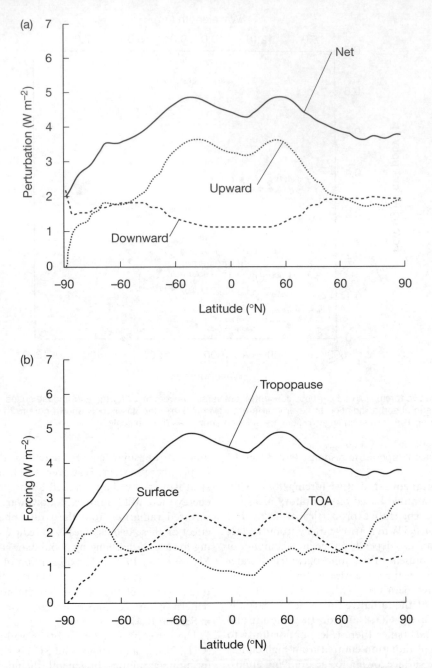

Fig. 7.3   (a) Latitudinal variation in the mean annual, zonally averaged instantaneous radiative forcing arising from a doubling in the atmospheric concentration of CO_2. The dotted line shows the decrease in upward infrared radiation at the tropopause, the dashed line shows the increase in downward infrared radiation at the tropopause, and the heavy solid line shows the total forcing. (b) Comparison of the net radiative forcing at the tropopause (the solid line here and in (a)) with the change in radiation at the top of the atmosphere (TOA, dashed) and at the surface (dotted) when the CO_2 concentration is doubled. The global mean values of the changes shown here are as follows: upward flux at the tropopause, 2.86 W m^{-2}; downward flux at the tropopause, 1.45 W m^{-2}; downward flux at the surface, 1.35 W m^{-2}; upward flux at the top of the atmosphere, 1.96 W m^{-2}. Based on zonally averaged AGCM flux data kindly provided by Jeff Kiehl.

CFCs are well-mixed gases within the troposphere (so their concentration in the troposphere is essentially the same everywhere), but their concentration rapidly decreases with height above the tropopause because they are destroyed only in the stratosphere. Alternative assumptions concerning how the CFCs are distributed with height in the stratosphere cause an uncertainty of about ±15% in the radiative forcing (Hansen *et al.*, 1997a). This also applies to some extent to the replacements for CFCs (see Table 13 of Christidis *et al.*, 1997). Although N_2O is also destroyed primarily in the stratosphere and has a lifetime comparable to that of CFC-12 (Table 2.7), alternative vertical profiles within the stratosphere cause an uncertainty in the forcing of only 2% (Minschwaner *et al.*, 1998).

7.2 Radiative forcing due to a decrease in stratospheric ozone

Computation of the radiative forcing due to changes in stratospheric O_3 is more complicated than for a change in atmospheric CO_2 concentration. First, a decrease in the amount of stratospheric O_3 increases the penetration of solar energy into the troposphere but decreases the downward emission of infrared radiation. Second, the decrease in downward infrared radiation depends strongly on the vertical profile of the O_3 change, with O_3 decreases just above the tropopause having the greatest effect on downward infrared emission. As noted in Section 5.6, the uncertainty in the trend in global mean total ozone from 1970 to 1994 is ±25%. Unfortunately, there is even greater uncertainty concerning the vertical distribution of O_3 changes than there is for the change in the total amount. Third, the stratospheric temperature response has a significant effect on the adjusted radiative forcing, changing the sign of the forcing from positive to negative. The negative forcing arises because of the substantial decrease in stratospheric temperature that occurs with O_3 depletion. The temperature decrease in turn is due to the central role of O_3 at present in heating the stratosphere by absorbing ultraviolet radiation.

Recent estimates place the global mean forcing due to the decrease in stratospheric O_3 at about –0.2 to –0.3 W m^{-2}. Forster and Shine (1997) estimated the forcing during 1979–1996 to be –0.22 ± 0.03 W m^{-2} using SBUV total O_3 trends, with the loss in total

column O_3 accommodated by a uniform reduction in a 7 km thick layer immediately above the tropopause. Figure 7.4 shows how the computed forcing varies with latitude. The direct effect of the decrease in ozone is an increase in the penetration of solar energy into the troposphere (a warming effect, shown by the light solid line in Fig. 7.4). The subsequent cooling of the stratosphere decreases the downward flux of infrared radiation (dotted line) by an even larger amount, giving an overall cooling effect (heavy solid line). Forster and Shine (1997) obtained a second estimate for the global mean forcing of –0.17 ± 0.03 W m^{-2} using SAGE profile data, that are available down to a height of 15.5 km, and making various assumptions for the changes between 15.5 km and the tropopause. Hansen *et al.* (1997a) estimated global mean forcings, for the combined stratospheric and tropospheric ozone changes between 1979 and 1994, of –0.20 W m^{-2} using a combination of TOMS and SAGE data, and –0.28 W m^{-2} using a combination of SBUV and SAGE data. Figure 7.5 compares the vertical profiles of global mean O_3 change during 1979–1994 for the two cases. The changes applied to tropospheric O_3 are somewhat arbitrary. Since the increase in tropospheric O_3 has a heating effect, the forcing due to changes in stratospheric O_3 alone would be even larger than given by Hansen *et al.* (1997a).

There is an additional factor that complicates the radiative forcing arising from changes in stratospheric ozone, which is not important for other forcings and is not accounted for in the above-mentioned estimates: the surface–troposphere forcing for changes in stratospheric ozone depends on how the tropopause is defined (Forster *et al.*, 1997). A common definition of the tropopause is the height at which the vertical temperature gradient (or *lapse rate*) drops to a value of 2 K km^{-1}. However, the important distinction for the surface climate response is between regions that are well mixed and those that undergo a purely radiative adjustment to heating perturbations. The boundary between these two regions occurs about 4 km below the tropopause as defined above, and will be referred to as the *radiative tropopause*. For changes in the concentration of CO_2 or CFCs (chlorofluorocarbons), there is little difference in the change in the net upward flux between these two heights, so the height at which the forcing is computed does not matter. However, for changes in stratospheric ozone, there is a significant difference in the perturbation at these two heights after (but not

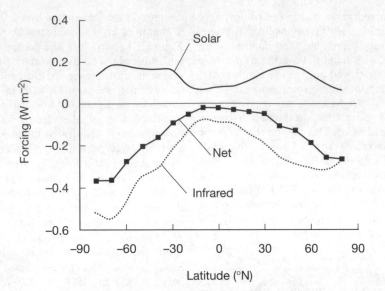

Fig. 7.4   Latitudinal variation in the mean annual, zonally averaged radiative forcing due to the change in total column O_3 for the period 1979–1991 as calculated by Forster and Shine (1997). The total O_3 reduction is based on SBUV data and is accommodated by a uniform percentage reduction in a 7 km thick layer immediately above the tropopause. Light solid line: increased penetration of solar radiation into the troposphere; dotted line: reduced downward emission of infrared radiation after adjustment of stratospheric temperature; heavy solid line: net forcing.

before) the adjustment in stratospheric temperature is allowed for. For a uniform 50% reduction in stratospheric O_3, Forster *et al.* (1997) find that the radiative forcing is two-thirds smaller at the radiative tropopause than at the usual tropopause (-0.11 W m^{-2} instead of -0.32 W m^{-2}).

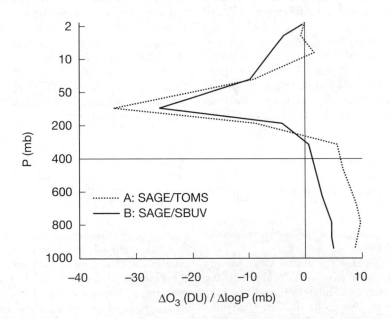

Fig. 7.5   Comparison of the global mean of the SAGE/TOMS and SAGE/SBUV profiles of O_3 change during 1979–1994, used by Hansen *et al.* (1997a). Redrafted based on data kindly provided in electronic form.

7.3   Radiative forcing due to an increase in tropospheric ozone

As discussed in Section 5.6, we have only sketchy information concerning how tropospheric ozone has changed since the Industrial Revolution, and not particularly good information on the global distribution of changes during the last 20–30 years. Thus, estimates of the increase in tropospheric O_3 must be based on AGCMs that include the full suite of relevant chemical reactions, and emissions, transport and removal of O_3 precursors. The radiative forcing is determined by comparing the radiative fluxes for simulations of present and pre-industrial conditions, with no changes in surface or tropospheric temperatures allowed. Apart from the uncertainties associated with the representation of tropospheric chemistry and transport, there are additional uncertainties in estimating both present and pre-industrial emissions of O_3 precursors.

In evaluating the calculated forcings, the first step is to see if the computed increases in the concentration of ozone since the Industrial Revolution agree with the evidence summarized in Chapter 5.6. Table 7.2 shows the zonally averaged percentage increase in surface ozone concentration or in total tropospheric ozone at 30°S and 45°N as computed with five different chemical models. As discussed in Section 5.6, there is evidence that the ozone concentration in the lower troposphere roughly doubled in the SH over the Atlantic Ocean and increased by a factor of 5 in the NH over the Atlantic Ocean and Europe. Zonally averaged changes were undoubtedly smaller. The

chemical models simulate zonally averaged increases in surface or column ozone at 45°N of 50–170% (i.e., up to almost a tripling), and increases of 30–70% at 30°S. These changes are consistent with the evidence cited in Section 5.6.

Figure 7.6 shows the latitudinal variation in the forcing due to the increase in tropospheric ozone as computed using four of the models listed in Table 7.2, while Table 7.3 compares the global and hemispheric mean forcings for a number of models. The radiative forcing due to increases in tropospheric O_3 involves both an increase in the trapping of longwave heating (responsible for about 80% of the forcing) and in an increase in the absorption of solar radiation (responsible for about 20% of the forcing). There is uncertainty by almost a factor of 2 in the global mean forcing (from 0.23 to 0.55 W m^{-2}), but less difference in the computed NH:SH forcing ratio (attention is drawn to the NH:SH forcing ratio here and in the next section because of its importance in detecting anthropogenic effects on climate and in setting limits on how big the aerosol forcing could be, as discussed in Section 11.4). There are strong regional variations in the computed forcing, with peak forcings in excess of 0.8 W m^{-2} over the central and eastern USA in July, and in excess of 0.9 W m^{-2} over central and eastern Europe and western Russia in July according to van Dorland *et al.* (1997). The main sources of uncertainty arise from uncertainty in

Table 7.2 Percentage increase in zonal mean tropospheric ozone from pre-industrial times to the present day as simulated by various atmospheric chemistry models. Results are given for either the increase in surface concentration, or the increase in the total amount of ozone in the troposphere.

Reference		SH, 30°S	NH, 45°N
Surface concentration			
Hauglustaine *et al.* (1994), July		35%	120%
	January	45%	75%
Forster *et al.* (1996)		30%	60%
Levy *et al.* (1997)		35%	90%
Tropospheric column			
Roelofs *et al.* (1997)		70%	170%
van Dorland *et al.* (1997), July		33%	84%
		37%	52%

Table 7.3 Hemispheric and global mean radiative forcing (W m^{-2}) due to the increase in tropospheric ozone since pre-industrial times, as computed with various atmospheric chemistry and radiative transfer models. Results by Forster *et al.* (1996) are given for two different chemistry models but the same radiative transfer model, while results by Berntsen *et al.* (1997) are for two different radiative transfer models but the same simulated change in O_3.

Reference	Global	NH	SH	NH:SH
Hauglustaine *et al.* (1994)	0.55	0.66	0.48	1.4
Forster *et al.* (1996)	0.51	0.61	0.42	1.5
	0.30	0.35	0.24	1.5
Roelofs *et al.* (1997)	0.42	0.51	0.33	1.6
van Dorland *et al.* (1997)	0.38	0.50	0.26	1.9
Berntsen *et al.* (1997)[a]	0.28	0.36	0.20	1.8
	0.31	0.40	0.23	1.8
Haywood *et al.* (1998)	0.32[b]	0.43	0.21	2.0

[a] Global mean adjusted forcings (their "cloudy" case) are given. The instantaneous forcings are about 0.07 W m^{-2} larger.
[b] This is the adjusted radiative forcing. The instantaneous forcing in this case is 0.03 W m^{-2} larger.

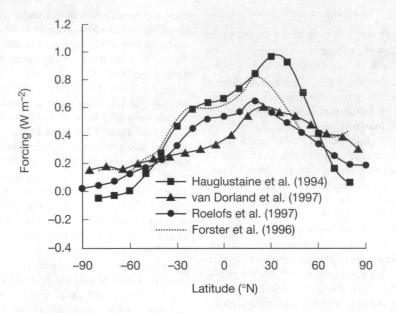

Fig. 7.6 Latitudinal variation in the mean annual, zonally averaged radiative forcing due to the estimated increase in tropospheric O_3 since the beginning of the Industrial Revolution, as computed by Hauglustaine *et al*. (1994), Forster *et al*. (1996), Roelofs *et al*. (1997) and van Dorland *et al*. (1997).

the change in total tropospheric O_3, and in the vertical distribution of the changes (which strongly affects the long-wave radiative forcing). If the true increase in tropospheric ozone is greater than simulated by the models, then the true forcing would be larger as well. Indeed, Portmann *et al*. (1997) believe that the effects of biomass burning in producing ozone have been generally underestimated. They estimated that the increase in ozone in the tropics alone (30°S–30°N) contributed 0.1–0.4 W m^{-2} to the global mean forcing, and that the total forcing could be 0.4–0.7 W m^{-2}.

7.4   Radiative forcing due to anthropogenic aerosols

The computation of the radiative forcing due to anthropogenic aerosols is undoubtedly the most difficult and uncertain of all the anthropogenic forcings. First, there is uncertainty by easily a factor of 2 concerning just how much the concentration of key aerosols has increased since pre-industrial times. Second, there is even greater spatial variability in the changes in aerosol concentration than for tropospheric ozone. Third, and as previously noted (Section

2.5), the aerosol forcing involves both direct and indirect effects. The direct effect involves the reflection of solar radiation by the aerosol particles themselves, while the indirect effect involves modifications in the properties and/or lifespan of clouds. Fourth, the direct effect depends on the size distribution of the aerosol particles (which is somewhat uncertain), on the presence of impurities, and when impurities are present, on the geometry of the impurities within the main aerosol substance. The indirect effect is also complicated, depending as it does on microphysical processes in clouds and the presence of alternative sources of cloud condensation nuclei.

Simulated aerosol loadings

The first step in computing the aerosol forcing is to compute the increase in the atmospheric aerosol loading due to anthropogenic emissions of aerosols or aerosol precursors. The vast majority of work to date on aerosol forcing has focused on sulphate aerosols, with a recent tendency also to consider the effect of admixed soot particles (black carbon). Organic carbon aerosols have also recently been considered. Sulphate aerosol consists of two major components: dilute sulphuric acid droplets (typically 75%

H_2SO_4, 25% H_2O) and droplets of ammonium sulphate ($(NH_4)_2SO_4$) solution (Chýlek et al., 1995). Sulphate aerosol can be largely $(NH_4)_2SO_4$ near the Earth's surface, but is essentially all H_2SO_4 in the upper troposphere and stratosphere. This is a result of the fact that SO_2 is not at all soluble in water, so it can be transported into the upper atmosphere and oxidized to SO_4^{2-}, whereas NH_4^+ is highly soluble and gets washed out before rising as high. The efficiency of H_2SO_4 in scattering radiation is about 50% greater than that of $(NH_4)_2SO_4$ per unit of sulphate mass, but its effectiveness in scattering radiation backwards – which is most important for the climatic effects of aerosols – is only about 15% greater (Boucher and Anderson, 1995). The scattering efficiency of both aerosol types depends significantly on the extent to which the aerosol contains water. So-called "dry" sulphate aerosols contain some water, and the mass of water in the aerosol – and hence the total aerosol mass – increases with atmospheric relative humidity. Carbon aerosols, in contrast, do not contain water. Since the optical effect of aerosols is directly related to the mass of aerosol present, the sulphate aerosol mass should be scaled upward to account for at least the presence of NH_4^+ in the aerosol, in order to facilitate a meaningful comparison with the carbon aerosol mass.

Table 7.4 compares the global, hemispheric, and land and sea average of sulphate, black carbon, and organic carbon aerosol loadings as computed using a number of different aerosol chemistry models embedded in AGCMs. The sulphate aerosol loading is given by multiplying the model sulphate loadings by the ratio of $(NH_4)_2SO_4$ to SO_4 molecular weights (132/96 = 1.38), which is the same as that of $H_2SO_4 + H_2O$ to H_2SO_4 in a 28:72 H_2O:H_2SO_4 mixture. Plate 8 shows the latitudinal variation in the sulphate aerosol loading, given as the mass of $(NH_4)_2SO_4$ per unit area, for pre-industrial (natural) conditions, and the additional loading due to anthropogenic emissions, for selected models. Since different workers adopted different total global emissions of sulphur, the original loadings have all been scaled to correspond to natural emissions of 28.0 Tg S per year and anthropogenic emissions of 69.0 Tg S per year. In spite of this scaling, differences by up to a factor of 4 occur in the zonal mean loading at any given latitude, while differences of up to a factor of 2 occur in the global and hemispheric mean loadings.

Figure 7.7 compares the variation with latitude in the black carbon and organic carbon aerosol loadings as computed by various researchers. Finally, Figure 7.8 compares the geographical variation in the

Table 7.4 Comparison of anthropogenic sulphur, black carbon (soot), and organic carbon (non-absorbing) aerosol loadings as computed by various atmospheric chemistry models embedded in AGCMs. Results were computed from grid-point data that were kindly provided by the lead authors, and have been separately averaged over the NH and SH and over land and ocean. The original grid-point sulphur loading data in all cases were linearly scaled to correspond to a global emission of 69.0 Tg S, so that the differences between the two sets of results shown below are not due to differences in the assumed total emission. The total anthropogenic S emissions assumed by the various workers are as follows: Chin et al. (1996), 67.4 Tg S; Langner and Rodhe (1991), 70.0 Tg S; van Dorland et al. (1997), 67.4 Tg S; Pham et al. (1996), 94.4 Tg S; Lohmann et al. (1999), 72.5 Tg S; Rasch et al. (1999), 67.0 Tg S.

Reference	NH Land	NH Ocean	NH Mean	SH Land	SH Ocean	SH Mean	Global mean
Sulphur loading (mg $(NH_4)_2SO_4/m^2$)							
Chin et al. (1996)	3.20	1.01	2.87	0.61	0.28	0.35	1.61
Langner and Rodhe (1991)	5.57	1.96	3.36	1.75	0.35	0.58	1.97
Rasch et al. (1999)	6.21	3.69	4.67	2.74	1.43	1.65	3.16
van Dorland et al. (1997)	8.37	3.27	5.25	3.49	0.68	1.16	3.20
Pham et al. (1996)	9.01	3.62	5.72	1.41	0.64	0.78	3.25
Lohmann et al. (1999)	11.28	6.76	8.52	2.70	1.34	1.57	5.04
Black carbon loading (mg C/m^2)							
Cooke and Wilson (1996)	1.48	0.52	0.89	0.83	0.15	0.26	0.58
Liousse et al. (1998)	0.85	0.45	0.61	0.63	0.17	0.25	0.43
Lohmann et al. (1999)	1.06	0.56	0.75	0.57	0.15	0.22	0.49
Organic carbon loading (mg C/m^2)							
Liousse et al. (1998)	5.27	3.03	3.92	5.76	1.50	2.24	3.07
Lohmann et al. (1999)	5.69	2.98	4.04	5.28	1.26	1.94	2.99

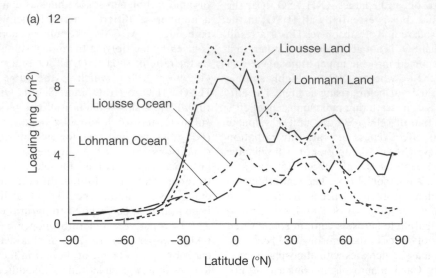

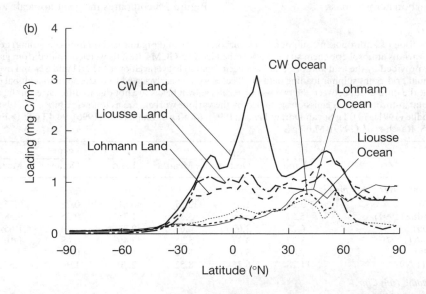

Fig, 7.7 Latitudinal variation in the zonally averaged loadings over land and ocean for (a) organic carbon, and (b) black carbon. Based on grid point data kindly provided to the author by Catherine Liousse and Johann Feichter, corresponding to simulations described in Liousse *et al.* (1996) and Lohmann *et al.* (1999), respectively. Additional black carbon results from Cooke and Wilson (1996) are also shown as the curves labelled "CW".

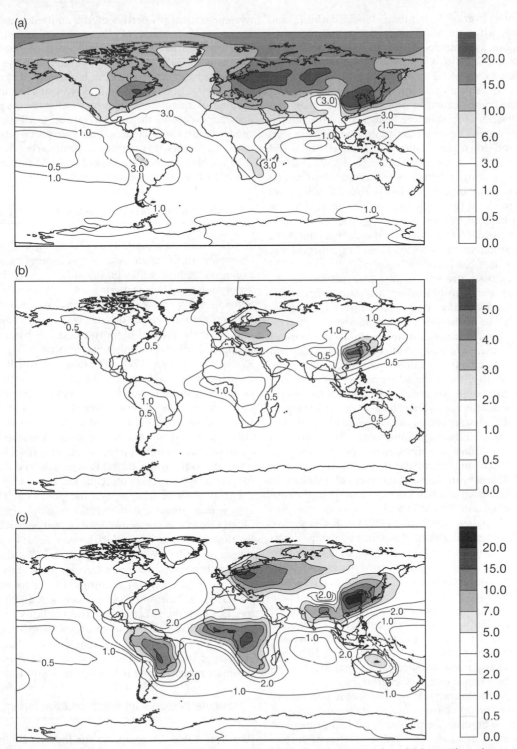

Fig 7.8   Annually average loading of (a) sulphate aerosol, (b) black carbon aerosol, and (c) organic carbon aerosol, due to estimated present-day anthropogenic aerosol emissions as simulated by Lohmann *et al.* (1999) using a single AGCM. Sulphate aerosol loadings are given in mg of $(NH_4)_2$ SO_4/m^2, while black and organic carbon aerosol loadings are given as mgC/m^2. Based on grid point data kindly provided by Johann Feichter.

annually averaged sulphate, black carbon, and organic carbon aerosol loadings due to anthropogenic emissions, as computed by Lohmann *et al.* (1999).

There is a marked spatial variation in the atmospheric loading of all the anthropogenic aerosols due to the concentration of emissions in certain regions and the limited extent to which aerosols or their precursors can spread before being removed from the atmosphere. The largest sulphate loadings are seen over and downwind of the industrialized regions of North America, eastern Europe, and China. Black carbon loadings are highest over eastern Europe and China, while organic carbon aerosol loadings are high in these two regions and over central South America and Africa, where significant biomass burning occurs. Organic carbon loadings tend to be comparable to sulphate loadings, but substantially greater than black carbon loadings.

Since there are few, if any, observations of the total column aerosol loading, it is hard to determine the validity of any given model. Instead, comparisons can be made with a limited number of observations of surface concentrations. In all the sulphur models, there are many grid points where modelled and observed surface sulphate concentrations agree to within 10–20%, but in all the models there are also many grid points where discrepancies of a factor of 2 or greater also occur. Kasibhatla *et al.* (1997) report that such discrepancies are widespread even after adjusting their model to improve the fit to observed aerosol loadings over Europe and North America. Their model and others also do poorly in simulating the differences in the seasonal cycle of sulphate at many grid points, which might be related to the substantial uncertainties in how clouds process SO_2 into SO_4^{2-}. Langner and Rodhe (1991) and Chin *et al.* (1996) presented a limited number of comparisons between modelled and observed sulphate concentrations in the free atmosphere, above the surface layer. In both models there are instances of agreement to within a factor of 2, but also instances of much larger discrepancies. A similar pattern is reported for the black and organic carbon simulations: there are some grid points where the model and observed surface concentrations agree to within 10–20%, but there are also many instances where there are discrepancies by a factor of 2–4 or even larger.

We do not know with any reasonable uncertainty what the correct emission factors are for the emission of carbon aerosols from either fossil fuel or biomass combustion. In models, the radiative effect of carbon aerosols is proportional to the average emission factor for the various emission sources times the average optical properties of the emitted aerosols. Direct measurement of the radiative effects of aerosols emitted from a coal-burning plant in Germany by Bond *et al.* (1998) indicates that the correct radiative effect, in this case, is an order of magnitude smaller than would be inferred using the usual method of calculation. The reason for the discrepancy appears to be the inadvertent use, in all current model studies of black carbon aerosol effects, of emission factors that pertain to total carbon rather than the black carbon fraction of only the fine particles (Charlson, personal communication, 1998).

Sulphate aerosols, direct radiative forcing

Aerosols can be classified as absorbing and non-absorbing. In the latter case, only scattering of solar radiation occurs, while in the former case, some absorption as well as scattering occurs. Pure sulphate aerosols are very weakly absorbing, whereas soot aerosols are strongly absorbing. When soot is mixed with sulphate, the extra absorption is 2–2.5 times as great as the absorption that would occur from soot alone (Chýlek *et al.*, 1995). The extent to which soot is mixed with sulphate can therefore dramatically alter the direct radiative forcing of both sulphate and soot aerosols. The determination of aerosol radiative forcing is somewhat simplified by the fact that the adjustment of stratospheric temperatures in response to an increase in anthropogenic aerosols (which are restricted to the troposphere) is negligible and therefore not a factor in computing the forcing.

Part A of Table 7.5 compares the global and hemispheric mean direct and indirect forcings for pure sulphate aerosols as computed by different researchers for the present day. The global mean direct forcing ranges from $-0.28\,\mathrm{W\,m^{-2}}$ to $-0.66\,\mathrm{W\,m^{-2}}$. The differences in the computed direct forcings are due to differences in the aerosol loadings, in the assumed scattering coefficient for sulphate aerosols, and in the way in which the effects of relative humidity on the aerosol particle size and hence optical properties are treated. Pan *et al.* (1997) present a comprehensive analysis of these and other sources of uncertainty in the direct radiative forcing by sulphate aerosols.

Sulphate aerosols, indirect radiative forcing

Part B of Table 7.5 compares the indirect sulphate aerosol forcing as computed by AGCMs, with the results grouped in three different ways: using the

Table 7.5 Summary of mean global, NH, and SH present-day radiative forcings (Wm^{-2}) due to pure anthropogenic sulphate aerosols as computed by different researchers. The last column gives the assumed anthropogenic emission (TgS/yr).

Reference	Global	NH	SH	SH:NH	Emission
A. Direct radiative forcing					
Boucher and Anderson (1995)[a]	−0.28	−0.47	−0.11	0.23	72.3
Myhre *et al.* (1998b)	−0.32	−0.55	−0.08	0.15	67.0
Feichter *et al.* (1997)	−0.35	−0.55	−0.13	0.24	72.5
van Dorland *et al.* (1997)	−0.36	−0.59	−0.13	0.22	67.4
Haywood *et al.* (1997)	−0.38	−0.60	−0.15	0.25	72.3
Chuang *et al.* (1997)	−0.43	−0.71	−0.15	0.21	77.9
Kiehl *et al.* (1999)	−0.56	−0.81	−0.30	0.37	67.0
Charlson *et al.* (1991)	−0.59	−1.07	−0.11	0.10	72.3
Schult *et al.* (1997)[b]	−0.66	−0.95	−0.37	0.39	72.5
B. Indirect radiative forcing					
Using the Hadley Centre AGCM but different S–CCN relationships[c]					
Hegg (1994)	−0.5	−0.7	−0.3	0.4	72.3
Boucher and Lohmann (1995)	−0.6	−0.8	−0.4	0.5	72.3
Jones *et al.* (1994)	−1.5	−2.0	−1.8	0.9	72.3
Using the Boucher and Lohmann (1995) S–CCN relationship but different AGCMs[d]					
Hadley Centre AGCM	−0.6	−0.8	−0.4	0.5	72.3
LMD AGCM	−1.1	−1.7	−0.4	0.2	72.3
ECHAM AGCM	−1.1	−1.6	−0.6	0.4	72.3
With alternative assumptions concerning cloud microphysics[e]					
Standard Case[f]	−1.0	−1.3	−0.7	0.5	69.0
Allow change in LWC	−1.4	−2.0	−0.8	0.4	69.0
+ Alternative parameterization of cloud fraction	−4.8	−6.5	−3.1	0.5	69.0
+ Alternative parameterization of conversion from cloud droplets to rainfall	−2.2	−2.9	−1.5	0.5	69.0
Other estimates					
Kiehl *et al.* (1999)	−0.40	−0.65	−0.16	0.25	67.0
Feichter *et al.* (1997)	−0.86	−1.12	−0.60	0.53	69.0
Chuang *et al.* (1997)[g]	−0.62	−0.91	−0.34	0.37	77.9
	−1.24	−1.74	−0.75	0.43	77.9

[a] This and all other direct forcing estimates assume no admixture of soot with S.

[b] Results are for a cloud-free atmosphere. Multiply by (1.0 – cloud fraction) (about 0.5) for comparison with the other results shown here.

[c] Results given below were computed by Jones and Slingo (1996) using the S–CNN relationship given in the reference listed in the first column.

[d] Forcings were computed by Boucher and Lohmann (1995).

[e] As computed by Lohmann and Feichter (1997) using the ECHAM AGGM.

[f] S aerosols affect the number of CCN and hence the droplet radius but not the cloud water content. This is the same as in all the other cases shown above.

[g] The lower of the two forcing estimates assumes that 85% of SO_2 is converted to SO_4^{2-} inside cloud droplets, while the higher forcing estimate assumes that 65% is oxidized in this way.

same AGCM but different S-cloud condensation nuclei (CCN) relationships, using the same S–CCN relationship but different AGCMs, and using the same AGCM and S–CCN relationship but allowing for differing sulphate–cloud interactions and alternative parameterizations of cloud fraction and cloud microphysics. The global mean indirect forcing ranges from –0.4 Wm^{-2} to –4.8 Wm^{-2}. Different AGCMs have different amounts and distributions of clouds, and compute cloud optical properties in different ways, so that the effect for a given aerosol loading and S–CCN relationship varies by a factor of 2 as the AGCM that is used changes. The use of alternative functional forms for the S–CCN relationships introduces a factor of 3 uncertainty, but uncertainty in the cloud parameterization with a given

S–CCN relationship gives an uncertainty of a factor of 5, which is greater than the uncertainty arising from the choice of AGCM.

The first two sets of indirect forcing results shown in Table 7.5 assume that a change in the concentration of CCN leads to a change in the mean droplet radius but not in the cloud water content or lifetime. The last set of results shows the effect of alternative assumptions, as computed by Lohmann and Feichter (1997) using the ECHAM AGCM. The first line is the standard case, where only cloud droplet size changes. The second line allows for a reduction in the rate of conversion from cloud droplets to rainfall as the droplet size decreases; this increases the mean cloud water content by about 17%, extends the cloud lifetime, and increases the indirect aerosol forcing by about 40% (from $-1.0\,\mathrm{W\,m^{-2}}$ to $-1.4\,\mathrm{W\,m^{-2}}$). The standard scheme computed the cloud fraction based on relative humidity only; allowing the cloud fraction to depend on relative humidity and cloud water content (which is now allowed to change) increases the forcing even further (to $-4.8\,\mathrm{W\,m^{-2}}$). However, use of an alternative empirical parameterization of the conversion from cloud droplets to rainfall reduces the forcing to $-2.2\,\mathrm{W\,m^{-2}}$. Pan *et al.* (1998) present further analysis of the sources of uncertainty in the indirect forcing by sulphate aerosols.

NH:SH forcing ratio

The last column in Table 7.5 gives the ratio of SH:NH radiative forcing. The SH direct forcing ranges from 0.1 to 0.4 of the NH direct forcing, while the SH indirect forcing ranges from 0.2 to 0.9 of the NH indirect forcing. This information might be of use in narrowing the uncertainty in the total forcing, as discussed in Section 11.4.

Black carbon aerosols mixed with sulphate aerosols

On a mass basis, black carbon aerosols are much more effective than sulphate aerosols in creating a radiative forcing; according to Haywood and Ramaswamy (1998), the average black carbon forcing is $1.85\,\mathrm{W\,m^{-2}}$ for an aerosol loading of $1.0\,\mathrm{mg\,m^{-2}}$, compared to $-0.46\,\mathrm{W\,m^{-2}}$ per $\mathrm{mg\,m^{-2}}$ of sulphate aerosol loading. There are two reasons for this. First, scattering aerosols (sulphate) are less effective in perturbing the radiative balance than absorbing

aerosols, because some of the scattered radiation is simply scattered in the forward direction. Second, the radiative forcing of scattering aerosols over clouds is very small, whereas the radiative forcing due to absorbing aerosols over a highly reflective cloud is greater than for the cloud-free region.

There are three ways in which the radiative effect of a combination of sulphate and soot aerosols can be computed. The first is to assume that the soot aerosol particles occur inside the sulphate particles, and to compute the optical properties of the combined particle by using a volume-weighted average of the separate sulphate and carbon refractive indices. This is referred to as *internal mixing*. The second approach is to assume that separate sulphate and carbon aerosols exist, and to compute the effects of multiple scattering among these independent but co-occurring aerosol particles. This is referred to as *external mixing*. The third approach is to separately compute the radiative forcing for an atmosphere containing only sulphate aerosols, and for an atmosphere containing only black carbon aerosols, and to add the results. This is referred to here as *independent mixing*. When either external or independent mixing is performed, the aerosol absorption coefficient that is used should be an effective coefficient rather than the real coefficient appropriate for isolated black carbon, to take into account the greater effectiveness of a given mass of black carbon in absorbing radiation when it is inside a sulphate particle. External mixing and independent mixing give essentially the same results because a given solar photon is unlikely to encounter both a sulphate aerosol and a black carbon aerosol, so that the effects of externally mixed aerosols are largely independent of one another. Further technical details are given in Box 7.1.

Table 7.6 shows the forcings computed for sulphate and black carbon aerosols alone, and for various ways of mixing sulphate and black carbon aerosols. Haywood *et al.* (1997) computed regionally varying sulphate–soot forcing by assuming a globally uniform soot:SO_4 ratio of 0.075 (this is meant to account for black carbon emissions from fossil fuel sources only). Even though the soot:SO_4 mass ratio was assumed to be uniform, the geographical patterns of soot forcing (a heating effect) and sulphur forcing (a cooling effect) were substantially different. This is because the soot forcing depends much more strongly on surface albedo than does the sulphate forcing, and is more positive the higher the surface albedo. Consequently, the soot offsets a larger frac-

Table 7.6 Summary of global, NH, and SH mean present-day direct radiative forcings ($W\,m^{-2}$) due to anthropogenic sulphate aerosols, black carbon aerosols, and various mixtures of sulphate and black carbon aerosols, as computed by different researchers.

Reference	Global	NH	SH	SH:NH
Haywood et al. (1997)				
Pure sulphate	−0.38	−0.60	−0.15	0.25
External mixture[a]	−0.18	−0.26	−0.10	0.38
Internal mixture	−0.02	0.00	−0.04	—
Schult et al. (1997)[b]				
Sulphate alone, January	−0.69	−1.02	−0.36	0.35
July	−0.63	−0.87	−0.39	0.45
Black carbon alone, January	0.51	0.92	0.10	0.11
July	0.34	0.57	0.11	0.19
External mixture, [c] January	−0.16	−0.01	−0.31	31
July	−0.26	−0.25	−0.27	1.08
Haywood and Ramaswamy (1998)				
Sulphate alone	−0.82	−1.40	−0.24	0.17
Black carbon alone	0.40	0.64	0.16	0.25
Independent mixture[d]	−0.42	−0.76	−0.08	0.11
Myhre et al. (1998)				
Sulphate alone	−0.32	−0.55	−0.08	0.15
Black carbon alone	0.16	0.28	0.03	0.11
External mixture	−0.16	−0.26	−0.05	0.19
Internal mixture	0.10	0.19	0.01	0.05

[a] A C:S ratio of 0.075 is assumed everywhere for this case and for the internal mixture case.

[b] Results are for a cloud-free global atmosphere. Under cloudy conditions, the sulphate aerosol forcing is weaker and the carbon aerosol forcing is stronger than for clear skies, so the net global mean forcing will be less negative than indicated here.

[c] Computed through multiple scattering calculations involving physically separated but intermixed sulphate and carbon aerosol particles.

[d] Computed as the linear sum of the independently computed sulphate and carbon aerosol forcings.

tion of the sulphate forcing over continents than over the oceans, and since the sulphate aerosol forcing is greater over continents than over oceans and greater in the NH than in the SH, the inclusion of soot aerosol reduces the geographical differences in the forcing (as well as reducing the global mean forcing).

Schult *et al.* (1997) allowed for a regionally varying S:C ratio by combining the sulphate field of Langner and Rodhe (1991) with the black carbon field of Cooke and Wilson (1996). The latter includes emissions from both fossil fuels (estimated to be 8.0 TgC/yr) and biomass burning (estimated to be 5.5 TgC/yr). Their global and annual mean soot:SO_4 ratio is 0.12. As seen from Table 7.6, the resulting sulphate–soot aerosol forcing is *larger* in magnitude in the SH than in the NH. Haywood and Ramaswamy (1998) adopted a similar approach, except that they used the sulphate field of Kasibhatla *et al.* (1997). This yielded a global mean C:SO_4 ratio of 0.32, which is quite high. On the basis

of the global mean loading from Liousse *et al.* (1996), they scaled the local black carbon aerosol loadings (from Cooke and Wilson, 1996) by a factor of 0.46, giving a global mean C:SO_4 ratio of 0.15. In this simulation and that of Haywood and Ramaswamy (1998), the combined aerosol forcing is strongly negative (in excess of –3 $W\,m^{-2}$) over much of the central and eastern USA and western Europe, and is positive (in excess of 1 $W\,m^{-2}$) over central Africa. In the case of Schult *et al.* (1997), the combined forcing is also positive over the former Soviet Union, but is negative according to Haywood and Ramaswamy (1998).

The way in which the sulphate and soot aerosols are assumed to be mixed also has a significant effect on the results. For the two cases in which results for both external and internal mixing are available, the net forcing with internal mixing is significantly less negative than for external mixing, or is positive. As discussed in Box 7.1, external mixing as currently

Box 7.1   Optical properties of sulphate and black carbon aerosol particles

When radiation is intercepted by an aerosol particle, it can be either absorbed or scattered (its original direction altered). Solar radiation travelling in the single, original direction is referred to as *direct beam* radiation, and both absorption and scattering reduce the intensity of direct beam radiation. The scattered radiation is referred to as *diffuse beam* radiation, and can be scattered with a component either in the forward direction or in the backward direction. The sum of absorption and scattering is referred to as *extinction*. The effect of a collection of aerosol particles in depleting direct beam radiation can be represented as the area of a disc, per unit mass of aerosol, that would block out the same amount of direct beam radiation as is blocked out by the aerosols. This is referred to as the *extinction cross-section* (k_e); when referring to depletion of direct beam radiation by absorption or scattering alone, the terms *absorption cross-section* and *scattering cross-section* are used. The three parameters usually used to characterize the optical properties of a collection of aerosols are the extinction cross-section, the *single scattering albedo* (the ratio of the scattering to extinction cross-sections), and the *asymmetry factor* (which is related to the fraction of scattering into the forward direction). These optical properties depend on the chemical composition, size, and shape of the aerosol particles. For spheres, these optical properties can be computed from the specified statistical distribution of particle sizes and the refractive index of the material composing the aerosol particle. For mixtures of sulphate and black carbon, the optical properties also depend on the geometry of the mixture (Chýlek *et al.*, 1995). The optical properties vary markedly (but generally smoothly) with the wavelength of radiation under consideration, as illustrated by Haywood and Ramaswamy (1998).

Pure, dry sulphate and ammonium sulphate aerosols have k_e values of $3.6\,m^2/g$ and $5.0\,m^2/g$, respectively, at a wavelength of $0.5\,\mu m$. At most wavelengths in the solar part of the spectrum, the extinction is due entirely to scattering. These aerosols can be characterized as spheres, so shape is not a source of uncertainty. However, these aerosols are hydrophilic, meaning that they attract water and therefore grow in size with increasing atmospheric relative humidity (RH). This in turn greatly increases k_e; at an RH of 92%, k_e for ammonium sulphate is about $20\,m^2/g$! The dependence of sulphate aerosol optical properties on RH is reasonably well known, but different AGCMs produce different RHs throughout the atmosphere. As discussed in Section 5.5 and in Chapter 10

(Box 10.1), sizeable errors in model-simulated RHs are common. This in turn will lead to errors (of perhaps 10% or more) in the simulated direct radiative forcing by sulphate aerosols.

Black carbon aerosols are hydrophobic (water-repelling), so they are not influenced by variations in RH. However, the k_e (and single scattering albedo) depend on the particle shape, size, density, and detailed chemical composition. All of these parameters are usually unknown, and vary from case to case. For spherical particles with a radius of $0.1\,\mu m$ and a density of $2.0\,g/cm^3$, k_e is $4.5\,m^2/g$, while for a density of $1.0\,g/cm^3$, it is about $9.0\,m^2/g$. For plausible changes in the chemical properties of black carbon, the k_e could be as large as $18\,m^2/g$. These extinction cross-sections (all of which were kindly provided to the author by P. Chýlek) pertain to isolated black carbon particles. Regardless of the extinction cross-section for isolated particles, the value will increase by a factor of 2.0–2.5 if the black carbon particle is inside a sulphate or water droplet.

As noted in the main text, the combined effect of black carbon and sulphate aerosols can be computed assuming external mixing or internal mixing. All of the external mixing calculations cited here use an extinction cross-section of about 9–$10\,m^2/g$, which corresponds to the minimum value that should be appropriate for black carbon embedded inside sulphate aerosols. The external mixing assumption thus crudely accounts for the fact that the BC and sulphate aerosols are not externally mixed in reality by using an effective k_e. However, the correct effective k_e may very well be much larger than has been used. If this is so, then the heating effect of BC aerosols will be larger, and the net cooling effect of combined BC-sulphur aerosols will be smaller than computed using the external mixing approach. When the optical properties are computed by explicitly accounting for the internally mixed natural BC and sulphate aerosols (internal mixing), the procedure is to volume-average the refractive indices of BC and sulphate and to directly compute the absorption and scattering cross-sections for the mixture. This causes errors in k_e that are typically about 10% compared to more accurate methods, but the absorption cross-section can be up to three times the correct value (Chýlek, 1988, 1999). Thus, the internal mixing approach will tend to give too weak a cooling or too strong a heating – which is opposite to the likely error in the external mixing approach. This is consistent with the differences shown in Table 7.6, where results using both approaches are given. In any case, the calculated cooling effect of sulphate, once the presence of black carbon is taken into account, is rather small.

calculated might underestimate the heating effect of black carbon aerosols, while internal mixing might overestimate the heating effect. This is consistent with the fact that the radiative forcings given in Table 7.6, as computed with internal mixing, are less negative (or more positive) than those computed with external mixing.

Soot aerosols inside cloud water droplets would significantly alter the optical properties of clouds (in the same way that they alter the optical properties of sulphate aerosols), as shown some time ago by Chýlek et al. (1984). However, none of the results cited above include the effects of soot mixed in cloud water droplets.

Organic carbon, ammonium nitrate, and dust aerosols

Apart from sulphate and black carbon aerosols, the other anthropogenic aerosols include organic carbon aerosols, ammonium nitrate (NH_4NO_3) aerosol, and human-induced increases in the dust content of the atmosphere. Liousse et al. (1996) and Lohmann et al. (1999) computed global distributions for organic and black carbon aerosol, and their hemispheric and global mean loadings are shown in Table 7.4 along with the black carbon loadings of Cook and Wilson (1996). Tegen and Fung (1995) computed a global distribution of anthropogenic dust. Hansen et al. (1998) estimated global mean radiative forcings of $-0.22\,W\,m^{-2}$ and $-0.12\,W\,m^{-2}$ for organic carbon and dust, respectively. For both aerosols, broad areas of both positive and negative radiative forcing occur.

The ammonium nitrate aerosol forcing, like that of pure sulphate, is negative everywhere or almost everywhere. Van Dorland et al. (1997) computed peak regional forcings of $-0.8\,W\,m^{-2}$ over Europe in July, $-0.6\,W\,m^{-2}$ over the USA in July, and $-0.4\,W\,m^{-2}$ over Asia in January. The computed NH mean annual forcing, however, is only $-0.06\,W\,m^{-2}$ – about one-tenth of their computed sulphate forcing. The computed forcing in the SH is negligible. Van Dorland et al. (1997) caution that the nitrate aerosol loadings, used to calculate the forcing, are highly speculative.

Geographical patterns of aerosol forcing

Plate 9 shows the variation with latitude in mean annual direct radiative forcing due to sulphate aerosols as computed by various researchers, while Fig. 7.9 shows the computed indirect radiative forcing. There is a marked variation in the forcing with latitude, which reflects the marked variation in the loading that was shown in Plate 8. The direct forcing tends to be much larger over land than over ocean, while the indirect forcing is stronger over ocean than over land at most latitudes. This is due to some combination of a greater percentage cover of low-level stratus clouds over the oceans and greater susceptibility of cloud optical properties to sulphate aerosols over the ocean. Figure 7.9 compares the contributions of black carbon and sulphate aerosol forcing to the zonally averaged net forcing as computed by Haywood and Ramaswamy (1998). Plate 10 shows the geographical variation of the direct and indirect sulphate forcings as computed by Kiehl et al. (1999).

7.5 Indirect radiative forcing for greenhouse gases

Like aerosols, some greenhouse gases have both direct and indirect radiative forcings. The direct radiative forcing is the perturbation in the downward radiation at the tropopause due to the presence of the gas itself, and was discussed above. The indirect radiative forcing arises from changes in the concentrations of *other* greenhouse gases due to the emission of the first gas, and occurs through atmospheric chemistry. The most important indirect forcings arise from changes in the amount of stratospheric ozone and in the atmospheric lifespan of methane.

Depletion of stratospheric ozone results in an increase in the amount of UV radiation penetrating into the troposphere, which in turn leads to an increase in the concentration of OH. Reaction with tropospheric OH is the main mechanism for removing methane, HFCs (hydrofluorocarbons), and HCFCs (hydrochlorofluorocarbons) from the atmosphere. Tropospheric OH also plays a role in regulating the amount of tropospheric O_3. Bekki et al. (1994) calculated that the decrease in methane concentration resulting from the increase in OH associated with depletion of stratospheric O_3 will add another 30–50% to the negative forcing caused by depletion of ozone alone. However, most of the destruction of methane occurs in tropical regions, while the largest ozone depletion occurs at high latitudes, so this indirect forcing due to ozone depletion depends on the latitudinal variation in the ozone losses.

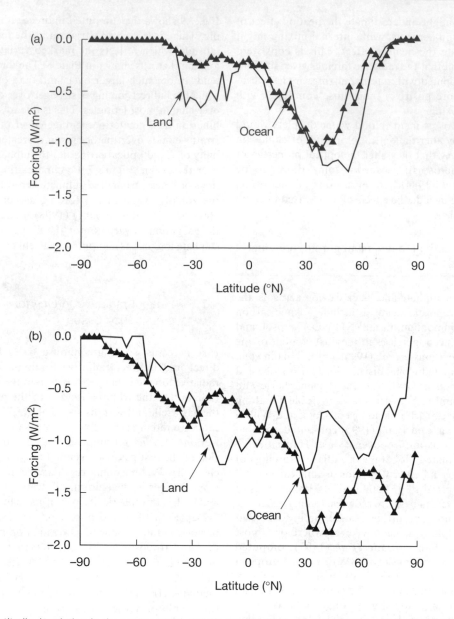

Fig. 7.9 Latitudinal variation in the mean annual, zonally averaged indirect sulphate aerosol forcing as computed by (a) Kiehl *et al.* (1999), and (b) Feichter *et al.* (1997). Based on grid point data kindly provided by Jeff Kiehl and Johann Feichter.

A second indirect forcing due to loss of stratospheric ozone arises from the fact that OH is involved in converting SO_2 to sulphate aerosols. An increase in OH would presumably lead to more sulphate aerosol and an increase in cloud reflectivity for the same reasons that increased anthropogenic emissions of SO_2 are expected to have these effects (see Box 2.3). Toumi *et al.* (1994) calculated that this indirect forcing could be 40–800% of the direct forcing due to depletion of stratospheric ozone, although Rodhe and Crutzen (1995) have challenged this result. This indirect forcing is subject to many of the same uncertainties associated with the indirect forcing arising from anthropogenic emissions of sulphur.

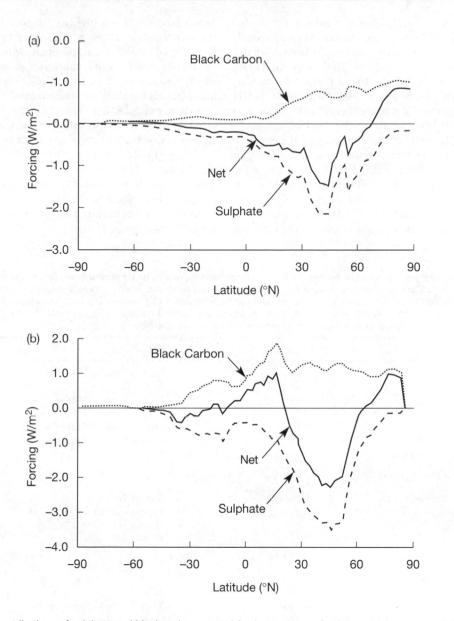

Fig. 7.10 Contributions of sulphate and black carbon aerosol forcing to the net forcing over (a) ocean, and (b) land, as computed by Haywood and Ramaswamy (1998). Based on grid point data kindly provided by Jim Haywood.

An indirect negative forcing is associated with the buildup of tropospheric O_3, as tropospheric O_3 promotes the production of OH. As previously noted, OH is central to the removal of CH_4 and halocarbons, and the conversion of SO_2 to SO_4^-. The reduction in the lifetime of CH_4 alone is thought to offset about 30% of the forcing due to the buildup of tro-

pospheric O_3 which, as noted in Section 7.3, is estimated to be at least 0.4–$0.7\,\mathrm{W\,m^{-2}}$.

There are several potentially important indirect radiative forcings associated with CH_4 emissions. First, the emission of a pulse of CH_4 would reduce the atmospheric concentration of OH, thereby increasing the atmospheric lifetime of the much larger amount of CH_4 already present in the atmosphere. This effect can only be assessed using models of tropospheric chemistry, and results from seven different groups presented in Prather *et al.* (1995, Section 2.10.2.3) indicate that a 1% increase in CH_4 concentration leads to a 0.17–0.35% decrease in OH concentration (the chemical feedback factor f is –0.17 to –0.35). The slower removal of the background CH_4 partly offsets the decay of the emission pulse, and the net effect appears as a slower decay or *adjustment time* of the initial pulse. The ratio of the adjustment time to the lifetime is given by $1/(1 + f)$, and ranges from 1.23 to 1.62 for the f values given above. Second, oxidation of CH_4 (by reaction with OH) leads to the production of ozone. The extent of ozone production is a highly nonlinear function of the NO_x concentration, which varies by at least a factor of 100 within the troposphere. Most of the O_3 production occurs in the mid- to upper troposphere, where the addition of O_3 is most effective as a greenhouse gas. The calculated production of O_3 is highly uncertain, but Albritton *et al.* (1995, Section 5.2.5.2) estimate that the heating effect of the CH_4-induced O_3 production is $25 \pm 15\%$ of the direct heating effect of CH_4. Finally, the oxidation of CH_4 produces water vapour. This is of no consequence in the troposphere, but it is an important enough contributor to stratospheric water vapour that it adds another 5% or so to the direct heating effect of methane according to Albritton *et al.* (1995).

Table 7.7 Summary of the radiative forcing due to changes in the composition of the atmosphere between the pre-industrial period and 1995. The direct radiative forcing due to methane is based on the observed increase, and so includes the effect of the change in the OH concentration induced by the anthropogenic emission of methane. The indirect methane forcing due to tropospheric O_3 is the contribution of the methane increase to the inferred increase in tropospheric O_3, so it is partly redundant with the tropospheric O_3 forcing that is shown further down in the table.

Forcing	Magnitude (W m^{-2})	References
CO_2	1.40	Data in Tables 2.7 and 7.9
CH_4, direct	0.44	
tropospheric O_3	(0.11)	Lelieveld *et al.* (1998)[a]
stratospheric H_2O	0.02	
N_2O	0.14	Data in Tables 2.7 and 7.9
CFCs	0.32	Data in Tables 2.7 and 7.9
Other well-mixed GHGs	0.05	Data in Table 7.1
Total for well-mixed GHGs	2.37 ± 0.24[b]	
Increase of tropospheric O_3	0.3 to 0.7	Section 7.3
Loss of stratospheric O_3	–0.2 to –0.3[c]	Forster and Shine (1997), Hansen *et al.* (1997a, b)
	–0.1[d]	Forster *et al.* (1997)
Total GHGs	2.1 to 3.2	
Combined S and C aerosols, direct effect	–0.1 to –0.4	Table 7.6
S aerosol, indirect effect	–0.5 to –2.2	Table 7.5, excluding extremes
Organic carbon aerosols	–0.22	Hansen *et al.* (1998)
Anthropogenic dust	–0.12	Hansen *et al.* (1998)
Nitrate aerosols	–0.03	van Dorland *et al.* (1997)
Total aerosols	–0.8 to –3.0	

[a] As an indication of the uncertainty, note that Minschwaner *et al.* (1998) computed an adjusted direct forcing for 1750–1992 of 0.55 W m^{-2}.
[b] ±10% uncertainty has been arbitrarily added to the total forcing.
[c] Based on the radiative forcing at the conventionally defined tropopause, for the period 1979–1994.
[d] Obtained by scaling the above forcing by a factor of 0.33 in order to given the forcing at the radiative tropopause. See text for clarification.

Table 7.8 Sources of uncertainty in the major anthropogenic forcings. The magnitude of the uncertainties is classified as follows: Very Minor, <10%; Minor, 10–30%; Modest, 30–100%; Major, greater than a factor of 2.

Forcing	Source of uncertainty	Magnitude
Increase in CO_2	Magnitude of increase Absorptive capability Overlap with other gases Baseline temperatures Cloud distribution Adjustment in T_{str}	Very minor
Decrease in stratospheric ozone	Total overhead change Vertical profile of change Adjustment of T_{str}	Modest
Increase in tropospheric O_3	Total overhead change Vertical profile of change	Modest
Direct effects due to increase in sulphate and black carbon aerosol in the troposphere	Vertically integrated mass Chemical composition Size distribution Vertical profile Effects of RH Presence of soot impurities Method of S:soot mixing	Major Minor Modest Minor for pure sulphate Modest for black carbon Minor Major Modest

Indirect effects due to increase in sulphate aerosol in troposphere	Aerosol number concentration (ANC)
	Relationship between ANC and
	    concentration of cloud
	    condensation nuclei (CCN)
	Presence of other CCN
	Distribution of sensitive clouds
	Way in which aerosols mix with water
	Relationship between CCN and cloud
	    radiative properties
	Relationship between CCN and cloud lifetime
	Presence of soot
		Modest to major

7.6   Summary of estimated radiative forcings in the 1990s and associated uncertainties

Table 7.7 presents the estimated radiative forcings due to changes in the atmospheric composition from the pre-industrial period up to 1995, while Table 7.8 lists the major sources of uncertainty in the forcing due to increases of CO_2, decreases in stratospheric O_3, increases in tropospheric O_3, and increases in tropospheric aerosols. The uncertainties in the radiative forcing for other well-mixed GHGs except CH_4 are comparable in origin and importance to that shown in Table 7.8 for CO_2. The total GHG forcing is estimated to be 2.1–3.2 W m^{-2}, while the total aerosol forcing is estimated to be –0.8 to –3.0 W m^{-2}. The direct sulphur + black carbon forcing is estimated to be 0.1 to –0.4 W m^{-2} (a slight cooling effect), while the indirect sulphate forcing is estimated to be –0.5 to –2.2 W m^{-2} if extreme estimates are excluded.

In the preceding section we discussed the direct and indirect heating effects of a given CH_4 *emission pulse*. One of these is the increase in the atmospheric lifespan for all pre-existing atmospheric methane. This effect, however, is implicitly accounted for in the *observed* CH_4 buildup that is used to compute the CH_4 forcing and so, in the present context, is counted as part of the direct CH_4 forcing. The other major indirect forcing due to CH_4 is from the induced buildup of tropospheric O_3, but since the O_3 forcing is included in Table 7.7 based on the estimated observed buildup of O_3, this indirect forcing need not

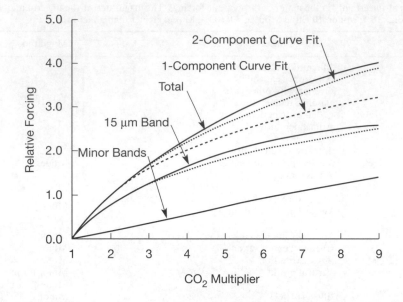

Fig. 7.11 Variation in the relative instantaneous radiative forcing with the CO_2 increase (given as a fraction of the pre-industrial concentration). Shown is the radiative forcing due to the 15 μm band alone, due to the minor bands alone, and due to all bands. Also given is the curve fit using Eq. (7.1) with $R_F = 1.0$ (1-component curve fit) and using the sum of Eqs (7.1) and (7.2) (2-component curve fit).

be included in the summation. Most of the uncertainty in the total GHG forcing is due to uncertainty in the forcing due to changes in O_3, but the preceding discussion suggests that the total GHG forcing is more likely to be near the upper end of the 2.1–3.2 W m^{-2} range than at the lower end.

7.7   Simple relationships between concentration and forcing

It is possible to develop simple mathematical relationships between GHG concentrations and radiative forcing, or between anthropogenic aerosol emissions and radiative forcing. To do so requires making use of the results of detailed radiative transfer calculations, along with simulations of the spatial patterns of changes in concentration for O_3 and aerosols. The commonly used relationships were summarized in Harvey *et al.* (1997), and are shown here in Table 7.9. The dependence of radiative forcing on the concentration of well-mixed GHGs varies as follows for different gases:

- The radiative forcing for CO_2 varies with $\ln(C/C_0)$, where C_0 is the starting concentration. Thus, the forcing tends to saturate as CO_2 concentration increases.
- The radiative forcing for CH_4 and N_2O varies as $\sqrt{C} - \sqrt{C_0}$. Thus, the forcing increases less rapidly than the increase in concentration, but is not as close to saturation as for CO_2.
- The radiative forcing for halocarbons varies linearly with the concentration.

These differences in the *shape* of the variation of radiative forcing with concentration are due entirely to the different initial concentrations and the extent to which the various molecular absorption bands are already saturated. The absolute forcings themselves also depend on the inherent strengths of the various molecular absorption bands, and where these bands occur within the wavelength spectrum (as discussed in the introduction to Section 7.1). Table 7.9 is complemented by Table 2.2, which compares the relative heat-trapping ability of different greenhouse gases on a molecule-by-molecule basis, starting from present-day concentrations.

Table 7.9 Relationship between the global mean radiative forcing due to changes in greenhouse gas concentration and aerosol emissions, based on results of detailed radiative transfer calculations in regionally resolved models. The forcings involving O_3 and aerosols are subject to considerable uncertainty. All formulae are taken from Harvey *et al.* (1997), with the exception of the halocarbons and SF_6, which are based upon updated data in Table 7.1; and stratospheric O_3, which is from Daniel *et al.* (1995). See the main text for a more accurate formulation for CO_2.

Constituent	Method for computing forcing ($W m^{-2}$)
CO_2	$\Delta Q = 3.75 \ln(C(t)/C_0)/\ln(2)$
CH_4	$\Delta Q = 0.036(\sqrt{C(t)} - \sqrt{C_0})$ – (correction for overlap with N_2O), where C and C_0 are in ppbv and $C_0 = 700$ ppbv
N_2O	$\Delta Q = 0.14 (\sqrt{C(t)} - \sqrt{C_0})$ – (correction for overlap with CH_4), where C and C_0 are in ppbv and $C_0 = 280$ ppbv
CFC-11	$\Delta Q = 0.300 C(t)$
CFC-12	$\Delta Q = 0.386 C(t)$ where C is in ppbv
Other halocarbons, SF_6	$\Delta Q = \alpha C(t)$, where α is given in column 4 of Table 7.1
Stratospheric water vapour	$\Delta Q = 0.05 \, \Delta Q_{CH_4 - pure}$
Tropospheric ozone	$\Delta Q = 8.62 \times 10^{-5} \Delta C_{CH_4}$ for O_3 formation due to CH_4 buildup ΔQ associated with O_3 formation due to emissions of other gases can ramp up to the assumed 1990 value
Loss of stratospheric ozone	$\Delta Q = (\Delta Q_{1990}/\Delta E_{1990 - 1980})\Delta E_{t-1980}$, up to 1990 $\Delta Q = \Delta Q_{1990} + 0.6(\Delta Q_{1990}/\Delta E_{1990 - 1980})\Delta E_{t - 1990}$, after 1990 where $\Delta E_{t_2 - t_1} = E_{t_2} - E_{t_1}$, $E_t = n_{cl}(t) + \alpha n_{br}(t)$, $n_{cl}(t)$ is the Cl loading in the stratosphere at time t, $n_{br}(t)$ is the Br loading in the stratosphere at time t.
Sulphate aerosols, direct forcing	$\Delta Q = e(t)/e_{1990} \, \Delta Q_{dir,1990}$ where $\Delta Q_{dir,1990} = -0.3 \, W m^{-2}$ and $e_{1990} = 69 \, Tg \, S \, yr^{-1}$
Sulphate aerosols, indirect forcing	$\Delta Q = \dfrac{\lvert \log(1 + e(t)/e_{nat}) \rvert}{\lvert \log(1 + e_{1990}/e_{nat}) \rvert} \Delta Q_{indir,1900}$ where $\Delta Q_{indir,1990} = -0.8 \, W m^{-2}$ and $e_{nat} = 42 \, Tg \, S \, yr^{-1}$
Biomass burning aerosols	$\Delta Q = $ ramps to $-0.2 \, W m^{-2}$ in 1990, and is held constant thereafter

Notes:

The climate forcing due to loss of stratospheric ozone does not include effects of ozone loss on tropospheric chemistry.

$\Delta Q_{CH_4 - pure}$ is the methane forcing before correction for overlap with N_2O.

$C(t)$ and $e(t)$ refer to concentrations and anthropogenic emissions of the gas in question at time t, while C_0 is the pre-industrial concentration.

The correction to the CH_4 forcing due to overlap with N_2O is given by $f(C_{CH_4}(t), C_{N_2O}(0)) - f(C_{CH_4}(0), C_{N_2O}(0))$, while the correction to the N_2O forcing due to overlap with CH_4 is given by $f(C_{CH_4}(0), C_{N_2O}(t)) - f(C_{CH_4}(0), C_{N_2O}(0))$, where $C_{CH_4}(t)$ and $C_{N_2O}(t)$ are the concentrations at time t, $C_{CH_4}(0)$ and $C_{N_2O}(0)$ are the initial concentrations, and $f(x,y) = 0.47 \ln (1.0 + \alpha_1 (x * y)^{0.75} + \alpha_2 (x * y)^{1.52})$, $\alpha_1 = 2.01 \times 10^{-5}$, and $\alpha_2 = 5.31 \times 10^{-15}$. This formulation comes from Shine *et al.* (1990).

The sulphate aerosol indirect forcing depends on the natural sulphur emission, e_{nat}, which until recently was assumed to be $42 \, Tg \, S$ per year. Recent estimates place it around $28 \, Tg \, S$ per year or less, as given in the caption to Plate 8.

The expression for the CO_2 radiative forcing given in Table 7.9 is written in a form such that the coefficient (3.75) gives the forcing ($W m^{-2}$) assumed for a CO_2 doubling, after taking into account the adjustment of stratospheric temperature. Myhre *et al.* (1998) computed essentially the same forcing (3.71 W m^{-2}), but also taking into account the absorption of solar radiation by the CO_2 near-infrared bands. The uncertainty in the forcing for a doubling is about $\pm 0.25 \, W m^{-2}$. Because the forcing depends on the ratio of the altered to the initial CO_2 concentration, the forcing for a CO_2 doubling does not depend on what the assumed starting concentration is. The main contributor to the CO_2 radiative forcing is the absorption band centred at $15 \, \mu m$ (see Fig. 7.1), which falls in the logarithmic regime. For very large increases in CO_2 concentration, other absorption bands which fall between the square root and linear regimes rather than in the logarithmic regime – become important.

Figure 7.11 shows the radiative forcing due to the 15 μm band and the minor bands, relative to the total forcing for a doubling of the pre-industrial concentration, as the CO_2 concentration varies from pre-industrial to nine times the pre-industrial concentration. These curves are based on data published by Augustsson and Ramanathan (1977). Also shown as dashed or dotted lines are various curve fits to the data. The 15 μm band curve is well approximated by the function

$$\Delta Q_{15} = R_F \ln(R)/\ln(2) \qquad (7.1)$$

where R is the CO_2 multiplier and R_F is the forcing due to the 15 μm band as a fraction of the total forcing, when CO_2 is doubled. The minor band curve is almost perfectly approximated by the function

$$\Delta Q_{\text{minor}} = (1 - R_F)(R^n - 1.0)/(2^n - 1.0) \qquad (7.2)$$

where $n = 0.775$. When the total radiative forcing is fitted to a logarithmic function (the curve labelled "1-Component Curve Fit"), the total forcing is noticeably underpredicted for CO_2 concentrations three or more times the pre-industrial concentration. A much better fit is obtained when Eqs (7.1) and (7.2) are added, giving the "2-Component Curve Fit" shown in Fig. 7.11. If the total absolute instantaneous radiative forcing for a CO_2 doubling is $\Delta Q_{2\times}$ and the reduction in the total forcing due to the adjustment of stratospheric temperature is ΔQ_{adj}, then the variation of the adjusted radiative forcing with CO_2 concentration is given by

$$\Delta Q_{CO_2} = (R_F \Delta Q_{2\times} - \Delta Q_{\text{adj}}) \ln(R)/\ln(2)$$
$$+ (1 - R_F)\Delta Q_{2\times}(R^n - 1.0)/(2^n - 1.0) \quad (7.3)$$

This expression assumes that the stratospheric cooling that gives rise to ΔQ_{adj} is caused entirely by the 15 μm CO_2 band, with no contribution from the minor bands. The reason for expecting this to be a good approximation is given in Box 10.1 of Chapter 10. As a result, the minor bands make a somewhat larger percentage contribution to the adjusted forcing than they do to the instantaneous forcing.

The expression given in Table 7.9 for the buildup of tropospheric O_3 accounts for the effect of CH_4 emissions only; ozone increases due to emissions of

CO, hydrocarbons, or NO_x are not accounted for. The expression given in Table 7.9 for the radiative forcing due to the loss of stratospheric O_3 is taken from Daniel et al. (1995). It is based in part on the result, obtained with AGCM models, that the forcing varies close to linearly with the global mean loss of O_3 as long as the shape of the vertical loss profile does not change. The O_3 loss in turn is driven by the increase in the effective concentration of chlorine and bromine atoms in the stratosphere. The effective concentration takes into account the fact that bromine atoms are a factor α more effective in depleting ozone than are chlorine atoms, where $\alpha = 30$–100. The stratospheric chlorine and bromine concentrations of each compound are approximated by the concentrations in the troposphere three years before, times a factor that takes into account the fraction of a given compound that is destroyed within the stratosphere (thereby liberating its chlorine and bromine atoms). The tropospheric concentrations in turn can be computed using a simple box model and scenarios of emissions of chlorine- and bromine-containing compounds. The radiative forcing at all times is scaled by the assumed forcing in 1990, which is rather uncertain (-0.1 to $-0.3\,\text{W}\,\text{m}^{-2}$).

The direct forcing due to sulphate aerosols can be assumed to vary linearly with the sulphate loading and hence close to linearly (Section 8.11) with the total anthropogenic sulphur emission (the sulphate concentration responds essentially instantly to changes in emission because the atmospheric lifespan of sulphate is only a few days). The major uncertainty is what the current forcing is, but once this forcing is specified – based on the results summarized in Table 7.5 or Table 7.6 – the direct forcing will scale close to linearly with the projected variation in total emissions. If the assumed forcing takes soot aerosols into account, then possible changes in the soot:SO_4 emission ratio should be allowed for. This is probably best done through a separate consideration of soot emissions and soot forcing. The indirect sulphate aerosol forcing is also highly uncertain, but should saturate in a manner close to that implied by the logarithmic dependence in Table 7.9. The projected future forcing then depends on the forcing assumed for 1990, future emissions, and the magnitude of the assumed background or natural emissions.

As noted in Section 6.5, the climate sensitivity can be directly prescribed (and altered) in simple models such as the 1-D upwelling–diffusion model. In AGCMs, on the other hand, the climate sensitivity is an internally computed quantity (although, as we shall see in Section 9.1, it is dependent on uncertain and in some cases rather arbitrarily chosen model parameters). Because the climate sensitivity is internally predicted rather than prescribed in AGCMs, it can and sometimes does turn out to be different for different forcings (see Section 3.7). When the radiative forcings computed according to the formulae given in Table 7.9 are used to drive simple climate models, on the other hand, there is no choice but to use the same climate sensitivity for all forcings. This problem could be easily circumvented by adjusting

the real radiative forcing to take account of differing climate sensitivities – in effect, applying an "effective" radiative forcing. However, there is no agreement at present as to the extent to which the sensitivity differs for differing forcings, and in most cases the expected differences in the climate sensitivity are overwhelmed by uncertainty concerning the baseline sensitivity for a CO_2 doubling – a subject we will address in Chapter 9. First, however, we need to understand the processes that determine the buildup in the concentrations of various GHGs, and the essentially instantaneous relationship between aerosol emission rates and aerosol concentrations. This is the subject of the next chapter.

CHAPTER 8

Response of the carbon cycle and other biogeochemical cycles: translating emissions of GHGs and aerosols into concentrations and radiative forcing

During the 1980s an average of 5.5 ± 0.5 Gt (giga-tonnes, or billions of tonnes) of carbon were released to the atmosphere every year as CO_2 due to the burning of fossil fuels, with an additional 1.6 ± 1.0 Gt C per year due to tropical deforestation and other land use changes (Schimel *et al.*, 1996). Regrowth of forests on abandoned agricultural land in the NH mid-latitudes might have absorbed 0.5 ± 0.5 Gt C per year, to give a range in total net anthropogenic emissions of 4.6–8.6 Gt C per year. Atmospheric CO_2 increased at the rate of 3.3 ± 0.2 Gt C per year during this time, which implies that 1.1–5.5 Gt C per year, or about one-third to two-thirds of total anthropogenic emissions, are quickly removed from the atmosphere at present. This carbon is thought to be removed primarily through absorption by the oceans and by an increase in the rate of global photosynthesis due to the stimulatory effect of the 25% increase in atmospheric CO_2 which has occurred since the Industrial Revolution. Recent analysis of oceanic and atmospheric data indicate that the likely rate of oceanic uptake is 2.0 ± 1.0 Gt C per year (Chapter 8.9). Thus, a consideration of the overall CO_2 budget implies that the terrestrial biosphere is absorbing an extra 0–4.5 Gt C per year.

Analysis of the changing ratios of different carbon isotopes in the atmosphere reduces this uncertainty somewhat. These ratios are affected by combustion of fossil fuels and a reduction in the mass of the terrestrial biosphere, and indicate that there has been little change in the total mass of the biosphere in recent decades (Siegenthaler and Oeschger, 1987). This implies that the undisturbed biosphere has been absorbing an amount of carbon comparable to the net release as a result of deforestation + regrowth on abandoned land. This could be a rate of absorption as large as 2.6 Gt C per year. Inasmuch as the global net primary productivity (or NPP, equal to plant

photosynthesis minus plant respiration) is on the order of 50–60 Gt C per year, the stimulation required to balance the carbon budget is no more than about 5%, which will be difficult to detect through direct measurement, particularly since part of the increased photosynthesis is likely to go into an increased mass of roots and soil carbon.

The biosphere and oceans, because they act to remove anthropogenic CO_2, are referred to as carbon "sinks". In order to project atmospheric CO_2 concentrations associated with future CO_2 emissions, it is important to know the relative importance of the terrestrial biosphere and oceanic sinks today, and how each sink will respond in the future. These sinks can be expected to change both as a result of continuing injections of CO_2 from human activities into the atmosphere, and as a result of the climatic changes – temperature and precipitation in particular – induced by the very buildup of CO_2 and other greenhouse gases. The terrestrial biosphere sink is not infinite, and there is widespread agreement that it is of a temporary nature because terrestrial ecosystems (including soils) do not have an unlimited capacity to absorb carbon. Furthermore, the uptake of CO_2 by the terrestrial biosphere is not irreversible. On the other hand, the oceanic sink can be regarded as almost infinite and close to irreversible, although it too could weaken considerably. The CO_2 that is not removed by these (or any other) sinks will accumulate in the atmosphere.

The broad workings of the present-day global carbon cycle were introduced in Chapter 2. In this chapter we discuss the many ways in which the carbon cycle might respond to further increases in atmospheric CO_2 concentration and to concurrent climatic change. We then discuss the techniques used to estimate what the current rates of uptake by the terrestrial biosphere and ocean are, and review the resultant estimates. In the case of the terrestrial bio-sphere sink, there is the additional issue of attribution – the estimated sink strength during recent decades could be due to the direct effect of higher atmospheric CO_2 in stimulating photosynthesis, but it could also be due, at least in part, to the effect of fertilization by nitrogen pollution or to the effect of short-term climatic variability. We close this chapter with a discussion of the cycling of other GHGs and aerosols in response to anthropogenic emissions.

8.1 Response of the terrestrial biosphere to higher atmospheric CO_2 in the absence of climatic change

The important processes involving the terrestrial component of the carbon cycle are photosynthesis, respiration by plants themselves for growth (growth respiration) and to maintain the living plant processes (maintenance respiration), and respiration by micro-organisms that decompose plant litter (heterotrophic respiration). The decomposition of plant litter releases nutrients that can be used to support new plant growth. Plant litter that is particularly resistant to decomposition can accumulate in the soil, locking up nutrients but also storing carbon on a long-term basis. Plant photosynthesis is usually stimulated by higher atmospheric CO_2 concentration – a phenomenon sometimes referred to as *CO_2 fertilization*. Higher CO_2 concentration can also enhance or inhibit rates of maintenance respiration. An increase in photosynthesis implies an increased transfer of carbon from the atmosphere to the biosphere, thereby tending to decrease atmospheric CO_2 concentration and increasing carbon storage on land. An increase in respiration has the opposite effect. Carbon dioxide enters the leaves of plants for use in photosynthesis through small openings called *stomata*. An increase in the CO_2 concentration allows smaller stomatal openings, thereby reducing the loss of water through evapotranspiration. This can indirectly affect the carbon balance by extending the active growing season where water is a limiting factor, and by changing the rate of decomposition of soil organic matter by reducing the drying of soils in summer. In the following subsections we discuss changes in photosynthesis, changes in respiration, and interactions with nutrient cycles and the water balance that can be expected in response to changing atmospheric CO_2 concentration.

Initial, direct effects on the rate of photosynthesis

Photosynthesis in a given species of plant occurs through one of three different chemical pathways. In one pathway, an intermediate compound with three carbon atoms occurs, so this pathway is referred to as the C_3 pathway, and plants using it are called C_3 plants. In the second pathway, referred to as the C_4 pathway, an intermediate compound with four carbon atoms occurs. The third pathway is not common and so is not of concern to us here. All trees and most temperate- and high-latitude grasses are C_3, while tropical grasses are C_4. Photosynthesis is a chemical reaction that is powered by sunlight and catalysed by the ribulose biphosphate carboxylase (rubisco) enzyme when a CO_2 molecule attaches to it. However, if an O_2 molecule attaches to the rubisco enzyme, then respiration can occur in the presence of sunlight. This respiration – known as *photorespiration* – undoes part of the effort of photosynthesis, thereby reducing the net productivity of the plant. The ratio of photosynthesis to photorespiration thus depends on the ratio of CO_2 to O_2 molecules bombarding the rubisco enzyme (and on the reaction probabilities). In C_4 plants, which evolved around or prior to 8 million years ago (Ehleringer *et al.*, 1991; Cerling *et al.*, 1998) as the atmospheric CO_2 concentration fell, the CO_2 concentration around the rubisco enzymes is kept at a higher level through an internal pumping mechanism, thereby inhibiting photorespiration.

An increase in the concentration of atmospheric CO_2 will increase the probability of a CO_2 molecule binding with the rubisco enzyme in C_3 plants, thereby reducing the rate of photorespiration and increasing net photosynthesis. This effect is negligible in C_4 plants because they already have a high CO_2 concentration around the rubisco enzymes. Hundreds of experiments with dozens of C_3 plant species indicate that plant growth increases by 30–40% (and sometimes much more) when the atmospheric CO_2 concentration is doubled (Idso and Idso, 1994). However, the differences between C_3 and C_4 plants that were identified above do not fully explain some of the observed responses to higher atmospheric CO_2. Bazzaz (1990) reports experiments on a natural early successional community in which a C_4 species shows a greater increase in biomass than an adjacent C_3 species. Poorter (1993) reviews other cases where a large response to elevated CO_2 has been found in C_4 plants.

Most of the experiments measuring the effect of higher atmospheric CO_2 on plants were performed in controlled and somewhat artificial greenhouse settings, although in recent years an increasing number of experiments have been performed in which the CO_2 concentration in outside air is increased (by piping CO_2 through the site and releasing it at a controlled rate) and the response of plants under closer-to-natural conditions observed. These latter experiments are referred to as "free-air CO_2 enhancement" experiments or FACE experiments. However, only a few studies with elevated CO_2 have used wild plants growing in natural ecosystems, and only a handful (discussed below) have lasted more than two growing seasons.

In assessing the implications of the observed stimulation of photosynthesis by higher atmospheric CO_2 for natural ecosystems worldwide and in the long term, a bewildering number of factors and interactions need to be taken into account. First, the stimulatory effect of higher CO_2 depends on the concurrent value of a wide range of other environmental parameters, which vary in time and space. Second, the "law of diminishing returns" is at work regarding the effect of increasing CO_2 on photosynthesis. Third, the long-term stimulation of photosynthesis by higher CO_2 is generally much less than the immediate effect. Fourth, interactions between plants and soil micro-organisms, and in the partitioning of the additional products of photosynthesis among different plant parts, could greatly amplify or diminish the effect of higher CO_2 on the rate of photosynthesis. Fifth, changes in the rate of *senescence* (leading to leaf-drop from deciduous trees) and in the relative abundance of different species within a plant community due to the unequal stimulatory effect of higher CO_2, could significantly alter the overall stimulation at the scale of an ecosystem. Underlying all of these considerations is the need to distinguish between changes in the rate of photosynthesis and changes in the net storage or accumulation of carbon in ecosystems, as it is the latter which is important to the buildup of CO_2 in the atmosphere. These issues will now be briefly discussed.

Role of environmental variables

With regard to the role of other environmental variables, reviews by Gifford (1992) and Idso and Idso (1994) indicate the following relationships for the short-term (a few weeks to months) response of plants to higher CO_2:

- the percentage enhancement in growth is greater under water-limited conditions than when water is not in short supply;
- the percentage enhancement in growth is greater the warmer the temperature, but below 12°C for carrot and radish, an increase in CO_2 concentration reduces plant growth;
- the percentage enhancement in growth can be higher, smaller, or unaltered under conditions of low nutrient availability compared to conditions of high nutrient availability;
- the percentage enhancement in growth can be smaller or larger under conditions of low light compared to high light; and
- the percentage enhancement in growth under polluted conditions is greater than in non-polluted conditions.

It should be noted that, although the percentage enhancement of growth is often greater under stressed conditions (low light, low nutrients, low water, or high pollution) than under unstressed conditions, the absolute increase in the rate of growth is usually smaller because the initial growth rate in ambient CO_2 is smaller. Nevertheless, photosynthesis is able to increase under higher CO_2 even when other factors (such as light, water, and nutrients) are limiting, because higher CO_2 lets the plant make more efficient use of limited resources.

Law of diminishing returns

The short-term response of plant photosynthesis shows a tendency to saturate; that is, each successive increase in atmospheric CO_2 has a progressively smaller stimulatory effect on the rate of photosynthesis. Figure 8.1 shows the average enhancement in short-term plant growth as CO_2 concentration increases, for resource-limited or stressed conditions, and for non-resource-limited and stress-free conditions. For unstressed and non-resource-limited conditions, an increase in CO_2 concentration from the present concentration by 300 ppmv enhances growth by about 40%, a further 300 ppmv increase causes a further 20% increase, but a further 800 ppmv increase is required to cause a further 20% increase, at which point no further increase in growth is observed. For plants under stressed conditions or in which resources are limiting, the percentage enhancement is much greater and saturation occurs more slowly, but in most cases the absolute enhancement is smaller.

Biochemical downregulation of the photosynthetic response

Many, but not all, of the plant species that show a marked increase in the rate of photosynthesis upon initial exposure to higher CO_2 show a gradual reduction in the stimulatory effect of higher CO_2 over time (Bazzaz, 1990). This reduction in the initial stimulation of photosynthesis is referred to as *downregulation*. Bazzaz *et al.* (1993) report that the percentage enhancement in plant mass under doubled CO_2 for six temperate-latitude tree species after three years of growth is only one-third to two-thirds of the enhancement seen after one or two years. Downregulation occurs for two known reasons. First, the plant might not be able to use the additional products of photosynthesis because plant growth is still strongly limited by environmental constraints. Second, the concentration of the rubisco enzyme – which catalyses the photosynthetic reaction – decreases under higher CO_2. Since about 25% of plant nitrogen is found in the rubisco enzyme, this means that a smaller amount of nitrogen is required for a given rate of photosynthesis. This in turn implies that the *nutrient utilization efficiency* (NUE) increases as downregulation occurs. As a result, ecosystem-scale productivity might not be as adversely affected by downregulation as is the productivity of an individual leaf or plant, particularly if a greater density of plant growth occurs.

Interactions with soil microfauna and soil nutrients

A number of feedbacks involving nutrient cycling could significantly alter the direct or initial effects of higher CO_2. Zak *et al.* (1993) suggest, on the basis of field data from nutrient-poor soils, that increased plant productivity due to higher CO_2 leads to an increase in the release of root-derived material that can support microbial activity. This in turn causes an increase soil micro-organism biomass and in rates of nutrient turnover, thereby amplifying the initial stimulation of plant photosynthesis in the short term. In contrast, Diaz *et al.* (1993) find that in nutrient-rich soils an increase in the release organic matter by roots results in greater nutrient sequestration by an expanded soil microflora and a nutritional limitation of plant growth. The net effect of interactions

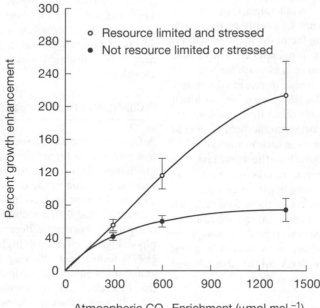

Fig. 8.1 Average percentage enhancement in plant growth as the magnitude of the enhancement in atmospheric CO₂ concentration changes. Results are shown for two sets of conditions: (i) for plants with less than adequate light, water, and nutrients, or experiencing stress induced by high levels of salinity, air pollution, or temperature (o); and (ii) for plants experiencing none of the above (●). Reproduced from Idso and Idso (1994).

between plant productivity, soil micro-organisms, and the availability of nutrients might therefore depend on the initial nutrient content of the soil.

A second interaction might occur through changes in the chemistry of plant litter. Stimulation of photosynthesis is widely believed to increase the C:N ratio in plant litter, which in turn might reduce the availability of N for plants by reducing decomposition rates, thereby eventually reducing the photosynthetic response of plants to higher CO_2. However, a number of other factors, whose response to increasing CO_2 is unknown, also influence organic matter decomposition rates (Gifford, 1992). Among these are changes in the relative amounts of specific carbon compounds. If this feedback and the feedback identified by Zak et al. (1993) both operate, then it is possible that the effect of nutrient feedbacks could be of opposite sign between the short term and long term.

Finally, increased carbon production in plants that contain nitrogen-fixing bacteria appears to increase rates of nitrogen fixation (Bazzaz, 1990). An increased supply of carbon to free-living nitrogen-fixing bacteria could also result in an increase in the rate of nitrogen fixation. This could clearly amplify the beneficial effects of high CO_2 in ecosystems where nitrogen fixation is an important process. Gifford (1994) suggests that, as a result of the above, nitrogen can be a limiting factor for plant growth on a short time scale (seasonal) but not on a longer time scale when photosynthetic rates increase for whatever reason. As with other processes increasing the supply of nutrients in an available form, the extent to which the increased nutrients are taken up by soil micro-organisms would be important to the final change in plant productivity and hence in carbon storage.

To conclude this subsection, the complexity of plant–soil–nutrient interactions is so great, and there are so many opposing negative and positive feedbacks, that the net effect of these feedbacks on carbon storage – and hence on the buildup of atmospheric CO_2 – is likely to vary from ecosystem to ecosystem and possibly even within a given ecosystem. The net feedback might also change over time.

Changes in the partitioning of plant production

Changes in the partitioning of plant production among different plant components as overall productivity increases can also strongly modulate the net increase in plant biomass. If plants respond to an increase in the rate of photosynthesis by increasing the proportion of net productivity going to fine roots (which are short lived) in order to maintain the same C:N ratio in plant tissues (by taking up more nutrients), then the increase in carbon storage due to higher rates of photosynthesis can be greatly reduced (Post et al., 1992; Luo et al., 1996) or completely eliminated (Norby et al., 1992). This is because the steady-state size of a given carbon pool is given by the product of input flux and the turnover time, as shown in Box 2.5.

Changes in senescence

Changes in the timing of senescence and of photosynthetic rates during senescence can also have significant effects on the end-of-season carbon gain. In experiments involving seedlings of five temperate forest species, McConnaughay et al. (1996) found that seedlings grown under elevated atmospheric CO_2 concentration had faster senescence and lower photosynthetic rates midway through senescence than seedlings grown under present CO_2. As a result, the end-of-season carbon gain was *smaller* for seedlings grown under higher CO_2. This is clearly too small a sample, and too short an experiment, to draw any meaningful conclusions other than that the effects of higher CO_2 during the entire annual cycle need to be considered, and not just during the rapid-growth phases.

Changes in relative species abundances

A further important consideration in the net ecosystem response to higher atmospheric CO_2 is the effect of changes in the relative abundance of different plant species due to the differential effects of higher CO_2 on different species. In simulations with a field-calibrated model that includes forest dynamics (i.e., competition between different species) as well as the physiological effects of higher CO_2, Bolker et al. (1995) found that allowing for changes in species abundances enhanced the increase in carbon storage due to higher CO_2 by about 30% over a period of 50–150 years. However, co-occurring changes in climate could require shifts in the distribution of species, resulting in a temporary (centuries) reduction in carbon storage owing to the lag between the decline of maladapted species and the arrival of new species at a site (see Section 8.2).

Decreased stomatal conductance

During photosynthesis, the stomata need to be open to permit more CO_2 to enter the leaf; this opening is associated with a loss of water from leaves. With a higher atmospheric CO_2 concentration, a smaller stomatal opening is required for a given inflow of CO_2, which in turn reduces the water loss. In a survey of experiments on 23 tree species, Field *et al.* (1995) found an average decrease in stomatal conductance of 23% in response to a doubling of atmospheric CO_2. The *water use efficiency* (WUE), or ratio of carbon uptake to water loss, therefore increases. In cases where the availability of water limits growth, this can increase the amount of photosynthesis over the course of a growing season. This effect is applicable to both C_3 and C_4 plants. An increase in WUE can also lead to a reduced drawdown of soil moisture during the summer growing season, which in turn can enhance the rate of decomposition of soil organic matter (if soils were initially too dry) or reduce the rate of decomposition (if soils were already too moist).

With downregulation of photosynthesis, the stomata can close further, since less carbon dioxide is required. However, this will result in an even larger increase in WUE (in parallel with the increase in NUE, noted above). Apart from its impact on the ecosystem carbon balance, the decrease in stomatal conductance (with or without the effects of the downregulation of photosynthesis) can have a significant effect on surface and surface-air temperature, and hence on the changes in these temperatures as the global climate changes in response to increasing CO_2 concentration. These effects will be discussed in Chapter 10.

Changes in the rate of respiration

A doubling of atmospheric CO_2 concentration reduces the rate of growth and maintenance respiration per unit of tissue dry mass by about 20% on average (Drake *et al.*, 1996a), although opposite effects have also been observed (Ryan, 1991; Wullschleger *et al.*, 1994). A suppression of respiration has been reported for leaves, roots, stems, and even soil bacteria. Part of this effect might arise from the decrease in protein and N concentration in tissues grown under higher CO_2, which reduces the amount of energy required for growth and maintenance. There is still considerable debate as to whether there is a true direct suppression of respiration by elevated CO_2 concentration.

Multi-year experiments on natural plant communities

The response of *in situ*, natural plant communities to artificial increases in the atmospheric CO_2 concentration, carried out over a period of several years, should reflect the integrated effect of the initial changes in the rate of photosynthesis and respiration, in stomatal conductance, and of many of the feedbacks discussed above. The results of a number of such experiments have been published in the last few years, and will be briefly reviewed here. These experiments involved tundra, estuarine marsh, and grassland ecosystems.

The tundra experiment involved placing environmentally controlled greenhouses over undisturbed tussock tundra at Toolik Lake, Alaska. This is a system in which one would expect both the low growing season temperatures and the extremely low available nutrient supplies each to restrict the response of plant growth to elevated CO_2. At present this tundra site is a net source of CO_2 to the atmosphere, probably because of the short-term effect of recent warming. Imposition of doubled atmospheric CO_2 concentration had a substantial effect during the first growing season, despite the temperature and nutrient limitations, shifting the site from a source of about 4 g C/m²/day to a sink of about 2 g C/m²/day. However, by the third year the stimulatory effect had entirely disappeared (Oechel *et al.,* 1994), thereby demonstrating near-total downregulation of the initial photosynthetic response. In contrast, experiments on an *in situ* estuarine marsh in Maryland (USA) indicate productivity increases of 30% at the canopy scale after eight years' exposure to higher CO_2 (Drake *et al.,* 1996b).

The third whole-ecosystem study involves a natural annual grassland at Jasper Ridge, California, on severely nutrient-limited serpentine soil. During the third growing season in which a doubled CO_2 concentration was maintained, Hungate *et al.* (1997) found a 25–37% increase (depending on soil type) in the sum of plant, detrital, and microbial carbon mass compared to controls with no enhancement of the CO_2 concentration. These are all actively overturning carbon pools, in which the new steady-state biomasses will adjust quickly to changes in the carbon input. The resistant soil carbon pool, in contrast, will take much longer to achieve its eventual increase in carbon mass. After three years, the increase in the carbon content of this large pool was 2–7%, giving increases in total carbon content of 8–9%.

The fourth whole-ecosystem study is an experiment fortuitously performed by nature. A species-rich plant community grows in a 1-hectare, bowl-shaped "CO_2 spring" near Siena, Italy. Geothermal release of CO_2 maintains daytime CO_2 concentrations of 500–1000 ppmv. Immediately adjacent to the spring area, similar plant communities grow on similar soils and in the same climate, but at about 355 ppmv. Körner and Miglietta (1994) report finding no evidence that plants in the CO_2 spring grow faster, flower earlier, or become larger than plants outside the spring. However, they did find that plants affected by the high CO_2 concentration did contain higher carbohydrate levels and lower nitrogen concentrations. More recently, Hättenschwiler *et al.* (1997) found, on the basis of tree-ring data from this site, that annual growth was larger during the first 25 years in trees that grew under higher CO_2 than in those that grew under ambient CO_2, but not thereafter, so that the net effect of higher CO_2 was to reach a given biomass sooner.

Needless to say, these whole-ecosystem experiments are so few in number, so restricted in scope (i.e., excluding forest ecosystems), and still so short in duration (except for the CO_2 spring) that they are not very helpful in anticipating the global-scale effects of higher atmospheric CO_2 on terrestrial carbon storage. They do nevertheless confirm the reality, under natural conditions, of some of the feedbacks identified above.

Evidence from tree rings

It has sometimes been stated that long records of tree-ring width provide evidence of an enhancement in the rate of growth of mature trees in recent decades that is linked to the increase in atmospheric CO_2 concentration. This evidence is reviewed by Luxmore *et al.* (1993). According to their review, the growth enhancement can be explained by climatic change in some cases. In other cases the supposed CO_2 effect is implausibly large (being greater than that observed in seedlings), while in yet other cases large growth enhancements are seen at various times during the pre-industrial era, prior to the increase in atmospheric CO_2. For these reasons, Luxmore *et al.* (1993) conclude that tree rings do not provide clear evidence of a CO_2 fertilization effect.

8.2   Effect of climatic change on the terrestrial biosphere sink

In the preceding section we discussed some of the myriad ways in which an increase in atmospheric CO_2 can directly alter the net flow of carbon into or out of terrestrial ecosystems. Changes in climate, induced by the increase in CO_2 and other GHGs, will also alter the net uptake of CO_2 by the terrestrial biosphere, either by changing the rates of photosynthesis and respiration in healthy ecosystems, or by stressing existing ecosystems to the point that they collapse and are replaced by new ecosystems. During the transition from one forest ecosystem to another, forest dieback and an increase in the frequency of forest fires might occur, which would inject additional CO_2 into the atmosphere. If, under the new climate, there is a greater proportion of ecosystems with a low density of carbon storage than under the current climate, there will be a long-term as well as a transient release of CO_2 to the atmosphere.

Direct effect of higher temperatures

Higher temperatures tend to increase plant maintenance respiration and the respiration of litter and soil organic matter. Growth respiration, on the other hand, is not believed to be affected by higher temperatures. Maintenance respiration and respiration by soil micro-organisms are generally assumed to increase with temperature using the so-called Q_{10} formulation, whereby the respiration rate R is given by

$$R = R_0 (Q_{10})^{\Delta T/10} \qquad (8.1)$$

where R_0 is the initial rate of respiration and Q_{10} is the factor by which respiration increases for a 10°C warming. Most researchers assume a Q_{10} of 2.0. According to Gifford (1994), this is based on short-term experiments (only hours in duration), and the longer-term increase in whole-plant respiration could be less than half as sensitive to increasing temperature. In the long term, Gifford found that the ratio of whole-plant respiration to whole-plant photosynthesis, averaged over 24 hours, is unchanged.

Large amounts of carbon are stored in the soils at high latitudes, which are also the regions that are expected to experience the greatest warming (Section 10.1). This raises the prospect of significant releases of carbon from these regions (Oechel *et al.*, 1993). Tropical regions are expected to experience smaller warming and have much less carbon stored in soils. However, the absolute respiration fluxes are larger than at higher latitudes and respiration increases most rapidly with increasing temperature in warm conditions, so the increase in the carbon flux could be greater from tropical soils than from polar soils during the 21st century (Townsend *et al.*, 1992).

Mitigating factors

Two factors could greatly reduce the loss of plant and soil carbon due to warmer temperatures. These are (i) an increase in the availability of nutrients as a result of increased respiration of soil organic matter, and (ii) a possible effect of warmer temperatures in suppressing the downregulation of the photosynthetic response to higher CO_2. As noted in the previous section, the initial stimulation of photosynthesis seen when CO_2 is doubled tends to decrease after a few years, and can completely disappear. This is probably due in part to the inability of plants to use the increased products of photosynthesis. However, warmer temperatures increase the demand for photosynthetic products, and are expected to reduce the long-term downregulation of photosynthesis. A sustained increase in the rate of photosynthesis implies a sustained increase in the flow of organic matter into the soil, thereby at least partially offsetting the effect of an increase in the rate of soil respiration.

Soil respiration is the process by which soil micro-organisms digest plant carbon and burn it for their own energy needs, releasing certain nutrient-containing compounds in the process. This increases the availability of nutrients to the plant roots, so any increase in the rate of soil respiration would tend to increase photosynthesis rates under nutrient-limited conditions (independently of any increase due to higher CO_2 and warmer temperatures). This would further reduce the loss of soil organic matter, since there would be a larger flow of litter into the soil. The net effect of temperature increases on terrestrial carbon storage, after accounting for possible nutrient cycle feedbacks, could be small or even beneficial (Pastor and Post, 1988; McGuire *et al.*, 1992). Simulations by Rastetter *et al.* (1992) using models of temperate forest and tundra ecosystems indicate that, in the new steady state, the increase in living biomass carbon more than offsets the decrease in soil carbon due to warming of temperate forests and tundra, although this result may depend on the extent to which the mobilized nutrients are lost from the

ecosystem or taken up by soil micro-organisms. McKane *et al.* (1995) reach a similar conclusion in their analysis of carbon dynamics in a moist tropical forest. The increase in biomass + soil carbon as the rate of respiration of soil carbon increases can be understood based on the difference in C:N ratio of soil and plant carbon (Gifford, 1994). Since a typical C:N ratio in soil organic matter is 12 but that of biomass is 60, each unit of soil carbon lost will lead to five extra units of biomass, assuming that all the mineralized N is taken up by plants. In the case of tropical forests, P is more of a limiting nutrient than N, but the disparity in the C:P ratio between plant and soil carbon (>3000 and <20, respectively) is even greater than for N, so that only a small fraction of the P released from increased decomposition of soil C need be taken up by plants to produce an increase in total carbon storage (assuming that other factors, such as availability of light, are not restricting above-ground carbon storage).

Along a similar vein, Waelbroeck *et al.* (1997) conclude that the century time-scale effect of warming in permafrost regions is to promote net carbon uptake due to the release of nutrients from thawing permafrost, even though the short-term effect is to release carbon because of a temperature-induced increase in the rate of respiration of soil carbon. Oechel *et al.* (1993) had suggested that Arctic ecosystems could switch from a net carbon sink to a net carbon source due to an increase in the rate of respiration induced by warming temperatures, and further, that this had already happened at the site in Alaska that they examined. Billings *et al.* (1982) also foresaw large increases in the respiration of soil carbon, but due to a lowering of the water table caused by thawing of permafrost. The analysis of Waelbroeck *et al.* (1997) suggests that, whatever the reason for an increase in soil respiration, the increased availability of nutrients could greatly diminish or even reverse the effect on carbon storage in the long term.

The mitigating effects of increased rates of nutrient recycling and/or suppression of photosynthetic downregulation are demonstrated in the previously discussed experiments on tussock tundra in Alaska (Oechel *et al.*, 1994). As noted in Section 8.1, the strong stimulation of photosynthesis observed in the first year following a doubling of atmospheric CO_2 completely disappeared by the third year. However, when a doubling of atmospheric CO_2 at the experimental site was combined with a 4 K increase in the air temperature, the CO_2 fertilization effect in the third year was about two-thirds of the effect observed in the first year.

In short, it is hard to see how warming alone could cause a major pulse of carbon from the terrestrial biosphere to the atmosphere. However, an increase in the frequency of drought in the interiors of some continents is expected as the climate warms, for reasons that are well understood (Section 10.2). This could lead to large-scale dieback of the current vegetation, and, as discussed next, is the major reason for concern regarding a temporary carbon pulse as ecosystems are reorganized.

Transient effects

A potentially significant terrestrial biosphere–climate feedback could occur if and when climatic change becomes large enough to induce changes in the species composition at a given site. The lag between dieback of maladapted ecosystems and their replacement by new ecosystems as climate changes could lead to significant decadal-to-century time-scale decreases in terrestrial carbon storage, even in cases where the net effect of CO_2 and climatic changes would be to increase carbon storage once the species distribution is fully adjusted to the new climate (Solomon, 1986; Pastor and Post, 1988). Calculations by Dixon *et al.* (1994) indicate that the average flux over the next century due to climate-induced forest dieback could be as high as 4.2 Gt C yr^{-1}, and would be further increased to as high as 6.0 Gt C yr^{-1} if climate-induced expansion of agriculture into forested areas is included. However, potential CO_2-fertilization effects are excluded from these calculations. Furthermore, an increase in atmospheric CO_2 increases the optimal temperature for plant growth (Idso and Idso, 1994), so that the need for ecosystems to shift poleward as the climate warms will be reduced. Nevertheless, it is clear that, with unrestrained GHG emissions and global warming, there is a growing risk that what is probably a net carbon sink at present could significantly weaken and possibly become a net carbon source.

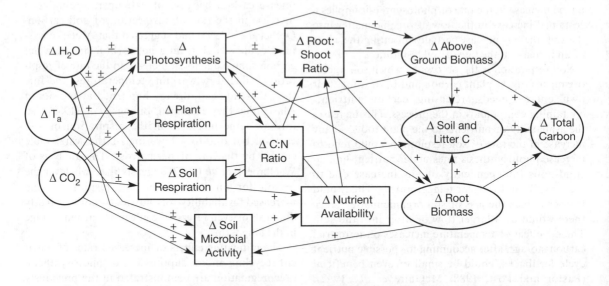

Fig. 8.2 Feedback loops involving the effect of changes in temperature, atmospheric CO_2, and soil moisture on terrestrial carbon storage. Plus signs indicate that an increase in the upstream quantity leads to an increase in the downstream quantity, and vice versa for a negative sign. If the product of all the signs in a given loop is positive, the result is a positive feedback. Not all branches from a given box will necessarily be operative at the same time at a given site. For example, a change in nutrient availability could change the root to shoot ratio or the rate of photosynthesis, but not necessarily both. Not included in this diagram are further feedbacks arising from changes in the species composition at a given site, or lags between the dieback of current species and their replacement by new species. Reproduced from Harvey (1996a).

8.3 Summary of feedback loops involving the terrestrial biosphere

Most of the feedbacks discussed in Sections 8.1 and 8.2 are summarized in Fig. 8.2. The structure of this figure is as follows: changes in three driving variables – temperature, atmospheric CO_2, and soil moisture – lead to changes in three primary processes – photosynthesis, plant respiration, and soil respiration. These changes ultimately lead to changes in above-ground biomass, in the amount of soil and litter carbon, and in root biomass, which sum to give the change in total carbon at a given site. Some of the linkages between the primary processes and the carbon storage components are direct, but others operate through changes in intermediate variables, such as the root:shoot ratio, the C:N ratio, microbial activity, and the availability of nutrients.

The plus and minus signs on the arrows joining the components shown in Fig. 8.2 indicate the sense of the causal relationship. Plus signs indicate that an increase in the upstream quantity leads to an increase in the downstream quantity, and vice versa for a minus sign. If the product of all the signs in a given loop is positive, the result is a positive feedback. Not all branches from a given box will necessarily be operative at the same time at a given site. For example, a change in nutrient availability could change the root to shoot ratio or the rate of photosynthesis, but not necessarily both. Not included in Fig. 8.2 are further feedbacks arising from changes in the species composition at a given site, or arising from lags between the dieback of current species and their replacement by new species.

The response of the terrestrial biosphere to increasing CO_2 concentration and climatic change involves a number of competing feedbacks whose relative strengths could very well change over time. Thus, even if the biosphere is serving as a net sink of CO_2, as appears to be the case at present, there is no justification for extrapolating this sink indefinitely into the future or in assuming that it will grow with increasing atmospheric CO_2 concentration. Nutrient limitations and transient effects on maladapted ecosystems in particular could lead to a significant weakening of the terrestrial sink over a period of several decades. Continued destruction of tropical rainforests, to the extent that a large fraction of the present terrestrial sink is in the tropics, will also limit the size of the future terrestrial sink.

8.4 Estimation of the current rate of uptake of anthropogenic CO_2 by the terrestrial biosphere

As noted in the introduction to this chapter, a consideration of the known anthropogenic emissions of CO_2, the observed rate of atmospheric increase, the estimated rate of uptake of CO_2 by the oceans, and the associated uncertainties, implies that the terrestrial biosphere (excluding the effects of land use changes) needs to be absorbing up to 4.5 Gt C yr^{-1} in order to balance the global carbon budget. This upper limit of 4.5 Gt C yr^{-1} is comparable to what would be expected based on a simple-minded extrapolation of the short-term response of plants to higher CO_2 as measured under controlled conditions in greenhouses or in the field. These results imply an increase in photosynthesis of 30–40% for a doubling of CO_2, and hence an increase of 8–10% – or 5–6 Gt C yr^{-1} – due to the 25% increase in the atmospheric concentration of CO_2 that has occurred so far. A more refined estimate is provided by Gifford (1994), who used a globally aggregated biosphere model driven by global mean changes in temperature and CO_2 to determine the current rate of uptake. Assuming a long-term increase in carbon storage of 10–40% for a CO_2 doubling and plausible lower and upper limits for the effects of temperature on photosynthesis and respiration, he obtained a net biomass + soil carbon uptake today of 0.5–4.0 Gt C yr^{-1}. These results largely span the range needed to balance the carbon budget. However, various additional observations, to be discussed next, can be used to show that the current rate of absorption by the terrestrial biosphere is substantially less than the upper limit of 4.5 Gt C yr^{-1} that is obtained from a consideration of the global-scale budget. This in turn provides the clearest evidence that the 30–40% stimulation of photosynthesis found in short-term experiments cannot be simply extrapolated to the entire natural biosphere.

In Section 2.11 it was explained that gases (such as CO_2) that have an atmospheric lifetime much longer than the time required for complete mixing within the atmosphere have close to a uniform concentration. There are, nevertheless, small spatial (and seasonal) variations in the concentration of CO_2 that can be used to infer the geographical distribution and magnitude of the various sources and sinks of CO_2. For example, the current CO_2 concentration is about

3 ppmv greater north of 30°N than it is south of 30°S. To infer sources and sinks from CO_2 concentration data requires reversing the usual analysis procedure: instead of trying to compute concentrations from emissions and transport processes, one starts with concentrations and works backwards. This kind of problem is called an *inverse problem,* and the process of finding the solution is called *inversion.* A major difficulty with inverse problems is that the solution can sometimes be very sensitive to errors in the input data.

Carbon occurs as two stable isotopes (varieties): 99% as ^{12}C and about 1% as ^{13}C (carbon also occurs as a radioactive isotope, ^{14}C, but ^{14}C accounts for only one carbon atom in about 10^{12}). Photosynthesis preferentially uses ^{12}C over ^{13}C, thereby enriching the remaining atmospheric CO_2 in ^{13}C, while the burning of fossil fuels and biomass releases ^{13}C-depleted CO_2 into the atmosphere. The absorption of CO_2 by the oceans, in contrast, entails relatively little discrimination between different isotopes. The $^{13}C/^{12}C$ ratio of a sample is expressed as $\delta^{13}C$, where

$$\delta^{13}C = \frac{(^{13}C/^{12}C) - (^{13}C/^{12}C)_0}{(^{13}C/^{12}C)_0} \times 100\permil$$

with $(^{13}C/^{12}C)$ pertaining to the sample and $(^{13}C/^{12}C)_0 = 0.0112372$ pertaining to a standard. Atmospheric $\delta^{13}C$ data have been used in two different ways to estimate the rate of absorption by the terrestrial biosphere: in the first way, the annual variation in the global mean isotope ratio has been used in conjunction with annual changes in the global mean atmospheric CO_2 concentration and in global fossil fuel emissions to deduce the separate total global absorption of CO_2 by the terrestrial biosphere and the oceans. In the second approach, the spatial pattern in atmospheric $\delta^{13}C$ is used to infer the geographical distribution of sources and sinks, and to differentiate between the terrestrial and oceanic sinks. In both approaches, the inferred terrestrial biosphere sink is the net effect of all changes in the terrestrial biosphere, including losses due to deforestation and regrowth of forests on previously cleared and abandoned land. This sink shall henceforth be referred as the *terrestrial biosphere exchange,* with a positive exchange indicating a net emission of carbon from the terrestrial biosphere.

Two estimates of the recent terrestrial biosphere exchange have been published based on what are supposed to have been interannual variations in the global mean $\delta^{13}C$ and CO_2 concentration. The CO_2 variations (combined with fossil fuel emission data) determine what the total biosphere + ocean exchange must be, while $\delta^{13}C$ data determine how this total must be partitioned between the biosphere and oceans. Keeling *et al.* (1995) deduced an average biospheric exchange from 1978 to 1994 of 0.21 Gt C yr^{-1} (a positive exchange being a CO_2 source), while Francey *et al.* (1995) deduced an average exchange from 1982 to 1992 of about 1.5 Gt C yr^{-1}. The discrepancy between the two results is largely due to the use of $\delta^{13}C$ data from different monitoring stations in the two studies. Both groups deduced significant interannual variations in the net exchanges, but disagree significantly during the period of overlap.

Global mean atmospheric CO_2 and $\delta^{13}C$ variations have also been used to infer the variation in the net exchange of carbon between the terrestrial biosphere and the atmosphere during the past 200 years or so. Figure 8.3 shows the net exchange since 1850 as reconstructed by Siegenthaler and Oeschger (1987). The atmospheric $\delta^{13}C$ variations were inferred from measurements of the isotope ratio of tree rings. Of particular interest is the inference that the net exchange has been very close to zero since about 1980, which supports the exchange estimate of Keeling *et al.* (1995), cited above. Also shown in Fig. 8.3 is the global net emission due to changes in land use, as calculated by Houghton (1998) and previously shown in Plate 1(b). The difference between the two gives the required terrestrial biosphere sink.

Table 8.1 shows the global mean net biospheric (and oceanic) exchanges, as deduced by Ciais *et al.* (1995) based on an analysis of the geographic distribution of atmospheric $\delta^{13}C$. Analyses were performed separately for 1990, 1991, and 1992. There are significant differences from one year to the next that could be related to the substantial problem of error propagation in the inversion methodology. That portion of the interannual differences that is real might reflect climate-driven differences in net primary productivity and in ocean–air CO_2 fluxes.

Table 8.1 Global terrestrial biosphere and oceanic sinks (Gt C yr^{-1}) for the years 1990, 1991, and 1992 as deduced by Ciais et al. (1995) based on inversion of atmospheric $^{13}C/^{12}C$ data.

Year	Terrestrial biosphere sink	Oceanic sink	Total sink
1990	2.2	1.0	3.2
1991	1.8	2.4	4.2
1992	1.4	3.1	4.6

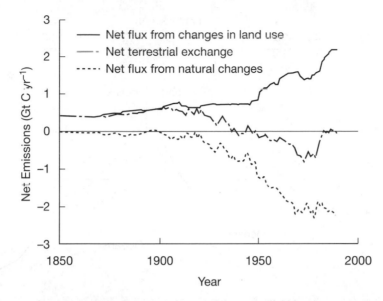

Fig. 8.3   Net exchange of carbon between the atmosphere and the terrestrial biosphere as computed by Siegenthaler and Oeschger (1987), the net flux to the atmosphere due to land use changes as computed by Houghton (1998) and previously shown in Plate 1(b), and the difference between the two. Based on data kindly provided by R.A Houghton.

Figure 8.4 shows the latitudinal variation in the deduced net terrestrial (and oceanic) carbon exchange for 1992. This and previous analyses based on the geographical variation in atmospheric $\delta^{13}C$ agree that there is a large sink in NH mid-latitudes. Inverse calculations by Tans *et al.* (1990), Ciais *et al.* (1995) and Fan *et al.* (1998) place most of the mid-latitude sink on land rather than in the oceans. Indeed, Fan *et al.* (1998) calculate that most of the NH mid-latitude sink is due to uptake by forests in North America (1.7 ± 0.5 Gt C yr^{-1}), with Eurasia–North Africa contributing only 0.1 ± 0.6 Gt C yr^{-1}. The CO_2 source shown in Fig. 8.4 for tropical regions ($16°S–16°N$) is smaller (1.5 Gt C yr^{-1}) than expected from oceanic outgassing and tropical deforestation, which implies some absorption of CO_2 by the undisturbed tropical biosphere. This is supported by Fan *et al.* (1998), who deduced a net tropical land source of only 0.2 ± 0.9 Gt C yr^{-1}.

The regional terrestrial sources and sinks inferred from these inverse calculations include the direct effects of human-induced land use changes (regrowth of forests on abandoned agricultural land in the NH mid-latitudes; deforestation and concurrent regrowth of forests on rapidly abandoned land in the tropics). The difference between the regional net source as deduced from inverse calculations, and the regional net source due to land use changes as estimated from land use and other ground-based data, would pre-sumably reflect the regional CO_2 fertilization effect. However, as explained below, the CO_2 fertilization effect is only one of three factors that can explain the difference between land use-related fluxes and the net biospheric exchange of carbon.

8.5   Causes of the recent absorption of CO_2 by the terrestrial biosphere

Two factors other than CO_2 fertilization and regrowth of forests on abandoned land could be causing the terrestrial biosphere to act as a sink at present: fertilization of terrestrial ecosystems (primarily forests) with nitrogen from NO_x emissions (which also contribute to acid rain), and the effects of recent decadal-scale variability in climate.

Fertilization by deposition of NO_x has been estimated to have caused a global carbon sink of $0.5–1.0$ Gt C yr^{-1} (Melillo *et al.,* 1996). This is comparable to lower estimates of the magnitude of the terrestrial biosphere sink needed to balance the global carbon cycle. If a large part of the current sink is due to NO_x pollution, and this pollution decreases in the future – or the adverse side effects of the associated cumulative acid deposition increase – then the terrestrial biosphere sink will become weaker in the future.

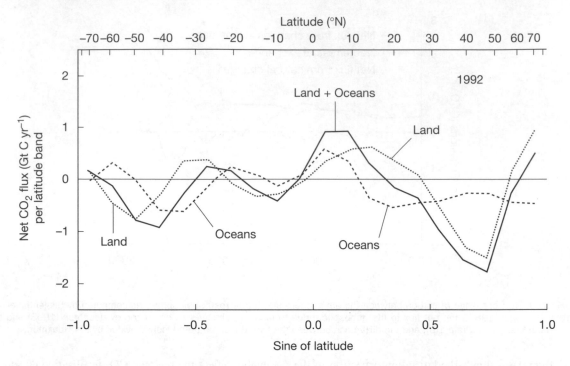

Fig. 8.4 Latitudinal variation in the net exchange of carbon in 1992 between the atmosphere and the terrestrial biosphere, the oceans, and the sum of the two as deduced from $^{13}C/^{12}C$ data by Ciais *et al.* (1995). Redrawn from Ciais *et al.* (1995).

There is some evidence that decadal fluctuations in climate in the recent past have affected the net storage of carbon in the terrestrial biosphere. Dai and Fung (1993) used simple empirical correlations between climatic variables (temperature and precipitation) and ecosystem variables (NPP and soil respiration), combined with historical variations in temperature and precipitation on a 2.5° × 2.5° latitude–longitude grid, to estimate changes in terrestrial carbon storage due to climatic variability during the period 1940–1988. At low latitudes, variations in precipitation have been more important than variations in temperature. Both NPP and soil respiration are computed to have increased in decades with greater precipitation, with little net effect on carbon storage in the tropics. A total carbon sink of 20 ± 5 Gt was computed for the period 1940–1984 (0.59 ± 0.15 Gt C yr^{-1}), caused largely by temperature-driven increases in NPP in middle and high latitudes. This sink, which does not include possible effects of CO_2 and N fertilization, is about one-third of the sink required to balance the carbon budget for middle-of-the-road estimates of land use emissions and the oceanic uptake of CO_2.

As noted in the preceding section, inverse calculations imply that there is a net terrestrial biosphere sink in NH mid-latitudes. This sink would include the effect of CO_2 and N fertilization, climatic change, and regrowth of forests on previously deforested land. Subtracting the carbon uptake due to regrowth of forests would then give the net effect of the first three factors. However, the carbon sink due to forest regrowth itself would include the effect of CO_2 and N fertilization, and of climatic change. Thus, it will be very difficult, if not impossible in practice, to accurately separate land use-related and non-land use-related effects on the net terrestrial biosphere exchange in mid-latitudes. This is of great significance, since a number of NH mid-latitude countries seem intent on using terrestrial biosphere sinks, for which they can supposedly claim credit, to meet their CO_2 emission obligations under the Kyoto Protocol.

Houghton (1996) presents evidence that part of the mid-latitude NH terrestrial biosphere sink is indeed due to CO_2 and/or N fertilization of regrowing forests, and not just due to normal regrowth following abandonment of previously cleared land. He has compared the expected sink based on changes in land use, and based on direct measurements of the rate of growth of new forests. The former method gives the expected sink in the absence of fertilization effects, and amounts to 0.0 ± 0.5 Gt C yr^{-1}. The second method includes any effects of fertilization (and of climatic change), and yields an estimated sink of 0.8 ± 0.4 Gt C yr^{-1}. The implied fertilization–climate effect, 0.8 ± 0.9 Gt C yr^{-1}, is about half of the total terrestrial biosphere sink that is required to balance the global carbon budget. The net sink of 0.8 ± 0.4 Gt C yr^{-1} is also only half of that inferred by Fan et al. (1998) from inversion of atmospheric data (1.7 ± 0.5 Gt C yr^{-1}), but both estimates are probably more uncertain than is implied by the quoted uncertainty range.

The remainder of the carbon sink could be in the tropics and due to CO_2 fertilization effects. Recall, from Section 8.4, that a sink is expected in the tropics due to the fact that the source deduced by inverse calculations (0.2 ± 0.9 Gt C yr^{-1}) is much less than expected based on estimated emissions due to changes in land use (1.6 ± 1.0 Gt C yr^{-1}). However, changes in climate have not caused a significant sink in the tropics according to Dai and Fung (1993), and NO_x deposition is also not likely to be important in the tropics. This leaves a tropical CO_2 fertilization effect of 0.0–3.0 Gt C yr^{-1}. Thus, CO_2 fertilization effects in the tropics could easily equal or exceed the combined CO_2–N fertilization effect in mid-latitudes.

8.6   The oceanic CO_2 sink in the absence of climatic change

The major processes involved in the oceanic part of the carbon cycle were identified in Section 2.4 and illustrated in Fig. 2.7. These processes are gaseous exchange of CO_2 between the atmosphere and ocean; chemical equilibration between CO_2, CO_3^{2-}, and HCO_3^-; vertical transfers between the mixed layer and deeper ocean through the biological pump, advective overturning, convection, and diffusion; and the net burial of organic tissue and $CaCO_3$ in oceanic sediments. Owing to the action of the biological pump, the concentration of total dissolved inorganic carbon (DIC, the sum of $[CO_2]$, $[CO_3^{2-}]$, and $[HCO_3^-]$) in surface waters is about 10% less than in deep water, and since the atmosphere tends to equilibrate with the surface water, this causes the atmospheric CO_2 concentration to be much smaller than it would be in the absence of the biological pump.

Anthropogenic emissions of CO_2 lead to an increase in the atmospheric partial pressure of CO_2 (pCO_2), causing it to be greater than the average value in ocean surface water. As a result, there is a net flow of CO_2 from the atmosphere to the ocean. However, unlike the terrestrial biosphere, a higher concentration of CO_2 has almost no stimulatory effect on marine photosynthesis, so there will be no direct effect of an increasing CO_2 concentration on the strength of the biological pump. Thus, unless the strength of the biological pump changes for some other reason (such as a climate-induced change in the rate of upwelling of nutrients), the biological pump plays no role in the uptake of anthropogenic CO_2.

The easiest way to understand how the ocean responds to the increasing concentration of atmospheric CO_2 is to first consider the response of the oceanic mixed layer, without interaction with the deep ocean. We then consider the role of the deep ocean, deep ocean sediments, and – in Section 8.7 – the potential role of feedbacks between climate and the oceanic part of the carbon cycle.

Table 8.2 Disposition of carbon in an atmosphere–mixed layer system that mimics pre-industrial conditions, and the changes after increasing the carbon content of the mixed layer by 1%. The results shown here were calculated using the chemical equations given in the appendix to Peng et al. (1987), using $T = 293.15$ K, a total alkalinity of 2320 µequiv/kg, and a salinity of 34.87‰.

Component	Pre-industrial carbon content (Gt)	Pre-industrial pCO_2 (µatm) or concentration (µmol/kg)	Change in pCO_2 (µatm) or concentration (µmol/kg)
Atmosphere	593.6	278.4 µatm	27.0 µatm
Mixed layer HCO$_3^-$	521.1	1737.28 µmol/kg	31.75 µmol/kg
CO$_3^{2-}$	69.8	233.69 µmol/kg	−12.83 µmol/kg
CO$_2$	2.7	9.03 µmol/kg	0.88 µmol/kg
Total	593.6	1980.00 µmol/kg	19.80 µmol/kg

Response with the mixed layer alone

Consider the case of a single atmospheric box in contact with a well-mixed surface box (the mixed layer). We shall choose a depth for the mixed layer box of 66.9 m since, for pre-industrial conditions, this results in the atmospheric and mixed layer boxes having exactly the same amount of carbon. This depth is a reasonable global mean depth for the real mixed layer, whose depth varies from 20–30 m near the equator to 100–200 m at high latitudes, but the subsequent analysis is greatly simplified if the two boxes have the same initial carbon content. Table 8.2 presents data on the initial conditions for the two boxes, computed from the equations for chemical equilibrium given in the appendix to Peng *et al.* (1987) using $T = 293.15$ K. Both boxes have 594 Gt C, all as CO_2 in the atmosphere but split among dissolved CO_2 (about 0.46%), CO_3^{2-} (about 11.8%) and HCO_3^- (about 87.7%) in the mixed layer.

The partial pressure of CO_2 in the atmospheric box, $(pCO_2)_a$, is given by

$$(pCO_2)_a = \gamma M_a \tag{8.2}$$

where $\gamma = 0.469$ µatm/Gt and M_a is the atmospheric mass of CO_2 in Gt C. The partial pressure of CO_2 in the mixed layer, $(pCO_2)_{ML}$, is given by

$$(pCO_2)_{ML} = \frac{[CO_2]_{ML}}{\alpha} \tag{8.3}$$

where α is the solubility of CO_2 and $[CO_2]_{ML}$ is the concentration of CO_2 in the mixed layer as µmol/kg. In equilibrium the atmospheric and mixed layer partial pressures are equal, so we can write

$$[CO_2]_{ML} = \alpha(pCO_2)_a \tag{8.4}$$

The solubility depends on water temperature, decreasing from 0.0622 µmol kg^{-1} µatm^1 at $T = 274$ K to 0.0326 µmol kg^{-1} µatm^{-1} at $T = 293$ K. Using $T = 293$ K and a pre-industrial atmospheric pCO_2 of 278.4 µatm, Eq. (8.3) gives $[CO_2] = 9.08$ µmol/kg. Given an ocean surface area of 3.61×10^{14} m^2, a density for seawater of 1027 kg/m^3, and a mixed layer depth of 66.9 m, this concentration gives a CO_2 content of the mixed layer of $[CO_2] \times 0.2976$ Gt kg/µmol = 2.71 Gt C. This is 0.46% of the CO_2 in the atmosphere (and 0.46% of the total dissolved carbon in the mixed layer).

Now suppose that we add 10 Gt C to the atmosphere as CO_2. This increases atmospheric pCO_2 by 4.69 µatm, which in turn induces a flow of CO_2 into the mixed layer. We will consider three cases.

Case 1

No chemical reactions with the added CO_2 occur. It then follows from Eq. (8.4) that the change in mixed layer $[CO_2]$ is given by

$$\Delta[CO_2]_{ML} = \alpha\Delta(pCO_2)_a \tag{8.5}$$

The oceanic uptake is $\Delta[CO_2] \times 0.2976$ Gt kg/µmol = 0.046 Gt C or 0.46% of the added carbon, which is the same as the initial fraction of DIC as CO_2.

Box 8.1   Uptake of anthropogenic CO_2 by an atmosphere–mixed layer system

Consider a system consisting of a thoroughly mixed atmospheric box and an oceanic mixed layer box, which are in an initial equilibrium state with equal CO_2 partial pressures. The atmosphere pCO_2 is directly related to the mass of carbon in the atmosphere as CO_2, M_a, while the mixed layer pCO_2 depends on the $[CO_2]$ in water next to the air–sea interface. The amount of carbon in the mixed layer, M_{ML}, will depend on its depth, but assume that $M_{ML} = rM_a$. A carbon mass ΔM is injected into the atmosphere as CO_2. Let f_a be the fraction of the injected CO_2 that remains in the atmosphere when a new steady state is established, so $f_{ML} = 1 - f_a$ is the fraction that is taken up by the oceanic mixed layer. In the new steady state, the atmospheric and oceanic pCO_2s will have increased by the same amount, so that we can write

$$\frac{\Delta(pCO_2)_a}{(pCO_2)_a} = \frac{\Delta(pCO_2)_{ML}}{(pCO_2)_{ML}} \tag{8.1.1}$$

The fractional change in atmospheric pCO_2 is equal to the fractional change in atmospheric carbon, while the fractional change in mixed pCO_2 is equal to β times the fractional change in mixed layer carbon, where β is the buffer factor. Thus, we can write

$$\frac{f_a\Delta M}{M_a} = \beta\,\frac{(1 - f_a)\Delta M}{M_{ML}} \tag{8.1.2}$$

Using $M_{ML} = rM_a$, we obtain

$$f_a = \frac{\beta}{r + \beta} \tag{8.1.3}$$

For $r = 1$ and $\beta = 9.7$, which corresponds closely to the real oceanic mixed layer, $f_a = 0.91$ and $f_{ML} = 0.09$. Thus, the ocean ends up with only one-tenth as much new carbon as the atmosphere, due to the fact that $(pCO_2)_{ML}$ increases 10 times faster than $(pCO_2)_a$ as its carbon content increases.

Case 2

It is assumed that the carbon that enters the mixed layer distributes itself among dissolved CO_2, HCO_3^-, and CO_3^{2-} in the same proportions as originally present. In this case the percent increase in mixed layer pCO_2 will equal the percent increase in total dissolved carbon in the mixed layer, since all the carbon species increase by the same fraction. The final atmospheric and mixed layer pCO_2 values will be the same if 5 Gt C remain in the atmosphere and 5 Gt C are transferred to the ocean. Thus, the ocean takes up 50% of the added carbon.

Case 3

The chemical reactions given by Eqs (2.1)–(2.4) occur, but the added carbon distributes itself among the three dissolved carbon species as dictated by the chemical equilibrium constants. In this case, a greater proportion of the added carbon remains in the form of dissolved CO_2 than for the pre-existing mixed layer carbon, so that the percentage increase in mixed layer pCO_2 (which is given by Eq. (8.3)) is greater than the percentage increase in total dissolved carbon. In particular, an increase in DIC by 1% (19.8 μmol/kg) starting from pre-industrial conditions causes $[CO_3^{2-}]$ to decrease by 5.5% (12.8 μmol/kg), while $[HCO_3^-]$ increases by 1.8% (31.8 μmol/kg) and $[CO_2]$ increases by 9.7% (0.88 μmol/kg). The $[CO_3^{2-}]$ decreases as CO_2 is added to the mixed layer because it is consumed by reaction (2.4). These results are summarized in Table 8.2. The relationship between changes in pCO_2 and DIC can be written as

$$\frac{\Delta pCO_2}{pCO_2} = \beta \frac{\Delta DIC}{DIC} \qquad (8.6)$$

where β is called the *buffer factor* and is equal to the ratio of the percentage increase in mixed layer pCO_2 (or $[CO_2]$) to the percentage increase in DIC. Based on the results presented above, $\beta = 9.7$ for pre-industrial conditions. In Box 8.1 it is shown that the fraction of the added CO_2 that remains in the atmosphere after the new equilibrium is achieved is equal to $\beta/(1 + \beta)$. Thus, about 9.1 Gt C remain in the atmosphere and only 0.9 Gt C flows into the ocean. The percentage increase in mixed carbon is only one-tenth that of the atmosphere, but since mixed layer pCO_2 increases 10 times faster, the final changes in atmospheric and mixed layer pCO_2 are the same (which is required to re-establish equilibrium). In this case the mixed layer takes up only 9% of the added carbon.

The above analysis was for a relatively small addition of carbon to the atmosphere and thence to the mixed layer. As previously noted, the addition of CO_2 to the ocean and its conversion to HCO_3^- consumes CO_3^{2-}. As more CO_2 is added and $[CO_3^{2-}]$ drops, more of any subsequently added CO_2 remains as dissolved CO_2, so the mixed layer pCO_2 rises more quickly. That is, the buffer factor increases as CO_2 is added to the mixed layer. By the time the mixed layer carbon content has increased by 10%, $\beta = 14.1$. The increase in β makes it harder and harder to put more CO_2 into the mixed layer.

The final point to consider is how quickly the mixed layer can absorb carbon from the atmosphere. The flux of carbon from the atmosphere to the mixed layer depends on the difference in partial pressures, namely,

$$F_{CO_2} = k_{as}((pCO_2)_a - (pCO_2)_{ML}) \qquad (8.7)$$

where k_{as} is an exchange coefficient with a global mean value of about 0.07 mol m^{-2} yr^{-1} μatm^{-1} (k_{as} is often determined based on the observed rate of absorption of ^{14}C that was released into the atmosphere by the testing of nuclear bombs during 1952–1981). This implies that the one-way flux between the atmosphere and mixed layer, for equilibrium conditions, is about 20 mol m^{-2} yr^{-1}. The pre-industrial DIC of the mixed layer (and the pre-industrial CO_2 content of the atmosphere) is about 136 mol m^{-2}, so that the turnover time of mixed layer carbon with respect to the atmosphere or vice versa is about seven years. However, because of the buffer factor, the turnover time does not equal the adjustment time of the mixed layer to a perturbation. The equations governing the *perturbations* in atmospheric and mixed layer carbon content, ΔC_a and ΔC_{ml}, are

Table 8.3 Vertical carbon flux from the mixed layer to the deep ocean for steady-state pre-industrial conditions, and the perturbation in the downward flux during the first year after the sudden injection of 10 Gt C into the atmosphere. Fluxes were computed using the model of Harvey and Huang (1999), which has separate non-polar, NH polar, and SH polar regions. A negative flux is an upward flux.

Transfer process	Steady-state flux (Gt C yr^{-1})	Perturbation flux (Gt C yr^{-1})
Advection	0.07	0.002
Convection	−2.48	0.033
Diffusion	−7.97	0.425
Biological pump	10.38	0.000

$$\frac{d\Delta C_a}{dt} = k'(\beta\Delta C_{ml} - \Delta C_a) + Q \qquad (8.8)$$

and

$$\frac{d\Delta C_{ml}}{dt} = -k'(\beta\Delta C_{ml} - \Delta C_a) \qquad (8.9)$$

respectively, where $1/k'$ is the turnover time τ (see Box 2.2) and Q is the rate of carbon input to the atmosphere. The exponential time scale for adjustment after a pulse input is given by

$$\tau_{adj} = \frac{1}{k'(1 + \beta)} = \frac{\tau}{1 + \beta} \qquad (8.10)$$

Thus, the adjustment time is less than one year.

Role of the deep ocean

As the concentration of DIC increases in the ocean mixed layer, the downward fluxes due to advective and convective overturning will increase, since in both cases the sinking water parcels will have a greater DIC concentration than before. In addition, the upward diffusion of DIC will decrease, since the DIC concentration will now increase less rapidly with increasing depth (recall the variation with depth of the DIC concentration shown in Fig. 2.8). We can think of the change in diffusion as a downward diffusive flux from the mixed layer, in response to higher mixed-layer DIC, superimposed on the pre-existing upward flux. Thus, the excess carbon in the mixed layer gradually works its way into the deeper ocean.

Table 8.3 lists the steady-state carbon fluxes between the mixed layer and deep ocean as simulated by the model of Harvey and Huang (1999), and gives the perturbation in the downward flux during the first year after a sudden injection of 10 Gt C into the atmosphere. The dominant balance in the undisturbed steady state is between a downward flux due to the biological pump and an upward flux due to diffusion. Since the biological pump is unaffected by the absorption of CO_2 by the mixed layer, the main flux to respond to the perturbation is the diffusive flux. Thus, on a global basis the dominant transfer process is diffusion, although at high latitudes, convective mixing is dominant.

Diffusion is a result of random turbulent eddies that mix water parcels and, in so doing, transfer dissolved constituents (and heat) from regions of high

Box 8.2 Oceanic uptake of anthropogenic CO_2 after complete mixing

To estimate what fraction of anthropogenic CO_2 can eventually be taken up by the oceans, after thorough downward mixing of the anthropogenic perturbation has had a chance to occur, we can use Eq. (8.1.3) of Box 8.1. The amount of carbon in the oceans is about 64 times the pre-industrial carbon content of the atmosphere (38,000 Gt/592 Gt), so $r = 64$. Assuming for the moment that β remains fixed at 9.7, we obtain $f_a = 0.13$, so the ocean takes up 87% of the injected carbon. Thus, if 3800 Gt C of CO_2 are released, 500 will remain in the atmosphere and 3300 will be taken up by the ocean. This would increase the average DIC concentration by 8.6%, which would increase β to about 14. Using an average β of 12, our revised estimate of the oceanic uptake is 84% rather than 87% of the injected carbon. However, this assumes that the added carbon is eventually distributed uniformly throughout the ocean, which is not the case. Using the 1-D model of Harvey and Huang (1999), we find that about 1000 Gt C remain in the atmosphere, and the actual oceanic uptake is only 74% of the injected carbon. This is the amount of carbon that would be removed after 1000–2000 years; if 3800 Gt C were emitted during the next 100–200 years, the buildup of CO_2 during this period would be several times greater than the final accumulation of 1000 Gt C because of the bottleneck created by slow downward mixing of the anthropogenic perturbation below the mixed layer.

concentration to regions of low concentration. However, the intensity of turbulence in the oceans is not the same in all directions. This is because there is a layering or stratification in the ocean, with water of slightly lower density on top of water of slightly higher density. Surfaces of constant density are called *isopycnal surfaces*. Turbulence perpendicular to surfaces of constant density (*diapycnal* mixing) is suppressed, while turbulence along constant-density surfaces (*isopycnal* mixing) is not suppressed. Diapycnal mixing involves eddies with a typical size of 1 m, while along-isopycnal mixing involves eddies with a typical size of 50 km. As a result, the empirical diffusion coefficient – which is a measure of the strength of mixing – is factor of about 10^5 greater for along-isopycnal than for diapycnal mixing. This is important because isopycnal surfaces are not perfectly horizontal, but have slopes as large as 10^{-3} to 10^{-4} over broad regions of the mid- and high-latitude oceans. Consequently, vertical fluxes due to isopycnal mixing can be as large as or larger than the vertical flux due to diapycnal mixing. Further analysis can be found in Section 11.4 (Box 11.2) and in Harvey (1995) and Harvey and Huang (1999).

In spite of the contribution from isopycnal mixing, vertical diffusion is still a slow process. The time scale required for a perturbation to penetrate to a significant extent to a depth D is given by D^2/k, where k is the vertical diffusion coefficient. Given an effective k value of about $1.0 \times 10^{-4}\,m^2\,s^{-1}$ for DIC (see Harvey and Huang, 1999), the time scale is on the order of 100 years to penetrate 500 m. In contrast, the mixed layer equilibrates with the atmosphere on a time scale of about one year. As DIC penetrates into the deeper ocean, the mixed layer is able to absorb more CO_2 from the atmosphere, but because the rate of penetration below the mixed layer is so slow, downward mixing serves as a bottleneck on the oceanic uptake of anthropogenic CO_2. Again, the exception is in regions of deep convection, where carbon added to the mixed layer is rapidly mixed through several hundred to thousands of metres of water.

Complete mixing of the CO_2 that would be released from combustion of a large part of the recoverable fossil fuel resource (3800 Gt or 10% of the total carbon storage in the ocean) through the entire volume of the oceans would dilute the DIC of the mixed layer enough to allow the oceans to eventually absorb about 75–80% of the carbon added to the atmosphere, as shown in Box 8.2. The time required for most of this absorption to occur is 1000 years. However, partial dissolution of $CaCO_3$ sediments would allow the ocean to absorb much of the remaining CO_2, as explained below.

Role of oceanic sediments

As discussed in Section 2.4, biological productivity in the mixed layer produces a steady rain of $CaCO_3$ (calcium carbonate) particles. Most of these are in the form of calcite. The solubility of ocean water with respect to calcite increases with pressure and hence with depth, and the top 1–4 km of the ocean (depending on location) is supersaturated with respect to calcite, while the deep ocean is unsaturated. Figure 8.5(a) shows the geographical varia-

tion in the depth of the boundary between saturated and unsaturated water (the saturation horizon). Calcite particles landing on the ocean bottom at depths shallower than the saturation horizon tend to be preserved, while particles landing at greater depths will dissolve. Thus, today, calcite is found primarily in sediments shallower than the saturation horizon but not in the deepest sediments. Figure 8.5(b) shows the percentage of $CaCO_3$ in sediments at the sea floor.

As noted above, as ocean water absorbs CO_2, $[CO_3^{2-}]$ decreases. This decreases the degree of supersaturation of water in the upper ocean with respect to $CaCO_3$. As CO_3^{2-}-depleted water penetrates into the deep ocean, the boundary between supersaturated and unsaturated water will rise, allowing the deepest $CaCO_3$ sediments to begin dissolving. As $CaCO_3$ dissolves, CO_3^{2-} is restored to the surrounding water. As this water mixes back up to the surface, the ocean can absorb more CO_2 from the atmosphere. The amount of $CaCO_3$ sediment that can dissolve is limited by the fact that, as dissolution occurs, the non-carbonate sediment is left behind, and this sediment eventually forms a capping layer thick enough to prevent further dissolution of carbonate sediments. Nevertheless, it is expected that enough carbonate sediment can dissolve to allow the oceans to take up about half of the remaining 20–25% of anthropogenic CO_2, so that the oceans ultimately take up 85–90% of CO_2 emitted into the atmosphere. However, several thousand years will be required for this to happen. Enhanced chemical weathering on land would take up the remaining 10–15%, but over a time period of about 200,000 years (Archer et al., 1997).

8.7 Effect of climatic change on the oceanic sink

Climatic change accompanying an increase in atmospheric CO_2 concentration could alter the oceanic sink either through changes in ocean surface temperature, through changes in the operation of the biological pump, or by altering the oceanic pool of dissolved organic carbon (DOC). Such changes would alter the subsequent atmospheric CO_2 concentration and associated climatic change.

(a)   Saturation Horizon

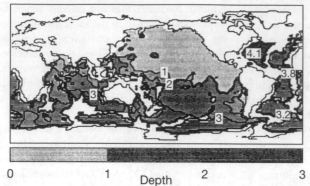

0   1   Depth   2   3

(b)   Dry Weight Percent CaCO$_3$

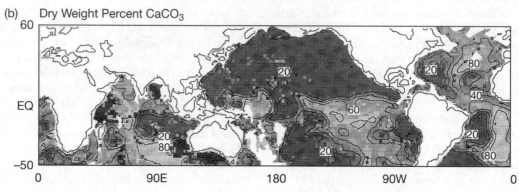

Fig. 8.5   (a) Depth of the calcite saturation horizon in the world oceans, and (b) percentage of CaCO$_3$ in ocean sediments at the seafloor. Reproduced from Archer (1996a) and Archer (1996b), respectively.

Ocean temperature feedback

Temperature directly affects the pCO$_2$ of sea water in two ways. First, the partitioning of total DIC between dissolved CO$_2$, CO$_3^{2-}$, and HCO$_3^-$ depends on temperature. Second, and more importantly, the solubility decreases with increasing temperature. Thus, according to Eq. (8.3), the pCO$_2$ of surface water will increase as temperature increases, even with fixed [CO$_2$]. The net effect of these two temperature dependencies is such that, for water with a fixed chemical composition (including fixed DIC), the pCO$_2$ of surface water initially at 20°C changes according to

$$\frac{1}{pCO_2}\left(\frac{dpCO_2}{dT}\right)\Bigg|_{DIC} = 0.043\,K^{-1} \qquad (8.11)$$

That is, pCO$_2$ increases by 4.3% for each 1 K (or 1°C) warming.[1] However, as pCO$_2$ increases, CO$_2$ will flow from the ocean to the atmosphere. This increases atmospheric pCO$_2$ and undoes some of the initial increase in the mixed layer pCO$_2$. At some point the two pCO$_2$s will become equal, at which point no further net transfer of CO$_2$ will occur. As shown in Box 8.3, the final increase in atmospheric (or mixed layer) pCO$_2$ is given by

$$\frac{\Delta(pCO_2)_a}{(pCO_2)_a} = \frac{\left(\frac{\Delta pCO_2}{pCO_2}\right)\Big|_{DIC}}{1 + \beta/\alpha} \qquad (8.12)$$

---

[1]   The 4.3% increase in pCO$_2$ arises from a 1.5% increase in the concentration of dissolved CO$_2$ (at the expense of HCO$_3^-$) and a 2.7% decrease in solubility.

Box 8.3 Feedback between temperature and oceanic mixed layer pCO_2

Assume that we have an initial steady state in which atmospheric and mixed layer pCO_2s are equal. The instant that a warming of the mixed layer is imposed, the mixed layer pCO_2 increases (by 4.3% per degree Celsius of warming) but atmospheric pCO_2 is unchanged. This causes CO_2 to flow from the mixed layer to the atmosphere, causing atmospheric pCO_2 to increase and the mixed layer pCO_2 to decrease from its elevated value. Eventually the two pCO_2s converge, and no further net flux occurs. The fractional change in atmospheric pCO_2 in the new steady state is equal to the fractional change in atmospheric CO_2 content. That is,

$$\frac{\Delta(pCO_2)_a}{(pCO_2)_a} = \frac{\Delta M_a}{M_a} \qquad (8.3.1)$$

The change in mixed layer pCO_2 in the new steady state is equal to the increase that occurred the instant the temperature increases, plus the change that occurred after CO_2 began flowing to the atmosphere. The latter change, in percentage terms, is equal to the buffer factor times the percentage change in mixed layer DIC, and is negative (a decrease). Thus, for the new steady state, we can write

$$\frac{\Delta(pCO_2)_{ML}}{(pCO_2)_{ML}} = \left.\left(\frac{\Delta pCO_2}{pCO_2}\right)\right|_{DIC} + \beta\frac{\Delta M_{ML}}{M_{ML}} \qquad (8.3.2)$$

Given that $\Delta(pCO_2)_a/(pCO_2)_a = \Delta(pCO_2)_{ML}/(pCO_2)_{ML}$, $\Delta M_{ML} = -\Delta M_a$, and writing $M_{ML} = rM_a$, we obtain

$$\frac{\Delta(pCO_2)_a}{(pCO_2)_a} = \frac{\left.\left(\dfrac{\Delta pCO_2}{pCO_2}\right)\right|_{DIC}}{1 + \beta/r} \qquad (8.3.3)$$

When $r = 1$ (i.e., an initial mixed layer carbon content equal to that of the atmosphere), and for $\beta = 10$, the final increase in pCO_2 is only 1/11 of the increase in mixed layer pCO_2 that occurred the instant that the temperature increased. This is because, as soon as enough CO_2 has flowed to the atmosphere to increase the atmospheric pCO_2 by 1/11 of the initial imbalance, the mixed layer pCO_2 will have decreased by 10/11 of the initial imbalance, thereby bringing the two pCO_2s back to a common value. If the mixed layer is r times as deep, then the reduction in mixed layer DIC and hence in pCO_2 for a given outflow of CO_2 will be only $1/r$ as large, so that almost r times as much carbon must flow to the atmosphere, giving a correspondingly larger final increase in atmospheric pCO_2.

where $(\Delta pCO_2/pCO_2)|_{DIC}$ is given by Eq. (8.11), and α is the ratio of the carbon content of that part of the ocean that exchanges with the atmosphere to the atmospheric carbon content. If only a 66.9 m deep mixed layer is involved in carbon exchange, $\alpha = 1$ and $\Delta pCO_2/pCO_2 = 0.00404$. That is, pCO_2 increases with increasing temperature about 10 times more slowly than for fixed DIC (see Box 8.3 for a detailed explanation). If the entire ocean depth contributes equally to carbon exchange, then $\alpha = 64$ and $\Delta pCO_2/pCO_2 = 0.0373$, which is close to $(\Delta pCO_2/pCO_2)|_{DIC}$. In reality, the added carbon is not mixed uniformly through the ocean (due to the solubility pump) and, for the 1-D model of Harvey and Huang (1999), $\Delta pCO_2/pCO_2 = 0.0343$. This corresponds to an effective α of 38. During the early part of the oceanic response to CO_2 emissions, relatively little of the ocean will be involved in exchanging carbon with the atmosphere, so the effective value of α will be small and the temperature–pCO_2 feedback will be weak. As more of the ocean contributes an efflux of CO_2 to the atmosphere in response to the increase in surface water pCO_2, the temperature feedback will become stronger. However, this occurs over a period of centuries, during which time atmospheric GHG concentrations and hence surface temperatures may have fallen from their peak values (depending on the future emission scenario). This would mitigate the effect of the temperature–pCO_2 feedback.

As a quantitative example, consider a cumulative emission of 4000 Gt C, which would be sufficient to double the pre-industrial CO_2 concentration after a period of about 1000 years (having fallen from much higher interim levels). If the global mean ocean surface temperature warming for a CO_2 doubling is 2°C and $\Delta pCO_2/pCO_2 = 0.035$, then the atmospheric carbon content will be 1267 Gt rather than 1184 Gt (a difference of 7%), but the change in carbon content compared to pre-industrial times will be 675 Gt rather than 592 Gt – a difference of 14%. If the temperature change (prior to any pCO_2 feedback) is 4°C instead of 2°C, the CO_2 buildup is enhanced by 28% rather than 14%. In either case, the extra CO_2 buildup would induce further warming that would induce yet further CO_2 buildup, giving eventual increases in excess of 30% and 15% (assuming a linear system response). One can thus conclude that temperature–pCO_2 feedback is likely to be important on long time scales if a significant fraction of the recoverable fossil fuel resource is used and if the climate sensitivity is high.

Changes in the biological pump

The biological pump could change in response to a change in climate for at least three different reasons: First, a change in the input of iron or other micronutrients from land areas, due to a change in winds or in aridity, could alter biological productivity; second, a change in temperature could alter the production of $CaCO_3$ by changing the proportion of calcareous and silicious micro-organisms in surface waters; and third, climatic change could lead to a change in the rate of mixing between the mixed layer and deep ocean, thereby changing the rate of supply of nutrients to the mixed layer. The first and second potential causes of a change in the biological pump are somewhat speculative (Falkowski *et al.* 1998), so our attention here will be directed to the third potential cause.

One of the more robust features of coupled atmosphere–ocean climate models is that warming of the climate at rates comparable to that anticipated during the 21st century will induce an at-least-temporary reduction in the intensity of the large-scale, thermohaline overturning in the oceans (Washington and Meehl, 1989; Mikolajewicz *et al.,* 1990; Gough and Lin, 1992; Manabe and Stouffer, 1994; Harvey, 1994). A reduction of 30–50% in the thermohaline overturning by the time CO_2 doubles is a common and quite credible model result. The reduction in the thermohaline overturning as the climate warms is a result of some combination of (i) the greater penetration of heat in polar downwelling regions than in lower latitude upwelling regions, which reduces the meridional density gradient in the upper ocean and thereby weakens the overturning, and (ii) a freshening of polar regions due to a poleward shift in the mid-latitude precipitation belt or due to increased runoff from melting polar ice caps, which also weakens the meridional density gradient.

The possibility of a significant reduction in the intensity of oceanic overturning has prompted a number of studies of the effect of changes in oceanic overturning either on the steady-state atmospheric CO_2 concentration, or on the transient variation of atmospheric CO_2 concentration in response to anthropogenic emissions of CO_2. Two ways in which changes in ocean circulation will affect atmospheric CO_2 are: (i) by altering the net vertical transfer of DIC associated with the sinking of cold water in polar regions and the upwelling of CO_2-rich water from depth elsewhere; and (ii) by altering the availability of nutrients in the surface layer and hence in the rate of biological production and export of particulate carbon from the surface layer. However, as reviewed by Harvey (1999), results from a variety of ocean carbon cycle models disagree as to the magnitude (and even the sign) of the effect of a reduction in thermohaline overturning on atmospheric CO_2. The recent results of Maier-Reimer *et al.* (1996) and Sarmiento and Le Quéré (1996) suggest that a 50% reduction in the intensity of thermohaline overturning would reduce the rate of uptake of anthropogenic CO_2 by the oceans by about 10%.

Insight into the reasons for the different responses of ocean carbon cycle models to a reduction in the rate of thermohaline overturning can be gained through a simple perturbation analysis, which is presented in Box 8.4. This analysis indicates that the most important parameters governing the response of atmospheric CO_2 to a change in thermohaline overturning are (i) the sensitivity of the biological pump to changes in the availability of nutrients, and (ii) the sensitivity of the vertical diffusive flux of DIC to changes in the vertical gradient of DIC. The latter depends on the vertical diffusion coefficient for DIC. The analysis presented in Box 8.4 suggests that the effect of changes in thermohaline overturning on the buildup of atmospheric CO_2 should be rather small.

Changes in the pool of dissolved organic carbon

Another potential feedback involves the dissolved organic carbon (DOC) pool in the ocean, whose size has recently been estimated to lie between 685 and 1800 Gt C (Druffel *et al.,* 1992; Martin and Fitzwater, 1992; Hansell and Carlson, 1998). DOC consists of organic carbon particles that are substantially smaller than the particulate organic carbon that can be measured through standard techniques. However, its turnover time is extremely slow – in excess of 6000 years for some components (Mopper *et al.,* 1991).

The rate-limiting step for the removal of a large fraction of DOC is photochemical degradation. Climate-induced changes in the DOC respiration flux are likely to be very small, and an even smaller fraction (10–15%) of the respired carbon would make its way into the atmosphere. Thus, there would seem to be no way in which changes in the DOC pool could exert a significant feedback on climate.

8.8 Impulse response for CO_2 and carbon isotopes

Our understanding of the response of the oceans and of the terrestrial biosphere to the injection of CO_2 into the atmosphere, in the absence of climate–carbon cycle feedbacks, can be summarized by running the oceanic and/or terrestrial components of a carbon cycle model and examining what happens when a pulse of CO_2 is suddenly added to the atmosphere. The variation through time in the amount of the injected carbon that remains in the atmosphere is called the *impulse response*.

Figure 8.6(a) shows the impulse response when the oceans alone absorb the injected CO_2, when the terrestrial biosphere alone absorbs the injected CO_2 (assuming a middle estimate of the CO_2-fertilization effect), and when the oceans and terrestrial biosphere act together to absorb the injected CO_2. These responses were computed using the carbon cycle model of Harvey and Huang (1999) for a sudden injection of 1 Gt C. For these calculations, no dissolution of $CaCO_3$ sediments was allowed to occur. The results shown in Fig. 8.6 are similar to those obtained by many other models. Focusing first on the ocean-only case, the first 20% of the carbon is removed in less than 10 years. Successive increments take progressively longer to be removed. The final 18% is not removed at all, which is equivalent to giving it an infinitely long time constant. Thus, unlike other GHGs, the removal of a pulse of CO_2 is not characterized by a single time constant. Rather, the time constant for the removal of the remaining CO_2 increases as more CO_2 is removed. When the terrestrial biosphere alone absorbs CO_2, the rate of removal is initially comparable to that due to the ocean alone. However, the biosphere absorption ceases around 40 years after the emission pulse, followed by a weak return flow of carbon to the atmosphere. This is a result of an increase in the respiration flux as respirable carbon accumulates in the soil pools. When the terrestrial biosphere and ocean act together, there is a markedly faster initial decrease in the atmospheric CO_2 pulse then when either act alone, but little difference in the total carbon uptake after 200 years.

Also shown in Fig. 8.6(a) is the rate of decay of a pulse of N_2O, which is governed with a single exponential time constant of 120 years (Table 2.7). Comparison with the impulse response for CO_2 underlines the fact that the removal of CO_2 is governed by multiple time constants, being considerable faster than the 120-year N_2O time constant at the beginning, and being considerable slower after about 200 years. Furthermore, the amount of the N_2O pulse that remains in the atmosphere asymptotically approaches zero, whereas the CO_2 pulse does not do so for all practical purposes (complete removal requiring on the order of 200,000 years). This latter difference creates significant difficulties when it comes to creating a single index for intercomparing different GHGs (Section 13.5).

Figure 8.6(b) shows the amount of ^{12}C and ^{13}C remaining (or entering) the atmosphere if the pulse input consists entirely of ^{13}C. Interestingly, the decay of a perturbation in the $^{13}C:^{12}C$ ratio is much faster than the decay in the total amount of carbon in the atmosphere, and after several hundred years only 2–3% of the ^{13}C pulse remains. The isotope ratio decreases much faster than the pulse of total carbon because, as ^{13}C diffuses into the ocean, some of it is replaced by ^{12}C which diffuses out (this oceanic ^{12}C can be thought of as being dislodged by the ^{13}C). The same is true if we consider a pulse consisting entirely of ^{14}C. As a result, the atmospheric ^{14}C content has recovered quickly from the injection of ^{14}C due to the atmospheric testing of nuclear bombs. This behaviour has led some to erroneously conclude that anthropogenic emissions of CO_2 should also be removed rapidly (with a time constant of around 10 years). If this were true, the steady-state buildup of anthropogenic carbon would be much smaller, given the relationship between the time constant for removal and concentration developed in Box 2.5. This in turn would imply that the current buildup of atmospheric CO_2 cannot be largely due to human emissions. As noted in Box 4.1, the evidence linking the CO_2 increase since the Industrial Revolution to human emissions is overwhelming.

Box 8.4 Analysis of the effect of a collapse of the thermohaline overturning on atmospheric CO$_2$

Considerable insight into the effect of a change in thermohaline overturning on atmospheric pCO$_2$ can be obtained through a perturbation analysis. Such an analysis is applicable to any ocean model. Since the strength of the biological pump is believed to be limited by the supply of nutrients from below the mixed layer, the first step is to analyse the change in the concentration of nutrients in the mixed layer. Here, we will assume that dissolved phosphate is the limiting nutrient, although the analysis is applicable to any combination of limiting nutrients that is supplied from the deep ocean.

Consider a two-box representation of the ocean, where the upper box is the mixed layer and the lower box represents the next few hundred metres below the mixed layer, where most of the dissolution of organic matter occurs and which therefore serves as the source region for upwelling nutrients and carbon. Let h_u and h_l be the thicknesses of the two boxes, with $h_u = 67$ m and $h_l = 1000$ m. Let P_u and P_l be the phosphate concentrations in the upper and lower boxes, respectively, and let C_u and C_l be the corresponding concentrations of DIC. The biological, advective, and diffusive phosphate fluxes per time step are given by

$$F_{bp} = \rho h_u \left(\frac{P_u}{P_u + P_h} \right) P_u \tag{8.4.1}$$

$$F_{ap} = -\rho w (P_l - P_b) \Delta t \tag{8.4.2}$$

and

$$F_{dp} = -\frac{\rho k}{\Delta z} (P_l - P_u) \Delta t \tag{8.4.3}$$

respectively, where ρ is the density of sea water, Δz is the distance between the midpoints of the two boxes, Δt is the time step length, and P_b is the phosphate concentration in sinking polar water. In nature, P_b is greater than the mean mixed layer value (P_u) due to convective mixing with phosphate-rich deep water. Thus, P_b will lie somewhere between P_u and P_l, with P_b being closer to P_l the stronger the convective mixing. This suggests that we can parameterize P_b as

$$P_b = P_u + a_1 (P_l - P_u) \tag{8.4.4}$$

where a_1 is a measure of the importance of convection. Empirically, $a_1 = 0.43$.

The biological flux is formulated using what is called a Michaelis–Menton relationship, where P_h is the phosphate concentration at which photosynthesis proceeds at half the maximum possible rate. Changes in P_u and P_l will be such that total phosphate is conserved. That is,

$$h_u \Delta P_u + h_l \Delta P_l = 0 \tag{8.4.5}$$

The fluxes of DIC are given by

$$F_{bc} = R_{CP} F_{bp} \tag{8.4.6}$$

$$F_{ac} = -\rho w (C_l - C_b) \Delta t \tag{8.4.7}$$

and

$$F_{dc} = -\frac{\rho k}{\Delta z} (C_l - C_u) \Delta t \tag{8.4.8}$$

where R_{CP} is the C:P ratio in organic matter (known as the Redfield ratio) and C_b is the DIC in sinking polar water. The latter is greater than C_u owing to the fact that colder water has a lower pCO$_2$ for a given DIC, so that CO$_2$ flows into the water from the atmosphere as it flows poleward and cools. Thus, C_b can be parameterized as

$$C_b = C_u(1 + \gamma) \tag{8.4.9}$$

where γ is an empirically determined coefficient equal to 0.08 for present-day conditions. Changes in DIC are subject to the same constraint as changes in phosphate, namely,

$$h_u \Delta C_u + h_l \Delta C_l = 0 \tag{8.4.10}$$

Appropriate values of the parameters in the above equations were obtained by Harvey and Huang (1999) by fitting a vertically resolved model to observational data. The appropriate values are $k = 1.5\,\mathrm{cm^2\,s^{-1}}$, $w = 2.0\,\mathrm{m\,yr^{-1}}$, and $P_h = 0.8\,\mathrm{\mu mol\,kg^{-1}}$.

In the initial steady state prior to a change in circulation, and in the final steady state after the nutrient distribution has adjusted to the change in circulation, the net flux of phosphate to the deep ocean is zero, so the change in the net flux is zero. The net flux depends on both P_u and P_l, so we can write

$$\Delta F_{ep} + \left(\frac{\partial F_{bp}}{\partial P_u} + \frac{\partial F_{ap}}{\partial P_u} + \frac{\partial F_{dp}}{\partial P_u} \right) \Delta P_u + \left(\frac{\partial F_{bp}}{\partial P_l} + \frac{\partial F_{ap}}{\partial P_l} + \frac{\partial F_{dp}}{\partial P_l} \right) \Delta P_l = 0 \tag{8.4.11}$$

where ΔF_{ep} is the "externally" imposed change in the net phosphate flux to the deep ocean, and ΔP_u and ΔP_l are the steady-state changes in P_u and P_l, respectively. When the ocean overturning rate is suddenly cut in half, ΔF_{ep} equals the negative of half the initial advective flux. Since $\Delta P_u = -(h_u/h_l)\Delta P_l$ and $\partial F_{bp}/\partial P_l = 0$, we can solve for ΔP_u as

$$\Delta P_u = - \frac{\Delta F_{ep}}{\dfrac{\partial F_{bp}}{\partial P_u} + \dfrac{\partial F_{ap}}{\partial P_u} + \dfrac{\partial F_{dp}}{\partial P_u} - \dfrac{h_u}{h_l}\left(\dfrac{\partial F_{ap}}{\partial P_l} + \dfrac{\partial F_{dp}}{\partial P_l} \right)} \tag{8.4.12}$$

This expression shows that the change in P_u depends on the magnitude of the initial advective flux and on the derivatives of all the fluxes with respect to concentration. Using the flux formulations given above, $\partial F_{bp}/\partial P_u = 62{,}800\,\mathrm{kg\,m^{-2}}$, $\partial F_{dp}/\partial P_u = 11{,}000\,\mathrm{kg\,m^{-2}}$, $\partial F_{ap}/\partial P_u = 2860\,\mathrm{kg\,m^{-2}}$, $\partial F_{dp}/\partial P_l = -11{,}000\,\mathrm{kg\,m^{-2}}$, and $\partial F_{ap}/\partial P_l = -2860\,\mathrm{kg\,m^{-2}}$. Thus, $\partial F_{bp}/\partial P_u$ dominates the denominator in Eq. (8.4.12) for the biological pump formulation used here.

A similar expression can be derived for ΔC_u, namely,

$$\Delta C_u = - \frac{\Delta F_{ec} + R_{CP}\Delta F_{bp}}{\dfrac{\partial F_{ac}}{\partial C_u} + \dfrac{\partial F_{dc}}{\partial C_u} - \dfrac{h_u}{h_l}\left(\dfrac{\partial F_{ac}}{\partial C_l} + \dfrac{\partial F_{dc}}{\partial C_l} \right)} \tag{8.4.13}$$

where the subscript c refers to DIC. The change in the downward biological flux is given by the change in the downward phosphate flux times the C:P Redfield ratio, and is independent of C_u. It therefore appears in the numerator of Eq. (8.4.13) as an additional fixed forcing, while the derivative $\partial F_{bc}/\partial C_u$ vanishes from the numerator. Since the remaining derivatives are much smaller than $\partial F_{bp}/\partial C_u$, one might expect that C_u will be considerably more sensitive to changes in the overturning strength than is P_u. However, since the ratio of DIC to phosphate in the ocean is relatively constant (and close to the Redfield ratio), there will be a strong tendency for F_{ap} and F_{ac} to be in the same direction. If F_{ap} and F_{ac} are both upward (<0), for example, then ΔF_{ep} and ΔF_{ec} will both be downward (>0) if w decreases, and ΔP_u and hence ΔF_{bp} will be negative (a decrease). Thus, ΔF_{bp} will be opposite in sign to ΔF_{ec}, and the partial cancellation between these two terms reduces the change in C_u. Furthermore, ΔC_u can be opposite in sign to ΔP_u and opposite to the direct effect of the change in upwelling (ΔF_{ec}). Thus, even though the net advective fluxes of DIC and phosphate are both upward, a reduction in the overturning intensity can lead to a decrease in P_u and an increase in C_u, as found in the model simulations of Harvey (1999). Whether P_u and C_u are directly or inversely correlated as w changes will, in general, depend on the chosen Redfield ratio, the initial magnitudes of the advective fluxes, and the magnitude of $\partial F_{bp}/\partial P_u$.

8.9 Estimation of the current rate of uptake of anthropogenic CO_2 by the oceans

The current rate of absorption of anthropogenic CO_2 by the oceans can be estimated by six different methods: (i) using uncalibrated 3-D OCGMs; (ii) using calibrated 1-D models; (iii) through direct estimation of the global air–sea CO_2 flux based on the observed difference between the mixed layer and atmospheric pCO_2; (iv) based on the observed change in the ^{13}C distribution in the oceans; (v) based on changes in the amount of O_2 in the atmosphere; and (vi) based on concurrent variations in the global mean atmospheric CO_2 concentration and $^{13}C/^{12}C$ ratio, combined with data on fossil fuel CO_2 emissions and the $^{13}C/^{12}C$ ratio for these emissions. These six methods are briefly described below, then a synthesis of the results is given.

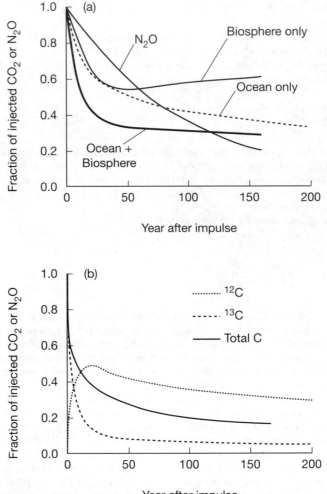

Fig. 8.6 (a) The impulse response for CO_2 taking into account removal by the ocean only, by the terrestrial biosphere only, and by the ocean and terrestrial biosphere acting together. Also shown is the impulse response for N_2O. (b) Impulse response for total carbon and for ^{13}C when a pulse consisting entirely of ^{13}C is added to the atmosphere. Also given is the increase in atmospheric ^{12}C as ^{13}C enters the ocean and dislodges some of the oceanic ^{12}C. All CO_2 results were computed using the carbon cycle model of Harvey and Huang (1999), while the N_2O impulse corresponds to an N_2O lifetime of 120 years.

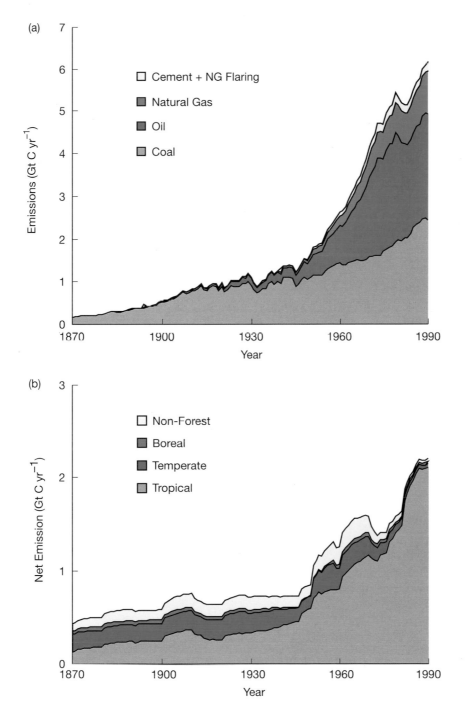

Plate 1 (a) Variations in CO_2 emissions (Gt C yr^{-1}) from 1870 to 1990 due to the combustion of coal, oil, natural gas, and the sum of the three. (b) Estimates net emissions of CO_2 (Gt C yr^{-1}) from 1870 to 1990 due to land use changes involving tropical, temperate, and boreal forest; grasslands; and the sum of the four. Fossil fuel and cement emission data are from Maryland *et al.* (1998) and were obtained from the web site http://cdiac.csd.ornl.gov, while land use emission estimates are from Houghton (1998) and were kindly provided in electronic form.

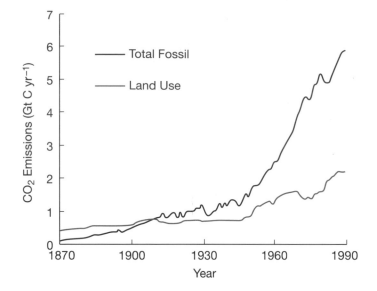

Plate 2 Comparison of global total fossil fuel CO_2 emissions (Gt C yr^{-1}) and global emissions due to changes in land use, the latter as estimated by Houghton (1998).

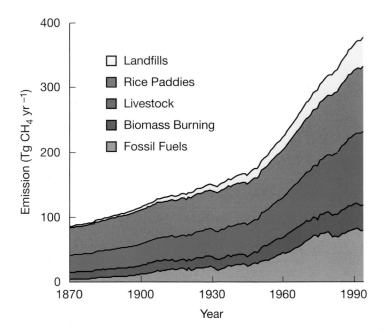

Plate 3 Historical variation in anthropogenic CH_4 emissions (Tg CH_4 yr^{-1}) from 1870 to 1990, as estimated by Stern and Kaufmann (1996a) and given at the web site http://cdiac.esd.ornl.gov/ftp/trends.ch4_emis/ch4.dat.

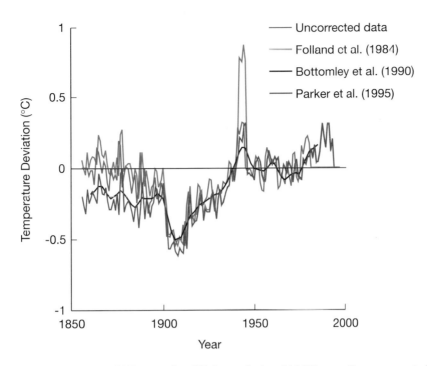

Plate 4 Uncorrected, global mean NMAT anomalies (°C) (green line) and NMAT anomalies as corrected by Folland *et al.* (1984) (blue line), Bottomley *et al.* (1990) (purple line), and Parker *et al.* (1995) (red line). Based on data files kindly provided by the UK Meteorological Office. The available Bottomley *et al.* (1990) data were smoothed.

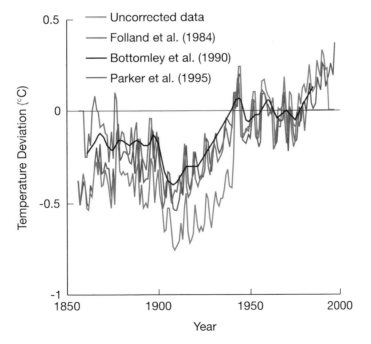

Plate 5 Variation in uncorrected global mean SST anomalies (°C) and as corrected by Folland *et al.* (1984), Bottomley *et al.* (1990), and Parker *et al.* (1995). Based on data files kindly provided by the UK Meteorological Office. The available Bottomley *et al.* (1990) data were smoothed.

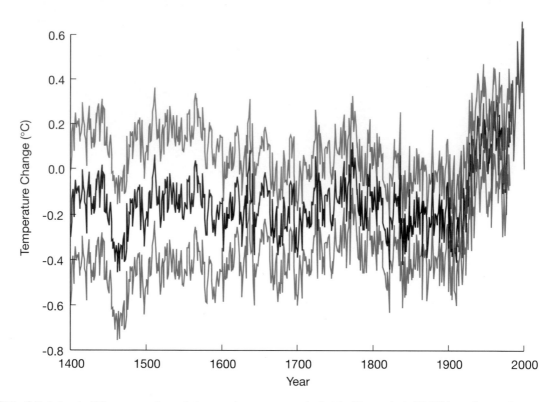

Plate 6 Variation in NH mean surface air temperature, as reconstruted by Mann *et al.* (1998) based on various proxy climate indicators. Also shown are the ±2 standard deviation error bars (light blue lines) and the observed NH mean temperature variation since 1900 (red line). Based on data obtained from the National Oceanographic and Atmospheric Administration (NOAA) Paleoclimatology web site http://www.ngdc.noaa.gov/paleo/pubs/mann1998.

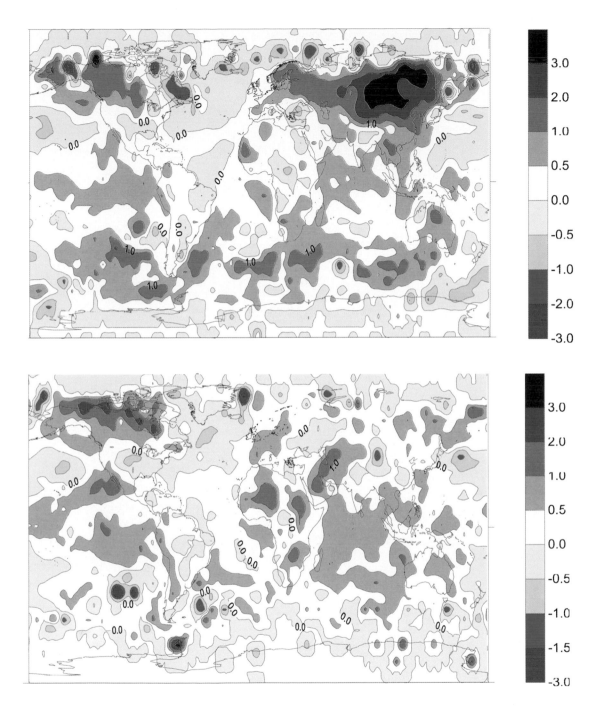

Plate 7 Change in mean annual and seasonal surface temperature (°C) from 1950–1959 to 1990–1998. Based on data obtained from the Climate Research Unit of the University of East Anglia (web site http://www.cru.uea.ac.uk).

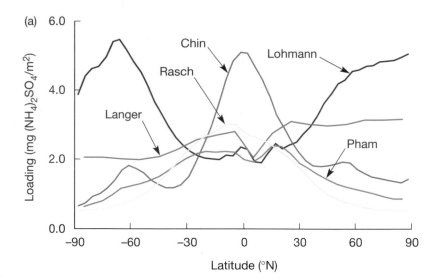

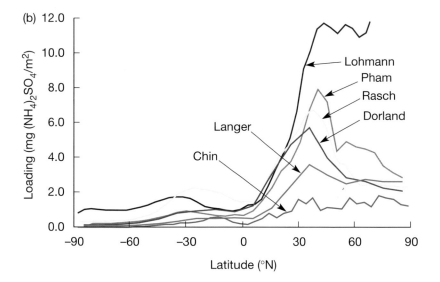

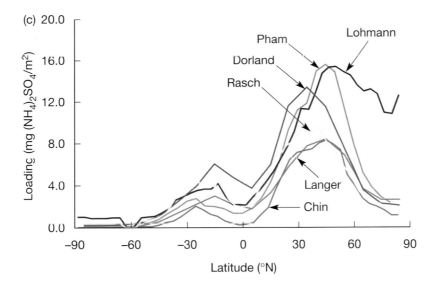

Plate 8 Latitudinal variation in the sulphate aerosol loading as computed by various researchers. Shown are (a) the zonal mean loading due to natural emissions alone, and the zonally averaged loading over (b) ocean and (c) land due to anthropogenic emissions alone. The total global natural emissions used for the simulations are as follows: Chin et al. (1996), 29.3 Tg S/yr; Lohmann et al. (1999), 25.5 Tg S/yr; Rasch et al. (1999), 15.5 Tg S/yr; Langer and Rodhe (1991), 28.0 Tg S/yr; Pham et al. (1996), 31.1 Tg S/yr. The total anthropogenic emissions used are as follows: Chin et al. (1996), 67.4 Tg S/yr; Lohmann et al. (1999), 72.5 Yg S/yr; Rasch et al. (1999), 67.0 Tg S/yr; Langner and Rodhe (1991), 70.0 Tg S/yr; Pham et al. (1996), 94.4 Tg S/yr; van Dorland et al. (1997), 67.4 Tg S/yr. In preparing this figure, the original loadings were scaled such that all of the loadings shown here correspond to a natural emission of 28.0 Tg S/yr and an anthropogenic emission of 69.0 Tg S/yr. Based on grid point data kindly by one of the authors of each of the papers cited above.

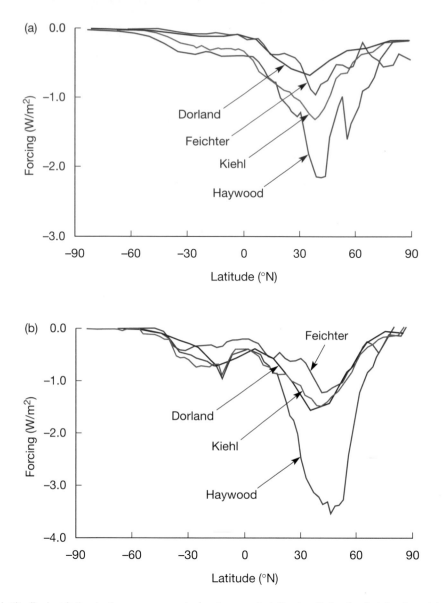

Plate 9 Latitudinal variation in the mean annual, zonally averaged direct radiative forcing due to sulphate aerosols over (a) ocean, and (b) land, as computed by various researchers. The curves labelled "Kiehl" here correspond to the curves labelled "Rasch" in Plate 8, and are based on Kiehl *et al.* (1999). The references corresponding to the other curves are van Dorland *et al.* (1997), Haywood *et al.* (1997), and Feichter *et al.* (1997). Based on grid point data kindly provided by the lead authors of these papers.

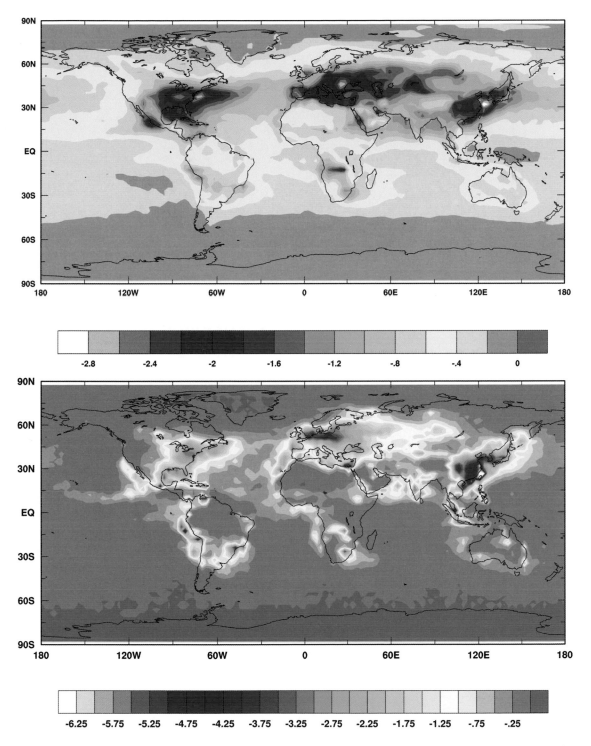

Plate 10 Geographical variation in (a) the direct radiative forcing due to sulphate aerosols, and (b) the indirect radiative forcing due to sulphate aerosols, as simulated by Kiehl *et al.* (1999). Reproduced from Kiehl *et al.* (1999).

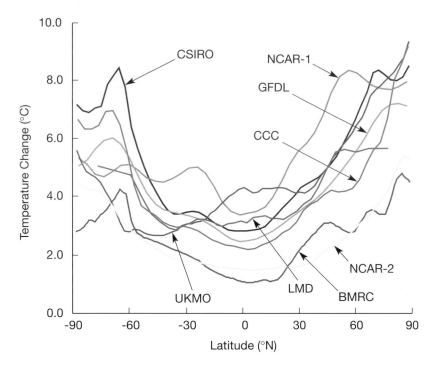

Plate 11 Variation with latitude in the zonally average steady-state surface-air temperature change for a CO_2 doubling, as simulated by various AGCMs. All results are based on data files directly provided to the author by the research institutes shown on this figure, but correspond to the following publications: CCC, Boer *et al.* (1992); CSIRO, Watterson *et al.* (1997), their Mark-2 version; GFDL, Manabe *et al.* (1991); LMD, Le Treut *et al.* (1994); NCAR-1, Washington and Meehl (1993); NCAR-2, Jeff Kiehl (personal communication, 1998); UKMO, Johns *et al.* (1997).

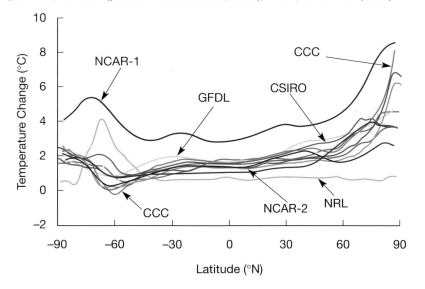

Plate 12 Variation in the zonally averaged surface-air temperature change at the time of CO_2 doubling, as obtained by a variety of coupled AOGCMs when driven by CO_2 increasing at 1% per year (compounded). These results were originally prepared for the CMIP2 project (Meehl *et al.*, 1997), and were kindly provided by Curt Covey with the permission of the participating modelling groups. Only selected cases have been labelled, to avoid clutter. See also the web site http://www-pcmdi.gov/cmip.

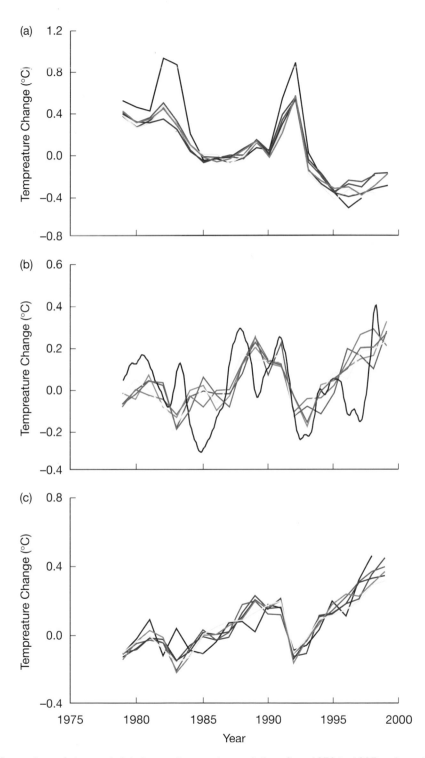

Plate 13 Comparison of observed global mean temperature variations from 1970 to 1995 and as simulated by Hansen *et al.* (1997c) for (a) the stratosphere, (b) the troposphere, and (c) the surface. For the troposphere, the MSU-2D data as corrected by Wentz and Schabel (1998) are shown.

Using uncalibrated OGCMs

The rate of uptake of anthropogenic CO_2 depends on the rate of CO_2 exchange across the air–sea boundary, and the rate of downward mixing by diffusion, convection, and oceanic circulation. The former depends on the exchange coefficient k_{as} (appearing in Eq. (8.7)), whose correct value is highly uncertain (Liss and Merlivat, 1986). However, except in regions of convection, the choice of k_{as} has little effect on the computed uptake of anthropogenic CO_2 because the rate-limiting step is downward mixing below the mixed layer. The downward mixing can be simulated using OGCMs, but the circulation field and other mixing processes in OGCMs are highly dependent on uncertain parameters (such as the subgrid-scale vertical diffusion coefficient).

The accepted procedure is to use an OGCM to simulate the concurrent absorption of anthropogenic CO_2 and of ^{14}C produced from the atmospheric testing of nuclear bombs. The oceanic uptake of ^{14}C can be characterized by two parameters: the total amount or inventory of bomb ^{14}C in the ocean, and the average depth to which the ^{14}C has penetrated. The oceanic ^{14}C inventory depends largely on k_{as}, while the ^{14}C penetration depth depends largely on the vertical mixing processes. Thus, the oceanic inventories of bomb ^{14}C and anthropogenic CO_2 are controlled by different dominant processes: k_{as} in the case of bomb ^{14}C, and downward mixing in the case of anthropogenic CO_2. Thus, the correct simulation of the bomb ^{14}C inventory does not imply that the uptake of CO_2 is correct. However, the ^{14}C penetration depth and the oceanic uptake of CO_2 are controlled by the same set of processes (vertical mixing). Thus, if the penetration depth for ^{14}C is too small, the oceanic uptake of CO_2 will also be too small. A better estimate of the real uptake of anthropogenic CO_2 can be obtained by multiplying the OGCM uptake by the ratio of the observed to model-simulated penetration of bomb ^{14}C.

The main sources of error in this procedure are: (i) the assumption that vertical mixing is the only factor controlling the uptake of CO_2, an assumption that is not entirely correct because k_{as} is important in regions of convection (as discussed above); (ii) the fact that until very recently, an estimate of the oceanic uptake of bomb ^{14}C existed only for 1973, so in scaling the model results based on data for this one year, it is assumed that the fractional error is the same at all times; and (iii) the fact that the estimated ^{14}C penetration depth is subject to an uncertainty of

$\pm 20\%$ (Jain et al., 1996). This being the case, it seems best to use the comparison of model and observed ^{14}C penetration depths only as an indication of the likely direction of the model errors.

Table 8.4 gives the uptake of anthropogenic CO_2 averaged over 1980–1989 as computed by three OGCMs and by the 2-D ocean carbon cycle model of Stocker et al. (1994). The computed uptake ranges from 1.5 to 2.1 Gt C yr^{-1}. In all cases except one (where ^{14}C penetration depth results are not published), it appears that the models are tending to underestimate the uptake of anthropogenic CO_2 (in the case of the Hamburg OGCM, the uptake is judged to be too small based on errors in the simulated temperature gradient in the upper ocean).

Using calibrated 1-D models

A second approach to estimating the uptake of anthropogenic CO_2 is to use simple models in which the mixing coefficients have been previously calibrated to replicate the observed pre-industrial distribution of various oceanic tracers, and the changes in tracer distributions since the start of anthropogenic interference. This approach yields a range of 1.8–2.3 Gt C yr^{-1} for the oceanic uptake during the 1980s (Table 8.4). Since one of the parameters used for a priori calibration of the simple models is the bomb ^{14}C penetration depth, it is not surprising that this approach gives results that are similar to those obtained using OGCMs with a posteriori calibration.

Based on direct computation of the air–sea CO_2 flux

The third approach is to directly measure the global air–sea CO_2 flux. This flux depends only on the measured mixed layer and atmospheric pCO_2 and on the air–sea exchange coefficient, k_{as} (see Eq. (8.7)); the observed mixed layer pCO_2 at any given time includes the net effect of downward mixing in the ocean up to that time, so mixing processes therefore do not need to be explicitly considered. The difficulties with this approach are (i) the mixed layer pCO_2 varies strongly with location and season (particularly at high latitudes), so that a large number of pCO_2 measurements are needed; and (ii) the magnitude of the air–sea exchange coefficient is quite uncertain, and also varies strongly with location and season. The air–sea exchange coefficient depends on wind

Table 8.4 Estimates of the rate of uptake of anthropogenic CO_2 by the oceans during the 1980s. See Table 6.1 for definitions of the acronyms GFDL and LODYC.

Method	Rate of oceanic uptake (Gt C yr^{-1})	Reference
Uncalibrated 3-D and 2-D models (1980–1989)		
Hamburg OGCM	>1.5	Orr (1993)
LODYC OGCM	2.1	Orr (1993)
GFDL OGCM	>1.9	Sarmiento *et al.* (1992)
2-D model	>2.1	Stocker *et al.* (1994)
1-D models with a priori calibration using multiple tracer data		
Box-diffusion model	2.32	Siegenthaler and Sarmiento (1993)
Upwelling–diffusion model	2.15	Siegenthaler and Joos (1992)
Upwelling–diffusion	1.84	Harvey and Huang (1999), 1980–1989
Direct estimation		
	1.1–2.4	Sarmiento and Sundquist (1992)
Analysis of ^{13}C penetration		
	2.1 ± 0.9	Heimann and Maier-Reimer (1994)
Analysis of O_2 data		
	≈2.0	Keeling *et al.* (1996)
Analysis of atmospheric CO_2 and $^{13}C/^{12}C$ variations		
	1.1	Francey *et al.* (1995), 1982–1992

speed, wave height, and the vertical stability of the atmosphere next to the surface. Estimates of the global mean value based on a combination of wind speed data and laboratory measurements (e.g., Heimann and Monfray, 1989; Boutin and Merlivat, 1991) differ by about a factor of 2 from estimates based on the oceanic inventory of bomb ^{14}C and natural ^{14}C (e.g., Sarmiento *et al.,* 1992; Harvey and Huang, 1999) – 0.035 mol m^{-2} yr^{-1} µatm^{-1} in the first case, and 0.07 mol m^{-2} yr^{-1} µatm^{-1} in the second case. Part of the discrepancy between the two estimates of k_{as} could be due to the effect of dissipating bubbles in enhancing the gaseous exchange in natural conditions (Watson, 1994). Given an estimate of the current global air–sea CO_2 flux (averaged over year-to-year fluctuations), subtraction of the pre-industrial air-sea flux gives the perturbation in the flux, which equals the oceanic uptake of anthropogenic CO_2. Tans *et al.* (1990) were the first to attempt this approach and derived an oceanic uptake of only 0.2–0.8 Gt C yr^{-1}, using exchange coefficients having a global mean value of 0.067 mol m^{-2} yr^{-1} µatm^{-1}. However, the discovery of a number of errors in their method (including the assumption that the pre-industrial flux was zero) led to a revised estimate of 1.1–2.4 Gt C yr^{-1} (Sarmiento and Sundquist, 1992). The revised estimate overlaps the range

deduced using ocean carbon cycle models. Note, however, that what agreement there is depends on using the same high k_{as} as used in 1-D and 3-D ocean models.

Based on oceanic ^{13}C data

The release of ^{13}C-depleted carbon from the oxidation of fossil fuels and biomass has caused the atmospheric $\delta^{13}C$ ratio to decrease during the past two centuries. The penetration of this anomaly into the ocean in principle provides another constraint on oceanic uptake of CO_2. Three approaches have been used to estimate the oceanic uptake of CO_2 based on either (i) the change in the oceanic inventory of ^{13}C (Quay *et al.* 1992), or (ii) the disequilibrium between atmospheric $\delta^{13}C$ and ocean mixed-layer $\delta^{13}C$ (Tans *et al.* 1993), or (iii) the constraint imposed by the expectation that the rate of penetration of CO_2 and $^{13}CO_2$ into the ocean should be the same (Heimann and Maier-Reimer, 1994). All three methods begin with global mass balance equations for total CO_2 and for $^{13}CO_2$, but they partition the global pools differently and make use of different, partly substitutable observed data. If the total set of input data used by all three methods is internally consistent, then all three methods should give identical results. Since the *a priori* values of the input data contain errors and are therefore not internally consistent, the three methods as initially applied give widely divergent results. However, adjusting the input data within the limits of their uncertainty yields an input dataset such that all three methods give an identical oceanic CO_2 uptake of $2.1 \pm 0.9 \, \text{Gt C yr}^{-1}$ for the time period 1970 to 1990 (Heimann and Maier-Reimer, 1994). This range agrees with that obtained by other methods, cited above.

Using atmospheric O_2 data

Absorption of anthropogenic CO_2 by the oceans has no effect on the atmospheric supply of O_2, while absorption by the terrestrial biosphere (through enhanced photosynthesis) produces O_2. If the amount of atmospheric O_2 falls faster (slower) than it is consumed by the burning of fossil fuels, there must be a net terrestrial source (sink) of CO_2 to (from) the atmosphere. Measurement techniques have recently become precise enough to permit determination of trends in the amount of O_2 in the atmosphere, and preliminary results indicate that the northern land biota and global oceans each absorbed about $2.0 \, \text{Gt C yr}^{-1}$ during 1991–1994. The O_2 data also imply that the tropical land biota were neither a strong source nor a strong net sink of CO_2 (Keeling *et al.*, 1996). Given estimated emissions from tropical deforestation of $1.6 \pm 1.0 \, \text{Gt C yr}^{-1}$ (Schimel *et al.*, 1996), this implies a comparable uptake by undisturbed tropical forests due to the effects of CO_2 fertilization. These results are consistent with the results based on the inversion of atmospheric $\delta^{13}C$ data, previously shown in Fig. 8.4, which indicate that the tropical biosphere is not a large net source or sink of CO_2. The O_2 concentration can also be measured in air trapped in firn in polar ice caps, and concurrent measurements of the O_2 and CO_2 concentration imply that the terrestrial biosphere was neither a source nor a sink of CO_2 over the period 1977 to 1985 (Battle *et al.*, 1996).

Using concurrent global mean atmospheric CO_2 and $\delta^{13}C$ data

As discussed in Section 8.4, Francey *et al.* (1995) and Keeling *et al.* (1995) used data on the concurrent rate of change in atmospheric CO_2 concentration and $^{13}C/^{12}C$ ratio to decompose interannual variations in the total CO_2 sink into variations in the terrestrial biosphere and oceanic components of the sink. Only Francey *et al.* (1995) ventured an estimate of the absolute oceanic carbon uptake (in addition to estimating interannual variations), and obtained a mean oceanic sink for 1982–1992 of $1.1 \, \text{Gt C yr}^{-1}$.

Synthesis

Based on the evidence cited above, it can be concluded that the average rate of absorption of anthropogenic CO_2 by the oceans during the 1980s was $2.0 \pm 1.0 \, \text{Gt C yr}^{-1}$.

8.10   Atmospheric methane

Compared to CO_2, the cycle of atmospheric methane is simple. The three main processes that need to be considered are (i) the effect of climatic change on natural methane sources (primarily wetlands) and on the rate of removal of CH_4 from the atmosphere; (ii) the potential release of methane from a water–methane compound known as clathrate, which is found in terrestrial permafrost at depths of several hundred metres and in continental slope sediments worldwide; and (iii) the feedback between the methane concentration and its own lifespan. The primary removal process for methane is oxidation with OH to form CO_2, so the decay of atmospheric methane is an additional source of atmospheric CO_2. Soils appear to absorb about 3–10% of total emissions (Prather *et al.*, 1995).

Higher temperatures could increase CH_4 emissions from wetlands both by enhancing metabolic rates (Whalen and Reeburgh, 1990) and by enhancing primary productivity (Whiting and Chanton, 1993), but a drop in the water table could lead to decreases in CH_4 production. The computation of methane release from destabilization of methane clathrates requires modelling the downward penetration of heat into oceanic sediments and in permafrost, and doing so in a fairly disaggregated manner (such as on a $1° \times 1°$ latitude–longitude grid). Harvey and Huang (1995) carried out such an exercise but concluded that the potential release of methane from clathrates due to a warmer climate is not likely to be large.

The rate of removal of methane by oxidation with OH depends on the concentration of OH and on the chemical reaction rate, which depends on temperature. The concentration of OH depends on the total emissions of CH_4 (and CO); as emissions increase, the OH concentration decreases and the lifespan of CH_4 increases (this is the first indirect heating effect of CH_4 emissions that was described in Section 7.5). However, as the climate warms, the water vapour content in the atmosphere increases, which acts to increase the concentration of OH and thereby reduce the lifespan of CH_4. The greater chemical reaction rate as temperatures increase will also act to reduce the lifespan of CH_4. As noted in Section 4.2, hypothetical temperature-induced changes in CH_4 emissions from wetlands, and temperature effects on the rate of oxidation, can explain the year-to-year variations in the rate of growth in atmospheric CH_4 that were observed during the 1980s and 1990s.

Once the natural methane flux has been estimated, and anthropogenic fluxes added, the atmospheric methane concentration can be modelled as

$$\frac{dC}{dt} = F - \frac{C}{\tau} \tag{8.13}$$

where F is the total methane source flux, and τ is the atmospheric lifespan of methane. It can be updated periodically based on changes in the concentration of methane and in temperature, using the results of 3-D atmospheric chemistry models (as discussed in Section 6.3). However, calculations of future atmospheric CH_4 concentration have generally considered the feedback between CH_4 concentration and τ, but not between climate and τ.

Harvey and Huang (1995) tested the impact of a hypothetical feedback between climate and the natural methane flux, assuming that the natural flux increases according to Eq. (8.1) using $Q_{10} = 2.0$. They found that a feedback of this magnitude increased the global radiative forcing and hence temperature response by only 5%. To the extent that higher methane fluxes from wetlands are caused by higher primary productivity rather than warmer temperatures, there would be an associated removal of CO_2 from the atmosphere, so feedbacks involving CO_2 and CH_4 would be partially offsetting. Similarly, conditions that would maximize CO_2 emissions from high-latitude peatlands as climate warms (namely, a lower water table) would tend to minimize CH_4 emissions (Gorham, 1991), so that large positive feedbacks involving both CO_2 and CH_4 would not occur simultaneously.

8.11   Other greenhouse gases and aerosols

The atmosphere can be treated as a well-mixed reservoir for other greenhouse gases, and the rate of removal is linearly proportional to the concentration of the gas in the atmosphere. Thus, concentrations of these gases are also governed by Eq. (8.13), with no feedback between the concentration of the given gas and its own lifespan. Using Eq. (8.13), the concentration at some future time $t_1 + \Delta t$ depends on the concentration at time t_1 and on the current rate of emission. Future concentrations thus depend on the *history* of previous emissions.

In the case of aerosols, the atmospheric lifespan is so short (days) that the current concentration can be regarded as depending only on the current emissions, and not in any way on past emissions. As noted in Section 7.4, most estimates of the total worldwide anthropogenic emissions of SO_2 generally fall within ±5% of the central estimate of about 70 Tg S for 1985. As noted in Section 2.5, about 80% of the emitted SO_2 is converted (oxidized) to SO_4^{2-} under present-day conditions, with about 60% of the conversion done by reaction with H_2O_2 (hydrogen peroxide) and the rest by reaction with OH and O_3. As SO_2 emissions increase, there is proportionately less H_2O_2 available for conversion of SO_2 to SO_4^{2-}, so the fractional increase in the atmospheric SO_4^{2-} loading will be smaller than the fractional increase in SO_2 emissions. For a 50% *decrease* in SO_2 emission, the sulphate loading decreases by 40–43% over the dense source regions (Misra *et al.*, 1989). On a larger scale, the H_2O_2 limitation on the conversion from SO_2 to SO_4^{2-} is presumably smaller (since SO_2 will be less concentrated than over the source regions), so the variation of sulphate loading with SO_2 emissions is presumably even closer to linear. However, if both SO_x (primarily SO_2) and NO_x emissions are reduced by 50%, Misra *et al.* (1989) found the decrease in sulphate loading to be less than 30% in the source regions. This is because, with less NO_x, there is more H_2O_2 available to convert SO_2 to sulphate. Once SO_2 is converted into sulphate aerosol, it is distributed by winds and turbulent mixing, and is removed by rainout and dry deposition. As noted in Section 7.4, there is a wide variation in the sulphate loading as computed by different models, even when given the same emissions.

8.12 Summary

This chapter has highlighted the complexity of processes involved in the response of the carbon cycle to anthropogenic emissions of CO_2, and the differences between the responses of atmospheric CO_2 and other greenhouse gases to anthropogenic emissions. Carbon dioxide differs from other GHGs in that it continuously cycles between a number of "reservoirs" or temporary storage depots (the atmosphere, land plants, soils, ocean water, and ocean sediments). All of the other GHGs (except water vapour, which is not directly influenced by human emissions) are removed by either chemical or photochemical reactions within the atmosphere. For this reason, the rate of removal of non-CO_2 GHGs can be characterized by a single time constant (which, in the case of CH_4, can change over time), and the concentration following a pulse emission will eventually decay to zero. The rate of removal of CO_2, in contrast, cannot be characterized by a single time constant. Rather, different portions of a pulse emission can be thought of as being removed from the atmosphere at different rates, with 10–15% of the emitted carbon requiring on the order of 10,000 years to be removed, and another 10–15% or so requiring on the order of 200,000 years to be removed.

Feedbacks between climate and the carbon cycle have the potential to alter the rate of removal of anthropogenic CO_2 from the atmosphere during the next few hundred to 1000 years. Current knowledge indicates that the potential change in the ocean CO_2 sink due to circulation changes or warmer temperatures is likely to be small, and that the effect of potential methane–wetland and methane–clathrate feedbacks is also small. However, the present terrestrial carbon sink could significantly weaken with continuing fossil fuel emissions, and there is a risk of a transient dieback of forests as climatic zones shift, giving a large CO_2 flux to the atmosphere for a century or longer. There will be a tendency for CO_2 and CH_4 feedbacks involving thawing of permafrost to be negatively correlated; that is, if one is comparatively strong, the other will be comparatively weak, and vice versa. Changes in the strength in ocean overturning also have partly compensating effects on atmospheric CO_2, by altering the rate of upwelling of nutrients and of CO_2-rich deep water. This minimizes the effect of such changes. Terrestrial nutrient feedbacks appear to limit changes in carbon storage due to both temperate enhancement of respiration and CO_2 enhancement of photosynthesis. However, these conclusions must be tempered by the knowledge that the climate system is being moved by human intervention into a state which is unprecedented in recent geological history, and current models could be omitting key processes which, if included, might give a different CO_2 buildup in response to anthropogenic emissions than expected. Scenarios of the expected CO_2 buildup are presented in Chapter 13, but first, we continue our process-based discussion of the climate system response to human emissions of GHGs.

CHAPTER 9

Climate sensitivity

The term "climate sensitivity" was defined in Section 3.6 as the ratio of the steady-state change in the global and annual mean surface air temperature to the global and annual mean radiative forcing. Only the fast feedback processes, involving atmospheric water vapour and temperature structure, clouds, and seasonal ice and snow, are accounted for in this definition. As discussed in Section 3.7, the climate sensitivity is believed to be approximately constant for radiative forcing perturbations of up to a few $\mathrm{W\,m^{-2}}$, and to be largely independent of the specific combination of forcings that add up to a given total forcing when changes in the concentration of different well-mixed greenhouse gases, in non-absorbing aerosols, and in solar luminosity are combined. Thus, if the climate sensitivity is known, the steady-state global mean temperature response can be scaled up or down as the magnitude of the total forcing changes. A commonly computed indicator of climate sensitivity is the global mean temperature response for a doubling in the atmospheric concentration of CO_2.

As discussed in Section 3.3, climate feedbacks affect the climate sensitivity by either adding to or subtracting from the initial radiative perturbation; positive feedbacks add to the initial perturbation, thereby provoking further temperature change, while negative feedbacks subtract from the initial perturbation. In principle, the climate sensitivity can be estimated by measuring the strengths of the major individual feedback processes from observations, or by attempting to simulate these feedback processes in 3-D AGCMs. The problem with the observational approach is that it is rarely possible to directly observe the feedback processes that determine climate sensitivity. The problem with computer models is that they are simplifications of nature, and not nature itself.

There are two other ways in which the climate sensitivity could be estimated: one is based on data from past climates, and the other is based on the observed temperature changes during the past 140 years. In order to use paleoclimatic data, we obviously need to be able to accurately determine how past temperatures differed from the present, but we also have to be able to identify the factors responsible for past climate changes, so that the associated global mean radiative forcing can be computed. The climate sensitivity is then given by the ratio of the computed change in global mean temperature to the computed global mean radiative forcing. In order to use historical temperature changes, we have to be able to identify all of the important radiative forcings, but we also have to take into account the effect of the oceans in delaying the surface warming.

In this chapter, these four ways of assessing climate sensitivity are examined. We begin with an analysis of climate feedback processes as simulated by AGCMs, as this affords a comprehensive overview of the processes involved in determining climate sensitivity. We then turn to observational evidence for the subset of feedback processes that can be observed directly, namely, those involving water vapour and its distribution within the atmosphere. Fortunately, these feedback processes turn out to be the most important for climatic sensitivity, so the limited observational constraints are quite useful. We conclude this chapter with a discussion of inferences based on paleoclimatic and historical temperature changes.

9.1 Assessments based on simulations with 3-D AGCMs

As explained in Section 3.3, the Earth's surface and atmospheric temperatures tend to adjust themselves so that the absorbed solar energy is balanced by the heat energy (infrared radiation) emitted to space. The increased heat trapping due to an increase of GHG concentrations requires that the mean temperature at which the Earth radiates heat to space increases so that a balance between absorbed solar energy and emitted heat energy can be restored. The required temperature change, ΔT, is given by

$$\Delta T = \frac{\Delta R}{\lambda} \tag{9.1}$$

where ΔR is the radiative forcing and λ, the radiative damping, is given by

$$\lambda = \frac{\mathrm{d}F}{\mathrm{d}T} - \frac{\mathrm{d}Q}{\mathrm{d}T} \tag{9.2}$$

where F is the global mean emission of infrared radiation to space and Q is the global mean absorption of solar radiation. The more rapidly the net emission of radiation to space increases with increasing surface temperature (that is, the larger the radiative damping), the less the surface temperature must increase to restore balance, and so the smaller the climate sensitivity. For a blackbody emitting the same amount of infrared radiation to space as the Earth, the radiative damping parameter is $3.76\,\mathrm{W\,m^{-2}\,K^{-1}}$, so the climatic response to a CO_2 doubling would be about $1.0\,\mathrm{K}$ (Section 3.4). This response would be modified by various climate feedbacks operating in the real Earth.

The feedbacks that affect the climate sensitivity were identified in Section 3.5. They are changes in the amount and vertical distribution of water vapour in the atmosphere; changes in the temperature lapse rate; changes in the amount, horizontal and vertical distribution, and radiative properties of clouds; and changes in the areal extent of seasonal snow and sea ice. The importance of individual feedbacks can be determined by calculating the contribution of each feedback to $\mathrm{d}F/\mathrm{d}T$ and $\mathrm{d}Q/\mathrm{d}T$ (the $\partial F/\partial I_i(\mathrm{d}I_i/\mathrm{d}T)$ and $\partial Q/\partial I_i(\mathrm{d}I_i/\mathrm{d}T)$ of Section 3.3). The only models with the potential to reliably simulate the individual feedback processes are three-dimensional AGCMs with high spatial resolution and a diurnal cycle. However, the complexity of the processes involved in these feedbacks and the need for rather coarse resolution in AGCMs means that results obtained from AGCMs are still rather uncertain. Some of the complexities are outlined below, followed by an overview of key results obtained from AGCMs.

Processes affecting climate sensitivity and their representation in AGCMs

The amount and distribution of water vapour depend on evaporation of surface water, horizontal and vertical transport of moisture by large-scale winds and small-scale eddies, upward movement by dry convection and within convective clouds, detrainment of moisture from cloud tops, dissipation of non-precipitating clouds, partial re-evaporation of precipitation, and the downward movement of dry air between rising convective columns at low latitudes. At present, only simplified representations of these processes are included in AGCMs, owing mainly to the relatively coarse resolution of these models (from many tens to several hundred kilometres) compared with the scales of the most important processes (micrometres to kilometres).

As explained in Section 3.5, the radiative impact of a given change in cloud properties or in cloud amount depends on the location of the cloud and on the time of year and day when the changes occur. The initiation of cloud depends on large-scale motions, coupling to the surface boundary layer (which feeds moisture into the overlying air), and the spatial distribution and intensity of surface–air fluxes of heat and moisture. Positive or negative feedback loops can be created between cloudiness and the surface energy budget in some regions (e.g., Miller and Del Genio, 1994). Once clouds have been initiated in a model, a number of cloud properties and processes need to be simulated, including:

- the liquid and ice water content of clouds and their horizontal variability within AGCM grid cells;
- the mean water droplet or ice crystal sizes and the distribution of sizes around the mean size;
- the optical properties of clouds for a given mean particle size and distribution – something that is complicated by the potential presence of impurities (such as soot) and is particularly difficult to predict for ice crystals because of their irregular shape;
- the fall velocity and hence lifespan of water droplets and ice crystals;
- the cloud geometry (thickness, patchiness);
- the detrainment of cloud moisture and mixing with surrounding air, which is especially important to the formation of cirrus anvils from cumulus clouds;
- the formation of precipitation and the partial re-evaporation of falling precipitation; and

- convective-scale downdraughts and their interaction with the planetary boundary layer.

Most of these processes and properties are now incorporated in at least some AGCMs, on either a routine or an experimental basis (see the reviews by Del Genio (1993) and Dickinson *et al.* (1996), and references therein). However, these processes are of necessity treated in a highly simplified and in some cases a rather *ad hoc* manner because most of them occur at scales well below the resolution of global models. This introduces the potential for substantial error in the simulated changes in cloud properties, locations, and amounts.

The correct simulation of changes in land snow-cover and their net radiative effect requires models with high spatial and temporal resolution on account of potential interactions with clouds and because of the spatial heterogeneity of the land surface. In the case of sea ice, high resolution is desirable because of the important role of ice motions to the distribution and thickness of sea ice, but there are special problems in applying ice stress–strain relationships observed in the laboratory or at the scale of an ice floe to that of even the smallest AGCM or OGCM grid cell (Overland *et al.*, 1995).

Climate sensitivity in the absence of snow, ice, and cloud feedbacks

The combination of changes in the amount of water vapour, in the vertical distribution of water vapour, and in the lapse rate in cloud-free regions is referred to as the "clear-sky" feedback. In cloud-covered regions, additional feedbacks arise from changes in the height, thickness, water and ice content, and microphysical properties of clouds, but changes in the amount and distribution of water vapour below clouds with an emissivity of 1.0 will have no feedback effect (since the only radiation leaving the atmosphere in this case is that originating from the cloud tops or higher). As discussed in Section 3.3, the aggregate effect of the feedback processes can be quantified as $dF/dT - dQ/dT$. Cess *et al.* (1990) estimated the infrared and solar feedback derivatives for clear skies, cloudy skies, and in the global mean for 19 different AGCMs (the global mean feedback derivatives are not the simple area average of the clear- and cloudy-sky feedback derivatives, as the global mean also includes the effect of changes in the amount of cloud and its

geographical distribution). This was done by imposing a uniform 4 K change in sea surface temperature (SST), allowing the AGCM and land surface modules to equilibrate to the new SSTs, and determining the changes in solar and infrared radiative fluxes at the top of the atmosphere. This is not a perfect analogue for a climatic change experiment, in which an external forcing is applied and the atmospheric, land surface, and sea surface temperatures are allowed to adjust, because in the latter case the change in SST would not be uniform. As a result, the changes in atmospheric circulation and hence the changes in the distribution of water vapour and clouds – and the resulting feedbacks – would be different from those of the idealized experiments performed by Cess *et al.* (1990). Furthermore, the experiments were run with the solar radiation fixed with the geographical distribution appropriate for July, referred to as a "perpetual July" experiment. This greatly reduces the computational expense of the experiments, and also largely eliminates feedbacks arising from changes in snowcover (sea-ice feedback having already been eliminated because a fixed sea-ice distribution was specified).

In spite of these shortcomings, the experimental design of Cess *et al.* (1990) provides considerable insight into the separate infrared and solar feedback processes operating in AGCMs, and in the differences between different AGCMs. As discussed in Section 3.3, the infrared feedback derivative, dF/dT, can be decomposed as

$$\frac{dF}{dT} = \frac{\partial F}{\partial T} + \Sigma_i \frac{\partial F}{\partial I_i} \frac{dI_i}{dT} \tag{9.3}$$

where $\partial F/\partial T$ is given by the derivative of the Stefan–Boltzmann Law (Eq. (3.1)) evaluated at the effective radiating temperature, T_e. Cess *et al.* (1990) present estimates of the mean clear-sky values of dF/dT and F for each of the 19 models, from which T_e and hence $\partial F/\partial T$ can be computed. The infrared effect of all other feedback processes, $\Sigma_i(\partial F/\partial I_i)$ (dI_i/dT), can then be computed as a residual using Eq. (9.3). The estimates of dF/dT and dQ/dT (which is also given by Cess *et al.*, 1990) can be combined to give λ using Eq. (9.2), and the steady-state ΔT resulting from clear-sky feedback processes alone can be estimated using Eq. (9.1).

The results of such an analysis are presented in Table 9.1 for four AGCMs that represent extreme values either for the overall clear-sky feedback or for the separate infrared and solar components. As one

Table 9.1 Contribution to the clear-sky radiative damping parameter and resulting climate sensitivity (excluding surface albedo and cloud feedbacks) for four different AGCMs, based on data presented in Cess *et al.* (1990). Column 2: Global mean clear-sky infrared emission to space (F, W m^{-2}); Column 3: Global mean effective radiating temperature (K) derived from F; Column 4: $\partial F/\partial T$, computed from Eq. (3.10); Column 5: the contribution of all other feedback processes to infrared radiative damping; Column 6: the total infrared contribution to radiative damping; Column 7: the solar contribution to radiative damping; Column 8: the radiative damping parameter (given by the sum of columns 4, 5, and 7); Column 9: the global mean temperature response (°C) for a CO_2 doubling, assuming an adjusted forcing of 3.75 W m^{-2} and the radiative damping given in Column 8. The units in Columns 4–8 are W m^{-2} K^{-1}.

1 Model	2 F	3 T_e	4 $\partial F/\partial T$	5 $\Sigma(\partial F/\partial I_i)\mathrm{d}I_i/\mathrm{d}T$	6 $\mathrm{d}F/\mathrm{d}T$	7 $-\mathrm{d}Q/\mathrm{d}T$	8 λ	9 ΔT
ECMWF	273	263.5	4.14	−1.68	2.46	−0.74	1.75	2.14
OSU/IAP	277	264.4	4.19	−1.52	2.67	−0.18	2.40	1.56
ECHAM	250	257.8	3.88	−1.28	2.60	−0.47	2.13	1.76
GISS	253	258.5	3.91	−1.98	1.93	−0.00	1.92	1.95

See Table 6.1 for definitions of the acronyms used here.

might expect, there is very little variation in $\partial F/\partial T$, and the variation that does occur is due to the slightly different initial atmospheric temperatures among the four models. This term represents the negative feedback due to the direct coupling between temperature and the emission of infrared radiation. The net effect on infrared emission of all other feedback processes, $\Sigma_i(\partial F/\partial I_i)(\mathrm{d}I_i/\mathrm{d}T)$, is a positive feedback; that is, as temperature increases it gets harder to emit radiation to space ($\Sigma_i(\partial F/\partial I_i)(\mathrm{d}I_i/\mathrm{d}T) < 0$), and the radiative damping is reduced. These processes include an increase in the total amount of water vapour in the atmosphere, changes in the vertical distribution of water vapour, and changes in the atmospheric lapse rate. The clear-sky solar feedback is also positive, and arises in part from an increase in the absorption of near-infrared radiation by water vapour. The last column of Table 9.1 gives the steady-state change in global mean temperature that would occur if these feedback processes were applicable to cloudy skies as well as to clear skies. The expected temperature change ranges from 1.56°C to 2.14°C.

Zhang *et al.* (1994) decomposed the infrared feedback into separate contributions from changes in temperature ($\partial F/\partial T$), atmospheric water vapour, and atmospheric lapse rate for three different versions of the NCAR[1] AGCM in which convective clouds were parameterized differently. The results of their analysis are summarized in Table 9.2. The water vapour feedback is strongly positive and the lapse rate feedback is modestly negative, with the strongest lapse rate feedback occurring in association with the strongest water vapour feedback.[2] This implies that the model version with the largest increase in upper tropospheric moisture also has the largest increase in upper tropospheric temperature (that is, the largest decrease in the lapse rate). The negative global mean lapse rate feedback found by Zhang *et al.* (1994) is probably the result of a positive lapse rate feedback at high latitudes (i.e., increasing lapse rate) and a negative lapse rate feedback at low latitudes (i.e., decreasing lapse rate), as this is what Ramanathan (1977) found in the GFDL AGCM.

A critical question concerning the clear-sky feedback is the extent to which the water vapour content of the upper troposphere increases. As explained in Section 3.1, water vapour in the upper troposphere is particularly effective as a greenhouse gas. At low latitudes the increase in water vapour in the upper troposphere could depend on the way in which convective clouds are treated. This issue is discussed more extensively in Section 9.2, but here we examine the behaviour of four different AGCMs. Del Genio *et al.* (1991) compared two different cumulus convection schemes in the GISS AGCM, both of which account for subsidence-induced drying between cumulus towers, and found that the entire model troposphere moistens as the climate warms. Indeed, for

---

[1] This and other acronyms denoting different AGCMs are defined in Table 6.1.

[2] Recall, from Section 3.5, that the lapse rate feedback is negative if the lapse rate decreases as the climate warms, so that upper tropospheric temperatures increase faster than for constant lapse rate. This causes a faster increase in the emission of infared radiation to space, so the surface temperature does not need to warm as much in order to restore radiative balance.

a 4 K globally uniform ocean surface warming, relative humidity (RH) is constant near the surface but increases in the global mean by 5% of saturation at the 200 mb level. The UKMO GCM also accounts for the drying effect of subsidence between cumulus clouds and also produces moistening throughout the troposphere as the climate warms, but not by enough to maintain constant RH; instead, RH decreases by an average of about 2–3% of saturation between 550–250 mb (Mitchell and Ingram, 1992). In the NCAR AGCM, version 2 (CCM2), RH decreases throughout the tropical troposphere in response to a 4 K uniform ocean surface warming (by up to 20% of the initial RH), but increases in the upper troposphere outside the tropics (Zhang *et al.*, 1994; their figure 16). Qualitatively similar behaviour occurs in the GFDL AGCM in response to a quadrupling of atmospheric CO_2, at least over continents (Wetherald and Manabe, 1995; their figure 6). Finally, Colman *et al.* (1997) report decreases in upper tropospheric RH in the BMRC AGCM in regions where strong increases in convection occur.

The reason why the GISS model gives an increase in upper tropospheric RH as climate warms, while the other models give a decrease in RH, appears to be that greater sublimation of ice crystals that are detrained from cumulus anvils occurs in the GISS model (Hansen, personal communication, 1998). In a warmer climate, more water is pumped into the upper tropical troposphere by convection. In all models, a new steady-state balance will be achieved in which essentially all of the extra upward moisture transfer will eventually fall out as precipitation (some of the extra moisture might diffuse downward). In the GISS model, the new steady state requires a greater increase in upper tropospheric moisture than

in other models. In the other models, the required increase in upper tropospheric moisture is smaller because more of the anvil water immediately falls as precipitation. Even where RH decreases in these models, however, absolute humidity increases.

As noted above, Zhang *et al.* (1994) found that versions of the NCAR model with a larger increase in upper tropospheric water vapour also had a larger increase in upper tropospheric temperature. This is reasonable, since all of the extra moisture that is pumped into the upper troposphere as the climate warms will condense and release latent heat, but it also leads to a higher steady-state water vapour content. Thus, as moisture pumping increases, both upper tropospheric temperature and absolute humidity increase, and these changes have opposing effects on the infrared emission to space. Thus, the overall variation in λ is smaller than if these changes were not positively correlated (Table 9.2).

There is another dimension to this issue, however, involving the *precipitation efficiency*. This refers to the fraction of upward moisture transfer that immediately falls as precipitation, with the rest detraining from convective columns and moistening the upper troposphere. If, as upward moisture pumping increases, a smaller fraction goes into moistening the upper troposphere and more into immediate precipitation, then the net effect will be a less positive change in upper tropospheric humidity. If the precipitation efficiency increases enough for upper tropospheric moisture to decrease in spite of increased moisture pumping, then the positive water vapour feedback due to the increase in total atmospheric water vapour would be considerably weaker. The primary effect of increased moisture pumping would then be the negative lapse rate feedback. Current

Table 9.2 Clear-sky radiative damping (λ) for three versions of the NCAR AGCM in which convective clouds are parameterized in different ways. λ is not exactly equal to the sum of the individual components because the different feedback processes are not entirely independent of one another. The last column gives the global mean surface air temperature response (°C) for doubling of atmospheric CO_2, given the computed radiative damping and assuming an adjusted forcing of 3.75 W m^{-2}. From Zhang *et al.* (1994).

| Model | $\partial F/\partial T$ | $(\partial F/\partial I_i)dI_i/dT$ | | $-(dQ/dT)$ | Sum | λ | ΔT |
		Water vapour	Lapse rate				
CCM2	4.20	−2.18	0.56	−0.19	2.39	2.34	1.60
MAA	4.21	−1.93	0.26	−0.17	2.36	2.30	1.63
NPBL	4.23	−1.95	0.26	−0.17	2.37	2.31	1.62

CCM2: Original model, using a mass flux convection scheme coupled to an explicit atmospheric boundary layer.
MAA: The mass flux scheme is replaced with a moist adiabatic adjustment scheme.
NPBL: Same as MAA except that the explicit boundary layer is removed.

AGCMs make rather arbitrary moistening assumptions, and some do not explicitly consider precipitation efficiency. The major convective schemes used in AGCMs and the related assumptions concerning precipitation efficiency are outlined in Box 9.1. Calculations by Sun and Lindzen (1993; their figure 20) indicate that the time required for raindrops to form decreases as temperature increases, which implies that the precipitation efficiency should increase.

To sum up: the effect on climate sensitivity of changes in upper tropospheric water vapour and temperature, when both increase as overall water vapour pumping increases, is minimal because such changes have opposing effects. This diminishes the importance of changes in upper tropospheric water vapour due to changes in moisture pumping. However, if the precipitation efficiency increases as moisture pumping increases, there will be less cancellation of the negative lapse rate feedback by the positive water vapour feedback. Thus, the near-cancellation between the effects of changes in lapse rate and of water vapour feedback as moisture pumping increases, found by Zhang *et al.* (1994), might not be a general result. It is also clear that the "clear sky" radiative feedbacks in AGCMs depend critically on the way in which cloud processes are treated, and are thus coupled to the cloud processes and associated radiative feedbacks – a subject we turn to next.

Role of cloud feedbacks

Another important source of uncertainty concerning climate sensitivity is the role of cloud feedbacks. Table 9.3 lists the feedback effect of various cloud changes, if they were to occur in association with a warming of the climate. Clouds reflect solar radiation, which has a cooling effect. They also absorb infrared radiation emitted from the surface and re-emit their own radiation, but the amount re-emitted is smaller than the amount absorbed because the tops of clouds are colder than the underlying surface. The higher the cloud, the colder the cloud top and the greater the reduction in net infrared emission to space. High clouds thus tend to have a net warming effect because the reduction in infrared emission to space is greater than the extra reflection of solar radiation. Low clouds, in contrast, have a net cooling effect. Thus, an increase in the amount of high cloud as the climate warms usually serves as a positive feedback, while an increase in the amount of low cloud

serves as a negative feedback. Similarly, an increase in the height of existing clouds, because this makes them colder and reduces the emission of radiation to space, has a warming effect on climate. Stratus clouds have an emissivity of 1.0 and so absorb all of the radiation emitted from the surface. An increase in the water content of these clouds increases the albedo but has no effect on infrared emission to space, so there is a net cooling effect. Cirrus clouds, on the other hand, have an emissivity less than 1.0. An increase in water content increases the albedo and the emissivity, so both the cooling and heating effects of the clouds increase. Thus, an increase in the water content of cirrus clouds can have a net warming or a net cooling effect. Ice crystals tend to be much larger than water droplets and so are less reflective (for the same reason that larger droplets are less reflective than smaller droplets, as explained in Box 2.4), but they also have greater fall velocities. Thus, a transition from ice crystal to water droplet clouds makes clouds more reflective (a cooling effect) but presumably decreases the lifespan of clouds which, in most cases, will be a warming effect.

The directions of the cloud changes listed in Table 9.3 are given for illustrative purposes only. In some cases the direction of change that would occur in reality is unknown, as illustrated in the following examples. First, Senior and Mitchell (1993, their fig. 18) show that the direction and magnitude of the changes in cloud fraction depend on the way in which cloud amounts are parameterized. Second, we are reasonably confident that convective clouds will rise to greater heights in a warmer climate, but the extent to which this occurs depends on the choice of convective scheme (Cunnington and Mitchell, 1990). Third, Sinha and Shine (1994) show that implementation of a feedback between temperature and ice water content in cirrus clouds changes the sign of the feedback due to increasing cloud height from positive to negative. For fixed cloud height and for clouds cold enough to remain as ice crystal clouds, warmer temperatures lead to a greater ice water content, which serves as a positive feedback. However, if cloud height increases, this feedback is greatly weakened because there is a smaller increase in cloud ice content due to the fact that the cloud hardly warms at all. Fourth, it was widely believed that the optical thickness of liquid water clouds would increase as climate warms, owing to an increase in cloud droplet size (Somerville and Remer, 1984). Now it appears that concurrent changes in cloud thickness – which

Box 9.1 Parameterization of cumulus convection and its effects in AGCMs

Cumulus convection in nature involves rising motions in clusters of columns, each on the order of a few hundred metres or so in horizontal extent. Cumulus clouds, of up to a kilometre in horizontal extent, may consist of several such updraughts. These motions transfer heat, moisture, and mass vertically within the atmosphere, thereby significantly affecting the large-scale atmospheric properties. Since the spatial scale of cumulus convection is much smaller than the grid size of global-scale AGCMs (usually at least 200 km × 200 km), it is necessary to use a short-cut method or *parameterization* to account for the effects of cumulus convection in models. The parameterizations in use today fall into two categories, as described below.

Moist convective adjustment (MCA)

When the vertical temperature gradient (the lapse rate) is unstable, temperatures are adjusted so as to re-establish a stable profile. This involves an implicit mixing that is also applied to water vapour. Certain layers invariably become saturated, while other layers may have already been saturated before mixing. In either case, the excess water is assumed to fall out as rain. In a variant of this method, a grid cell is not required to be saturated for precipitation to occur, and mixing is assumed to involve only a portion of the grid cell. At the end of each time step, any remaining water in the convective column, and the excess heat, are assumed to be mixed laterally with the rest of the grid cell. There is no explicit consideration of compensating subsidence between convective columns. This scheme was first used by Manabe *et al.* (1965).

Penetrative convection (PC)

In the moist convective adjustment scheme described above, moisture will not be transferred higher than the top of the unstable region. In reality, rising air parcels will be able to penetrate to some extent into the overlying stable region, until the *equivalent potential temperature* of the surrounding air equals that of the parcel. As a result, moisture will be pumped to a greater height than using MCA. There are many variants of PC. It can be implemented assuming that there is a single rising parcel in a given horizontal grid cell that rises without dilution due to entrainment of surrounding air, or that rises with entrainment. If

there is entrainment, a fixed fractional or absolute rate of entrainment can be specified. Alternatively, an ensemble of rising parcels can be assumed to exist within each horizontal grid cell, with different members of the ensemble rising to different heights owing to the fact that they have different initial temperatures (Hansen *et al.*, 1983) or are assumed to experience different fractional rates of entrainment as they rise (Arakawa and Schubert, 1974). A PC parameterization can be implemented in an AGCM such that rising plumes can originate only from the planetary boundary layer (PBL), or can originate anywhere within the atmosphere (Ding and Randall, 1998). The initial mass flux in the rising plumes (prior to entrainment) must be somehow specified, based on either the mass or water vapour flux convergence within the PBL, and perhaps taking into account the local rate of evaporation. Entrainment as the parcels rise can be tied to the large-scale convergence above the PBL (Tiedtke, 1989). Localized downdraughts next to rising columns, and compensating subsidence between convective columns, may or may not be explicitly considered. Detrainment of air at the top of the rising column can be treated in a variety of different ways as well, including allowing for some sinking motion of formerly rising parcels before detrainment (e.g., Emanuel, 1991). In short, a large number of somewhat arbitrary choices have to be made concerning the parameterization of convective mixing in AGCMs, choices that can sometimes dramatically alter the model sensitivity to GHG increases.

Of particular interest here is the way in which cloud condensation is partitioned into the fraction that immediately falls out as precipitation (the *precipitation efficiency*), and a fraction that remains, is mixed with the surrounding air, and moistens it through re-evaporation. The fraction of falling precipitation that re-evaporates is also important. The most commonly used convection schemes in AGCMs are some variant of MCA (in eight out of 29 AGCMs surveyed by Gates, 1992), some variant of the Kuo (1974) scheme (nine out of 29 AGCMs), or some variant of the Arakawa and Schubert (1974) scheme (eight out of 29 AGCMs).

There is no explicit consideration of precipitation efficiency in the MCA scheme. In the Kuo (1974) scheme, the partitioning of condensation into precipitation and an increase in cloud liquid water content is parameterized as a function of the grid-scale relative humidity (RH). A lower RH results in less precipitation and hence greater eventual moistening of the layer in question, which tends to dampen changes in RH. In the schemes developed by Arakawa and Schubert (1974) and Emanuel (1991), precipitation efficiency is explicitly specified but as a fixed function of cloud height; there is no feedback between precipitation efficiency and the large-scale RH, although such a feedback could be easily added. Rennó et al. (1994a,b) present an analysis of the role of the specified precipitation efficiency and other microphysical parameters using a radiative–convective climate model.

Washington and Meehl (1993) computed the climatic sensitivity to a CO_2 doubling using MCA and PC schemes in the NCAR AGCM, and obtained global mean warmings of 4.0°C and 6.2°C, respectively. The much greater warming with PC was attributed to the warmer control climate and the greater increase in upper tropospheric moisture as the climate warms. Washington and Meehl (1993) suggested that PC without some feedback between cloud water content and optical depth (a negative climate feedback) is unrealistic. However, the feedback analysis by Cunnington and Mitchell (1990), using a prescribed change in SST in the UKMO AGCM, implies that PC pro-duces a smaller climate sensitivity than MCA. PC gives greater warming in the upper tropical troposphere than does MCA (4.5–5.0°C instead of 3.5–4.0°C), which results in a more negative lapse rate feedback. There is also a greater reduction in upper-level tropical cloudiness using PC, which is also a negative feedback (since upper-level clouds have a net warming effect). The greater strength of these two negative feedbacks using PC apparently overrode the stronger positive feedback due to greater upward transport of moisture using PC in the UKMO AGCM. Finally, Colman and McAvaney (1995) compared the control climate and climate sensitivity to a CO_2 doubling using the Kuo (1974) and Tiedtke (1989) PC schemes in the BMRC AGCM. Both schemes cause the tropical atmosphere to be 40–60% too dry below a height of 400–500 mb, and 20–40% too moist above this height. This implies that there is too much upward pumping of moisture in both schemes. Nevertheless, there was little difference in the model response to a CO_2 doubling, with the Kuo and Tiedtke schemes giving global mean warmings of 2.1°C and 2.2°C, respectively.

depend on changes in atmospheric stability in regions of subsiding motion – can override the relationship between temperature and optical depth that one would otherwise expect (Del Genio, 1996). Thus, the direction of the feedback between climate and cloud optical thickness is also uncertain.

Table 9.4 lists the changes in cloud amount and height that occurred in response to an increase in CO_2 in a number of simulations with AGCMs or AOGCMs. There is a general tendency for cloudiness to decrease everywhere, except for low clouds at high latitudes and for clouds near the tropopause at most latitudes. There are, however, many exceptions. Changes in cloud optical depth also occur in some AGCMs, but these changes are not given here because they depend on arbitrarily prescribed input assumptions. Changes in cloud amount and cloud height, on the other hand, depend on the interaction of several cloud and non-cloud processes as implemented in any given AGCM.

In order to accurately simulate cloud feedbacks, at least four conditions need to be satisfied: (i) the relevant feedback processes must be incorporated in the cloud parameterization; (ii) these processes have to be incorporated in a way that mimics nature; (iii) changes in the large-scale factors that influence clouds must also be correctly simulated, and this depends on the fidelity of all the other processes in the climate model; and (iv) the present-day distribution and characteristics of clouds must be accurately simulated. Because conditions (i) and (ii) must both be satisfied, the incorrect incorporation of a feedback process could result in worse results than if the feedback were neglected altogether. An example might be the incorporation of a strong temperature–optical depth feedback, when the true feedback is much weaker. Thus, more complex cloud parameterization schemes are not necessarily better than simpler schemes, unless the individual processes added to the complex scheme are separately validated against observations.

With regard to the last condition, the net global cloud feedback could be grossly in error if the initial relative amounts or locations of different kinds of clouds are wrong, even if the feedbacks within individual cloud types are accurately simulated. This is because the various feedback processes are of different strength and sign for different cloud types or for clouds at different altitudes. Zhang et al. (1994) show that, for models in which cloudiness increases near the tropopause, the net radiative effect of these changes depends on the initial height of the tropopause, while Cess et al. (1996) show that the simulation of too large a cloud albedo for the present

Table 9.3 Qualitative effect of various changes in cloud characteristics that could occur as the climate changes. Also given are references where the indicated effect is quantitatively assessed or included in climate simulations. The actual changes in cloud property that occur as the climate warms may be different from those indicated here, and sometimes differ from model to model.

Cloud property change	Effect on climate	Example reference
Decrease in amount of low cloud	Warming	Le Treut and Li (1991)
Increase in amount of high cloud	Warming[a]	Wetherald and Manabe (1988)
Increase in cloud height	Warming	Mitchell and Ingram (1992)
Increase in water content of stratus clouds	Cooling	Somerville and Remer (1984)
Increase in water content of cirrus clouds	Warming or cooling	Lohmann and Roeckner (1995)
Increase in ratio of water droplets to ice crystals	Cooling through enhanced reflectivity	Senior and Mitchell (1993)
	Warming through increased particle fall velocities	Senior and Mitchell (1993)

[a] This assumes that high clouds have a net heating effect. However, Del Genio et al. (1996) calculate a net cooling effect for high tropical cirrus clouds in the GISS AGCM.

climate results in too large a positive feedback when cloudiness decreases. Thus, accurate simulation of the cloud amounts, locations, and properties for the present climate is a necessary (but not sufficient) condition for reliable simulation of cloud feedbacks.

In the original feedback analysis of Cess et al. (1990), four of the models had sufficiently strong positive cloud feedbacks to more than double the global mean temperature sensitivity. Six years later, one model had a positive feedback strong enough to just double the climate sensitive; revisions to the other models with strong positive feedbacks resulted in substantially weaker positive feedbacks or a weak negative feedback (Cess et al., 1996). The data published by Cess et al. (1996) indicate that the effect of cloud feedbacks in the current generation of models ranges from a 27% reduction in climate sensitivity to a doubling of climate sensitivity, compared to the sensitivity expected based on the clear-sky radiative damping alone. Assuming that the clear-sky radiative damping has not changed from the 1990 analysis, the expected climatic response for a CO_2 doubling ranges from 1.3°C to 3.3°C. Lee et al. (1997) presented more detailed information concerning radiative damping due to different feedbacks in five versions of the NCAR AGCM that differ in the way in which cloud optical properties are computed. They presented the total infrared damping due to temperature changes but, using the methodology applied above to the clear-sky results of Cess et al. (1990), we can break this into contributions due to $\partial F / \partial T$ and due to changes in lapse rate. Results averaged over clear and cloudy skies are presented in Table 9.5. The inferred

lapse rate feedback is negative but is not particularly well correlated with the strength of the water vapour feedback. Cloud solar feedback can be weaker or stronger than the cloud infrared feedback, and either feedback can be of either sign. The expected climatic response for a doubling of CO_2 ranges from 1.53°C to 2.45°C. Recall that this range and the results of Cess et al. (1996), cited above, exclude ice and snow feedbacks, and are based on an idealized experimental design (a uniform change in SST and a perpetual July simulation).

The magnitude and even the sign of the net cloud feedback also depend on the spatial *pattern* of temperature change. In the feedback analyses discussed above, a globally uniform increase in SST was imposed. In reality, the surface warming will vary with latitude and with longitude, and both variations can cause a markedly different cloud feedback than for the cause of globally uniform warming. With regard to east–west variations, the critical region appears to be the tropical Pacific Ocean. Today, the eastern Pacific is relatively cold due to the upwelling of deep water, while the western Pacific is relatively warm. This temperature contrast drives an east–west circulation known as the Walker circulation, with rising motion over the western Pacific and sinking motion over the eastern Pacific. In the GFDL AOGCM, greater warming occurs in the eastern than in the western tropical Pacific Ocean, thereby reducing the temperature contrast and weakening the Walker circulation (Knutson et al., 1997). In experiments in which a globally uniform change in SST is imposed, Del Genio et al. (1996) found that the

Table 9.4 Changes in cloud amount and cloud height in response to a warming of the climate, as simulated by various AGCMs. See Table 6.1 for definitions of the model acronyms used here.

Model	References	Change in cloud amount			Mean cloud height	Type of experiment
		Total or low	Mid	High		
ARPEGE	Timbal *et al.* (1997)	Total and mid-level clouds decrease (with magnitude dependent on forcing, generally around 1–2%)		Positive except at ITCZ	Increased	Forced with $2 \times CO_2$ SST anomalies from ECHAM1 and UKMO models
BMRC	Colman and McAvaney (1995)	Low = +0.23%	–3.7%	–2.1%	High clouds occur higher	$2 \times CO_2$ equilibrium
BMRC	Colman and McAvaney (1997), Colman *et al.* (1997)	Total clouds = –0.45% for 0–2 K ΔSST		–1.2% for 0–2 K ΔSST mainly in tropics	Increased by about 30 m	Perpetual July, ±2 K ΔSST
CCC	Boer *et al.* (1992)	Total = –2.2% (–1.9% winter, –2.4% summer)		Positive except between ±30°		$2 \times CO_2$ equilibrium
CCM2	Zhang *et al.* (1994)	–0.5%	–3.5%	–3.5%		Perpetual July, ±2 K ΔSST
GFDL	Wetherald and Manabe (1988)	Cloud cover decreases in the upper troposphere at mid and low latitudes, except for an increase near the tropopause, especially at high latitudes. Low cloud cover increases at high latitudes				$2 \times CO_2$ equilibrium
GISS	Rind *et al.* (1995), Cess *et al.* (1996)	–2.3% global average; +0.1% polar		–0.3% global average; +0.4% polar		$2 \times CO_2$ equilibrium
LMD	Le Treut and Li (1991)	Cloud cover decreases almost everywhere at all heights, except that it increases near the tropopause at most latitudes				Perpetual July, ±2 K ΔSST
NCAR	Washington and Meehl (1996, 1989)	Increase at high latitudes		Increase		Transient $2 \times CO_2$
UKMO	Murphy and Mitchell (1995)	NH = –0.2% SH = +1.1% Increase in Arctic	NH decrease	NH = +0.3% SH = –0.7%		Transient $2 \times CO_2$, values given after 75 years

global mean cloud feedback was slightly negative. However, when greater warming in the eastern than in the western Pacific Ocean was allowed, the cloud feedback was positive. This is because the weakening of the Walker circulation in the latter case resulted in less anvil cloud in the western Pacific, which otherwise has a cooling effect. The warming effect of less anvil cloud in the western Pacific Ocean was only partly offset by increased anvil cloud in the eastern Pacific Ocean. As discussed on Section 11.2, these feedbacks could be different during a changing climate than for an unchanging, steady-state climate.

In closing this discussion of cloud feedbacks, it is worth returning to the issue of the linkage between clear-sky and cloud-radiative feedbacks. As previously noted, if upper tropospheric RH decreases as the climate warms, this implies less moistening and a weaker water vapour feedback compared to the case with constant RH. The decrease in RH also tends to reduce the average cloudiness in the upper troposphere (as in Colman *et al.*, 1997), and since high clouds generally have a warming effect, a reduction in their amount as the climate warms also acts as a negative feedback. Thus, if increased convective

Table 9.5 Analysis of the global mean radiative damping in the NCAR AGCM using five different parameterizations of cloud radiative properties, based on data presented in Lee *et al.* (1997). In the original model (CCM2), the optical properties of layer clouds are prescribed based on latitude, cloud altitude, and thickness. In CW, the amount of cloud water is explicitly computed and affects the cloud lifespan, but optical properties are still fixed. In CWRF, the cloud optical thickness depends on the liquid water content but the droplet radius is assumed to be constant. In CWRV, the cloud droplet radius is allowed to vary. In CWRI, the size of ice crystals in ice clouds is also allowed to vary. F is the global mean emission of infrared radiation to space ($W m^{-2}$), T_e is the effective radiating temperature (K), λ is the total global mean radiative damping ($W m^{-2} K^{-1}$), and ΔT is the steady-state change in surface air temperature (°C) for a CO_2 doubling, assuming an adjusted forcing of $3.75 W m^{-2}$.

Model	F	T_e	$\partial F/\partial T$	IR damping ($W m^{-2} K^{-1}$)				Solar damping ($W m^{-2} K^{-1}$)			λ	ΔT
				Lapse rate	Water vapour	Cloud	Net	Water vapour	Cloud	Net		
CCM2	246.1	256.7	3.83	0.72	−1.85	1.10	3.71	−0.21	−0.60	−0.90	2.81	1.33
CW	233.0	253.3	3.68	0.29	−1.51	1.04	3.36	−0.17	−1.01	−1.25	2.11	1.78
CWRF	240.3	255.2	3.77	0.56	−2.12	−1.04	1.24	−0.17	0.55	0.29	1.53	2.45
CWRV	243.2	256.0	3.80	0.48	−2.08	−0.86	1.42	−0.21	0.81	0.51	1.93	1.94
CWRI	244.6	256.4	3.82	0.46	−2.07	−0.83	1.46	−0.18	1.04	0.76	2.22	1.69

pumping is accompanied by a shift in the partitioning of the moisture flux towards greater immediate condensation and less moistening, then the effect of this shift on upper tropospheric moisture and cloudiness will be for both to act to reduce climate sensitivity (as long as upper tropospheric clouds have a net warming effect). On the other hand, if upward moisture pumping increases with no or little change in the partitioning, then both the negative lapse rate feedback and the positive water vapour feedback will increase as convective pumping increases. These have opposing effects on climate sensitivity.

Feedbacks involving land snowcover

A reduction in the extent of snowcover on land as climate warms is widely expected to serve as a positive feedback, contributing to a greater warming than in the global average in high latitude regions. This is because snow tends to have a higher albedo (reflectivity) than snow-free land, so a reduction in snowcover leads to an increase in the amount of solar radiation that is absorbed at the surface. This will lead to further warming and thus acts as a positive feedback. However, analyses by Cess *et al.* (1991) and Randall *et al.* (1994) indicate that a number of competing changes come into play when snowcover changes, and these changes can substantially weaken the positive snowcover feedback that would otherwise occur or can even change the snowcover feedback into a weak negative feedback.

First, changes in snowcover can induce changes in the amount, location, or optical properties of clouds. In particular, cloudiness tends to increase over snow-free regions, so the reduction in albedo at the top of the atmosphere (the planetary albedo) is much smaller than the reduction in surface albedo. It is even possible for the planetary albedo to increase if the clouds are more reflective than the former snowcover. Second, once the snow melts away, the surface temperature is free to rise above 0°C, so the emission of infrared radiation from the surface and thence to space can increase. This serves as a negative feedback. Third, the atmospheric lapse rate tends to increase after snow retreats (due to greater surface heating), and this serves as a positive feedback. Finally, the drying of the surface after the snow all melts reduces the water vapour content of the overlying atmosphere, which serves as a negative feedback. Overall, the short-wave feedback tends to be positive, while the long-wave feedback tends to be negative. Either feedback can be stronger, so the net feedback due to retreat of snowcover can be positive or negative.

Cess *et al.* (1991) and Randall *et al.* (1994) quantified snow feedbacks using the same methodology as described above for the evaluation of clear-sky and cloud-related feedback processes, except that they imposed a uniform 4 K change in SST in a perpetual April simulation rather than in a perpetual July simulation. This experiment was performed for model versions with fixed and freely computed snowcover, and the radiative changes compared. Cess *et al.*

(1991) presented results for 17 different AGCMs, and found that the effect of snowcover feedback ranged from a reduction in climate sensitivity by 10% to almost doubling the climate sensitivity. That is, the *global mean* temperature response can as much as double, with even larger changes (by up to a factor of 3–4 or more) in regions where snow retreats. However, snowcover feedbacks are expected to be strongest during transitional seasons, so the mean annual climate feedback in these models is certainly smaller than indicated by the perpetual April experiments.

Feedbacks involving sea ice

Changes in the areal extent of sea ice can also exert a strong feedback effect on climate by changing the radiative balance. Changes in the area and thickness of sea ice also alter the fluxes of latent and sensible heat between the ocean and atmosphere, which has a strong effect on the seasonality of temperature change at high latitudes. However, except through possible interactions with cloud amount, these changes will not significantly alter the planetary radiative balance and so will not noticeably affect the climate sensitivity. We will defer discussion of the impact of changing air–sea heat fluxes to Section 10.1, in which regional and seasonal patterns of climatic change are discussed, and focus here only on direct and possible indirect radiative feedbacks.

The simulation of sea ice in global-scale climate models is a particularly challenging problem (Randall *et al.,* 1998). Sea ice involves vertical growth (at the base of the ice), lateral growth (into cracks or "leads" created by divergent ice motions), surface and basal melting, and lateral melting (from leads in summer). The surface albedo of sea ice depends on its thickness and the possible presence of snowcover or surface melt ponds; the way in which the surface albedo is computed generally has a dramatic effect on the simulation of sea ice for the present climate (Shine and Henderson-Sellers, 1985), and on the sensitivity of sea ice to changes in climate (Meehl and Washington, 1990). Ice motions, caused by surface winds and ocean currents, have a significant effect on the distribution of ice thicknesses within a given region, on the occurrence of leads and larger areas of open water, and on the spatial extent of sea ice. There are a number of important interactions between the thermodynamics of sea ice (processes related to

freezing and melting) and the dynamics of sea ice (processes related to ice motions). However, only six of the 16 coupled climate models discussed in Gates *et al.* (1996) included sea ice dynamics; in the other 10, only sea ice thermodynamics are explicitly considered, with dynamical effects included only through the specification of a minimum lead fraction.

Pollard and Thompson (1994) found that, in simulations with ice dynamics, the ice was less compact for the control simulation than for the case without sea ice dynamics. Thus, the total area of sea ice was smaller, and because of this, the temperature–albedo feedback was weaker. This reduced the global mean equilibrium warming for a doubling of CO_2 from 2.27°C to 2.06°C, with much larger effects at high SH latitudes in winter and spring. Watterson *et al.* (1997) also found that the introduction of sea ice dynamics reduces the global mean climate sensitivity, in their case from 4.8°C to 4.3°C using the CSIRO AGCM. The reduced sensitivity with sea ice dynamics is consistent with the more general result that the climate model sensitivity tends to be smaller the smaller the initial amount of sea ice, all else being equal. The weaker high-latitude feedback in the presence of ice dynamics implies a smaller polar amplification of the response, with important implications for the change in soil moisture in mid-latitudes (as discussed in Section 10.2).

The simulated high-latitude climatic change and hence the global mean climate sensitivity are also quite sensitive to the initial ice thicknesses. The initial sea ice thickness in turn is very sensitive to the heat flux from the underlying water to the base of the ice, which is related to the poleward heat flux simulated for the present climate. Rind *et al.* (1995) found that, when the GISS model was adjusted so that the sea ice thicknesses (which were initially too great) match observational estimates, the global mean warming for a CO_2 doubling increased from 4.17°C to 4.78°C. This was due to the fact that, when initial ice thicknesses in the model match those observed for the present climate, it is easier for the ice to retreat when the CO_2 concentration is increased.

Simulation results with AGCMs coupled to mixed layer-only ocean models

The feedback analyses discussed above shed considerable light on the relative importance of different feedback processes in AGCMs, and serve to indicate the major sources of uncertainty in the climate sensitivity of these models. However, the feedbacks were evaluated in the above analyses by imposing a globally uniform change in SST, and so will not be the same as when an AOGCM is allowed to respond with regionally varying temperature changes to a radiative perturbation. Thus, we must still fall back on simulations of steady-state temperature change as obtained by AGCMs that have been coupled to an ocean model.

Although a number of research teams have developed coupled AOGCMs, these models are not practical for evaluating the steady-state response to an increase of CO_2. This is because integrations on the order of 1000 years in duration are required to reach close to steady state, and this is prohibitively expensive. Since the feedback processes that determine climate sensitivity operate in the atmosphere or at the surface, an alternative is to couple the AGCM to an ocean model consisting of the mixed layer only. This permits steady state changes to be attained after only a few decades' simulation. The disadvantage is that horizontal heat transport in the oceans, which depends on the 3-D overturning circulation, cannot be simulated. Instead, the oceanic heat transports required to give the present-day variation of temperature with latitude are prescribed. These are then held constant as the climate changes. Spatial variations in the warming can develop, in response to regional variations in the feedback strengths, but the spatial patterns in the new steady-state climate would undoubtedly have been different in an AOGCM simulation due to changes in the horizontal transport of heat by the oceans. Not only would this affect regional patterns of climatic change, it would also probably affect cloud feedbacks and hence the global mean climate responsiveness to a doubling of atmospheric CO_2.

Table 9.6 list the steady-state, global mean change in surface air temperature obtained with coupled AGCM-mixed layer models for a doubling of atmospheric CO_2. In most cases, the global mean warming for a CO_2 doubling is 2.0°C to 4.0°C. This is consistent with the feedback analyses discussed above.

9.2   Direct observations of the water vapour feedback

The preceding discussion has highlighted the complexity of the processes that determine climate sensitivity in nature, and of the difficulties and uncertainties associated with trying to simulate these processes – and their interactive effects – in computer climate models. The increase in the amount of water vapour in the atmosphere as climate warms is projected to be the single largest feedback, and to be strongly positive. Cloud feedbacks could also be important, but the net feedback could be either positive or negative. However, the importance of cloud (and surface albedo) feedbacks depends on the overall strength of the water vapour feedback. This is because, if the water vapour feedback is weak, the absolute effect of the other feedbacks will be small, as previously shown in Box 3.2. The primary factors driving the overall increase in atmospheric water vapour are the global mean warming and its latitudinal distribution. The primary factors governing the vertical distribution of changes in water vapour at low latitudes are likely to be how the intensity and height of convection change as the climate warms, and how the partitioning of convective moisture pumping between upper tropospheric heating and

Table 9.6 Global mean change in surface air temperature (ΔT, K) for a doubling of atmospheric CO_2 as obtained by various AGCMs coupled to a slab ocean.

Model	ΔT	Reference
BMRC	2.1	Colman and McAvaney (1995)
Genesis	2.1	Pollard and Thompson (1994)
	2.3	Pollard and Thompson (1994)
UKMO	2.5	Johns *et al.* (1997)
	2.8	Murphy and Mitchell (1995)
GLA	2.6[a]	Sellers *et al.* (1996a)
	2.8[b]	Sellers *et al.* (1996a)
CCC	3.5	Boer *et al.* (1992)
GFDL	4.0	Manabe *et al.* (1992)
NCAR	4.0	Washington and Meehl (1989)
	4.6	Washington and Meehl (1993)
CSIRO	4.3	Watterson *et al.* (1997)
GISS	4.2[c]	Rind *et al.* (1995)
	4.8[d]	Rind *et al.* (1995)

Notes: The Genesis model is an outgrowth of the NCAR AGCM. See Table 6.1 for definitions of the acronyms used here.

[a] Standard model version.
[b] Version with reduced stomatal conductance in response to a higher atmospheric CO_2 concentration (see Section 10.3).
[c] Standard model version, having sea ice that is too thick for the present climate.
[d] Model version in which sea ice thickness for the present climate matches observational estimates.

humidification changes as convection changes. At middle and high latitudes, the changes in the vertical moisture transport by large-scale eddies (storms) are likely to be the dominant control on the vertical distribution of changes in water vapour.

The amount of water vapour and its vertical distribution can be directly observed at present. This provides the possibility of directly inferring the strength of the feedback between temperature and water vapour from observations. There are three different kinds of observations that shed light on the water vapour feedback: the relationship between surface temperature and the total amount of water in the atmospheric column (the precipitable water); the relationship between surface temperature and the vertical distribution of water vapour; and the relationship between specific processes (such as deep convection) that influence water vapour and the amount of water vapour in the upper troposphere.

The ability to assess the water vapour feedback from direct observations is crucial, because the only potentially serious objection to the viewpoint that the climate sensitivity to increasing GHG concentrations is large enough to matter centres around the water vapour feedback. Lindzen (1990) hypothesized that an increase in cloud convection due to overall warming would lead to a drying of the upper troposphere through the increase in subsidence between clouds that would be induced. Since water vapour in the upper troposphere is particularly effective as a GHG (for the two reasons discussed in Section 3.1), this could result in a significantly smaller net water vapour feedback, thereby substantially reducing the projected warming. This idea was pursued further by Sinha and Allen (1994) using a computer model of an individual convective column and the surrounding atmosphere. They concluded that an increase in the intensity of convection can cause the upper troposphere to become dryer in some height intervals, depending on the choice of model parameters which are themselves highly uncertain. For two intermediate cases they obtain peak drying by 10% centred at heights of 500 mb and 250 mb in response to a 1 K surface warming, with less drying at other heights in the upper troposphere and moistening close to that expected for constant relative humidity below a height of 700 mb. The model used by Sinha and Allen (1994) adopts one of the key assumptions of Lindzen's hypothesis – that all condensed water vapour in deep clouds falls to the ground as rain rather than partially evaporating and moistening the

upper troposphere – an assumption that was first questioned by Betts (1990) and later refuted by model-based analysis of observations (Sun and Lindzen, 1993). Relaxation of this assumption may very well cause the simulated upper troposphere drying to disappear. Sinha and Allen (1994) indicate that introduction of other processes can also substantially reduce the simulated drying, but that introducing yet further processes can bring it back.

Observations pertaining to the variation of atmospheric water vapour with surface temperature will now be discussed.

Variations in the total amount of water vapour

There is no doubt that the total amount of water vapour in the atmosphere will increase as the climate warms. This follows from very fundamental principles involving the fact that the driving force for evaporation from the ocean – the difference between surface and atmospheric vapour pressures – tends to increase as the surface temperature increases, combined with the fact that the atmosphere's ability to hold water increases as the air warms. As summarized in Section 5.5, observations made during recent decades indicate that the atmospheric water vapour content has indeed increased as the climate warmed. The relationship between surface temperature and precipitable water in the tropics is weak at the monthly and annual time scales, where concurrent changes in atmospheric dynamics complicate the picture, but is stronger at the decadal time scale (Gaffen et al., 1992). An increase in total column water vapour with increasing surface temperature will lead to an increase in heat trapping as temperatures increase unless a sufficiently strong downward shift in the moisture distribution occurs.

Interannual variations in the amount and vertical distribution of water vapour

As discussed in Section 5.5, the available rawindsonde observations are not reliable for determining long-term trends in the amount of water vapour in the upper troposphere, while satellite observations have not been available for long enough to determine long-term trends. However, the available datasets are adequate for assessing the spatial distribution of interannual changes in water vapour and their correlation with interannual changes in surface temperature.

Much of the interannual variability is related to the El Niño oscillation, which involves large-scale oscillations in surface temperatures in the tropical Pacific Ocean and a shift in the main region of convection between Indonesia and the eastern Pacific Ocean. Since the changes in surface temperature are associated with major shifts in the tropical circulation that are unique to the El Niño oscillation, the *local* correlation between temperature and humidity is obviously not going to be applicable to long-term, large-scale climatic change (the same is also true for correlations between temperature and humidity over the course of the seasonal cycle).

Sun and Oort (1995) have attempted to circumvent this problem by examining the correlation between atmospheric water vapour and surface temperature averaged over the *entire* tropics (30°S–30°N). Their analysis is based on rawindsonde data, and includes several stations within the very dry regions of the tropics (recall Fig. 5.22). Based on data for the period from May 1963 to December 1989, they found that the amount of water vapour in the tropical atmosphere increased with increasing temperature as follows:

- by 70–75% of the increase that would occur with fixed RH in the 1000–900 mb layer;
- by only 15% of the increase expected with fixed RH at 700 mb; and
- by 70% of the increase expected with fixed RH at 300 mb.

In the planetary boundary layer (1000–900 mb), the change in water vapour is most directly tied to the thermodynamically driven increase in the rate of evaporation. There is also a sizeable relative increase in the upper troposphere, related no doubt to an increase in the rate of detrainment of moisture from the tops of cumulus cloud columns. However, in the middle troposphere there is almost no increase in the amount of water vapour, which can be explained by an increase in the rate of subsidence of relatively dry air from the upper troposphere (even though this air becomes moister, it is still much drier than the air below).

Since AGCMs simulate close to constant RH at all heights as the climate warms, the results of Sun and Oort (1995) – assuming that they are applicable to long-term climatic change – imply that the water vapour feedback is too strong in AGCMs. This in turn could be due to the convective coupling between the moist surface layer and middle troposphere being too strong, so that water vapour increases enough at all heights to keep RH constant as temperatures increase (Sun and Held, 1996). Sun and Oort (1995) compared the effect of their observed water vapour variation on the outgoing emission of infrared radiation with that assuming constant relative humidity. Their results imply that the assumption of constant relative humidity increases $|\mathrm{d}F/\mathrm{d}T|$ by about $0.8\,\mathrm{W}$ $\mathrm{m}^{-2}\mathrm{K}^{-1}$ compared to the observed water vapour variations. Since this error is applicable only to cloud-free regions in the tropics, the error in the global mean $\mathrm{d}F/\mathrm{d}T$ is about $0.2\,\mathrm{W\,m}^{-2}\,\mathrm{K}^{-1}$. From Table 9.5, it can be seen that this error is only about 10% of the total water vapour feedback strength as computed by AGCMs with constant RH. However, as discussed above, the RH does decrease in the middle and/or upper tropical troposphere as the climate warms in some AGCMs, and in these AGCMs the error in $\mathrm{d}F/\mathrm{d}T$ will be even smaller.

Direct observations of processes governing upper tropospheric water vapour

An alternative to examining the correlation between changes in surface temperature and atmospheric water vapour is to examine the relationship between specific processes and atmospheric water vapour. Soden and Fu (1995) examined the relationship between deep convection and upper tropospheric water vapour as inferred from satellite data. As the indicator of deep convection, they used the frequency of occurrence of high, optically thick clouds in 2.5° × 2.5° latitude–longitude grid cells. As the indicator of upper tropospheric humidity, they used the emission of infrared radiation at a wavelength of 6.7 μm using the TOVS satellite system (see Section 5.5). This can be linked to the vertically averaged humidity in the 500–200 mb layer with an uncertainty of ±8%. Their analysis spans the period July 1983 to June 1990.

Figure 9.1 compares the frequency of deep convection (FDC) with the upper tropospheric humidity (UTH), as analysed by Soden and Fu (1995). Regions of strong convection are associated with regions of high upper tropospheric humidity (UTH), while regions of low convection are associated with low UTH. That is, there is a positive spatial correlation between UTH and FDC. There is also a positive local correlation through time everywhere in the tropics; that is, at essentially all grid points, the UTH humidity increases if the FDC increases *at that grid point*. The relevant question for climate sensitivity,

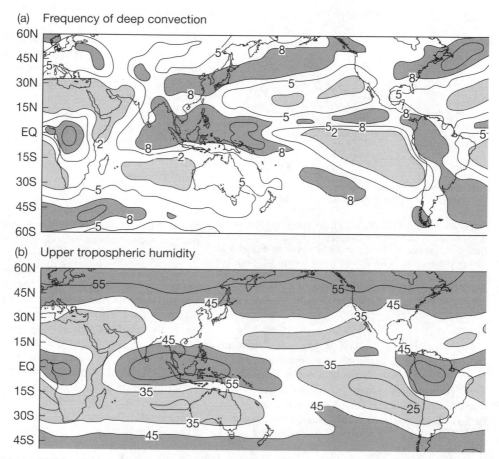

Fig. 9.1 Comparison of the annual average frequency of deep convection with the annual average upper tropospheric relative humidity for the period July 1983–June 1990, as inferred from satellite observations. Reproduced from Soden and Fu (1995).

however, is how the *areally averaged* UTH changes through *time* as the areally averaged FDC changes. Unfortunately, currently available data (including the data shown in Soden and Fu, 1995) are not reliable enough to answer this question clearly.

However, an alternative, and more physically meaningful, approach is to directly examine the relationship between clear-sky greenhouse trapping (GHT) and sea surface temperature, where GHT is simply the difference between the surface emission (σT^4) and the outgoing emission for clear skies at the top of the atmosphere (which can be easily measured by satellite). For the period April 1985 to December 1987, Soden (1997) found that the tropical-mean GHT increased in association with an increase in tropical-mean SST, as conditions

changed from La Niña to El Niño. This is illustrated in Fig. 9.2. GHT trapping increased in regions where deep convection increased, and decreased elsewhere (due to an increase in subsidence). There were marked intra-annual variations in GHT that are not reflected in changes in the tropical-mean SST. This must be related to changes in the spatial patterns of SST, which are known to be as important to the occurrence of deep convection as absolute temperatures (Zhang, 1993). Since it is not clear how the spatial patterns of SST will change as the climate warms or whether the patterns of temperature change will resemble El Niño patterns (Section 11.2), the correlation seen over the course of a La Niña–El Niño transition might not apply to a warmer climate.

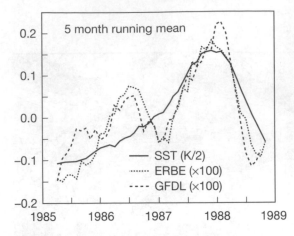

Fig. 9.2   Variation (times 100) from 1985 to 1989 in the greenhouse trapping as measured by the Earth Radiation Budget Experiment (ERBE) satellite observations and as computed by the GFDL AGCM when driven by observed sea surface temperature (SST) variations. Results are averaged over the region from 30°S to 30°N. Also given are the average SST variations divided by two. Reproduced from Soden (1997).

However, the GFDL AGCM has successfully simulated the overall variation in tropical-mean GHT from 1985 to 1989, in tropical-mean intra-annual fluctuations, and the local patterns of increasing and decreasing GHT when forced with observed changes in SST. The simulated tropical-mean variations are shown alongside the observed variations in Fig. 9.2. This model, like all AGCMs, produces a positive long-term water vapour feedback. Thus, although observations of intra-annual, longer-term, and spatial patterns of GHT do not directly tell us anything about the multi-decadal water vapour feedback, they can be used to test the fidelity of convective process embedded in AGCMs, the same processes that give rise to the long-term feedback in AGCMs. This test indicates that, *given the correct patterns of sea surface temperature change,* current AGCMs should do a reasonable job in simulating the water vapour feedback in tropical regions.

With regard to middle and high latitudes, Del Genio *et al.* (1994) find that, in AGCMs, eddies and mean meridional motions are the dominant factors influencing upper tropospheric moisture outside the tropics. Although their analysis is based on model simulations rather than on nature (for which adequate observations are lacking), the processes in question are fully resolved by AGCMs. This is in contrast to convective processes, which must be parameterized because they occur at scales smaller than the model grid resolution. Thus, we can have more confidence in these results than in model results concerning the role of convection. The analysis of Del Genio *et al.* (1994) indicates that the moistening effect of eddies and mean meridional motions would both increase as the climate warms. Thus, even if increasing convection were to dry the upper or middle troposphere in regions surrounding intense convection, this does not imply a drying of the upper or middle troposphere in the global mean.

Conclusion

The evidence reviewed above supports the conventional wisdom that the water vapour feedback is positive. Observations indicate that total precipitable water increases as the climate warms, which will exert a positive feedback. Observations of interannual variability indicate that the amount of water vapour also increases in the upper troposphere, although the increase in the middle troposphere is probably less than simulated by current AGCMs. Observations show that the clear-sky GHT, averaged over the tropics, increases as the tropical mean SST increases. AGCMs are able to simulate the observed variation when driven by observed changes in SST. This agreement indicates that AGCMs should do well in simulating the long-term water vapour feedback in the tropics if the simulatd patterns of SST change are correct.

In Section 5.5 it was noted that current models tend to simulate too much moisture in the very dry regions of the middle and upper tropical troposphere (corresponding to the descending branches of the Hadley Cells). We have seen here that the models might also simulate too much moistening in the middle troposphere as the climate warms, when averaged over the entire tropics. However, these two potential errors will cancel to some extent. This is because, if the initial moisture content is too high, the impact on the trapping of infrared radiation of a *given* increase in the amount of water vapour will be too small, and this will be offset to some extent if the increase in moisture content too large.

Thus, there is no reason to suspect that the water vapour feedback is other than positive, and of a magnitude sufficient to increase the climate sensitivity by at least 50% compared to the case with no feedbacks. Recall (from Section 3.1) that the climate sensitivity for a CO_2 doubling in the absence of any feedbacks is about 1.0°C. The water vapour feedback probably increases this to at least 1.5°C. Negative cloud feedbacks could reduce the climate sensitivity below 1.5°C, but ice and snow feedbacks would increase the sensitivity. Thus, we can accept 1.5°C as a likely lower limit for the climate sensitivity to a CO_2 doubling on the basis of the evidence reviewed so far.

9.3   Assessments based on paleoclimatic data

Since climate sensitivity is defined as the ratio of the global mean change in temperature to the global mean radiative forcing, it can be calculated empirically if these two quantities can be estimated for past climates. Hoffert and Covey (1992) did this using data for the Cretaceous Period, when the climate was substantially warmer than at present, and for the peak of the last ice age, when the climate was substantially colder. The climate sensitivity calculated in this way implies a climatic response for CO_2 doubling of 2.3 ± 0.9°C, which falls within and below the lower half of the 2.1–4.8°C range derived from recent AGCMs (Table 9.6). Interestingly, Hoffert and Covey (1992) deduced a similar sensitivity for a climatic warming (the Cretaceous Period) as for a climatic cooling (the last ice age).

Further, qualitative, evidence that the climate is responsive to changing atmospheric CO_2 concentration is found by examining climatic changes throughout geological history. In general, times of inferred high CO_2 concentration coincide with times of inferred warm climates, while times of low atmospheric CO_2 coincide with episodes of extensive glaciation (Berner, 1990). This implies either that atmospheric CO_2 variations do indeed exert a significant influence on climate, or that atmospheric temperature controls atmospheric CO_2 concentration at multi-million year time scales and longer. The latter contradicts the main competing geochemical models of the carbon cycle at this time scale (i.e., Berner *et al.*, 1983; Francois and Walker, 1992). These models have been successful in explaining key features of the observed geochemical record, and contain negative feedbacks between atmospheric temperature and CO_2 concentration but no mechanism whereby warmer temperatures would induce higher atmospheric CO_2 at multi-million-year time scales. Hence,

the correlation between atmospheric CO_2 and climate at multi-million-year time scales and longer is further evidence that the climate is sensitive to variations in CO_2 concentration.

9.4   Assessments based on the analysis of historical temperature changes

The analysis of global mean temperature changes during the 20th century could, in principle, provide an independent estimate of climate sensitivity. However, because of uncertainty in the magnitude of cooling effects associated with biomass and sulphur aerosols, the observed changes in global mean temperature cannot provide a meaningful constraint on climate sensitivity. This is compounded by uncertainty in the radiative heating or cooling effects associated with changes in O_3 and in the magnitude of solar radiation variations during the last 100 years. Hence, we must first deduce the relative importance of different forcing mechanisms. Constraining the aerosol forcing requires examining the spatial patterns of climatic change. We must also take into account the effect of the oceans in delaying the temperature response to radiative forcings. This problem is discussed more fully in Section 11.4, but it suffices to state here that the observed temperature changes during the past century are consistent with a climate responsiveness for a doubling of CO_2 of 1.0°C to perhaps 3.0°C *if* the effective climate sensitivity during a time-dependent change is the same as the equilibrium sensitivity. This is an assumption that will be fully addressed in Section 11.2.

9.5   Synthesis

The argument that increases in GHG concentrations of the magnitude expected during the 21st century will lead to a significant warming of the climate is based on very fundamental physical principles and observations. Increasing GHG concentrations trap heat, and the heat trapping can be calculating to within ±10% based on laboratory measurements. A trapping of heat must lead to an increase in temperature so as to restore radiative balance. The first benchmark that one can establish is the warming that would be required by a perfect radiator: about 1.0°C for a doubling of CO_2. The second benchmark arises from the observation that the amount of moisture increases throughout the troposphere as the climate warms. This exerts a strong positive feedback, and raises the expected warming for a doubling of CO_2 to at least 1.5°C. This is the second benchmark, albeit less certain than the first benchmark. Cloud feedbacks could further amplify the climate response or could diminish it. Combined with ice and snow feedbacks, this gives an overall range of 2.1°C to 4.8°C in current AGCMs.

Two tests to which the prediction of a CO_2 doubling sensitivity of this magnitude can be subjected are as follows: the ability to explain paleoclimatic data, and the ability to properly simulate observed temperature variations during the past 140 years or so. Both of these tests suggest a CO_2 doubling sensitivity within and somewhat below the lower half of the AGCM results. Figure 9.3 summarizes the climate sensitivities as inferred from the three lines of evidence discussed here. Since a sensitivity of 2.0–3.0°C is common to all three estimates, this can be taken as the most likely climate sensitivity.

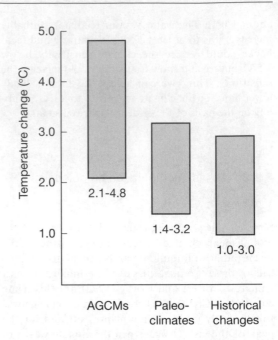

Fig. 9.3 The global mean equilibrium climatic change ("climate sensitivity") for a doubling of atmospheric CO_2 as given by AGCM simulations, paleoclimatic data, and historical temperature variations.

CHAPTER 10

The regional equilibrium response to a doubling of the atmospheric concentration of CO_2

In this chapter we examine the characteristics and geographical patterns of the climatic changes that are simulated by 3-D AGCMs to result from a doubling of atmospheric CO_2. These climatic changes are computed by first running a model with the present (or near-present) concentration of CO_2 until it has settled into a statistically constant climate, then suddenly increasing the concentration of CO_2 and allowing the simulated climate to reach a new, statistically constant state. In the vast majority of cases, a doubling of the atmospheric CO_2 concentration is imposed. The difference between the two climates is then determined. In some cases a second simulation, without increasing the CO_2 concentration, is continued from the exact point where the CO_2 increase was applied. This simulation is referred to as the "control" simulation, and this is what is compared with the perturbed simulation. The resultant climatic change is often referred to as the "equilibrium" climatic change, although the term "steady state" climate change is preferable and will be used here interchangeably with the term "equilibrium" climatic change. In order to achieve a new equilibrium climate after only a few decades of simulated time, only the mixed layer of the ocean is included. If the AGCM were coupled to an OGCM, representing the full depth of the ocean, then about 1000 years of simulated time would be required before the new equilibrium would be achieved. This in turn would require a prohibitively large amount of computer time. Owing to the absence of the deep ocean in a mixed layer-only simulation, the horizontal heat transport by the ocean for the present climate has to be specified and then held constant as the climate changes.

The use of AGCM-mixed layer models, rather than AOGCMs, introduces two likely sources of error in the simulated geographical patterns of climatic change that can be expected at any given time. First, the geographical patterns of climatic change during the transition from one climate to another could be quite different from equilibrium patterns of climatic change. This is because the delay in climatic warming is expected to vary regionally due to differences in the rate of downward mixing of heat into the oceans. The resulting differences in surface warming could be further amplified by differences in cloud feedbacks. This is because the magnitude and even the sign of cloud feedback in a given region seem to depend on the regional patterns of climatic change, and not just on the overall magnitude of climatic change. Furthermore, changes in ocean circulation could occur abruptly rather than smoothly, and the sense of the change in ocean circulation during the transition to a warmer climate (a weakening) could be different from that of the equilibrium change. Second, the geographical patterns of equilibrium climatic change itself, as simulated by AGCM-mixed layer models and AOGCMs, are likely to differ. This is due to the fact that changes in the oceanic circulation and associated heat transport can occur in AOGCMs, but these are implicitly held constant in AGCM-mixed layer models. All of these issues will be discussed in the next chapter. Nevertheless, the steady-state response to a CO_2 doubling, as simulated by AGCM-mixed layer models, serves as a useful benchmark for comparing different AGCMs, for understanding the role of different processes in climatic change, and for gaining an appreciation of the potential magnitude of climatic change at the regional scale.

10.1   Temperature response

As discussed in Section 9.1, the majority of recent simulations with AGCMs give a global and annual mean surface air warming for a doubling of CO_2 of 2.0–4.0°C in equilibrium. However, a single number – the global average temperature change – masks substantial geographical and seasonal differences in the simulated temperature response. The main features of the simulated response patterns, that occur in all models to some extent, are as follows:

- a greater mean annual warming at high latitudes than at low latitudes at the surface and in the lower troposphere, also referred to as a *polar amplification* of the warming;
- a greater warming in winter than in summer at high latitudes, particularly over the oceans;
- a greater overall warming of the land surface than of the ocean surface;
- a greater warming of the surface air than of the surface, particularly over the oceans;
- greater warming in summer than in winter in those land areas where soils become sufficiently dry in the summer;
- a tendency for somewhat greater warming at night than during the day;
- a greater warming in the tropics than at middle and high latitudes in the upper troposphere;
- a greater warming in the upper troposphere than in the lower troposphere or at the surface in the tropics (so that the tropical lapse rate decreases);
- a greater warming at the surface and in the lower troposphere than in the upper troposphere at high latitudes (so that the polar lapse rate increases); and
- a cooling of the stratosphere.

These characteristics of the response of AGCMs are illustrated in Figs 10.1 and 10.2. Figure 10.1 shows the geographical pattern of surface air warming for December–January–February (DJF) and June–July–August (JJA) as simulated by three different AGCMs that have low, intermediate, or high climate sensitivity (the corresponding zonally averaged changes are compared with a larger set of models later, in Plate 11). Figure 10.2 shows the variation with latitude and height of the zonally averaged mean annual warming as simulated by representive AGCMs using two different parameterizations of convection.

Polar amplification of the surface response

The greater surface and near-surface warming at high latitudes is a result of (i) the feedback between temperature and the extent of snow and ice, and (ii) differences in the feedback involving atmosphere lapse rate. As temperatures warm, the extent of ice and snow decreases. This results in a reduction in the surface albedo which, in spite of complications arising from changes in clouds and other factors discussed in Section 9.1, tends to amplify the local temperature response. Since ice and snow cover are restricted to high latitudes, the high-latitude temperature response is amplified. The other factor is the lapse rate feedback. At high latitudes the lapse rate tends to increase owing to the stability of the high-latitude atmosphere, which restricts the warming to a shallow layer of the atmosphere. At low latitudes the average lapse rate is closely linked to the moist adiabatic lapse rate, which decreases as the climate warms. As discussed in Section 3.5, an increase in lapse rate as surface temperature warms acts as a positive feedback, while a decrease in lapse rate serves as a negative feedback.

Plate 11 compares the latitudinal variation in the zonally (east–west) averaged surface-air warming as simulated by several different AGCMs. All models produce a polar amplification of the surface temperature response relative to the tropical response, ranging from less than a factor of 2 (LMD) to a factor of 3 (CSIRO) or more (NCAR-2). Part of the reason for the difference is probably due to differences in the amount of ice and snow simulated for the present climate. Models that simulate more ice and snow are likely to have a stronger ice/snow–albedo feedback and hence a stronger polar amplification. As discussed in Section 9.1, the way in which sea ice is simulated can strongly affect the response of sea ice to warming, with the inclusion of sea ice dynamics tending to reduce the reduction in the areal extent of ice as the climate warms, and hence tending to reduce the strength of the ice–albedo feedback. Finally, there could be important differences among the models in the cloud feedback at high latitudes relative to low latitudes, particularly since feedbacks involving convective clouds – which are treated differently in different models – will be important at low latitudes but not at high latitudes.

An important consequence of the polar amplification of the surface temperature response is that the difference between polar and equatorial temperatures decreases (that is, the meridional temperature gradient decreases). As explained in Section 10.2, differences in the polar amplification have important consequences for the simulated changes in rainfall and soil moisture in both tropical and middle latitude locations. The degree of polar amplification is also important to potential changes in mid-latitude storminess, as discussed in Section 10.5.

All climate models predict substantially smaller polar amplification than implied by analysis of paleoclimatic data. Russian geologists in particular have attempted to reconstruct spatial patterns of climatic change that occurred at various times during the past 100 million years when the climate was warmer than at present. An English summary of this work is found in Borzenkova (1992). Kheshgi and Lapenis (1996) estimated error bars in the latitudinal profile of temperature change for the Holocene Climatic Optimum (5300–6200 years before present). Figure 10.3 shows the latitudinal temperature profile for the Holocene, with error bars, along with the latitudinal profile deduced by Hoffert and Covey (1992) for the Last Glacial Maximum and the Late Cretaceous. The middle estimate of the polar amplification is about a factor of 6 – substantially larger than the factor of 2–3 obtained by current climate models (as shown in Plate 11). At present it is not possible to say which approach is more likely to be correct.

Greater warming in winter than in summer at high latitudes

As long as ice or snow is present at the surface, the surface temperature cannot rise above 0°C. Additional heating of the surface will go into melting of ice and snow rather than raising the surface temperature. Where sea ice is thick enough to persist right through the summer in spite of the increased melting, the surface air warming will be strongly suppressed. However, as air temperatures begin to drop during autumn and winter, the thinner sea ice will permit a greater conductive flux of heat from the underlying ocean water (which cannot drop below the freezing point of sea water, about −1.8°C) to the atmosphere. This will reduce the normal seasonal cooling of the atmosphere, with the result that

late autumn and winter temperatures in a doubled CO_2 experiment can be more than 10°C warmer than for the control climate (Fig. 10.1(b,c)). The greater heat flux to the atmosphere is offset by a greater rate of freezing at the ice base than in the control climate, which releases the latent heat that ends up in the atmosphere. In effect, greater melting of sea ice in summer is offset by greater freezing of sea water in winter, so that the energy that went into the enhanced summer melting is released in winter. This suppresses the summer warming and enhances the winter warming.

Of course, if enough extra summer melting occurs to completely remove the ice, than a modest summer-time warming of the mixed layer and overlying atmosphere can also occur. This delays the formation of new ice in the autumn, producing a large warming in this season as well.

Greater warming of the land surface than of the ocean surface

An increase in atmospheric CO_2 leads to an increase in the downward flux of infrared radiation to the surface, partly as a direct result of the higher atmospheric emissivity but largely due to the increase in atmospheric temperature. The surface will respond by warming, to an extent determined by how fast the surface can increase the upward fluxes of latent and sensible heat and infrared radiation as it warms. This increase can be called the surface heat flux damping. Over land, the increase in latent heat flux is limited by the availability of water, but not so over the ocean. Thus, the surface heat flux damping is smaller for land, so a larger land surface warming is required. The air in immediate contact with the land surface also experiences greater warming. In the case of the ocean, more of the extra surface heating goes into evaporating water, so less warming occurs. Even if the extra water vapour is all condensed over the ocean, this occurs sufficiently high in the atmosphere that much of the latent heat that is released can be efficiently radiated to space. Thus, the warming of the coupled surface–troposphere system is less for oceans than for continents.

It can be seen from Fig. 10.1(a–c) that sea ice-covered regions in winter are an exception to the above generalization. In this case, the surface temperature response is driven by an increase in the upward heat flow through sea ice (as explained above), a process that operates in the oceans only.

(a) DJF, BMRC

(b) DJF, CCC

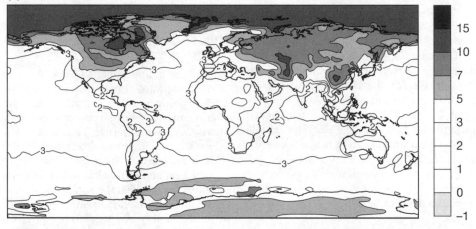

(c) DJF, NCAR

Fig. 10.1 Equilibrium change in surface air temperature (°C) following a CO_2 doubling, as simulated by three different AGCMs having low (BMRC) intermediate (CCC) and high (NCAR) climate sensitivity, and averaged over the months of December–January–February (DJF) or June–July–August (JJA). All results are based on data files kindly provided by one of the authors of the following publications, which also describe the model simulations: BMRC, Colman and McAvaney (1995); CCC, Boer *et al.* (1992); NCAR, Washington and Meehl (1993).

(d) JJA, BMRC

(e) JJA, CCC

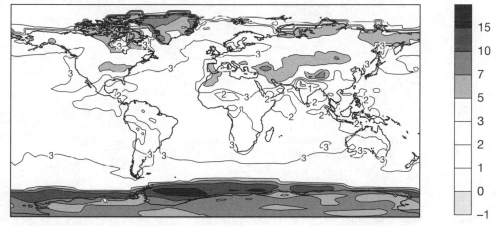

(f) JJA, NCAR

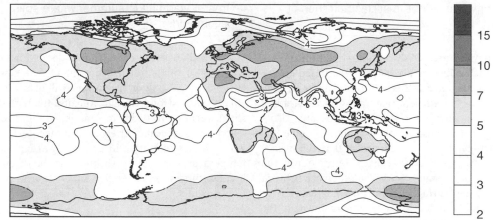

Fig. 10.1 Continued.

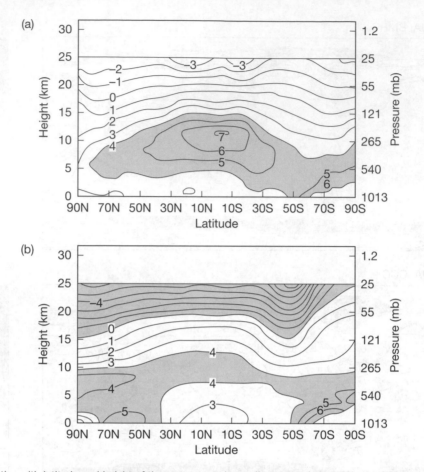

Fig. 10.2 Variation with latitude and height of the mean annual, zonally averaged change in temperature in response to a doubling of atmospheric CO_2 as simulated by representative AGCMs using (a) penetrative convection and (b) moist convective adjustment. Reproduced from Mitchell *et al.* (1990).

Greater warming of the surface air than of the surface

A tendency for greater warming of the atmosphere than of the surface can be anticipated based on the analysis presented in Box 3.1 of Chapter 3. The atmosphere is less efficient than the surface at ridding itself of extra energy by warming up, because of its lower emissivity. Consequently, one can expect the surface air temperature to increase more than the surface temperature, an expectation that is borne out by the results of AGCM simulations. The difference between the surface and surface-air warming is greatest over low- and mid-latitude oceans in summer, owing to the advection of warm air from the continents.

Greater warming in summer where soils become drier

Where soils become drier, the tendency for greater warming of the land surface is intensified, since the evaporative cooling of the surface will decrease. This effect can be strong enough to cause the largest seasonal warming to occur in summer in those regions where soils become drier in summer. Drier soils would be less effective in creating warmer surface temperatures in winter, should soils be drier then, because there is less energy available to heat the surface in winter.

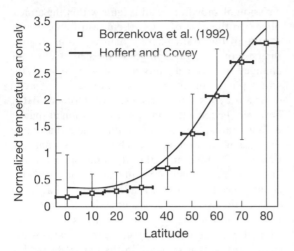

Fig. 10.3 The dependence of zonal-mean temperature changes on latitude. Shown are the normalized temperature response profile for the Holocene, as given by Borzenkova (1992) with error bars estimated by Kheshgi and Lapenis (1996), and Hoffert and Covey's (1992) fit to the latitudinal variation in temperature change between the present and the Last Glacial Maximum, and between the present and the Late Cretaceous. Reproduced from Kheshgi and Lapenis (1996).

Greater warming at night than during the day

In a number of AGCMs where the diurnal pattern of warming in response to a CO_2 increase has been examined, it is found that the nighttime surface warming is modestly greater than the daytime warming at most grid points (Cao *et al.*, 1992; Hansen *et al.*, 1995a; Watterson, 1997; Zwiers and Kharin, 1998). Consequently, there is a reduction in the diurnal temperature range. Stenchikov and Robock (1995) analysed the reasons for a diurnal variation in warming using a 1-D radiative convection model (see Section 6.1). In analysing the reasons for the diurnal variation in the reponse, it is appropriate to examine the radiative forcing and feedback fluxes at the surface (rather than at the tropopause). Stenchikov and Robock (1995) found that both the direct radiative forcing and the feedback fluxes are important to the diurnal response pattern.

As noted in Section 7.1, there is a small negative radiative forcing at the tropopause when CO_2 increases due to additional absorption of NIR (solar) radiation by CO_2 in the stratosphere. Absorption of NIR by CO_2 in the troposphere further reduces the direct forcing at the surface. The diurnal *variation* in the negative NIR surface forcing is comparable to the

variation in the infrared forcing, but opposes it: the strongest negative NIR forcing occurs around noon (up to $-1.5\,W\,m^{-2}$), when the positive IR forcing is also strongest. The net surface forcing is always positive at night, but can be positive or negative during the day, with the daytime forcing tending to be more negative the greater the incident solar radiation at the top of the atmosphere. Thus, the direct effect of the surface forcing is to cause greater surface warming at night than by day. This tendency is reinforced by an increase in the absorption of NIR due to the increase in atmospheric water vapour as the climate warms; this can reduce the radiative energy at the surface by $4-6\,W\,m^{-2}$ at noon. The combination of these two factors is such that, for tropical conditions, Stenchikov and Robock (1995) obtain a surface warming of 2.55°C around sunrise, but only 1.95°C at noon. This corresponds to a reduction in the diurnal temperature range by 0.6°C or 27% of the mean warming of about 2.25°C. For mid-latitude winter conditions, the reduction is about 18%. In AGCM simulations, the ratio of the reduction in diurnal temperature range to diurnal mean warming averaged over all land grid points is 0.26/6.3 = 0.041 (Cao *et al.*, 1992), 0.36/5.9 = 0.061 (Watterson, 1997), 0.30/3.52 = 0.085 (Zwiers and Kharin, 1998), and 0.44/2.80 = 0.16 (fixed clouds) or 0.64/3.56 = 0.18 (variable clouds) (Hansen *et al.*, 1995a). However, there is considerable spatial variability in these results, with the diurnal temperature range increasing rather than decreasing at some grid points, most probably due to a decrease in daytime cloudiness.

Greater warming at low latitudes in the upper troposphere

The greater warming of the upper troposphere at low latitudes is a consequence of the increase in latent heating by the condensation of water vapour, which in turn is a result of an increase in the upward pumping of water vapour by cumulus convection. As discussed in Section 9.1, not all of the extra water vapour pumping leads to local latent heating, as there is also an increase in the detrainment of moist air from cumulus anvils into the surrounding air. In addition, different AGCMs differ substantially in the extent to which cumulus pumping increases as the climate warms, depending on how cumulus convection is parameterized (Cunnington and Mitchell, 1990).

The increase in warming with increasing height in the tropics has two, and possibly three, important consequences. First, it represents a decrease in the lapse rate, which serves as a negative feedback on surface temperature warming (see Sections 3.5 and 9.1). Second, it tends to make the middle and upper tropospheric warming at low latitudes greater than at high latitudes. This in turn will increase the meridional temperature gradient in the upper troposphere, which is opposite to the decrease in meridional temperature gradient that occurs in the lower troposphere. As discussed in Section 10.5, the change in storminess in mid-latitude regions depends to a large extent on how the meridional temperature gradient changes, so the changes in the upper and lower troposphere have opposing effects. Third, the increase in the meridional temperature gradient in the upper troposphere is likely to reduce the ability of planetary waves to transport energy upward into the stratosphere, thereby contributing to a cooling of the stratosphere as the surface climate warms. This topic is discussed next.

Cooling of the stratosphere

The change in stratospheric temperature in response to an increase in CO_2 (and other GHGs) is governed by two factors: changes in the *radiative* energy balance, and changes in the *dynamical* heat transports. The latter involves the reduction in upward heat transport associated with planetary-scale waves that is expected to occur if the upper troposphere warms more at low latitudes than at high latitudes (Shindell *et al.*, 1998). This pattern of warming is driven by overall warming itself and the associated increase in tropical convection. Thus, to the extent that dynamical heating of the stratosphere does decrease as the climate warms, the increase in all GHGs will contribute to a cooling of the stratosphere. However, current AGCMs still do rather poorly in simulating planetary-scale wave interactions, so dynamically induced changes in stratospheric temperature are rather uncertain.

The effect on the stratospheric radiative energy balance of an increase in GHG concentrations is quite different for different GHGs. An increase in the concentration of any GHG increases the downward emission of radiation from the stratosphere to the troposphere (this has a cooling effect on the stratosphere, but contributes to the surface–troposphere radiative forcing). In addition, an increase in the concentration of any GHG reduces the upward

emission of radiation at all heights within the atmosphere. In the case of CO_2, the upward emission at the top of the stratosphere is reduced by *less* than at the base of the stratosphere, so the net effect of the changes in upward emission is also to cool the stratosphere, thereby reinforcing the effect of extra downward emission. In the case of all other GHGs, however, the reduction in upward emission at the top of the stratosphere is *greater* than the reduction at the base of the stratosphere. Thus, there is a net convergence of radiation within the stratosphere, which has a warming effect. As explained in Box 10.1, the convergence in the upward radiative flux is larger relative to the extra downward emission the less saturated the absorption lines are for the gas in question. Thus, for gases whose current concentrations are small enough that they are in the linear absorption regime (namely, SF_6 and the halocarbons; see Section 7.7), increasing concentration tends to warm the stratosphere (see Table 3.1). For increases in CH_4 and N_2O (which lie in the square-root absorption regime discussed in Box 10.1 and Section 7.7), the changes in upward and downward flux divergence almost perfectly cancel, the net result being a very slight stratospheric cooling tendency (see Table 3.1 again). Only CO_2 has a high enough current concentration to lie in the logarithmic absorption region and, as noted above, tends to cool the stratosphere. This cooling tendency due to infrared radiative fluxes is offset by about 10% due to extra absorption of solar radiation by CO_2. Finally, an increase in tropospheric O_3 also tends to cool the stratosphere, by reducing the upward emission of infrared radiation at the tropopause (Berntsen *et al.*, 1997).

According to calculations by Ramaswamy *et al.* (1996), the heating effect of increases in SF_6 and the CFCs offsets just over half of the CO_2 cooling effect over the period 1765–1990, but offsets all of the cooling effect over the period 1979–1990. Whether or not there is a net radiative cooling effect in the future depends on the relative increases in CO_2, SF_6, and the halocarbons. The cooling that has occurred during the last 20 years (see Sections 5.1 and 5.3) has been overwhelmingly caused by the concurrent decrease in stratospheric ozone (see Section 11.4), but this effect will diminish as the amount of ozone gradually recovers during the course of the 21st century (WMO, 1998).

The net cooling of the stratosphere, to the extent that it occurs and influences the temperature changes just below the tropopause, is likely to influence how

Box 10.1 Analysis of the effect of increases in different GHGs on the stratospheric radiative balance

The different effects of increases in different GHGs on the stratospheric energy balance, and hence on stratospheric temperatures, can be understood through the simple three-layer model in Figure 10.1.1. This model is an extension of the two-layer model shown in Fig. 3.1. The three layers are the surface layer, the troposphere, and the stratosphere, with temperatures T_s, T_1, and T_2, respectively. The troposphere and stratosphere have emissivities ε_1 and ε_2, respectively. The upward fluxes at the surface, the tropopause, and the top of the atmosphere (TOA) are indicated in Fig. 10.1.1 as F_s, F_{trop}, and F_{TOA}, respectively. The effective radiating temperature for the surface–troposphere system is the temperature of a blackbody that would produce the same emission at the tropopause. That is,

$$T_{eff} = \left(\frac{F_{trop}}{\sigma}\right)^{0.25} = (T_s^4 - \varepsilon_1(T_s^4 - T_1^4))^{0.25} \quad (10.1.1)$$

Since $T_1 < T_s$, T_{eff} will fall between T_s and T_1. If ε_1 is large, T_{eff} will be close to T_1, and since $T_1 < T_2$ (temperature increases with height in the stratosphere), T_{eff} can be less than T_2. Since F_{TOA} is given by

$$F_{TOA} = \sigma T_{eff}^4 - \varepsilon_2(\sigma T_{eff}^4 - \sigma T_2^4) \quad (10.1.2)$$

it follows that, if $T_{eff} < T_2$, then $F_{TOA} > F_{trop}$ – that is, there is a *divergence* of the upward radiative flux, which tends to cool the stratosphere (more radiation leaves than enters). An increase in ε increases the flux divergence and the associated cooling effect. This reinforces the cooling due to extra emission of downward radiation at the tropopause, which also occurs when ε is increased. Conversely, when ε is small, $T_{eff} > T_2$, so $F_{TOA} < F_{trop}$ – that is, there is a *convergence* of the upward radiative flux, which tends to heat the stratosphere. When ε increases, the convergence also increases, thereby offsetting to some extent the extra downward emission.

The above arguments can be applied to the radiative fluxes at individual wavelengths. If a

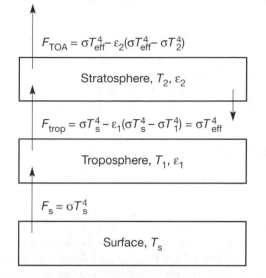

Fig. 10.1.1 A three-layer model

GHG is of low enough concentration that its absorption of radiation is weak (such that it falls in the linear absorption regime) and it absorbs at wavelengths where there is little other absorption, then $\varepsilon \ll 1$, $T_{eff} \approx T_s$, and an increase in the gas concentration causes a sufficiently strong additional convergence of the upward radiative flux that the net effect is to warm the stratosphere. This is the case for the CFCs, HCFCs, and HFCs. If the gas is of high enough concentration such that T_{eff} at the wavelengths where the gas absorbs is only moderately larger than T_2, then the cooling effect of increased downward emission will not be cancelled out, and an increase in the concentration of the gas will have a weak cooling effect on the stratosphere. This is the case for CH_4 and N_2O. If the initial gas concentration is higher still, then ε is sufficiently large that the upward flux in the stratosphere divergences at the wavelengths where the gas absorbs. An increase in gas concentration increases this vertical flux divergence, thereby reinforcing rather than counteracting the effect of increased downward emission, giving a strong cooling effect on the stratosphere. This is the case for CO_2.

peak tropical cyclone intensity changes as the climate warms, as discussed below in Section 10.6. Stratospheric cooling, whether caused by increases in GHG concentrations or by the depletion of stratospheric O_3, is also expected to delay the recovery of stratospheric O_3 (Shindell *et al.*, 1998).

10.2 Soil moisture response

The changes in soil moisture that occur as the climate warms are among the most important characteristics of climatic change. If present soil moisture levels can be maintained as the climate warms, the impacts of climate change on natural ecosystems and agriculture will be much less severe or, in many instances, decidedly positive. If, on the other hand, soils become drier as the climate warms, than the prospect of a warmer climate is much more serious. AGCMs project a general drying of soils as the climate becomes warmer, but disagree sharply concerning the extent of drying and concerning the geographical patterns of soil moisture change. Paleoclimatic data from climates warmer than at present, in contrast, have been interpreted as implying that a warmer climate will be associated with moister soils.

In order to provide some insight into the differences in soil moisture changes as projected by different AGCMs, and of the difficulty in making accurate projections, we begin this section with a brief outline of the surface hydrological schemes used in AGCMs. We then discuss the changes in soil moisture as simulated by recent AGCMs, with an emphasis on understanding the reasons for the differences between different models and how one can subjectively correct for known shortcomings in the models. Next, we discuss the fundamental principles that underlie changes in soil moisture and that are independent of the details of any given model. In this way, we gain insights that do not depend on the quantitative validity of any given AGCM. This section closes by reconciling the apparent disagreement between AGCM-based and paleoclimate-based projections of future changes in soil moisture.

Surface hydrology in nature and in AGCMs

The amount of moisture in the soil will change in response to changes in the inputs or withdrawals from the soil column. There are two major inputs to soil water in nature: infiltration of a portion of the rainfall reaching the land surface, and (at higher latitudes) infiltration of a portion of snowmelt. Rainfall reaching the land surface is subject to a number of processes: interception by the vegetation canopy and direct evaporation or drip; runoff when rainfall rates (minus interception) exceed the infiltration rate of the soil; and infiltration of the remainder. The partitioning of rainfall between interception, runoff, and infiltration depends on the amount, type, and characteristics of the vegetation, the rainfall intensity, the amount of water already in the soil, and the topography.

Withdrawals of soil moisture occur through evapotranspiration, direct evaporation from bare soils, and subsurface drainage. Evapotranspiration depends on surface temperature, atmospheric humidity, wind speed, and the stomatal and aerodynamic resistance to the movement of water. Stomatal resistance depends on the density and degree of opening of the plant stomata, the latter of which is affected by the availability of light and water (through their influence on the rate of photosynthesis) and on the ambient atmospheric CO_2 concentration. The aerodynamic resistance depends on the degree of turbulence or mixing of the atmosphere next to the surface, being smaller the greater the turbulence. This turbulence in turn depends on wind speed, the vertical stability of the atmosphere, and the surface roughness.

These rather complex land surface processes are treated in a highly simplified manner in AGCMs. A widely used scheme is referred to as the "bucket model". In this scheme, each land surface grid point is assumed to contain a bucket with a prescribed water-holding capacity. Some models use the same water-holding capacity everywhere, while others allow for geographical variation. When the water content of the soil is less than the water-holding capacity, rainfall and snowmelt enter the bucket. Runoff occurs only when the bucket reaches its capacity. The evaporation rate is equal to that which would occur for a saturated surface, times a factor that is smaller the less water there is in the bucket. Usually this factor decreases linearly with water content below some threshold water amount. A variant of the bucket model is to allow for a shallow soil moisture reservoir (with a water-holding capacity of around 1 cm) and a deeper reservoir. The evaporation rate in this case depends on the moisture content of the upper reservoir, not of the total soil reservoir, and the water content of the upper reservoir can respond quickly to precipitation events. A further variant is to allow some runoff to occur even when the soil reservoir (or upper reservoir) is not full, depending on the rainfall rate. The main thing that is lacking in the bucket model or any of its variants is the explicit representation of the vegetation canopy.

One of the simplest ways to incorporate a vegetation canopy is outlined in Box 10.2, which provides the foundation for the Biosphere–Atmosphere Transfer Scheme (BATS), developed by Dickinson (1984), and the *Si*mple *Bi*osphere Model (SiB) of Sellers *et al.* (1986). With an explicit canopy layer, separate temperatures are computed for the ground surface, canopy foliage, and canopy air. Since these three temperatures can differ by several degrees, the incorporation of a canopy layer can lead to significant changes in the simulated heat and moisture fluxes, and in the response of soil moisture to rainfall. Box 10.3 compares the computation of evaporation in AGCMs using BATS or SiB with the well-known and widely used Penman–Monteith equation.

Sato et al. (1989) compared surface air temperatures and the partitioning of net surface radiation into latent and sensible heat fluxes by the US National Meteorology Center GCM using the bucket model and SiB. The bucket model almost always substantially overestimates evaporation compared to observations, while the SiB agrees well with observations. This is illustrated in Fig. 10.4, which compares the bucket model and SiB with observations for an evergreen coniferous forest grid cell in British Columbia, Canada. The changes in surface evaporation using SiB lead to changes in modelled precipitation of 25–50% over parts of North America and Eurasia. A comparison of observed seasonal soil moisture variation in the former Soviet Union with that simulated by three AGCMs using the bucket model indicates that all three models simulate too much seasonal drying for the present climate (Vinnikov and Yerserkepova, 1991).

The overprediction of evaporation by the bucket model could cause too much drying of soils in summer in response to a warmer climate. On the other hand, if, as a result of error in the bucket model, present soils are too dry compared to observations, then the scope for additional drying is reduced and the simulated decrease in soil moisture might be too small. As will become even clearer below, a grid-point-by-grid-point analysis of the errors in the simulated soil moisture for the present climate is required in order to at least partially assess the validity of the simulated changes in soil moisture.

An important parameter in models such as BATS and SiB is the stomatal resistance, r_s. The smaller the stomatal pore size, the greater r_s and the less water loss through evapotranspiration that will occur. In nature, r_s is tied to the process of photosynthesis, since the stomata open to allow more CO_2 to enter the plant when photosynthesis rates go up. Photosynthetic rates in turn depend on climate-related variables such as the availability of sunlight and soil moisture, and on temperature. For a given rate of photosynthesis, r_s depends on the atmos-pheric CO_2 concentration, since a higher CO_2 concentration permits a smaller stomatal opening for a given inflow of CO_2. This in turn will tend to reduce water loss, but at the same time will increase leaf temperature (since there will be less evaporative cooling). The feedback between r_s, canopy temperature, and evapotranspiration is accounted for in models such as BATS and SiB, but until recently, the computation of r_s has not had a sound biophysical basis. A second version of the SiB model, SiB2, has been developed that explicitly computes rates of photosynthesis (Sellers et al., 1996a,b; Randall et al., 1996). This allows r_s to be directly parameterized in terms of the computed photosynthesis rate and atmospheric CO_2 concentration, and allows for the coupling between the carbon, water vapour, and energy fluxes. Sellers et al. (1997) provide a concise review of developments leading up to SiB2.

Snowmelt is also modelled very crudely in global climate models: if snow is present and the computed surface temperature is above 0°C, the temperature is reset to 0°C and the excess energy computed and used to melt snow. In most cases, no distinction is made between ground and snow temperatures or between ground and snowpack heat capacities. The processes of melting and refreezing as the snowpack ripens, which can lead to a sudden rather than gradual release of snowmelt once the snowpack collapses, are not represented. In most models, snowmelt is assumed to infiltrate into the soil unless the soil is saturated. In practice, relatively little infiltration will occur if the snowmelt is released suddenly or the soil is frozen. However, an improved scheme that treats snowmelt and related processes in reasonable detail has been developed by Lynch-Steigletz (1994) and Douville et al. (1995) for AGCMs, but CO_2 increase experiments using the new schemes have yet to be published.

Table 10.1 summarizes the land-surface schemes that have been used in recent CO_2-increase experiments. Many models still use the bucket model or one of its variants; in only one case has BATS been used, while results using SiB2 have only recently been published (and are discussed more fully in Section 10.3).

Results from AGCM simulations

The major hydrological changes that are simulated by AGCMs to result from a doubling of atmospheric CO_2 are as follows:

Box 10.2 Incorporation of a vegetation canopy in AGCMs

An increasing number of AGCMs allow for an explicit vegetation canopy between the soil surface and the atmosphere, and compute separate soil surface, foliage, and canopy air temperatures. They also allow for separate moisture fluxes from the soil to the canopy air, from leaves to the canopy air, and from the canopy air to the free air above the canopy. In this box we outline the principles involved in the computation of the heat and moisture fluxes when there is an explicit vegetation canopy. These principles form the foundation for the Biosphere–Atmosphere Transfer Scheme (BATS), developed by Dickinson (1984) and first used for CO_2 doubling experiments by Henderson-Sellers *et al.* (1995), and the *Simple Biosphere Model* (SiB) of Sellers *et al.* (1986).

Figure 10.2.1 illustrates the sensible and latent heat fluxes that will be considered here. The sensible heat flux from the ground to canopy air, from foliage to canopy air, and from canopy air to the air above the canopy can be written as

$$H_g = \rho c_p C_D U (T_g - T_{af}) \quad (10.2.1)$$

$$H_f = \rho c_p \sigma_f L_a \left(\frac{T_f - T_{af}}{r_a} \right) \quad (10.2.2)$$

and

$$H_a = \rho c_p C_D U (T_{af} - T_a) \quad (10.2.3)$$

respectively, where ρ is air density; c_p is the specific heat of air; C_D is a coefficient that depends on wind speed, atmospheric stability, and surface roughness; U is the near-surface wind speed; σ_f is the foliage fraction; r_a is the aero-dynamic resistance; and L_a is the leaf area index (the ratio of leaf area to ground area). The ground temperature T_g and canopy air temperature T_a are determined by computing the rates of change at the beginning of a time step and extrapolating the rate of change over a short time step (a *prognostic* approach). The canopy temperature T_{af} is determined by assuming that the canopy air has zero heat capacity, and deducing the T_{af} required so that there is no heat gain or loss from the canopy (a *diagnostic* approach). That is, one solves for T_{af} such that

$$H_a - H_f - H_g = 0 \quad (10.2.4)$$

which is equivalent to assuming that T_{af} adjusts instantly so as to make Eq. (10.2.4) true. This leaves the foliage temperature T_f, which is also determined diagnostically but in conjunction with the evaporative fluxes (as explained below).

The water vapour fluxes from the ground to canopy air and from canopy air to air above the canopy are given by

$$E_g = \rho C_D U (q_g - q_{af}) \quad (10.2.5)$$

and

$$E_a = \rho C_D U (q_{af} - q_a) \quad (10.2.6)$$

respectively, where q_g, q_{af}, and q_a are the specific humidities of the ground, canopy air, and air above the canopy (these are related to the vapour pressures as follows: $q_x = 0.622 e_x / P_a$, where P_a is the air pressure). The vapour flux from the canopy to canopy air can be divided into evaporation E_f^{WET} from the wet fraction f_{wet}, and transpiration E_{tr} from the dry fraction $(1 - f_{wet})$. These fluxes can be represented by

$$E_f^{WET} = f_{wet} \, \rho \sigma_f L_a \left(\frac{q_{sat}(T_f) - q_{af}}{r_a} \right) \quad (10.2.7)$$

and

$$E_{tr} = (1 - f_{wet}) \, \rho \sigma_f L_a \left(\frac{q_{sat}(T_f) - q_{af}}{r_a + r_s} \right) \quad (10.2.8)$$

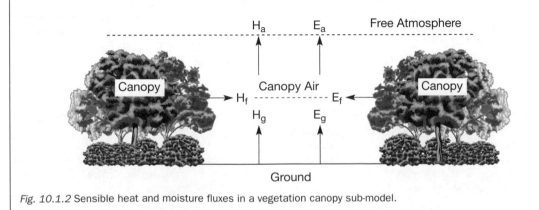

Fig. 10.1.2 Sensible heat and moisture fluxes in a vegetation canopy sub-model.

where $q_{sat}(T_f)$ is the saturation specific humidity at temperature T_f, and r_s is the resistance to the transfer of water vapour from inside to outside the leaf (which occurs overwhelmingly through the stomata). The total vapour flux from the canopy is $E_f = E_f^{WET} + E_{tr}$. As with temperature, q_g and q_a are determined prognostically, and q_{af} is determined diagnostically by solving

$$E_a - E_f - E_g = 0 \qquad (10.2.9)$$

which implicitly assumes that the canopy air has no moisture-holding capacity. The foliage temperature T_f, needed for the computation of both sensible and latent heat fluxes from the foliage to free air, is solved diagnostically by assuming the foliage to have zero heat capacity. That is, one solves for T_f such that

$$K_f^* + L_f^* - H_f - L_e E_f = 0 \qquad (10.2.10)$$

where K_f^* and L_f^* are the foliage short-wave and long-wave net radiation, respectively, and L_e is the latent heat of evaporation. The absorption of short-wave radiation can be determined from a multiple scattering computation assuming a horizontally extensive and uniform canopy with random or non-random leaf orientations (Dickinson, 1983).

To summarize the temperature and moisture computations used in the BATS and SiB models, surface and free air temperatures and specific humidities are determined prognostically, while foliage temperature, canopy air temperature, and canopy specific humidity are determined diagnostically using Eqs (10.2.10), (10.2.4), and

(10.2.9), respectively. These equations are coupled, since each of the three unknowns appears in two or more of the equations, and they are non-linear, which necessitates an iterative solution. The calculated vegetation canopy temperatures can differ by several degrees Celsius from the ground temperature.

In both the BATS and SiB models, runoff arises both from precipitation in excess of infiltration and by gravitational drainage from the lowest soil layer. Both models allow for storage and direct evaporation of intercepted water, with the maximum interception storage dependent on vegetation type and leaf area index. In both the BATS and SiB models, r_a depends on the roughness of the surface, which is parameterized in terms of vegetation type and the seasonally dependent leaf area index.

The surface resistance r_s in the BATS and SiB schemes assumes that factors representing leaf temperature, soil moisture, incident solar radiation and (in the case of SiB) atmospheric vapour pressure deficit can be multiplied together to obtain their interactive effect; whether this approach correctly captures the interaction between the controlling factors is uncertain. The individual factors in turn depend on parameters such as soil texture, plant type, and seasonal growth stage. In the bucket model, evaporation varies linearly with soil moisture content (if all other parameters are unchanged), whereas in BATS and SiB, r_s varies with soil moisture in a more physically realistic, non-linear manner.

- an increase in global mean evaporation from the oceans by 5–10% in most models and a corresponding increase in the global mean rate of precipitation, so that there is an overall intensification of the hydrological cycle;[1]
- a poleward shift in the mid-latitude belt of high rainfall associated with the westerly jet stream;
- an increase of precipitation in mid-latitudes in winter and increases or decreases in summer;

- a tendency for soils to become drier in summer in mid-latitudes;
- a tendency for the Asian and West African monsoons to become stronger; and
- a mixture of increases and decreases in soil moisture in summer in tropical regions.

Kellogg and Zhao (1988), Zhao and Kellogg (1988), and Meehl and Washington (1988) presented and analysed the differences between the geographical patterns of soil moisture change as simulated by various AGCMs for equilibrium CO_2-increase experiments. Papers by Mitchell and Warrilow (1987), Mitchell et al. (1987), Rind (1987, 1988), and Milly (1992) all contain useful insights concerning the ways in which the soil moisture changes simulated by AGCMs depend on the way in which the models are formulated. Rind et al. (1990) clarify the underlying physical principles that can be expected to drive soil

---

[1] Wild et al. (1997) reported results from a high-resolution version of the ECMWF AGCM, in which there was a slight *reduction* in the intensity of the hydrological cycle due to a reduction in surface wind strength. Present-day sea surface temperatures were prescribed for the simulation with present CO_2, and changes in SST from a low-resolution, coupled AOGCM were prescribed for the doubled-CO_2 run. The low-resolution model itself produced the normal intensification of the hydrological cycle.

Box 10.3 Relationship between the formulation of evapotranspiration in AGCMs and the Penman–Monteith formulation used in applied micrometeorology

In this box the relationship between the bulk aerodynamic formulation of the surface–air sensible and latent heat fluxes, which is the cornerstone of the computation of these heat fluxes in AGCMs, and the widely used Penman–Monteith equation is explained. At first glance, these two approaches appear to assume that completely different physics drive evaporation from the land surface. In this box it will be shown that, in fact, the two approaches are based on the same underlying physics.

The Earth's surface in general has a net radiative surplus (Q^*) that is balanced by upward fluxes of sensible (Q_H) and latent (Q_E) heat, and by a conductive heat flux into the surface (Q_G) which is very close to zero when averaged over a year. That is,

$$Q^* = Q_H + Q_E + Q_G \qquad (10.3.1)$$

The most direct driving force for the latent heat flux is the difference in vapour pressure at the surface–air interface and some distance above the interface. Similarly, the sensible heat flux is driven by the difference between the surface and near-surface atmospheric temperatures. For given surface–air temperature and humidity differences, the fluxes will depend on wind speed and a proportionality coefficient. This leads to the "bulk aerodynamic" formulation of the sensible and latent heat fluxes which, for a saturated surface, are given by

$$Q_H = \rho c_p C_D U (T_s - T_a) \qquad (10.3.2)$$

and

$$Q_E = \rho L_e C_D U (q_{sat}(T_s) - q_a) \qquad (10.3.3)$$

respectively, where T_a is the near-surface air temperature extrapolated from the lowest model atmospheric layer, T_s is the surface temperature, and the other terms are as defined in Box 10.2.

The computation of Q_H and Q_E from Eqs (10.3.2) and (10.3.3) requires knowledge of the surface temperature. Surface temperature is readily available for computer simulations of climate, but is not readily available from observations of the real world. Screen-level atmospheric temperature, however, is easily obtainable. The surface saturation vapour pressure e_s can be approximated in terms of the atmospheric saturation vapour pressure and surface–air temperature difference using a first-order Taylor series expansion:

$$e_s(T_s) \approx e_s(T_a) + \frac{de_s}{dT}(T_s - T_a) \qquad (10.3.4)$$

The latent heat flux for a saturated surface can be written as

$$Q_E = \frac{\rho c_p}{\gamma} \left(\frac{e_s(T_s) - e_a}{r_a} \right) \qquad (10.3.5)$$

where $\gamma = c_p P_a / 0.622 L_e$ is the psychrometric constant and r_a is the aerodynamic resistance. If $r_a = 1/UC_D$, then Eqs (10.3.5) and (10.3.3) are equivalent for saturated surfaces. Combining Eqs (10.3.2) (with UC_D replaced by $1/r_a$), (10.3.4), and (10.3.5) with the surface energy balance given by Eq. (10.3.1), one obtains

$$Q_E = \frac{(Q^* - Q_H)\Delta + \rho c_p (e_s(T_a) - e_a)/r_a}{\Delta + \gamma} \qquad (10.3.6)$$

where $\Delta = de_s/dT$. This is the Penman (1948) equation. It embodies the dependence of Q_H and Q_E on $T_s - T_a$ and $q_s - q_a$, respectively, as in the bulk aerodynamic equations used in climate models, but neither T_s nor q_s appears explicitly.

In the case of evaporation from a leaf, one can assume that the vapour pressure inside the leaf is at the saturation value $e_s(T_s)$ corresponding to the leaf surface temperature T_s (as with q_{sat} and T_f in Eq. (10.2.8) of Box 10.2), but that passage of the vapour to the outside of the leaf (primarily through stomata) encounters a surface resistance r_s. Once outside the leaf, the vapour is subject to the aerodynamic resistance r_a in its passage to the free air immediately above the canopy, where e_a applies. Monteith (1965) showed that the evaporative flux from a single leaf in this case is given by

$$Q_E = \frac{\rho c_p}{\gamma} \left(\frac{e_s(T_s) - e_a}{r_s + r_a} \right) \qquad (10.3.7)$$

This expression is equivalent to that used in climate models that have an explicit vegetation canopy (Eq. (10.2.8) of Box 10.2). However, as in the case of a saturated surface, explicit consideration of the surface temperature and $e_s(T_s)$ can be avoided through use of Eqs (10.3.1), (10.3.2), and (10.3.4), along with Eq. (10.3.7), to yield

$$Q_E = \frac{(Q^* - Q_H)\Delta + \rho c_p (e_s(T_a) - e_a)/r_a}{\Delta + \gamma \left(1 + \dfrac{r_s}{r_a}\right)} \qquad (10.3.8)$$

This is the Penman–Monteith equation or "combination model" for unsaturated surfaces.

The Penman–Monteith equation is generally interpreted as implying that evaporation is driven by two terms: the net radiation at the surface (minus the conductive heat flux into the ground), and the atmospheric vapour pressure deficit (VPD, $e_s(T_a) - e_a$). However, it is clear from the above derivation that the VPD is merely a *proxy* for the single physical driving term, $e_s(T_s) - e_a$, and that the net radiation only indirectly drives evaporation, to the extent that it serves to maintain the surface temperature in the face of evaporative cooling. The Penman–Monteith equation can nevertheless be used to understand the effect of warmer temperatures on evapotranspiration in AGCMs, since it is based on the same underlying physics (Eqs (10.3.1), (10.3.2), and (10.3.7)). In particular, if T_a increases with constant relative humidity, then both terms in the numerator of Eq. (10.3.8) will increase, and the evaporation rate increases (albeit by a smaller fraction due to the presence of Δ as an additive term in the denominator).

moisture changes in reality – principles that are independent of the details of any particular model. Papers published since then in which soil moisture changes are presented have largely involved transient simulations, the main exception being Wetherald and Manabe (1995). The qualitative lessons that can be gleaned from the above-mentioned papers are outlined below.

A general tendency for drying of mid-latitude soils in summer is expected based on very fundamental principles, and so can be regarded as a reliable model result (Rind et al., 1990)

Drought tends to occur in nature and in climate models when potential evapotranspiration exceeds precipitation. Potential evapotranspiration (E_p) is the evapotranspiration rate that occurs when water is freely available. It is driven by the difference between the saturation vapour pressure evaluated at the surface temperature, $e_{sat}(T_s)$, and the atmospheric vapour pressure (e_a). Actual evapotranspiration is equal to E_p times some resistance factor that depends on how depleted the soil moisture is. The relationship between saturation vapour pressure and temperature is given by the Clausius–Clapeyron equation, and is shown in Fig. 10.5. The relationship is highly non-linear, with a strongly concave upward shape. The non-linearity of the Clausius–Clapeyron equation is absolutely crucial to the response of the hydrological cycle and of soils to a warmer climate.

First, with equal warming of the land surface and of surface air over land, $e_{sat}(T_s)$ will increase faster than $e_{sat}(T_a)$ because T_s is warmer than T_a by 1–2°C. In the version of the GISS model analysed by Rind *et al.* (1990), both the land surface and surface air temperature over land warmed by about 4.5°C, which increased $e_{sat}(T_s) - e_{sat}(T_a)$ by 40%. If the surface-air relative humidity is constant (as is roughly the case in AGCMs), e_a will increase even more slowly than $e_{sat}(T_a)$ since $e_a = \mathrm{RH}\, e_{sat}(T_a)$, so that $\Delta e_a = \mathrm{RH}\, \Delta e_{sat}(T_a)$. In this case, $e_{sat}(T_s) - e_a$ will grow even more quickly as temperatures increase. Thus, E_p over land tends to increase. Two factors (not related to the non-linearity of the Clausius–Clapeyron equation) cause the increase in ocean evaporation to be much smaller than the increase in E_p over land. First, the land surface warms more than the ocean surface, so the increase in $e_{sat}(T_s)$ over the ocean is less than that over land. Second, air temperature over land tends to warm more than the ocean surface, and because this air is advected over the ocean, the air over the ocean warms more than the ocean surface, on average. This allows ocean air to hold more moisture, thereby causing a larger increase in e_a and inhibiting the increase in $e_{sat}(T_s) - e_a$ over the ocean. It also increases the stability of the atmosphere next to the surface, which reduces turbulent mixing and hence evaporation. The greater warming of the land than of the ocean surface, and of oceanic air than of the ocean surface, are consistent and intuitively reasonable results of climate models. The net result is that, in the GISS model, evaporation from the ocean (and global mean precipitation) increases by 11% on average while E_p from land increases by 30%. As result, E_p from land increases faster than precipitation over land, and soils tend to become drier on average.

The overall tendency for drier soils is exacerbated at low and middle latitudes by further effects related to the non-linearity of the Clausius–Clapeyron equation. Low and middle latitudes see the largest increase in E_p because they are warmer to begin with. In contrast, the largest increase in precipitation will tend to occur where the air is coldest and hence easiest to saturate (i.e., at high latitudes). Thus, increases in E_p will have a particularly strong tendency to exceed increases in precipitation at low and middle latitudes.

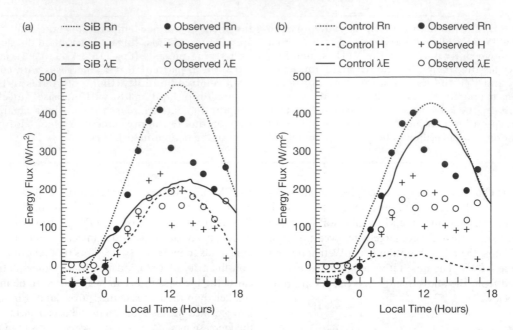

Fig. 10.4 Comparison of vertical sensible (H) and latent (λE) heat fluxes and of net surface radiation (Rn) as computed by the bucket model, by SiB, and as observed for a deciduous forest site in British Columbia, Canada. Reproduced from Sato *et al.* (1989).

Table 10.1 Treatment of surface–air heat fluxes and surface hydrology used in equilibrium CO_2-increase experiments with AGCMs (discussed in this chapter) and in transient experiments with coupled AOGCMs (discussed in Chapter 11). See Table 6.1 for definitions of the acronyms used here.

Model	Reference	Soil hydrology scheme
Equilibrium experiments		
BMRC	Colman and McAvaney (1995)	Bucket model
CCC	Boer *et al.* (1992)	Explicit vegetation canopy[a]
CSIRO	Watterson *et al.* (1997)	Explicit vegetation canopy[b]
GLA	Sellers *et al.* (1996c)	SiB2
GFDL	Manabe *et al.* (1991, 1992)	Bucket model[c]
GISS	Rind (1988)	Two-layer bucket model[d]
NCAR	Washington and Meehl (1989, 1993)	Bucket model
	Henderson-Sellers *et al.* (1995)	BATS
UKMO/Hadley	Early papers	Bucket model
Centre	Betts *et al.* (1997)	Multi-layer soil + vegetation canopy
Transient CO_2 increase experiments		
CSIRO	Gordon and O'Farrell (1997)	Explicit vegetation canopy
ECHAM	Cubasch *et al.* (1992)	Bucket model?
Hadley Centre	Murphy (1995)	Explicit vegetation canopy
NCAR	Washington and Meehl (1989)	Bucket model

[a] Has a single soil layer for moisture and geographical variation in vegetation and in soil properties.
[b] Has two layers for soil moisture, geographical variation in soil properties, canopy interception of moisture, and fast and slow runoff.
[c] Assumes a 15 cm soil water-holding capacity everywhere.
[d] Generation of runoff depends on soil moisture content and rainfall rate.
[e] Allows for interception storage, and fast and slow runoff, with one soil moisture reservoir.

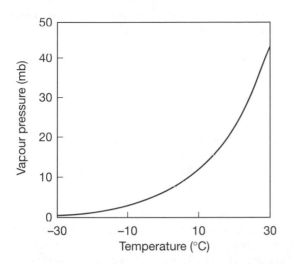

Fig 10.5 The relationship between temperature and saturation water vapour pressure, known as the Clausius–Clapeyron relation.

Changes in both mid-latitude and low-latitude precipitation depend on the polar amplification of the temperature response (Rind, 1987, 1988)

The greater the polar amplification of warming, the greater the decrease in the meridional temperature gradient that occurs and the greater the decrease in mid-latitude eddy activity. This leads to a reduction in water supply at mid-latitudes compared to the case with smaller polar amplification. For a given global mean warming, a greater polar amplification implies a smaller tropical warming and evaporation increase, which also decreases rainfall in mid-latitudes. The pattern of rainfall increases and decreases at low latitudes also depends on the degree of polar amplification, as this affects where increases or decreases in subsidence and rising motion occur.

The simulation of soil moisture conditions in late winter and early spring for the present climate is critical to the simulated change in summer soil moisture when CO_2 is increased

Meehl and Washington (1988) analysed the change in summer soil moisture for CO_2 doubling as predicted by the NCAR and GFDL models. The GFDL model generally simulated mid- to high-latitude soils as being saturated in winter at present, while the NCAR model simulated the soils as being unsaturated in winter. Both models predict an increase of winter

precipitation, which is lost as runoff in the GFDL model but leads to greater spring soil moisture in the NCAR model. Both models undergo about the same increase in spring-to-summer soil moisture loss for a doubled CO_2 climate. However, soils in the NCAR model receive a boost in their moisture content in early spring under a warmer climate, so the NCAR model projects little to no drying of soils in summer compared to present, while the GFDL model projects significant drying. On the other hand, the NCAR model simulates close to zero present-day summer soil moisture in mid-latitudes, such that there is very little scope for further summer soil moisture decreases. The correct soil moisture change is therefore likely to lie somewhere between the NCAR and GFDL model results.

The treatment of snowmelt also affects the change in summer soil moisture through its effect on spring soil moisture in the unperturbed simulation. Mitchell and Warrilow (1987) compared the effect on summer soil moisture of a CO_2 doubling using two versions of the UKMO GCM: in one, all snowmelt infiltrates the soil unless the soil is saturated, while in the second all snowmelt is assumed to produce runoff if either of the middle two soil layers is sub-freezing. In the first case, mid-latitude soil conditions for present CO_2 are simulated to be saturated in spring. The increase in winter precipitation and subsequent snowmelt under doubled CO_2 therefore does not increase initial soil water amounts prior to the period of enhanced evaporation in late spring and summer, so late summer soil moisture decreases when CO_2 is increased. In the second case, soils are not simulated to be saturated in spring for present CO_2, so that the winter precipitation increase under doubled CO_2 leads to greater spring soil moisture. This increase in spring moisture largely offsets the increased spring–summer evaporation, resulting in only a minor decrease in soil moisture by late summer. These two parameterizations represent extreme cases, but they demonstrate the importance of snowmelt modelling through its control on the late spring soil moisture in the unperturbed climate.

The way in which runoff is modelled affects the sensitivity of soil moisture to changes in climate

In the widely used bucket model, no runoff occurs in response to rainfall until the soil is saturated. Allowing for some runoff under unsaturated conditions (due to subsurface drainage) reduces the changes in soil moisture in response to climatic change (Milly, 1992). This is because, as the amount of soil moisture decreases, the runoff decreases, and this reduction acts to partially counteract the decrease in soil moisture. Gregory *et al.* (1997) allow for runoff from unsaturated soils, but still simulate significant (20–30%) decreases in summer soil moisture in NH mid-latitudes in a transient simulation.

A poleward shift in the mid-latitude rainfall belt could be a result, rather than a cause, of decreases in soil moisture

In early studies, the decrease in mid-latitude soil moisture was ascribed in part to a poleward shift in the mid-latitude rain belt. However, Wetherald and Manabe (1995) found that sometimes this shift occurs only in summer in their model, and it is not accompanied by a similar shift in the jet stream. They attribute the poleward shift in the rain belt to the fact that soils dry more on the equatorward side of the rainbelt than on the poleward side as the climate warms, which reduces subsequent evaporation. As a result, rainfall (much of which is recycled soil moisture) decreases on the equatorward side, thereby causing a shift in the axis of maximum precipitation.

Earlier snowmelt and an earlier onset of spring rains also contribute to decreases in summer soil moisture

Wetherald and Manabe (1995) find that these changes are the main reasons for a decrease in high-latitude soil moisture in summer, and are a contributing factor to the decrease in mid-latitude soil moisture. However, these effects will be mitigated to the extent that they lead to greater end-of-spring soil moisture.

Feedbacks between soil moisture and cloud cover tend to prolong and/or amplify the changes in spring and summer soil moisture

Kellogg and Zhao (1988), Meehl and Washington (1988), and Wetherald and Manabe (1988) all concluded that a positive soil moisture–cloud feedback operates in the models that they studied. A decrease in soil moisture leads to a reduction in the subsequent rate of evaporation and in near-surface relative humidity. This leads to reduced cloud cover, more solar radiation at the surface, and further drying. Opposite changes occur when soil moisture is initially greater. This feedback is consistent with find-

ings based on the analysis of seasonal anomalies that occur today (e.g., Atlas *et al.*, 1993).

Synthesis of AGCM results

Although AGCMs simulate widespread decreases in summer soil moisture, there are significant regional variations in the magnitude of the drying as simulated by any one model, and significant disagreements at the regional level between different models. The quantitative soil moisture changes simulated by current AGCMs are of low reliability. A good simulation of present amounts of ice and snow, precipitation patterns, and soil moisture is a *necessary* (but not *sufficient*) condition for an accurate simulation of the changes in soil moisture associated with a warmer climate. However, basic principles that are independent of the details of any given model provide several reasons for expecting an overall decrease in summer soil moisture at low and middle latitudes, and for expecting the decreases to become larger and more widespread the more the climate warms.

Implications from the study of past climates

In contrast to computer-based results, analyses of inferred moisture patterns associated with warm intervals in the geological past suggest that continents will become moister rather than drier as the climate warms. However, past warm climates are not suitable guides to the moisture conditions that would occur in association with future warming for at least two reasons: (i) as discussed by Crowley (1990), the seasonal solar radiation pattern (in the case of the mid-Holocene and last-interglacial warm periods) or polar ice cover (in the case of earlier warm periods) were quite different from that which will be associated with warming during the next few centuries, and these differences could readily produce a substantially different precipitation response to warming; and (ii) a significant part of the moisture response to past warm intervals probably involved gradual adjustments in soil properties that would not be involved in moisture changes during the 21st century (Lapenis and Shabalova, 1994). In particular, as climate warms and precipitation in coastal regions increases, the soil moisture-holding capacity is believed to increase. This increase allows for greater storage of rainfall after rainfall events, greater re-evaporation of the moisture, and greater precipitation further inland. However, the adjustment in soil moisture-holding capacity would occur over a period of several hundred years. Hence, data from past climates cannot be taken as evidence that a warmer world would lead to moister continents during the next century. Rather, simple principles indicate that, for fixed soil properties, increasing warming should lead to an increasing tendency for drying in continental interiors, as discussed above.

10.3 Impact of changes in stomatal conductance on the surface temperature and moisture response

In the preceding section the importance of explicitly including a vegetation canopy for the simulation of soil moisture, and changes in soil moisture, was stressed. A key parameter in models of the vegetation canopy is the stomatal resistance to evapotranspiration, r_s. The results surveyed in the preceding section do not allow for changes in r_s induced by an increase in the concentration of atmospheric CO_2; the soil moisture response in these simulations is entirely governed by the changes in precipitation and in the surface energy balance, and by feedbacks between soil moisture, temperature, and r_s. Increasing atmospheric CO_2 concentration will affect r_s in two ways: through the direct effect of atmospheric CO_2 concentration on r_s (higher CO_2 increases r_s by allowing smaller stomatal pore openings), and through the effect of higher CO_2 on the rate of photosynthesis (greater photosynthesis requires larger stomatal pore openings and hence smaller r_s). Changes in r_s will affect the canopy temperature, which will feed back on the rates of photosynthesis and evapotranspiration.

These feedbacks have been explored by Sellers et al. (1996c) using SiB2 coupled to the GLA AGCM. A doubling of atmospheric CO_2 with no change in climate increases the global mean rate of photosynthesis by 32% but decreases the canopy conductivity (equal to $1/r_s$) for CO_2 and water by 26%. That is, in spite of the greater photosynthetic demand for CO_2, the conductivity decreases. When these effects were allowed in combination with CO_2-induced changes in climate, the increase in r_s enhanced the global mean surface warming by only 0.2°C (from 2.6°C to 2.8°C). The greatest enhancement, 0.4°C, occurred in the tropics due to the greater evapotranspiration fluxes there. However, as discussed in Section 8.1,

downregulation of the photosynthetic response might occur, such that the long-term stimulation of photosynthesis is smaller than the initial stimulation. As an upper limit, Sellers et al. (1996c) assumed the downregulation to be strong enough to eliminate the increase in photosynthesis. This reduces the demand for CO_2, which allows a further reduction in the stomatal openings and a larger reduction of canopy conductance (by 35% now). This might be expected to cause yet further enhancement of the surface warming everywhere, but this happened only in the tropics, where the warming was enhanced by 0.9°C rather than by 0.4°C. This enhancement would be stronger still during midday. Elsewhere, changes in atmospheric circulation induced by the changes in surface temperature appear to have opposed the effect of increased r_s on the surface temperature change, since the warming enhancement with downregulation of photosynthesis was less than without downregulation.

The feedbacks involving r_s had a dramatic effect on the change in precipitation over land. With CO_2-induced climatic change but no direct effects of CO_2 on r_s, the average precipitation over land increased by 7.0%. With direct effects of higher CO_2 on photosynthesis and r_s, the increase in precipitation was 6.5%, but when maximum downregulation of photosynthesis was allowed, precipitation increased by only 3.5%. This can be attributed to reduced recycling of soil moisture due to less evapotranspiration associated with the lower canopy conductance. Unfortunately, Sellers et al. (1996c) do not indicate how CO_2-induced changes in r_s altered the impact of a warmer climate on soil moisture.

These experiments demonstrate important linkages between the response of the carbon cycle to higher atmospheric CO_2, and the surface temperature response to the heating effects of higher atmospheric CO_2. Other linkages could occur through the possible effect of higher atmospheric CO_2 on leaf area, which would affect not only the canopy-scale r_s but also the canopy albedo (see Betts et al., 1997).

10.4 Changes in the frequency of days with heavy rain

AGCMs generally show an increase in the proportion of days with heavy rainfall as the climate warms, particularly in the tropics (Kattenberg et al., 1996,

Section 6.5.6). Hennessy *et al.* (1997) found this to be the case even in places where the total number of rainfall days decreases. Gordon *et al.* (1992) found the increase in the intensity of daily precipitation to be a much clearer response of their model to warming than the changes in the total amount of precipitation. An increase in the intensity of rainfall will contribute to the widespread reduction in soil moisture obtained by AGCMs, since relatively more of the rainfall will be lost as runoff.

10.5 Changes in mid-latitude storms

Simulations with AGCMs differ widely concerning the effect of increasing CO_2 on the strength and frequency of mid-latitude storms (also known as extra-tropical cyclones). Branscome and Gutowski (1992), Stephenson and Held (1993), and Zhang and Wang (1997) report a decrease in storminess when CO_2 is doubled; Lambert (1995) finds an overall decrease in the frequency of mid-latitude cyclones but an increase in the frequency of intense cyclones, while Beersma *et al.* (1997) report opposite changes; and Hall *et al.* (1994) report an overall increase in storminess. These contradictory results arise because mid-latitude storminess depends on a number of factors which change in opposite ways as the climate warms, so that the net effect is uncertain. The meridional gradient of surface temperature tends to decrease (tending to reduce the jet stream speed and associated storminess), but the meridional gradient of upper tropospheric temperature tends to increase (which tends to increase storminess). Changes in the vertical temperature gradient (or lapse rate) also affect the occurrence of extra-tropical cyclones. Regional changes in the temperature gradient can be quite different from the zonally averaged change. For example, in the AGCM-mixed layer simulation of Hall *et al.* (1994), only slight warming occurs over southern Greenland, but strong warming occurs over western Europe. This increases the temperature gradient in the north-east Atlantic, and leads to a marked intensification and downstream shift in the Atlantic storm track. Stephenson and Held (1993), in contrast, obtain a weakening of the Atlantic storm track but no change in the Pacific storm track.

10.6 Changes in El Niños, monsoons, and tropical cyclones

Three of the most important features affecting weather and climatic variability in the tropics are the El Niño oscillation; the Australian, Asian, and west African monsoons; and tropical cyclones. The term El Niño refers to the periodic occurrence of warmer than average sea surface temperatures (SSTs) in the eastern equatorial Pacific Ocean. It is associated with colder than average SSTs in the western equatorial Pacific Ocean, greatly increased rainfall in the eastern equatorial Pacific and the adjacent parts of South America, and reduced rainfall in the western equatorial Pacific, south-east Asia, and northern Australia. Altered weather patterns occur in parts of the northern hemisphere mid-latitudes (Fraedrich *et al.*, 1992). The opposing temperature pattern (cold eastern Pacific and warmer than usual western Pacific) is referred to as La Niña. The El Niño drives a rising motion over the eastern Pacific Ocean and compensating descending motion over the western Pacific Ocean, which explains the rainfall anomalies in these regions that were noted above. Variations in the strength of the monsoons and in the occurrence of tropical cyclones are associated with El Niño–La Niña fluctuations. There are additional, direct links between the strength of the Asian monsoon and the occurrence of cyclones in the western Pacific and Indian Oceans. It is therefore appropriate to discuss the projected changes in these phenomena together, as they form an interlinked system.

Monsoons

The monsoons are driven by differences in the temperature contrast between continents and oceans. Moisture-laden surface air blows from the Indian Ocean to the Asian continent, and from the Atlantic Ocean into west Africa during the NH summer, when the land masses become much warmer than the adjacent ocean. Monsoon-like winds also blow from the Gulf of Mexico onto the south-central United States, and onto northern Australia during the SH summer. In winter, as the continents become colder than the adjacent ocean, higher surface pressure develops over the continents and the surface winds blow towards the ocean.

Climate models indicate an increase in the strength of the summer monsoons in response to increasing CO_2. This is a credible result, as there are three very simple reasons why it should occur:

- The continents tend to warm more than the oceans, for reasons that are well understood (see Section 10.1), and this differential warming will be greatest in summer. This increases the land–sea temperature contrast, which is the ultimate driver of the summer monsoon.
- In the case of the Asian monsoon, decreased snowcover on the Tibetan plateau – which can be expected to occur as the climate warms – has been observed today to be associated with a stronger summer monsoon (Dey and Bhanu Kumar, 1982; Dickson, 1984).
- With a warmer climate, the atmosphere can hold more water vapour, so that winds converging over the landmass will be able to supply more moisture.

The first two points pertain to the vigour of the summer monsoon circulation, while the third point pertains to the moisture transport associated with a given air flow.

As for specific results, we draw upon both equilibrium CO_2 doubling simulations using coupled AGCM–mixed layer models and a simulation using the Hadley Centre coupled AOGCM, in which the CO_2 concentration increases by 1% per year (compounded). The changes in the behaviour of the Asian monsoon in the latter case are analysed by Bhaskaran *et al.* (1995), based on a comparison of 70 years of the control run (with constant CO_2) and years 65–75 of the transient run (the CO_2 concentration having doubled by year 70). Ten years is not a particularly long sample, but is comparable to the

Table 10.2 Changes in the characteristics of the Asian monsoon associated with a doubling of atmospheric CO_2 concentration, based on simulations to date with AGCMs.

Monsoon characteristic	Change
Intensity of air flow	Increases in some areas
Position	Shifts north over India
Onset date	No detectable change
Seasonal mean precipitation	Generally increases by 10–20%
Interannual variability of mean seasonal precipitation	Increases by 25–100%
Proportion of days with heavy rainfall	Dramatic increase

sample length used by other researchers in comparing equilibrium simulations. The variables of greatest interest are the timing of the first monsoon rains, the intensity of the air flow, the position of the axis of maximum rainfall, the total seasonal precipitation, the interannual variability of total seasonal rainfall, and the proportion of days with heavy rainfall. The changes in these variables, as simulated by a variety of models, are summarized in Table 10.2 and are discussed below.

Bhaskaran *et al.* (1995) found that monsoon circulation over India intensified by about 10% and shifted north by about 10°. Seasonal mean rainfall increased from 2.53 ± 0.23 mm/day to 3.06 ± 0.46 mm/day. Thus, mean rainfall increased by 20% *but the interannual variability of rainfall doubled.* The average number of days per season with 2–3 mm/day rainfall decreased from about 55 to about 15, while the number of days with 4–5 mm/day increased from about 5 to about 60. Thus, there was a dramatic increase in the number of days with heavy rainfall at the expense of days with light rainfall. Finally, in the small 10-year sample, no statistically significant change in the onset date of the monsoon was detectable. Meehl and Washington (1993) found only a minor (6%) increase in precipitation associated with the Asian monsoon in a doubled-CO_2 climate, and a 25% increase in the interannual variability of precipitation. They cite observations showing an increase in the variability of the Asian monsoon in decades where temperatures are warmer on average. Suppiah (1995) obtained a 20% increase in precipitation by the Australian monsoon, but no change in interannual variability.

The quantitative changes in the monsoon, as simulated by AGCMs, seem to depend strongly on the model resolution. Hirakuchi and Giorgi (1995) compared the changes in the monsoon in east Asia (east of India) as simulated using the NCAR AGCM (having a $4.5 \times 7.5°$ latitude–longitude grid) and as simulated when output from the AGCM was used to drive a high-resolution (50 km grid) regional model. Unfortunately, the comparisons are based on five-year samples only. Nevertheless, the high-resolution model showed a distinctly greater increase in monsoon rainfall when CO_2 was doubled (by 11–26%, depending on location) compared to the low-resolution model (in which rainfall changed by −14% to +14%).

Two factors might counteract the expected tendency for the strength of the monsoons to increase, at least in some regions: (i) both the Asian and

Australian monsoons tend to be weaker during El Niños (Kiladis and Diaz, 1989; Evans and Allan, 1992; Kane, 1997), so if there is an increase in the occurrence or strength of El Niños, this could reduce the average strength of the monsoons; and (ii) the cooling effect of anthropogenic sulphate aerosols over the Asian landmass would tend to reduce the strength of the monsoon by reducing the land–sea temperature difference in summer (Meehl et al., 1996; Mitchell and Johns, 1997).

El Niños

Simulations with coupled AOGCMs, or with AGCMs using prescribed background SSTs and El Niño-like SST anomalies, indicate the following effects of a warmer climate:

- no change in the frequency of El Niños (Meehl et al., 1993a; Knutson et al., 1997);
- greater differences in precipitation and soil moisture between El Niño and La Niña years (Meehl et al., 1993a), no change in interannual variability of SST (Tett, 1995) or of rainfall (Smith et al., 1997), or a decrease in interannual variability of SST (Knutson et al., 1997); and
- very different patterns of change in sea level and upper tropospheric pressure outside the tropics in response to El Niños, compared to the changes that occur under the present climate (Meehl et al., 1993a).

In the three references cited above that used coupled AOGCMs (Meehl et al., 1993a; Tett, 1995; Knutson et al., 1997), the model-simulated El Niño-like variability in SST is about half that observed. The factors that limit present-day El Niño-like variability in these models may also be limiting their response to a warmer climate.[2] Smith et al. (1997), on the other hand, applied observed SST anomalies on present-day background temperatures and on warmer background temperatures associated with a prior CO_2-doubling experiment, and examined the response of their AGCM but without any feedback onto the SST anomalies.

---

[2] The main problem seems to be low resolution (a $4.5° \times 7.5°$ latitude–longitude grid in two cases, $2.5° \times 3.75°$ in the other case), since the El Niño-like variability matches the observed amplitude in a high-resolution version ($100 \text{ km} \times 33 \text{ km}$ at the equator) of the GFDL ocean model (Philander et al., 1992).

The conflicting results concerning changes in wet and dry extremes are due to several competing factors acting at once. The first factor arises from the non-linearity of the Clausius–Clapeyron equation. A given change in SST will have a larger effect on evaporation the warmer the base climate. The greater increase in evaporation in the eastern Pacific when an El Niño occurs in a warmer climate will lead to a greater increase in precipitation in this region, greater rising motion, and greater compensating subsidence in the western Pacific. Similar changes, but with the opposite geographical pattern, will occur during La Niña events. Hence, both wet and dry extremes will tend to be intensified as the climate warms. However, the increase in atmospheric stability at low latitudes that occurs in a warmer climate (owing to a decrease in the lapse rate) will tend to dampen the effects of the non-linear increase in evaporation. This was sufficient to neutralize the tendency for increased variability in the simulations of Smith et al. (1997).

Changes in the average east–west SST gradient in the tropical Pacific Ocean also influence the extent to which the amplitude of El Niño anomalies changes. However, there is no agreement at present as to how this gradient will change as the climate warms. A purely thermodynamic argument indicates that the greatest SST warming should occur where the increase in heat flux away from the surface as the surface warms (the surface heat flux damping) is lowest. Much of the heat flux damping is through an increase in the latent heat flux, which in turn is driven by the increase in surface vapour pressure as SST increases. Owing to the non-linearity of the Clausius–Clapeyron equation (shown in Fig. 10.5), saturation vapour pressure increases more slowly with increasing temperature where the SST is initially colder (i.e., in the eastern Pacific Ocean). This permits greater warming in the eastern than in the western tropical Pacific, thereby reducing the east–west temperature gradient. This is what occurs in coupled AOGCMs (Mitchell et al., 1995; Meehl and Washington, 1995; Knutson et al., 1997; Knutson and Manabe, 1998). Knutson et al. (1997) find that the reduction in the background east–west gradient reduces the amplitude of El Niño oscillations (although the background state becomes more El Niño-like than the present climate).

However, others have argued that the upwelling of cold water in the eastern tropical Pacific Ocean should result in smaller rather than greater SST warming in the eastern Pacific than in the western

Pacific (Seager and Murtugudde, 1997; Cane *et al.,* 1997; Liu, 1998). This in turn would increase the east–west SST gradient. If, as a result of the initial strengthening of the SST gradient, the trade winds that drive the upwelling were to increase, there would be stronger upwelling and an even greater suppression of warming in the eastern Pacific. The extent to which upwelling moderates the warming in the eastern Pacific also depends on how the temperature of upwelling water changes. Since the upwelled water originates in part at the surface in the subtropics, changes in the upwelling temperature will depend on how the subtropical SST changes. Seager and Murtugudde (1997) investigated all of these interactions using a high-resolution (ranging from about 30 km × 30 km to 100 km × 100 km) model of the tropical and subtropical Pacific Ocean (from 40°S to 40°N), coupled to a model of the atmospheric surface layer. They find that the latent heat flux damping is stronger in the subtropics than in the tropics (owing to a greater wind speed in the subtropics), so that SST warming is smaller in the subtropics. This in turn limits the warming of water that upwells along the equator in the eastern Pacific ocean, so that the east–west SST gradient increases as the climate warms. The reason for the difference between this result and the behaviour of coupled AOGCMs is not clear. It is possible that the response of the east–west temperature gradient – and the associated effects on El Niño – could be different during the *transition* to a warmer climate than for an equilibrium warming. This possibility is discussed in Section 11.2.

The lesson that emerges from the above considerations is that the details of the simulation of the present climate are important in determining how El Niño changes in a warmer climate, as are the large-scale spatial patterns of changes in temperature.

Tropical cyclones

There are a number of largely independent issues concerning the effect of a warmer climate on tropical cyclones (TCs). The impact of warmer temperatures on TCs has generated considerable disagreement, as exemplified by the exchange between Lighthill *et al.* (1994), Emanuel (1995), and Broccoli *et al.* (1995).

The first issue is the global area affected by TCs. Today, TCs do not form where ocean temperatures are colder than about 26°C. However, temperatures colder than 26°C do not inhibit TC formation. Rather, the 26°C isotherm corresponds to the begin-

ning of the trade wind inversion as one moves poleward, which does directly inhibit the formation of TCs. The equatorward limits of the trade wind inversion are not likely to change significantly as the climate warms, so the area affected by TCs is likely to remain about the same.

The second issue concerns the frequency of TCs. There are several AGCM-based studies of these issue (as reviewed by Henderson-Sellers *et al.,* 1998), and they give conflicting results: some indicate an increase in the frequency of hurricanes, and others indicate a decrease.

The third and fourth issues concern the average intensity of TCs and the intensity of the most severe storms that occur. The most recent simulations, in which a high-resolution hurricane prediction model is embedded in a global AGCM with prescribed SSTs, give an increase in the average strength of a sample of strong hurricanes by 5–12% under climatic conditions associated with a CO_2 doubling (Knutson *et al.,* 1998). However, observations of the present climate fail to show any correlation between average hurricane strength and SST, and enough processes are omitted from even high-resolution models that the predictions from such models must still be regarded as very tentative.

Current theories of tropical cyclone development indicate that the maximum potential intensity (MPI) of cyclones increases with increasing SST. The problem is that current theories omit a number of factors, all of which will tend to limit the increase in MPI with increasing SST (Henderson-Sellers *et al.,* 1998). In particular, the cooling effect of increasing evaporation of ocean spray as winds pick up could be a self-limiting process. The question can also be approached empirically. DeMaria and Kaplan (1994) find that the maximum hurricane strength (as opposed to the average hurricane strength) does increase with increasing SST. The problem here is that the temperature of the tropopause also decreases with increasing SST in the SST range where hurricanes occur (from the 26°C isotherm to the equator), and this also tends to increase the maximum hurricane intensity. Unless the correlation between tropopause temperature and SST as climate changes is the same as the spatial correlation observed at present, the observed SST–MPI correlation is inapplicable to a change in climate. Thus, it is not possible to say anything definite about the most destructive hurricanes, based on either theoretical or empirical arguments.

10.7 Linkage between uncertainties in climate sensitivity and changes in mid-latitude soil moisture

In closing this chapter, it is useful to return to the question of the climate sensitivity and the implications of low-latitude climate feedback processes for the impact of global warming on summer soil moisture. Part of the uncertainty over climate sensitivity involves the effect of convection on changes in the amount of water vapour in the upper tropical troposphere. A strong negative feedback due to a downward shift in the distribution of water vapour in this region (due to enhanced subsidence, as suggested by Lindzen, 1990) would reduce the global mean climate sensitivity by making the overall water vapour feedback less positive, but would have its largest impact at low latitudes. As a result, the polar amplification would increase. However, a greater polar amplification of the temperature response results in a greater decrease in mid-latitude summer soil moisture for a given warming in mid-latitudes (Rind, 1987, 1988). Thus, a smaller global mean climate sensitivity does not necessarily mean smaller impacts in the major mid-latitude food-producing regions of the world. The beneficial impacts of a smaller global mean warming could be offset to some extent by more severe impacts on soil moisture. The implications of any substantially revised estimates of climate sensitivity (whether lower or higher) depend on the specific reasons for the revised estimate and the associated changes in the geographical – and especially the latitudinal – pattern of climatic change.

10.8 Reliability of regional forecasts of climatic change

This chapter has emphasized the largest-scale patterns of climatic change expected from increasing GHG concentrations, such as the differences in the zonally averaged response at low, middle, and high latitudes; continental-scale changes in soil moisture; and general tendencies concerning changes in tropical and extratropical cyclones, interannual variability, and monsoon circulations. Many policy analysts and political decision makers, in contrast, are interested in climatic change at the national and subnational scales. However, the smaller the spatial scale of interest, the less reliable are the projections – something that seems to be true in many fields of human enquiry and not just in climate modelling. It is generally agreed that AGCMs and AOGCMs are quite unreliable at the scales of greatest interest to policymakers. Furthermore, it should be quite clear based on the evidence presented so far, and based on further evidence to be presented in Chapter 11, that it will be a very long time before reliable regional forecasts of climatic change can be made. Indeed, it might *never* be possible to make reliable regional-scale forecasts in advance. It is hard to see how the uncertainty in even the global mean forecast of temperature change can be reduced below a factor of 2, given the incredible complexity and multiplicity of interacting feedback processes.

This is not to say that regional forecasts are not without value. Regional forecasts provide an indication of how large the local climatic (and soil moisture) changes can be in association with a given global average warming. That is, they provide an indication of the spatial variability in changes. Even if the forecast change is wrong (in detail) in each and every single region (some regions warming or drying more in reality than forecast, other less), AGCMs and AOGCMs can still provide useful information for the development of GHG emission policies. This is because the range of regional changes seen in a given latitude band for a given model, or the range of changes simulated for a given region by a collection of models, gives an indication of the *risks* associated with climatic change. Policy development on the basis of estimated risks, rather than on the basis of specific regional forecasts (as are often asked for), has a much sounder scientific foundation.

10.9 Summary

This chapter has presented the main features of the equilibrium temperature and soil moisture response to a doubling of atmospheric CO_2, as simulated by state-of-the-art AGCMs. Although the quantitative results and regional details differ significantly from one model to the other, the large-scale features of the temperature response (as listed by the bullets in Section 10.1) and of the soil moisture response (as listed by the bullets in Section 10.2) can be readily understood in terms of basic principles. The discussion in this chapter has therefore emphasized the development of an intuitive understanding of the simulated response, and has stressed those features of the response that are largely independent of the details of any single model.

Many examples have been presented in which an accurate simulation of the present climate (and soil moisture) is a necessary (but, unfortunately, not a sufficient) condition for the reliable simulation of future changes. Nevertheless, it is possible to assess at least the direction of the likely errors in some of the model results and, in so doing, narrow somewhat the range of expected outcomes. This is especially the case for simulated changes in soil moisture; those models which simulate the largest decreases in soil

moisture seem to be exaggerating the response, while those models with minimal change in soil moisture seem to be underestimating the response.

Changes in other climatic elements of great interest and importance – mid-latitude storms, El Niños, and monsoons – seem to depend critically on the spatial patterns of simulated change. These in turn depend in part on the simulated changes in the ice and snow distribution and on many other details of the models. In addition, the spatial patterns of climatic change during the transition to a warmer climate are likely to differ in important ways from the equilibrium patterns of climatic change. The question of transient or time-dependent climatic change is addressed in the next chapter.

The transient climatic response and the detection of anthropogenic effects on climate

In the preceding chapter we examined the regional patterns of steady-state climatic change following a doubling of atmospheric CO_2, as simulated by AGCMs coupled to simple, mixed layer-only ocean models. A CO_2 doubling was used as a surrogate for the actual combination of GHG concentration changes giving rise to the same global mean forcing as for a CO_2 doubling. Since the focus in these experiments was on steady-state climatic changes, there was no need to consider the delay in surface warming caused by the mixing of heat into the deep ocean. To reduce the computational cost further, the models were subjected to a sudden doubling of the CO_2 concentration once a steady-state climate with the present (or near-present) CO_2 concentration had been reached.

Real climatic change involves the gradual response to gradually increasing concentrations of GHGs. The time-dependent response to a radiative forcing is referred to as the *transient response*. The transient response is relevant to a number of important issues, such as (i) determining the rate of future climatic change and hence the rate at which human societies and natural ecosystems will have to adapt (where possible); (ii) constraining estimates of the climate sensitivity by comparison of simulated global and hemispheric mean temperature changes (ΔTs) during the 20th century with observed temperature trends; (iii) estimating the relative importance of anthropogenic and natural forcing factors for climatic change during the past century, which is inextricably tied to the question of the climate sensitivity; and (iv) assessing the extent to which the patterns of temperature and rainfall change computed for a doubling of CO_2 would occur in association with smaller amounts of warming during the transition to a doubled-CO_2 climate.

In order to simulate the transient response to increasing GHG concentrations, the mixing of heat into the deep oceans has to be accounted for. In this chapter we examine the transient response. We begin with an explanation of the physical processes and ocean–climate feedbacks that are involved in the transient response, which are best elucidated with the aid of the 1-D upwelling–diffusion ocean model that was described in Section 6.1. We examine not only the surface temperature response, but also the warming of the deep ocean, since the latter is directly relevant to changes in sea level (which are discussed in Chapter 12). We then proceed to the global mean and regional patterns of transient climatic change as simulated by 3-D AOCGMs. Lastly, we compare temperature changes during the past 140 years as simulated by the 1-D upwelling–diffusion model and as observed, and during the past 20 years as simulated by one particular AGCM–ocean model. To make a valid comparison requires developing estimates of the impact of natural forcing factors (at the very least, volcanic activity and solar variability) and taking into account the cooling effect of aerosol emissions. In so doing, we develop indirect observational constraints on the magnitude of the aerosol cooling effect and on climate sensitivity.

11.1 Principles and ocean mixing feedbacks

In order to develop a solid understanding of the factors that govern the transient response, it is useful to begin by considering the response of a simple system consisting of a single, well-mixed box that represents the ocean mixed layer. We will then proceed to a consideration of mixing processes between the mixed layer and deep ocean in the absence of feedbacks between climate and oceanic mixing processes, and finally, we will consider the role of climate–mixing feedbacks. A similar approach was used in Chapter 8 (Sections 8.6 and 8.7) in explaining the oceanic uptake of anthropogenic CO_2.

Mixed layer-only response

To illustrate the factors governing the transient climatic response to an external forcing change, consider a planet that consists of an infinitely mixed, isothermal reservoir of thermal inertia R ($J\,m^{-2}\,K^{-1}$) and temperature T, governed by

$$R\frac{dT}{dt} = Q - F \qquad (11.1)$$

where Q is the global mean absorption of solar radiation and F is the emission of infrared radiation to space. The thermal inertia R is the amount of heat that needs to be added per square metre of horizontal area to raise the temperature of the reservoir by 1 K; the larger it is, the more slowly temperature will change. For a mixed layer of depth h, R is given by $\rho c_p h$, where ρ is the density of seawater ($1027\,kg/m^3$) and c_p is the specific heat of water ($4.186 \times 10^3\,J\,kg^{-1}\,K^{-1}$). If the planet is initially in a balanced state ($Q = F$), and a radiative forcing of magnitude ΔF is applied, the equation governing the change in temperature, ΔT, is

$$R\frac{d\Delta T}{dT} + \lambda\Delta T = \Delta F \qquad (11.2)$$

where $\lambda = dF/dT - dQ/dT$ is the radiative damping parameter (first introduced in Section 3.3, and extensively discussed in Section 9.1). The solution to this equation is

$$\Delta T(t) = \Delta T_{eq}(1 - e^{-t/\tau}) \qquad (11.3)$$

where ΔT_{eq} is the equilibrium or steady-state temperature change, and $\tau = R/\lambda$. A detailed derivation of Eqs (11.2) and (11.3) is presented in Box 11.1. When $t = \tau$, the temperature has changed by $1 - e^{-1} = 0.632$ of the final change, when $t = 2\tau$ the temperature warms by another 0.632 of the change that still had to occur at $t = \tau$, and so on, as ΔT asymptotically approaches ΔT_{eq}. τ is referred to as the *e-folding* time, and it depends on the thermal inertia *and* the radiative damping parameter. For a 70 m deep mixed layer and $\lambda = 2\,W\,m^{-2}\,K^{-1}$ (corresponding to about 2 K warming for a CO_2 doubling), $\tau \approx 5$ years.

Since $\Delta T_{eq} = \Delta F/\lambda$ and $\tau = R/\lambda$, it immediately follows that τ varies in direct proportion to the climate sensitivity. Thus, if the assumed climate sensitivity is

larger, the ultimate climatic change will be larger, but the approach to the final climatic change will be slower in relative terms. These differences will have competing effects on the absolute temperature change that occurs during the initial response. Figure 11.1 shows the effect of this competition, by comparing the transient response for climate sensitivities of 2 K and 4 K for a forcing of $3.75\,W\,m^{-2}$ (these curves are labelled as "Low Sensitivity, Low Forcing" and "High Sensitivity, Low Forcing", respectively). There is very little difference between the two cases during the first few years of the transient response. On the other hand, if the climate sensitivity is fixed and the final warming doubles due to a doubling of the forcing, then the temperature response at all points in time is multiplied by 2 (assuming that λ is constant as the climate changes and does not depend on the magnitude of the forcing); this case is also illustrated in Fig. 11.1, as the "Low Sensitivity, High Forcing" case.

If we were to consider the atmosphere rather than the mixed layer as a single, well-mixed box, then $\tau \approx 4$ months, since R for the atmosphere is about $1 \times 10^7\,J\,m^{-2}\,K^{-1}$ instead of $3 \times 10^8\,J\,m^{-2}\,K^{-1}$. However, the atmosphere and mixed layer are so tightly coupled through sensible and latent heat exchanges (as demonstrated in Box 3.1) that the atmosphere cannot warm significantly faster than the mixed layer without creating a large counteracting heat flux between the two; thus, the atmosphere and mixed layer warm together with a time constant governed by the mixed layer thermal inertia.

Effect of heat exchange between the mixed layer and deep ocean

The above analysis indicates that, if the mixed layer were to be isolated from the deeper ocean, it would respond to changes in external heating with an exponential time scale of 5–10 years, depending on the climate sensitivity. However, as soon as the mixed layer begins to warm, the heat flows between the mixed layer and deep ocean are altered, and this affects the subsequent warming of the mixed layer (and atmosphere).

The processes responsible for transferring heat between the ocean mixed layer and the deeper ocean were introduced in Section 2.2. These are advection, diffusion, and convection. The major advective movement in the global ocean is the large-scale, density-

---

Box 11.1 Derivation of Equations (11.2) and (11.3)

The governing equation for a planet consisting of a single mixed layer with temperature T is

$$R \frac{dT}{dt} = Q - F \qquad (11.1.1)$$

where Q is the global mean absorption of solar radiation, F is the emission of infrared radiation to space, and R is the thermal inertia (see the main text). Rather than dealing with the above equation, for the total temperature T and the total radiative fluxes, it is much more convenient to deal with temperature and radiative flux perturbations. To derive an equation governing temperature perturbations, we write

$$T = T_0 + \Delta T \qquad (11.1.2)$$

$$Q = Q_0 + \frac{dQ}{dT} \Delta T \qquad (11.1.3)$$

$$F = F_0 + \frac{dF}{dT} \Delta T \qquad (11.1.4)$$

where T_0 is the steady-state temperature and Q_0 and F_0 are the corresponding radiative fluxes. The temperature perturbation from the initial steady state, ΔT, arises in response to some external forcing ΔF. Adding ΔF to the right-hand side of Eq.

(11.1.1) and substituting Eqs (11.1.2)–(11.1.4) into Eq. (11.1.1), we obtain

$$R \frac{d(T + \Delta T)}{dt} = \left(Q + \frac{dQ}{dT} \Delta T \right) - \left(F + \frac{dF}{dT} \Delta T \right) \qquad (11.1.5)$$

However, dT_0/dt and $Q_0 - F_0$ equal zero (by the definition of steady state), so these terms can be subtracted from Eq. (11.1.5) to yield

$$R \frac{d\Delta T}{dt} = \frac{dQ}{dT} \Delta T - \frac{dF}{dT} \Delta T + \Delta F \qquad (11.1.6)$$

Upon rearranging Eq. (11.1.6) and making use of the definition of λ, Eq. (11.2) results. This equation is of the form

$$\frac{d\Delta T}{dt} + a\Delta T = b \qquad (11.1.7)$$

where $a = \lambda/R$ and $b = \Delta F/R$. Multiplication of Eq. (11.1.7) by the integrating factor e^{aT} and collapsing the resultant terms on the left-hand side yields

$$\frac{d}{dt} (e^{at}\Delta T) = be^{at} \qquad (11.1.8)$$

Integrating both sides of Eq. (11.1.8) from time 0 to time t, dividing through by e^{at}, substituting for a and b, and noting that $\Delta T(0) = 0$ (there is initially no temperature change) and $\Delta F/\lambda = \Delta T_{eq}$ (the equilibrium or steady-state temperature change), we obtain Eq. (11.3).

---

driven or thermohaline overturning. The thermohaline overturning consists of sinking to the ocean bottom in the North Atlantic Ocean, in the Weddell Sea, and elsewhere near Antarctica; lateral movement throughout the major ocean basins; and widespread upwelling. There are also shallow (<500 m) overturning cells that are driven by winds, the most important lying between 30°S and 30°N and consisting of upwelling near the equator, poleward flow in the mixed layer, sinking near 30°S and 30°N, and return flow in the 200–500 m layer. Figure 11.2 illustrates the east–west average of the vertical overturning in the Atlantic, Indian, and Pacific basins, and in the global mean. This circulation is too slow to be directly measured; the average upwelling due to thermohaline overturning is only 2–4 m/yr, and even the most intense wind-driven upwelling amounts to no more than about 50 m/yr (except possibly in isolated coastal areas). Rather, the overturning circulation is determined from 2-D or 3-D ocean circulation models that have been adjusted to replicate the

observed distribution of various water mass properties (such as temperature, salinity, DIC, and the concentrations of the ^{13}C and ^{14}C isotopes).

As discussed in Section 2.2, the net heat flux into the deep ocean in the steady state is zero. As the atmosphere warms in response to an increase in GHG concentrations, the downward flux of infrared radiation to the ocean surface will increase and the upward fluxes of sensible and latent heat will initially decrease. This creates an energy surplus for the mixed layer, so the mixed layer begins to warm. As the mixed layer warms, the downward diffusive heat flux increases and the intensity of convection and the associated upward transfer of heat from the deep ocean decreases. Since the warming is initially greatest for the surface layer and less for deeper water, the temperature of downwelling water will increase faster than that of upwelling water. This will reduce the net upward heat flux due to the thermohaline overturning. Thus, the changes in diffusive, convective, and advective heat transfer all result in a transfer of heat

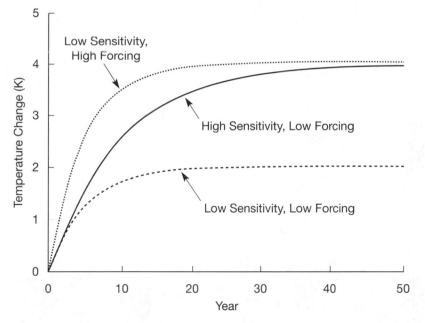

Fig. 11.1 Transient response of a one-box model to step-function external forcing changes of 3.75 and 7.5 W m⁻² applied at year 0. Results using a forcing of 3.75 W m⁻² are given for climate sensitivities for a CO_2 doubling of 2 K and 4 K, while results using a forcing of 7.5 W m⁻² are given for the lower climate sensitivity only.

into the deep ocean, which reduces the mixed layer energy surplus and slows down the subsequent mixed layer warming. As the deep ocean gradually warms up, the downward heat flux perturbations due to diffusion, advection, and convection will decrease, which allows further warming of the mixed layer, but at a rate equal to the rate of deep ocean warming. The time required for the deep ocean to reach, say, 90% of its final warming depends on the vertical profile of temperature change in the ocean, which in turn depends on a number of other factors.

To illustrate the factors that govern the deep ocean warming, the 1-D upwelling–diffusion model of Harvey and Schneider (1985) will be used. In this model the vertical diffusion coefficient (k), the global mean upwelling velocity outside the regions of bottom water formation (w), and the temperature at which polar water sinks from the mixed layer to the ocean bottom (θ_b) are all prescribed. If k and w are constant with depth, the steady-state temperature profile in this model is given by

$$\theta(z) = \theta_b + (T_{ml} - \theta_b)e^{-z/\tau_d} \qquad (11.4)$$

where $\tau_d = k/w$ is the thermocline scale depth, T_{ml} is the mean mixed layer temperature, θ_b is the temperature at which bottom water forms, $\theta(z)$ is the ocean

temperature at depth z, and z is measured downward from the base of the mixed layer. For $k = 3.6 \times 10^{-5}$ m² s⁻¹ and $w = 2.0$ m yr⁻¹, $\tau_d = 600$ m. Figure 11.3 compares the temperature profile given by Eq. (11.4) with the observed variation in global mean ocean temperature, using $\tau_d = 600$ m, $T_{ml} = 289$ K, and $\theta_b = 273$ K. These parameter values produce a good fit to the observations.

Now consider two cases when the climate changes. In the first case, θ_b is constant as the climate warms (which would happen to the extent that bottom water forms at a fixed distance from the sea ice margin, and the locus of bottom water formation merely shifts poleward as the climate warms). It then follows from Eq. (11.4) that the profile of steady-state temperature change is given by

$$\Delta\theta_{eq}(z) = (\Delta T_{ml})_{eq}\, e^{-z/\theta_d} \qquad (11.5)$$

From Eq. (11.5) it follows that the mean mixed layer warming is about $\frac{1}{7}(\Delta T_{ml})_{eq}$, for an ocean depth of 4000 m. Letting R be the ratio of polar sea to mean mixed layer warming, $R = 0.0$ in this case. In the second case, θ_b warms by the same amount as T_{ml} ($R = 1.0$), so that $\Delta\theta_{eq}(z) = (\Delta T_{ml})_{eq}$ at all depths, and the vertical mean warming equals $(\Delta T_{ml})_{eq}$. Hence, the mean deep ocean warming is seven times larger

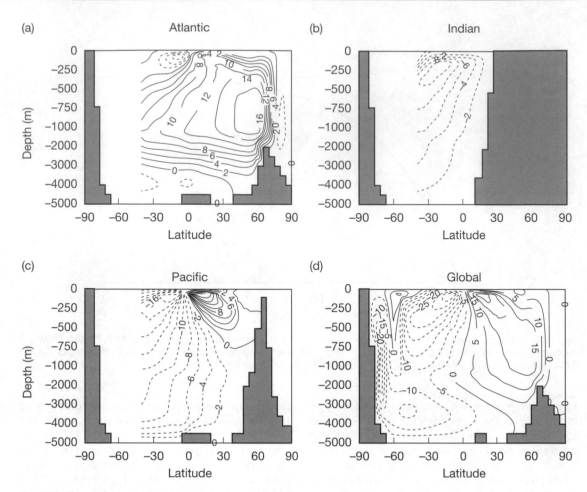

Fig. 11.2 Zonally averaged overturning circulation for (a) the Atlantic Ocean, (b) the Indian Ocean, (c) the Pacific Ocean, and (d) the global ocean. The numbers on the contour lines give the flow in Sverdrups (Sv, where $1\,\text{Sv} = 1$ million m^3/s) between the zero contour and the given contour line, with positive numbers indicating clockwise overturning and negative numbers indicating counter-clockwise overturning. Reproduced from Hovine and Fichefet (1994).

than for the first case. Changes in w would also dramatically alter the deep ocean temperature change profile, and will be considered later.

Figure 11.4 compares the transient response of the mixed layer and the mean deep ocean temperature to a sudden radiative forcing for the two cases considered above and for climate sensitivities of 2 K and 4 K for a CO_2 doubling.[1] Note that a logarithmic time scale is used. During the first two years the rate of change of mixed layer temperature is close to that expected for the mixed layer-only case, but the response becomes progressively slower as a net heat flux into the deep ocean develops. Increasing R from 0.0 to 1.0 delays the final mixed layer warming but has no effect on the magnitude of the eventual mixed layer warming (Fig. 11.4(a)). In contrast, changing R has a dramatic effect on the mean deep ocean warming (Fig. 11.4(b)). For the case with $R = 1.0$, most of the mixed layer warming has occurred by year 100, while most of the deep ocean warming (and associated sea level rise) has yet to occur. During the first 100 years, there is very little difference in the deep ocean warming for the two R cases.

---

[1] These climate sensitivities pertain to the atmospheric temperature response, but it is the mixed layer response that is shown in Fig. 11.4 and subsequent figures. As explained in Section 10.1, the mixed layer warms by less than the atmosphere.

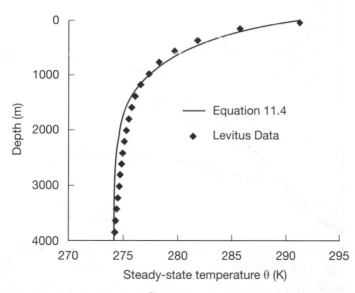

Fig. 11.3 The global mean ocean temperature profile as seen in observations (Levitus, 1982) and as computed in steady state by the 1-D upwelling–diffusion model using $\tau = 600$ m.

There is no longer a single response time constant for the mixed layer temperature response, as in the mixed layer-only case that was shown in Fig. 11.1. It is nevertheless instructive to examine the time required by the mixed layer to reach 63.2% of its steady-state response. These times are given in Table 11.1 for the two ocean model versions and for the two sensitivities, along with the e-folding times for the mixed layer-only case. The response time scale varies in direct proportion to ΔT_{eq} for the mixed layer-only case, but increases much more rapidly when the mixed layer is coupled to the deep ocean.

The 1-D upwelling–diffusion model is also a convenient tool for analysing the impact of alternative

choices for k and w on the transient response. Figure 11.5 shows the effect of halving and doubling the original values for k and w, both individually and together. An increase in k leads to a slower transient response, since a larger increase in the downward diffusion of heat occurs as the mixed layer warms when k is larger. An increase in w, in contrast, leads to a faster transient response. This is because, with larger w, the positive heat anomaly, that diffuses into the deep ocean, is partially advected back up into the mixed layer (Harvey and Schneider, 1985). Since the ratio k/w is constrained by the observed thermocline scale depth, k and w should be increased or decreased together. Since these changes have opposing effects, the uncertainty in the transient response due to uncertainty in the appropriate values of k and w is comparatively small.

To analyse the changes in mixed layer–deep ocean heat fluxes that can occur as the mixed layer warms,

Table 11.1 e-folding response times for the mixed layer in the mixed layer-only and the 1-D upwelling–diffusion model for cases in which $\Delta\theta_b = 0$ and $\Delta\theta_b = \Delta T_{ml}$, and for climate sensitivities for a CO_2 doubling of 2.0 K and 4.0 K.

Climate sensitivity	$\Delta\theta_b$	e-folding time
Mixed layer-only model		
2.0	n/a	5 years
4.0	n/a	10 years
Upwelling–diffusion model		
2.0	0	7.4 years
	ΔT_{ml}	9.0 years
4.0	0	20 years
	ΔT_{ml}	43 years

Table 11.2 Heat fluxes from the mixed layer to the deep ocean in the unperturbed steady state, and changes in these heat flux in the new steady state after applying a radiative forcing perturbation of 3.75 W m^{-2}.

Flux	Unperturbed value (W m^{-2})	Change (W m^{-2})
Convective	1.44	0.09
Diffusive	−1.96	−0.09
Advective	0.52	0.00

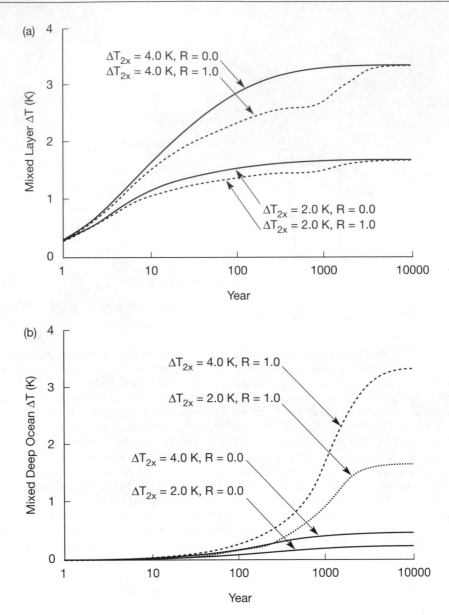

Fig. 11.4 Transient response of (a) the mixed layer temperature, and (b) the mean deep ocean temperature of the 1-D upwelling–diffusion model of Harvey and Schneider (1985) to a step-function external forcing change of 3.75 W m^{-2} applied at year 0. Results are given assuming no change in the temperature at which bottom water forms, and assuming that bottom water warms at the same rate as the mixed layer, in both cases for climate sensitivities of 2 K and 4 K.

it is more appropriate to use the quasi-1-D upwelling–diffusion ocean model of Harvey and Huang (1999), which has separate polar sea columns where deep convective mixing occurs. The temperature at which bottom water forms is freely predicted in this model rather than prescribed (which would have made it less convenient than the Harvey and

Schneider (1985) model for comparing the effect of alternative assumptions concerning θ_b). Table 11.2 shows the breakdown of the mixed layer–deep ocean heat fluxes for the pre-industrial steady-state condition, and for the new steady state after a CO_2 doubling. The principal balance is between downward heat transfer by diffusion and upward transfer by convective mixing and advection in a 3:1 ratio. As the mixed layer warms, a net heat flux to the deep ocean develops, which reduces the subsequent rate of mixed layer warming. At the same time, the net emission of radiation to space increases, and this also reduces the subsequent rate of mixed layer (plus atmosphere) warming. We shall call the change in heat flux to the deep ocean the *ocean heat flux damping,* since it has the same effect as radiative damping: to slow down the subsequent surface warming.

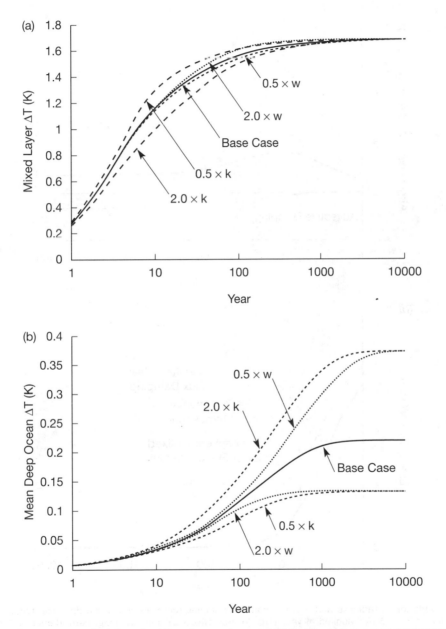

Fig. 11.5 Transient response of (a) the mixed layer temperature, and (b) the mean deep ocean temperature of the 1-D upwelling–diffusion model of Harvey and Schneider (1985) to a step function external forcing of 3.75 W m⁻² applied at year 0. Results are given for the base case values of the vertical diffusion coefficient (k) and the upwelling welocity (w), and for various multiples of these values.

Figure 11.6(a) shows the radiative and ocean heat flux damping following a sudden doubling of CO_2 (having an assumed forcing of $3.75\,W\,m^{-2}$). The radiative and ocean heat flux damping rise in tandem at the beginning, but then the ocean heat flux damping levels off and then gradually decreases to zero. This is because the subsurface ocean begins to warm in response to the downward heat flow,

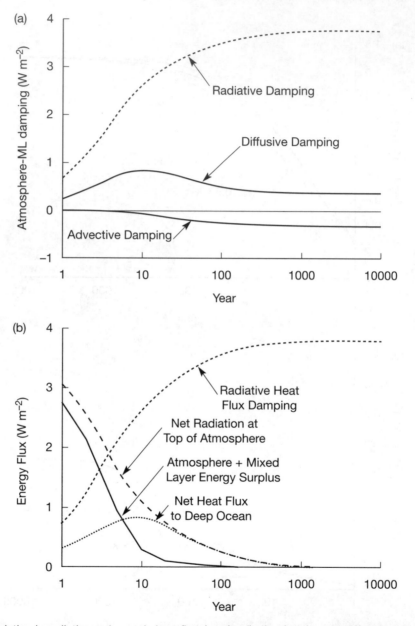

Fig. 11.6   (a) Variation in radiative and oceanic heat flux damping (both advective and diffusive) following a sudden external forcing of $3.75\,W\,m^{-2}$ applied at year zero. (b) Radiative heat flux damping, net radiation at the top of the atmosphere (equal to the external forcing minus the radiative damping), total heat flux from the mixed layer to deep ocean (equal to the sum of the advective and diffusive damping, shown in (a)), and the atmosphere + mixed layer energy surplus (equal to the difference between the preceding two items). All results were computed using the quasi-1-D upwelling–diffusion model of Harvey and Huang (1999).

thereby reducing the temperature difference between surface and subsurface water, which results in a gradual reduction in the downward heat flow. Meanwhile, the surface continues to gradually warm, which increases the radiative damping to space until it equals the external heating perturbation, at which point no further changes occur. Figure 11.6(b) shows the net radiation at the top of the atmosphere (equal to the radiative forcing minus the radiative damping), the total heat flow to the deep ocean, and the atmosphere–mixed layer energy surplus (equal to the net radiation at the top of the atmosphere minus the oceanic heat flux damping). After a few years, the net radiation is almost exactly equal to the net heat flow into the deep ocean, and the atmosphere + mixed layer energy surplus is close to zero. At this point, almost all of the net heat from the radiative surplus is being transferred into the deep ocean, and the subsequent atmosphere + mixed layer warming is very slow. This explains why, in Fig. 11.4, we see an initially rapid mixed layer warming and little deep ocean warming, followed by deep ocean warming with very slow mixed-layer warming.

The preceding analysis is for an idealized, step-function external forcing. This allows identification of the natural time constants in the system and of the key processes involved in the system response. However, the more practical case involves a gradually increasing external forcing. Figure 11.7 shows the transient response to a radiative forcing that increases linearly to $3.75 \, \mathrm{W \, m^{-2}}$ over a period of 70 years for climate sensitivities of 2 K and 4 K. Also shown are the climate responses that would occur if the temperature could instantly adjust to the changing forcing (the "instant equilibrium" response). Doubling the sensitivity doubles the instant equilibrium response at all times, but the realized warming increases by only about 50% during the period of increasing forcing. During the period of increasing forcing, the realized warming is about 80% of the equilibrium warming for low climate sensitivity and 66% for high climate sensitivity. That is, the warming at any given time as a fraction of the instantaneous response is smaller the greater the climate sensitivity. These responses can also be viewed in terms of the time lag between a given equilibrium response and the realized warming. This lag grows with time, to about 15 years for the low-sensitivity case and about 25 years for the high-sensitivity case.

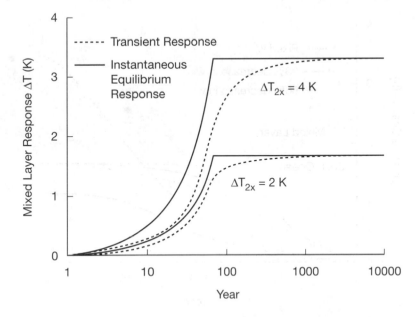

Fig. 11.7 Response of the mixed layer in the quasi-1-D upwelling–diffusion model of Harvey and Huang (1999) to an external forcing that increases linearly from $0.0 \, \mathrm{W \, m^{-2}}$ at year 0 to $3.75 \, \mathrm{W \, m^{-2}}$ at year 100. Results are shown for climate sensitivities of 2 K and 4 K for a forcing of $3.75 \, \mathrm{W \, m^{-2}}$. Also given are the temperature responses that would occur if the mixed layer temperature could instantaneously adjust to the changing forcing.

Impact of climate–ocean mixing feedbacks

All of the results shown above were obtained assuming that the strength of the thermohaline overturning does not change. As discussed in Section 8.7, one of the more robust results of coupled 2-D and 3-D atmosphere–ocean models is that the thermohaline overturning weakens as the climate warms, in some cases by 30–50 % by the time that the CO_2 concentration has doubled. Since the thermohaline overturning transports heat upwards, this can be expected to slow the warming of the mixed layer. This result was first demonstrated by Harvey and Schneider (1985), using the 1-D upwelling–diffusion model. Figure 11.8 shows the transient mixed layer and mean deep ocean temperature response as obtained by this model when the overturning circulation is assumed to weaken by 0%, 20%, or 40% per kelvin of mixed layer warming. The mixed layer response is slowed down, such that the warming after 70 years with decreasing w is only 80% (or 68%) of the case with fixed w. However, the decrease in w has no effect on the final warming in this model because the net heat flux to the deep ocean always eventually decreases to zero, so that the coupled atmosphere–surface temperature change is

entirely governed by the radiative damping to space. In nature, the weakening of the thermohaline overturning would lead to different regional patterns of climatic change. Since the global mean cloud and surface albedo feedbacks depend on the regional patterns of climatic change (as discussed in Section 9.1 and demonstrated below in Section 11.2), a climate–ocean circulation feedback would very probably have some effect on the global mean surface temperature response in reality.

The weakening of the thermohaline overturning does, however, result in a substantially greater warming of the deep ocean due to the fact that the thermocline scale depth (τ_d) increases. The mean deep ocean warming was illustrated in Fig. 11.8, while the vertical profile of warming for different cases is shown in Fig. 11.9. Figure 11.9(a) shows the steady-state temperature profile in the ocean before any climatic change, and after a 2 K surface warming with constant w and for w decreasing by 33% and 66% (θ_b is held fixed). Figure 11.9(b) shows the difference between the final and initial temperature profiles. When w decreases, the largest warming occurs in the 400–800 m layer and can be 2–3 times greater than the surface warming, whereas when w is fixed, the

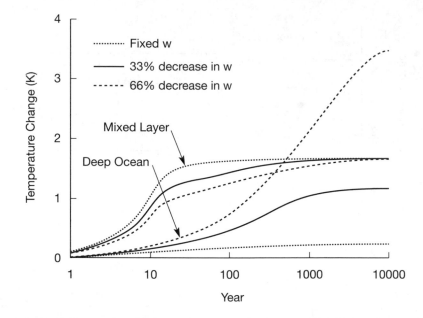

Fig. 11.8 Effect of a reduction in the thermohaline overturning on the transient response of mixed layer and mean deep ocean temperature in the 1-D upwelling–diffusion model, given a sudden doubling of CO_2 concentration at year 0. Results are given for cases in which the overturning decreases by 0%, 20%, or 40% per K of mixed layer warming (0%, 33%, or 66% total decrease).

warming decreases exponentially with increasing depth (as predicted by Eq. (11.5)).

Manabe and Stouffer (1994) found, using an AOGCM, that a doubling of atmospheric CO_2 provoked a 50% reduction in the strength of the thermohaline overturning. A quadrupling of atmospheric CO_2 had an even more dramatic effect, causing a 90% reduction in the thermohaline overturning that persisted until the end of their 500-year simulation.

Further insight into the circulation collapse is provided by experiments with the 2-D ocean circulation model of Harvey (1994). The occurrence of a collapse of the overturning circulation depends on the choice of the vertical diffusion coefficient, a parameter that is known to significantly influence the dynamics of OGCMs as well (e.g., Bryan, 1987). Furthermore, for cases in which a thermohaline collapse does occur, the thermohaline circulation is sud-

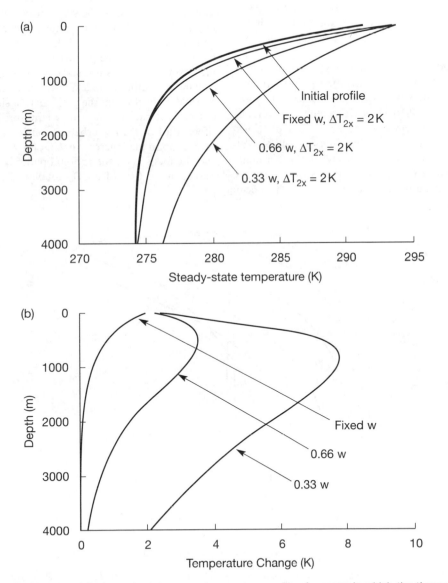

Fig. 11.9 (a) The initial and final steady-state ocean temperature profiles for cases in which the thermohaline overturning decreases by 20% or 40% per K of mixed layer warming (33% or 66% total decrease). (b) The corresponding profile of temperature change.

denly reinvigorated after a period of 700–1000 years. This behaviour is analogous to the sudden "flushes" obtained with OGCMs that undergo self-sustained, internally generated oscillations with periods of a few thousand years (e.g., Weaver *et al.,* 1991). The reinvigoration of the thermohaline circulation causes a large heat transfer from the deep ocean to the mixed layer that causes the mixed layer to temporarily overshoot its steady-state response. Thus, an irregular temperature response to GHG heating is possible, with the surface temperature response first being slowed down, then speeding up and overshooting its steady-state response, then decreasing to its final response.

Associated with the oscillatory variation in surface temperature described above are marked variations in the vertical profile of temperature change in the deep ocean. Figure 11.10 shows the globally averaged change in ocean temperature for the case described above involving an initial circulation collapse followed by a sudden invigoration. When the thermohaline overturning collapses, the thermocline deepens, leading to a subsurface maximum in the ocean

warming similar to that shown in Fig. 11.9(b). When vigorous thermohaline overturning resumes, heat is transferred to the surface, and a sharp cooling of the upper 2 km of the subsurface ocean occurs.

11.2   Regional patterns of projected transient climatic change

The problem posed by the very long simulation times required to reach a steady state or unchanging climate with coupled AOGCMs has been circumvented in two steps. First, the perturbation experiment with increasing CO_2 does not start from an equilibrium climate, but rather from a spin-up that still has some residual drift. Second, the perturbation simulation is not carried through to a new equilibrium climate either. Rather, the perturbation consists of a 1% per year compounded increase in CO_2 for 70 years or so, and the focus is on the regional patterns of climatic change at the time of a CO_2 doubling (year 70), not

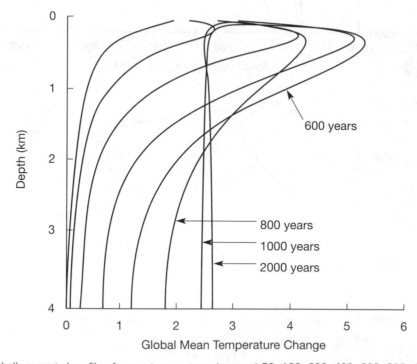

Fig. 11.10 Globally averaged profile of ocean temperature change at 50, 100, 200, 400, 600, 800, 1000, and 2000 years after applying a sudden radiative forcing of 6 W m^{-2} at year zero, for a case in which the thermohaline overturning collapses (between years 0–100) and then is suddenly reinvigorated shortly after year 600. Reproduced from Harvey (1994).

the $2 \times CO_2$ equilibrium change. The assumption is that the drift seen in the control climate (due to the fact that the spin-up simulation was not carried through to equilibrium) can be subtracted from the perturbation simulation to determine the climatic change that would have occurred if there had been no initial drift. This assumes that the model response is linear, which is probably a reasonably good approximation unless some critical threshold (involving ocean circulation or atmospheric dynamics) is crossed. Figure 11.11 shows the drift in the control climate, and the simulated change with the CO_2 concentration increasing by 1% per year, for the CSIRO AOGCM.

Manabe *et al.* (1991, 1992), Murphy and Mitchell (1995), and Gordon and O'Farrell (1997) all adopted this two-step approach. Gregory *et al.* (1997) focused on an analysis of the changes in mid-latitude soil moisture in the experiments of Murphy and Mitchell (1995). Among the important differences between the steady-state patterns of climatic change (found in experiments with AGCMs) and transient patterns of climatic change are:

- a less than average warming in the North Atlantic and Antarctic Circumpolar Ocean in the transient case, whereas the steady-state case shows a polar amplification (Manabe *et al.*, 1991; Murphy and Mitchell, 1995);
- a greater reduction of mid-latitude soil moisture in summer in the transient case than in the steady-state case because the ratio of land:ocean warming

is greater in the transient case, so there is less of an increase in evaporation from the oceans and thus less of an increase in rainfall over land (Manabe *et al.*, 1992);

- significantly different cloud feedbacks between the NH and SH in the transient case due to the greater warming in the NH than in the SH (Murphy and Mitchell, 1995).

Plate 11 shows the latitudinal variation in the zonally averaged surface air temperature response at the time of a CO_2 doubling, as computed by a number of AOGCMs when forced with CO_2 increasing at 1% per year (in most cases, a 10-year average centred around year 70 is given). This figure should be compared with Plate 10, which shows the equilibrium response to a CO_2 doubling. Some of the models show much less polar amplification of the transient temperature response than for the equilibrium response. This can be explained by the greater delay effect of the oceans at high latitudes, where the deepest mixing occurs. Many models show a minimum surface temperature response in the Antarctic Ocean, also due to deep mixing. Some models give a polar amplification in the NH by a factor of 2–3, which is comparable to the equilibrium pattern shown in Plate 10.

Figure 11.12 shows the geographical variation of surface air warming averaged over years 60–80 of the transient simulation of Manabe *et al.* (1991). Also shown is the steady-state warming for a CO_2 doubling as computed using an AGCM–mixed layer

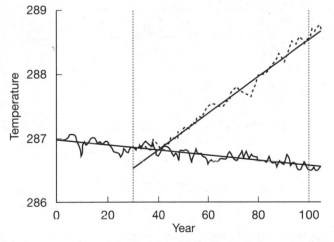

Fig. 11.11 Variation in global mean surface air temperature (K) for the control run of a 100-year simulation with the CSIRO AOGCM (solid line), and for the perturbation experiment, in which atmospheric CO_2 concentration increases by 1% per year (compounded) beginning in year 30 (dashed line). From Gordon and O'Farrell (1997).

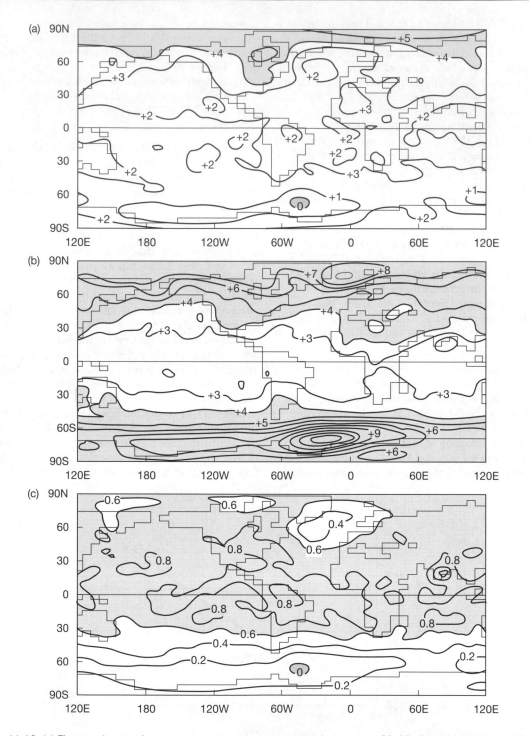

Fig. 11.12 (a) The transient surface air temperature change, averaged over years 60–80 of the AOGCM simulation of Manabe *et al.* (1991), in which atmospheric CO_2 increases by 1%/year starting at year 0. (b) The equilibrium response of surface air temperature as obtained with the same AGCM, but coupled to a mixed layer-only ocean. (c) The ratio of the transient to equilibrium temperature responses. Reproduced from Manabe *et al.* (1991).

model. The lowest panel shows the ratio of transient to steady-state warming, and highlights the limited warming in the North Atlantic and southern circumpolar regions (indeed, a slight cooling occurs near the Antarctic Peninsula). The slower warming in these regions is a result of (i) deeper penetration of heat by downwelling, and (ii) a reduction in the intensity of convection and the associated upward transfer of heat. The reduction in convection in turn is due to a reduction in the salinity of mid- and high-latitude surface water due to the poleward shift in the mid-latitude precipitation maximum.

Murphy and Mitchell (1995) found that, as a result of the greater penetration of heat into the deep ocean throughout the SH circumpolar region, the hemispheric mean warming of the surface air by year 70 (when CO_2 doubled) was only 1.1°C in the SH but 2.4°C in the NH. At the 300 mb level the interhemispheric difference was much less, with mean warmings of 2.37°C and 2.87°C in the SH and NH, respectively. As a result, the lapse rate decreased in both hemispheres and the atmosphere became more stable, which tended to suppress atmospheric convection, but the effect was much greater in the SH. In the NH, the amount of high cloud increased but the amount of middle and low cloud generally decreased, which served as a positive feedback. In the SH, the greater suppression of convection led to an increase in humidity in the lower atmosphere, so there was a pronounced increase in low-level cloud and a decrease rather than an increase in high cloud. These changes both serve as a negative cloud feedback. The occurrence of a positive cloud feedback in the NH and a negative cloud feedback in the SH served to amplify the interhemispheric difference in the surface temperature response. The transient response obtained by Murphy and Mitchell (1995) thus provides a clear example where the sign of the cloud feedback is different between transient and steady-state climatic change experiments. It also demonstrates the importance of the spatial patterns of climatic change to the overall cloud feedback.

These simulation results also show that the concept of near-constant climate sensitivity might not be valid for transient climate changes. Murphy (1995) computed the variation through time of the effective climate sensitivity for the transient response obtained by Murphy and Mitchell (1995), where the *effective climate sensitivity* is defined as the response to a CO_2 doubling that would occur if the feedbacks that occurred at various times during the transient were to persist until a steady-state climate had been reached (that is, if the initial values of the partial derivatives given in Box 3.2 were to persist). At year 15 of the transient simulation, the effective climate sensitivity is only 1.0°C. It reaches a climate sensitivity of about 2.8°C (equal to that obtained for a steady-state climatic change) by year 40 of the transient response, and remains close to this value until the end of the simulation (at year 70).

Another potentially important feature of the transient response to increasing GHG concentrations is asymmetry in the response of the western and eastern tropical Pacific Ocean. Today, the eastern Pacific Ocean is relatively cold due to the upwelling of cold water that originates at subtropical and higher latitudes, whereas the western Pacific Ocean is relatively warm. As discussed by Cane *et al.* (1997), when a quasi-uniform heating is imposed, the sea surface temperature will initially rise more slowly in the eastern tropical Pacific due to the upwelling of cold water. This, acting alone, will *increase* the east–west temperature difference and the strength of the trade winds, which will tend to strengthen the upwelling and may even cause a transient cooling in the eastern Pacific. This represents a positive dynamical feedback. The mechanism discussed here is a purely transient effect, however, since, as the subtropical and middle latitude source regions for the upwelled water warm, so will the sea surface temperature in the eastern tropical Pacific Ocean. Liu (1998) has analysed the factors controlling the east–west temperature gradient, and he also finds that the gradient should increase at least during the transient. The long-term change in the east–west gradient depends on the change in the *meridional* gradient of SST, as this determines the extent of warming of the source regions for the upwelling water. If extra-tropical warming is sufficiently limited, the east–west gradient could remain higher in the new steady-state climate, although the more likely result would seem to be a decrease in the east–west gradient (greater warming in the tropical eastern Pacific).

As noted in Section 10.6, different models give conflicting results concerning how the east–west SST gradient should change in equilibrium, with coupled AOGCMs consistently showing a decrease in the gradient. This decrease is also found in the transient simulations of Meehl and Washington (1996) and Knutson and Manabe (1998), appearing as it does by year 40–50 in experiments in which CO_2 increases by 1% per year. Alternative analyses also disagree concerning what has happened in reality: Cane *et al.* (1997) concluded that the eastern tropical Pacific cooled and the western Pacific warmed from 1900 to 1991, while Knutson and Manabe (1998) infer that the eastern Pacific warmed during this time span (see their figure 9). The observational data are not particularly good for inferring regional-scale temperature patterns in this region, and the inferred temperature trends require the use of rather elaborate statistical techniques.

If the effect of transient climatic warming is to increase the east–west temperature gradient, as suggested by Cane *et al.* (1997) and Liu (1998), then the characteristics of El Niño during the 21st century could be quite different from those suggested by equilibrium climatic change simulations (Section 10.6), with the amplitude of interannual variability increasing rather than decreasing.

Differences between the transient and steady-state changes in the east–west temperature gradient in the tropical Pacific could also lead to quite different cloud radiative feedbacks during the transient and in the steady state. Recall, from our discussion of cloud feedbacks in Section 9.1, that Del Genio *et al.* (1996) found the global mean cloud feedback to be slightly negative when uniform surface warming is imposed in the tropical Pacific Ocean, but that the net feedback is positive when there is greater warming in the eastern tropical Pacific than in the western tropical Pacific. If the transient warming in the eastern tropical Pacific is smaller rather than larger than in the western Pacific, then the net cloud feedback during the transient will be more strongly negative than in the steady state. This is another reason, in addition to that found by Murphy and Mitchell (1995), why the effective climate sensitivity during the transient might be different (and smaller) than the steady-state sensitivity.

11.3 Detection of anthropogenic effects on climate

Detection of anthropogenic effects on climate amounts to showing that the magnitude of climatic change during the past century is larger than what could be expected based on natural variability alone, and/or that the pattern of climatic change is inconsistent with the patterns expected due to natural variability but agrees with that expected for anthropogenic effects. To do this, one must first estimate the magnitude and pattern of natural variability. This can be done in three different ways: (i) based on model simulations, (ii) based on instrumental climatic data; and (iii) based on paleoclimatic data. Then one must estimate, statistically, the probability that the observed variation could have arisen by chance, given the estimate of natural variability. If it can be shown that the observed climatic change is not likely to have arisen entirely by chance as a result of internal variability, then the next step is to assess the relative importance of various anthropogenic forcings and of external natural forcings (volcanic and solar). This is referred to as *attribution.* It can be done by comparing the fit between the model-simulated and observed variation in global mean surface air temperature for various combinations of forcings, or the fit between model-simulated and observed spatial patterns for different forcings. Comparison of model-simulated and observed variations in global mean temperature requires taking into account the lag effect due to the uptake of heat by the oceans and requires specifying the climate sensitivity, whereas pattern-based analyses do not explicitly consider the oceanic uptake of heat or the climate sensitivity.

Estimates of natural variability

Natural variability in the climate system can be caused by variable external forcing, principally volcanic eruptions and variable solar luminosity, and through internal feedback mechanisms involving the atmosphere, oceans, and land surface (see Section 2.10). Here, we discuss attempts to estimate the magnitude of internal variability at the century time scale, and defer attempts to estimate past volcanic and solar variability to Section 11.4.

Simulations with coupled AOGCMs indicate that, *in models,* a warming of 0.5°C per 100 years cannot be caused by internal variability. In a 1000-year run using the GFDL AOGCM without any GHG or other external forcings, the longest that such a rate of warming was sustained is 60 years (Stouffer *et al.,* 1994). Figure 11.13 shows the variation in global mean surface air temperature in this run and in similar runs using the Hadley Centre and Hamburg AOGCMs. The Hadley Centre AOGCM has the largest interannual variability, and closely matches the observed magnitude and pattern of variability (Stouffer *et al.* 1999). Nevertheless, even the Hadley centre AOGCM fails to generate a warming comparable to that observed in the absence of anthropogenic forcing (Figure 11.13, right). Thus, to the extent that one believes the models, the failure of model-generated internal variability to give a warming comparable to that observed implies that the observed warming was not entirely caused by internal variability of the real climate system.

The reconstruction of NH average temperature during the past 600 years by Mann *et al.* (1998), shown in Plate 6, provides an alternative estimate of the magnitude of natural variability. The warming of the last 100 years clearly stands out as unusual, as it far exceeds the magnitude of temperature fluctuations experienced during the preceding five centuries. This reinforces the conclusion, based on model simulations, that the warming during the past century exceeds that which can be expected from natural variability alone.

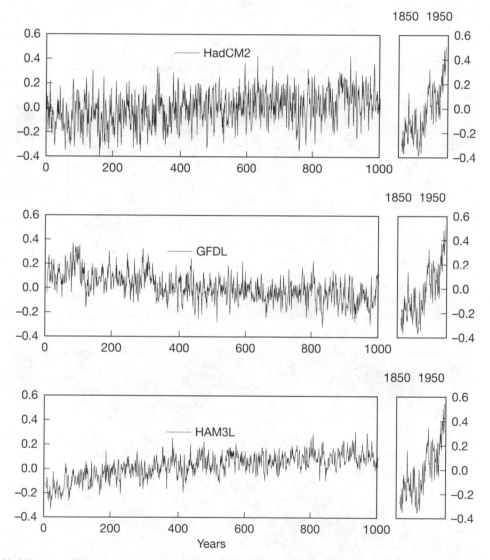

Fig. 11.13 Variation in globally and annually averaged surface air termperature (SAT) as simulated by (top) the Hadley Centre AOGCM, (centre) the GFDL AOGCM, and (bottom) the Hamburg AOGCM with no natural or anthropogenic radiative forcings. Also shown, to the right, is the observed variation in SAT. Reproduced from Souffer *et al.* (1999).

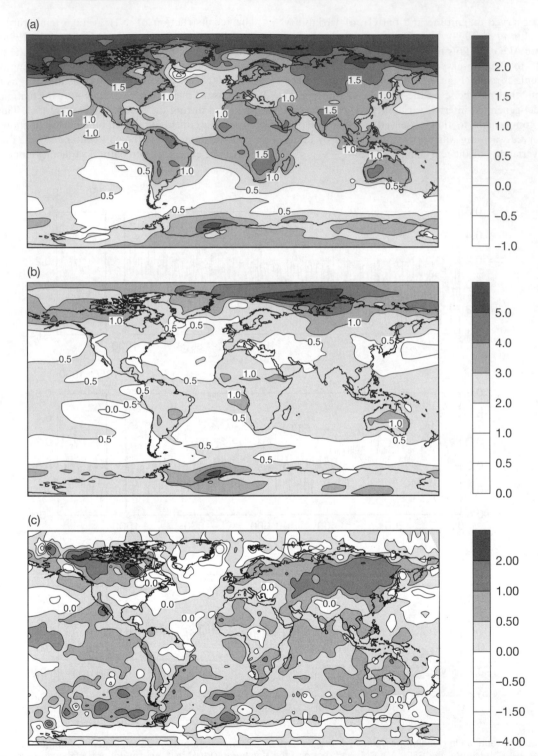

Fig. 11.14 Comparison of the change in annually averaged surface air temperature, from the period 1950–59 to 1990–98, as simulated by the Hadley Centre AOGCM with (a) greenhouse gases only or (b) greenhouse gases plus a hypothetical cooling effect due to sulphate aerosols, and (c) as observed. Model simulation data in (a) and (b) are each based on the average of four simulations and are relative to the control run. Model results were obtained from the IPCC Data Archive web site (http://ipcc-ddc.cru.uea.ac.uk), while observations were obtained from the Climate Research Unit of the University of East Anglia (http://www.cru.uea.ac.uk).

Pattern-based analyses

Analysis of the spatial patterns of climatic change can be used to identify at least part of the component of natural variability in the observed temperature record. Such analysis can also be used to try to detect specific anthropogenic causes of climatic change. Pattern-based detection of anthropogenic effects on climate requires comparing the pattern of temperature change for a specific combination of anthropogenic forcings with observed patterns. Early studies used CO_2 as the only anthropogenic forcing, while more recent studies have used CO_2 and aerosols (e.g., Hegerl *et al.*, 1997) or CO_2, aerosols, and depletion of stratospheric O_3 as forcings (e.g., Santer *et al.*, 1996; Tett *et al.*, 1996). When CO_2 and aerosol forcings are used, the pattern being compared with observations depends on the assumed relative magnitudes of the CO_2 and aerosol forcings. Since the CO_2 forcing is well known, the pattern depends on the assumed absolute aerosol forcing and its spatial distribution, both of which are highly uncertain (see Section 7.4). The magnitude and vertical profile of O_3 depletion are also uncertain (see Section 5.6), but for any reasonable ozone depletion profile, there is a clear temperature signature which can be easily distinguished from other anthropogenic or natural causes of climatic change (as discussed below).

There are a number of other problems that complicate the comparison of modelled and observed temperature patterns. Both the modelled and observed temperature patterns contain a component arising from random internal variability, so that even if the specified forcings are correct and the long-term, average response of the model is correct, there will be some mismatch between model and observations at any given time. Second, the observations contain errors or are limited by sparse coverage in some regions. Third, there will be real errors in the specified forcings and in the model response to the forcings. Fourth, some known or likely forcings have been omitted altogether: the indirect effects of sulphur aerosols, other aerosols, changes in tropospheric ozone, and land use changes (recall, from the discussion in Section 2.11, that land use changes are calculated to have caused regional and seasonal temperature changes of up to 1.5°C). Inclusion of changes in tropospheric O_3 would reduce the hemispheric asymmetry in the total forcing, thereby allowing a greater aerosol forcing (tropospheric ozone heating and aerosol cooling are both concentrated in polluted regions of the NH). Finally, in some cases the model climatic change patterns were taken from equilibrium simulations but compared with observed transient patterns of change (e.g. Santer *et al.*, 1996). This implicitly assumes that transient and equilibrium patterns are the same, and that the transient signal is stable over time. As a result of all these problems, the match between modelled and observed patterns found in pattern-based studies is seen only in a statistical sense, and the strength of the statistical fit increases irregularly over time (see Barnett *et al.*, 1998). Consequently, it is not possible to rigorously quantify the magnitude of the aerosol forcing, nor to quantify the climate sensitivity, based on the degree of fit between modelled and observed patterns.

When horizontal temperature patterns are used for detection and attribution of climatic change, the statistical fit between model and observed changes can be decomposed into the global mean change and the spatial deviation from the global mean change. Hegerl *et al.* (1997) found, in the case of the Hamburg AOGCM, that most of the statistical fit between modelled and observed changes is due to common global mean warming, rather than due to a high similarity between the detailed predicted and observed patterns. Furthermore, they found that the amplitude and pattern of the aerosol forcing cannot be convincingly distinguished from noise. The most convincing results from the analysis of Santer *et al.* (1996) also pertain to global- and hemispheric-scale temperature changes. In particular, they obtain a better fit between modelled and observed patterns when the sulphur aerosol forcing is scaled to a global mean value of $-0.7\,W\,m^{-2}$ rather than $-0.35\,W\,m^{-2}$, the weaker forcing reducing the difference between NH and SH warming. A weaker aerosol forcing also allows greater tropospheric warming relative to stratospheric cooling, thereby increasing the difference between tropospheric and stratospheric temperature changes. This implies that there is little gained in going from the analysis of global or hemispheric mean temperature changes to the analysis of two-dimensional horizontal patterns, at least at present.

Figure 11.14 compares the geographical pattern of mean annual temperature change between the periods 1950–59 and 1990–98 as observed and as simulated by the Hadley Centre AOGCM with forcing by GHGs alone, and with forcing by GHGs plus aerosols (the global mean aerosol forcing being $-0.6\,Wm^{-2}$ by 1990). In both the GHG-only and GHG+aerosol experiments, minimum NH warming is seen in the North Atlantic, downwind from eastern Europe, and downwind from China. These areas

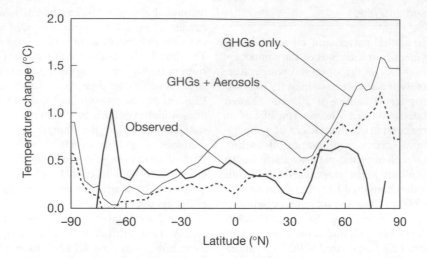

Fig. 11.15 Zonal average of the temperature changes shown in Fig. 11.14.

are downwind from the strongest sulphate aerosol loadings (Fig. 7.8) and crudely correspond to areas with minimal warming in the observations. However, the fit to the observed patterns of climatic change is not convincingly better when aerosols are included. This is consistent with the analysis of Hegerl et al. (1997), cited above. A clearer picture of the impact of aerosols is seen in Figure 11.15, which shows the latitudinal variation in the zonal average of the temperature changes shown in Figure 11.14. In the GHG-only experiment, the ratio of NH to SH warming is much larger than observed and the global mean warming is too large (0.59°C instead of 0.31°C). Inclusion of aerosols brings to NH:SH warming ratio closer to observed and gives essentially the same global mean warming (0.33°C) as observed. However, the agreement in the global mean warming masks the fact that the warming is still too large in the NH and too small in the SH. It should be stressed that these results pertain to only one particular model, a model which – according to Plate 11 – tends to produce a much higher NH:SH warming ratio than most other models for the CO2-doubling, equilibrium experiment. Also, the observations undoubtably contain a component due to internal natural variability that is not found in the model simulations. These results do, nevertheless, illustrate the point discussed above that the main improvement between modelled and simulated temperature changes when aerosols are included occurs at the global and hemispheric scales, not at the regional scale.

Observed temperature changes show a surface and tropospheric warming and a marked stratospheric cooling. Natural climatic variability in the GFDL AOGCM involves negligible changes in global mean stratospheric temperature (Vinnikov et al., 1996). Solar variability does not cause out-of-phase tropospheric and stratospheric temperature changes, and the CO_2 increase alone does not cause adequate stratospheric cooling (Tett et al., 1996; Vinnikov et al., 1996). Tett et al. (1996) obtained the best fit to the observed vertical structure of temperature changes using a combination of CO_2, aerosol, and stratospheric ozone changes in the Hadley Centre AOGCM, but the model shows a maximum in the tropical warming at 300 mb that is not seen in the observations.[2] This is illustrated in Fig. 11.16, which compares the difference between the 1961–1980 and 1986–1995 average temperatures in the observations and as computed by a transient run of the model. In a transient simulation spanning the period 1979–1995

---

[2] Tett et al. (1996) used an equivalent CO_2 increase rather than the mixture of real GHG increases. Since other GHGs do not cool the stratosphere at all or to the same extent as a CO_2 increase (see Table 3.1 and Section 10.1), the simulated stratospheric cooling will tend to be too large, except for the fact that the dominant cause of stratospheric cooling is the decrease in ozone, not the increase in CO_2 concentration. Since there is a large uncertainty in the ozone change, errors in the ozone change could easily have swamped the error arising from using an equivalent CO_2 increase.

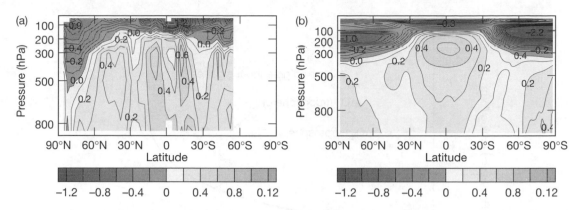

Fig. 11.16 Comparison of latitude–height variation in the change in zonally averaged atmospheric temperature from 1961–1980 to 1986–1995 as (a) observed, and (b) simulated in a transient run of the Hadley Centre AOGCM using CO_2 + aerosol + stratospheric O_3 radiative forcing. From Tett *et al.* (1996).

by Hansen *et al.* (1997c), the tropical upper tropospheric warming maximum is not present, in accord with observations. The bottom line is that only some combination of anthropogenic effects can explain the vertical structure of the observed temperature changes. However, one cannot rule out, on the basis of the observed vertical temperature structure alone, the possibility that the stratospheric cooling is due to anthropogenic effects but that the tropospheric warming is due to some other factor. The full significance of the successful simulation of the stratospheric cooling for climate sensitivity will be discussed later.

Analysis of the horizontal patterns of climatic change has also been used to identify at least one component of natural internal variability in the record of global mean temperature change. The global mean temperature shows a rapid rise between 1908 and 1946, then a minor reversal of the warming until about 1965 (recall Fig. 5.5). Schlesinger and Ramankutty (1994) show that this is largely the result of an oscillation in the North Atlantic Ocean and the surrounding areas. The peak-to-peak temperature variation associated with the oscillation is 0.19°C in the global mean (compared to a warming of 0.65 ± 0.05°C during the past 120 years) but 0.75°C in the North Atlantic Ocean. This oscillation is not likely to be a result of solar variability (the effect would be global) or white noise forcing of the ocean by the atmosphere (the effect would be ocean-wide). Rather, it is likely the result of an internal atmosphere–ocean oscillation of the type found in a variety of coupled atmosphere–ocean models (both 2-D and 3-D).

Granger causality argument

Statistical models can be constructed whereby the temperature change in a given place and at a given time is predicted based on the temperature changes on the immediately preceding time steps. Such statistical prediction models can be constructed with or without inclusion of a deterministic forcing term related to one or more known radiative forcings. When a statistical model is constructed to predict changes in NH temperature but with no anthropogenic forcings, a knowledge of the preceding ΔTs in the SH aids in the prediction of subsequent NH ΔTs, compared to predictions based on past temperatures in the NH alone (Kaufmann and Stern, 1997). SH ΔTs are thus said to "Granger-cause" the NH ΔTs. However, when anthropogenic CO_2 and aerosol forcings are explicitly included in the statistical regression model, there is no longer any separate predictability of NH ΔTs from SH ΔTs. This implies that the enhanced predictability of NH ΔTs using SH ΔTs in the former case is due to the fact that (i) aerosols retard warming largely in the NH, (ii) NH warming eventually occurs due to the heating effect of GHGs, and (iii) this heating effect is seen first in SH ΔTs, which therefore appear to "cause" the NH ΔTs when anthropogenic forcings are not explicitly included. NH ΔTs are not found to Granger-cause SH ΔTs. This provides evidence that observed hemispheric mean temperature changes are caused at least in part by combined GHG and aerosol forcing.

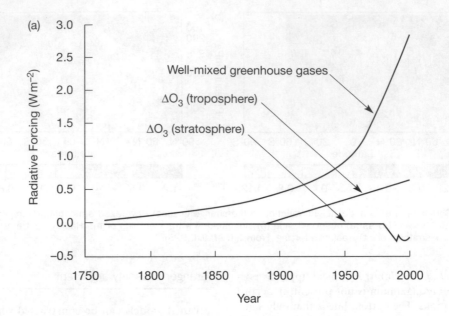

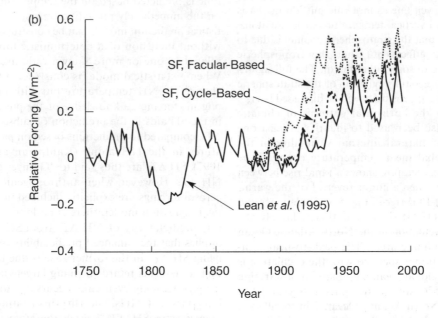

Fig. 11.17 Estimates of the globally and annually averaged radiative forcing from 1750 to the present: (a) due to changes in GHG concentrations; (b) due to solar variability, as estimated by Solanki and Fligge (1998) based on either the length of the solar cycle or the number of faculae (bright areas) seen on the sun (SF), and as estimated by Lean *et al.* (1995); (c) due to volcanic activity, based on time series of volcanic activity as reconstructed by Sato *et al.* (1993) and by Robock and Free (1995); and (d) due to anthropogenic aerosols. Two volcanic activity indices by Robock and Free (1995) are shown, one referred to as the Dust Veil Index (DVI), and the other a composite Ice Core–Volcano Index (IVI). For aerosols, three cases are shown, assuming that the total aerosol cooling in 1990 was either 20%, 40%, or 60% of the assumed GHG heating in 1990 of 2.71 Wm⁻².

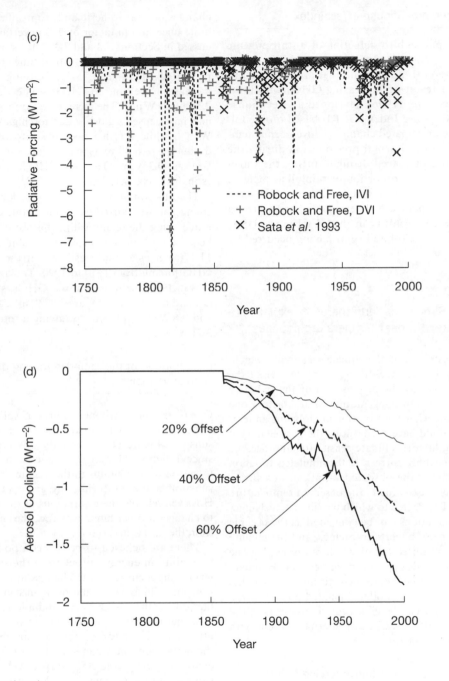

Fig. 11.17 continued

Enhanced greenhouse effect index

Another approach to detection of anthropogenic effects is to create an index which combines, in one number, several phenomena that are projected to occur as a result of increasing GHG concentrations. Karl *et al.* (1996) developed a Greenhouse Climate Response Index (GCRI) based on the following five anticipated changes: above normal temperatures, above normal precipitation during the cold season, increased drought during the warm season, a greater proportion of rainfall as extreme precipitation events, and reduced day-to-day variability in temperature. They find a positive trend during the 20th century in the GCRI in the USA, with only a 5% probability of having occurred by chance alone.

11.4   Simulating (and therefore explaining) recent past temperature changes

Given an estimate of the climate sensitivity (which, admittedly, can change during the transient), of the uptake of heat by the oceans, and of the radiative forcing due to anthropogenic and natural factors, one can calculate what the variation in global mean temperature should have been during the past century if the forcing factors, climate sensitivity, and oceanic uptake of heat are correct. If the simulated variation closely matches the observed variation, we can say that we have "explained" the observed temperature variation. The degree to which the simulated changes match observations can be compared as the oceanic uptake of heat, uncertain forcings, and the climate sensitivity are varied (all of which can be easily done with simple models). A range of acceptable climate sensitivities can be determined, given that some disagreement between model-simulated and observed climate change can be expected due to errors in the total forcing, the presence of internal variability in nature, or other factors.

Estimation of the radiative forcing due to greenhouse gases

The radiative forcings due to well-mixed GHGs can be accurately estimated (to within $\pm 15\%$) based on the known variation in their concentration. On the other hand, the present-day radiative forcing due to changes in stratospheric and tropospheric ozone is quite uncertain (a factor of 2–3 uncertainty), as discussed in Sections 7.2 and 7.3. The estimated radiative forcings from pre-industrial times to 1995 were presented in Table 7.7. The forcing due to the increase in well-mixed GHGs was estimated to be $2.37 \pm 0.24\,\mathrm{W\,m^{-2}}$. The increase in tropospheric ozone may have caused a global mean forcing as large as 0.7 $\mathrm{W\,m^{-2}}$, while the decrease in stratospheric ozone during the past 30 years may have caused a forcing as large as $-0.3\,\mathrm{W\,m^{-2}}$. The total GHG forcing in 1995 is estimated to be between 2.1 and $3.2\,\mathrm{W\,m^{-2}}$.

The time variation in the radiative forcings due to changes in well-mixed gases and ozone can be computed using the relationships that were discussed in Section 7.7 (and summarized in Table 7.9). Figure 11.17(a) gives a middle estimate of the variation in GHG forcing from 1750 to 1990. The assumed forcings in 1990 due to well-mixed GHGs, stratospheric O_3, and tropospheric O_3 are $2.37\,\mathrm{W\,m^{-2}}$, $-0.26\,\mathrm{W\,m^{-2}}$, and $0.6\,\mathrm{W\,m^{-2}}$, respectively, giving a total forcing of $2.71\,\mathrm{W\,m^{-2}}$.

Estimation of the radiative forcing due to solar variability

One of the most striking features of solar activity is the observed variation over time in the number of sunspots. Figure 11.18 shows the observed variation since 1500; an 11-year cycle is evident, as well as longer-term oscillations in the maximum number of sunspots achieved during any given 11-year cycle. However, relating these variations in sunspot activity to changes in solar luminosity (i.e., the energy output from the sun) is uncertain.

There are believed to be two components to the variability in energy output from the sun: a short-term component related to changes in the number of sunspots, which has been directly measured by satellites during the past 20 years; and a long-term variation that is derived indirectly from observations of other stars. These two components are referred to as the active sun and quiet sun components, and are denoted by ΔS_{act} and ΔS_{qs}, respectively. The active sun variability is caused by the passage of dark sunspots across the solar disc, which reduce the solar luminosity, and the passage of associated bright areas (faculae), which enhance the solar luminosity. The net result is an increase in solar luminosity with increasing sunspot number, with a luminosity variation of $\pm 1.0\,\mathrm{W\,m^{-2}}$ during the course of the 11-year

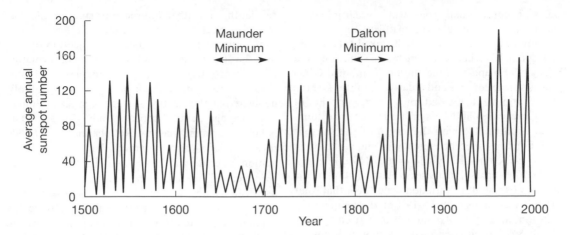

Fig. 11.18 Variation in the average number of sunspots visible during each year from 1500 to 1993. Reproduced from Bryant (1997).

sunspot cycle. Long-term variations are thought to be due to changes in the intensity of convective mixing between the deep interior and surface of the sun, and there is no reason to expect this to be tightly coupled to surface blemishes.

Possible indicators of ΔS_{qs} include the rate of solar rotation (faster rotation implies less deep convection and hence lower luminosity), the rate of decay of sunspots (which has varied by 25%), and the length of the sunspot cycle (which has varied from 9.9 to 12.1 years). The latter is inversely related to the sunspot decay rate. Unfortunately, it is not possible to quantitatively relate the sunspot decay rate to the intensity of convection. Hoyt and Schatten (1993) derived an index for ΔS_{qs} using a combination of five indicators (including those listed above). Based on various lines of indirect evidence, they assigned an absolute magnitude to the maximum variation in ΔQ_{qs} of 2–5 W m^{-2} during the preceding 100 years. Solanki and Fligge (1998) estimated that the variation in ΔS_{qs} between the Maunder Minimum (a period of almost no sunspots in the 17th century, seen in Fig. 11.18) and the present could be 2–8 W m^{-2}. However, when they fit the combined ΔS_{act} and ΔS_{qs} variation to direct observations since 1978, they are forced to limit the ΔS_{qs} variation to no more than 5 W m^{-2}. They obtain the best fit when they assume a variation of only 2 W m^{-2} – comparable to the short-term variation. Lean *et al.* (1995) estimated the long-term variation in solar luminosity using an approach similar to that of Hoyt and Schatten (1993), except that they based their ΔS_{qs} on the amplitude rather than the length of the

solar cycle. Figure 11.17(b) shows the variation of solar radiative forcing[3] as reconstructed by Lean *et al.* (1995), and for two alternative reconstructions by Solanki and Fligge (1998) beginning in 1874. There are some notable differences between the three reconstructions. However, all three time series agree in giving minimum solar luminosity around 1880 and a maximum in the 1930s or 1950s.

Another way to estimate the long-term variability of solar luminosity is to attempt to simulate the observed variation of global mean temperature during the past century, with the magnitude of the luminosity variation chosen so as to give the best statistical fit between the simulated and observed temperature variation. Simple models, such as the 1-D upwelling–diffusion model, are ideal for this task. The amplitude of solar luminosity variation determined in this way depends strongly on what other forcing factors are included when performing the climate simulation. If, for example, variation in solar luminosity is assumed to be the *only* factor that caused climatic change during the preceding century, then a rather large variation in solar luminosity (1% or 14 W m^{-2}) over the 20th century is deduced, capable of explaining over 90% of the variation in global mean temperature during the past century (Friis-Christensen and Lassen, 1991). However, such an

---

[3] The solar radiative forcing, ΔQ, is the change in the global average absorption of solar radiation due to changes in solar luminosity, ΔS. Thus, $\Delta Q = (1.0 - \alpha)\Delta S/4$, where α is the average reflectivity of the Earth.

analysis is not internally consistent, since it implicitly assumes that the climate was unaffected by substantial radiative forcing due to well-mixed GHGs, but did respond to radiative forcing by the sun. Kelly and Wigley (1992) analysed the effect of combined GHG (plus aerosol) forcing and solar variability. For the sake of argument, they adopted the assumption of Friss-Christensen and Lassen (1991) that solar luminosity is correlated with the length of the sunspot cycle. For two different estimates of the variation in solar cycle length, Kelly and Wigley (1992) deduce that GHG + aerosol increases explain 45–48% of the observed temperature variation since 1860, that solar variability explains only 13–19% of the temperature variation, and that solar luminosity varied by only 0.27–0.38%. This is supported by a thorough statistical analysis of temperature and solar variation by Schönwiese *et al.* (1994), who conclude that solar variability explains at most 15–20% of global mean temperature changes during the past century. Note that these results also leave room for a significant role for internal variability (causing up to 30% of the observed variation), which is consistent with the pattern-based analyses cited above.

Haigh (1994) reports that the spectral distribution of the change in solar radiation is important to the production of ozone; in particular, there is a greater relative variation in the ultraviolet flux than in the total solar flux. Using a two-dimensional model of atmospheric chemistry, he found that variation in UV radiation causes extra ozone production during peaks in the 11-year sunspot cycle which, by absorbing some of the solar radiation before it reaches the troposphere, significantly reduces the global mean surface–troposphere forcing. Poleward of 45°N in summer, the combined effect of an increase in solar luminosity and in stratospheric ozone is to *reduce* the solar flux to the troposphere. It is not known whether the spectral distribution of solar radiation changes at long time scales in the same way that it changes over the 11-year time scale. Nevertheless, it is clear that the correct surface–troposphere radiative forcing due to solar luminosity variations could be substantially smaller than that shown in Fig. 11.17(b).

Estimation of the radiative forcing due to volcanic activity

A number of workers have attempted to construct indices of volcanic activity during the past century or longer, based on information from known volcanic eruptions, measurements of atmospheric opti-

cal depth, and other indirect information. As discussed in Section 2.10, the climatic effects of volcanic eruptions arise through the injection of SO_x into the stratosphere and the subsequent formation of a hemispheric or global blanket of sulphate aerosols. As the sulphate aerosols are gradually deposited at the Earth's surface, a spike of higher than average acidity appears and is preserved in polar ice caps and glaciers. Robock and Free (1995) review various volcanic indices derived from information other than ice cores, as well as reviewing ice core records. Acidity or the sulphate concentration itself can be measured in ice cores. The disadvantage of relying on measured acidity is that not all acidity maxima are related to maxima in sulphate aerosols, while the disadvantage of relying directly on sulphate is that there are non-volcanic anthropogenic and biogenic sulphate sources that can obscure the volcanic signal. Local volcanic eruptions that are not climatically significant can produce large sulphate deposits, while mixing of deposited snow by wind can obscure the volcanic signal. Robock and Free (1995) compared eight extratropical NH ice core records, one tropical record, and five extratropical SH records. Sometimes nearby cores do not show the same peaks. In an attempt to remove random errors, Robock and Free (1995) prepared composite NH and SH volcanic indices for the period 1850–1990 by averaging all the ice core peaks in each hemisphere. Sato *et al.* (1993) prepared an independent volcanic time series based on estimated aerosol optical depths for the period 1883–1990 and based on volcanological evidence for the volumes ejected from known major volcanoes for the period 1850–1882. Both the Robock and Free (1995) and Sato *et al.* (1993) time series can be converted to estimates of global mean radiative forcing based on observations made after the 1991 eruption of Mt Pinatubo. Satellite and aircraft observations of the aerosols that were injected into the stratosphere after the recent eruption of Mt Pinatubo indicate that this eruption caused a global mean radiative forcing that peaked at a value of about $-3.0\,W\,m^{-2}$ about one year after the eruption, and then decayed to zero after a further two years (Hansen *et al.,* 1996b, their figure 12). From this and the importance of previous eruptions relative to Mt Pinatubo, the variation in global mean volcanic forcing can be computed.

Figure 11.17(c) compares the resulting global mean radiative forcing as estimated by Sato *et al.* (1993) and Robock and Free (1995). In the latter case, the forcing as estimated from an "Ice core–Volcano

Index" (IVI) is given, as well as from a previously developed "Dust Veil Index" (DVI). The largest forcing is due to the 1815 eruption of Tombora, with a peak annual mean forcing of $-6.5\,W\,m^{-2}$ according to the DVI index, and $-14.9\,W\,m^{-2}$ according to the IVI index. There are other notable differences between the three time series.

Constraining the sulphur aerosol forcing

As noted above, an internally consistent analysis of radiative forcings and climatic change indicates that solar variability can explain at most 15–20% of the temperature variation during the past century, with net anthropogenic (GHG + aerosol) forcing explaining about 50%. However, the implied climate sensitivity is highly dependent on the assumed aerosol forcing. As discussed in Section 7.4, the forcing due to sulphur + black carbon aerosols is extremely uncertain: after discarding the most extreme estimate, the combined global mean direct + indirect forcing could be anywhere between -0.4 and $-2.6\,W\,m^{-2}$. The cooling effect of other aerosols could bring the total aerosol cooling to -0.8 to $-3.0\,W\,m^{-2}$. Since the total forcing due to increases in GHGs is estimated to be between 2.0 and $3.2\,W\,m^{-2}$, the aerosol forcing would appear to be capable of offsetting most or all of the GHG heating in the global mean. Because of this large uncertainty, it is not possible to set a meaningful constraint on the climate sensitivity without some additional constraints on the magnitude of the aerosol forcing.

Fortunately, there are two observational constraints that can be used, at least in principle, to limit the magnitude of the aerosol forcing. These two constraints are (i) the difference in the warming between the NH and SH during the past century; and (ii) the difference between daytime and nighttime warming over land.

During the period 1900–1985, the SH warmed about $0.10 \pm 0.2°C$ more than the NH. Over the period 1964–1989, the SH warmed by 0.04 ± 0.05 more than the NH (Table 5.1). The greater SH warming occurred in spite of the greater land fraction in the NH, which tends to cause a greater hemispheric mean equilibrium warming (since land has a greater equilibrium sensitivity, as explained in Section 10.1) and a faster transient response. This difference (if real) can be explained by an aerosol cooling effect, which would be stronger in the NH owing to the greater concentration of anthropogenic aerosols. The observed magnitude of the global mean warming

and the interhemispheric difference in the warming together set a constraint on the climate sensitivity and on the magnitude of the aerosol cooling effect. If the climate sensitivity is large, then a large aerosol cooling effect is needed to avoid too large a global mean warming, but this in turn increases the difference in the warming between the hemispheres. Calculations performed by Wigley (1989) imply an aerosol forcing in the NH of $-0.7\,W\,m^{-2}$, or about one-sixth of the GHG heating perturbation in the global mean. However, Wigley assumed that aerosols have no cooling effects in the SH. As discussed in Section 7.4 (and shown in Table 7.5), more recent computations of the direct cooling effect of aerosols indicate that the SH mean forcing is 10–40% as large as the NH mean forcing and that the indirect forcing in the SH could be 20–90% of that in the NH. As noted above in Section 11.3, reasonable patterns of temperature change are simulated using an AGCM with a global mean (rather than NH mean) aerosol cooling of $0.7\,W\,m^{-2}$, owing, no doubt, to the fact that some of the cooling in the AGCM is in the SH. When the effect of black carbon aerosols is included in model simulations, the direct cooling effect is quite small or could even be a heating effect, and in any case, there is very little difference in the radiative forcing between the two hemispheres. The primary aerosol cooling effect would then be due to the indirect effect of aerosols on clouds, which tends to have less hemispheric asymmetry than the direct radiative forcing. This in turn would allow a greater aerosol offset of GHG heating than found by Wigley (1989). Another factor permitting greater aerosol cooling is the fact that the calculated forcing due to tropospheric O_3 is about 50% larger in the NH than in the SH (see Table 7.2). However, until the uncertainty in the interhemispheric asymmetry is reduced, it is not possible to say how large an aerosol forcing is permitted under this constraint.

The second potential constraint on the aerosol forcing arises from the observed difference in daytime and nighttime warming over land. As discussed in Section 5.2, minimum nighttime temperatures over land increased at a rate of $1.79°C$ per 100 years over the period 1950–1993, while maximum daytime temperatures increased at a rate of only $0.84°C$ per 100 years. Hansen et al. (1995a) tested the effect on diurnal temperature variation of a wide variety of heating perturbations, and found that the only mechanism capable of simultaneously generating changes in the mean temperature and in the diurnal amplitude comparable to that observed is localized anthropogenic aerosol and aerosol-induced cloud changes *in combination with* a large-scale warming factor, such as GHG increases. Their analysis implies that the combined aerosol–cloud cooling perturbation is about half as large as the anthropogenic greenhouse heating perturbation in the global mean.

In light of the above discussion, it seems reasonable to consider aerosol forcings that offset 20–60% of the 1990 heating due to GHGs. The temporal variation in the forcing up to (and after) 1990 can then be computed based on the variation in anthropogenic sulphur emissions, the assumed partitioning of the 1990 forcing between direct and indirect effects, and the relationships given in Table 7.9. Figure 11.17(d) shows the resulting variation in the aerosol forcing for cases in which the forcing reaches 20%, 40%, or 60% of the assumed 1990 GHG forcing of $2.71 \, \mathrm{W}$ m^{-2}, using the sulphur emission estimates given in Fig. 4.5, assuming the 1990 forcing to be one-third direct and two-third indirect, and assuming the background (natural) sulphur emission (e_{nat} in Table 7.9) to be $28 \, \mathrm{Tg \, S/yr}$. It is also assumed that the atmospheric sulphate loading varies linearly with the anthropogenic sulphur emission; according to the discussion in Section 8.11, this is a very good approximation except possibly when there are concurrent changes in NO_x emissions (as there were during this time period). However, this source of uncertainty is swamped by the uncertainty in the forcing for the reference year, 1990. The computed aerosol forcing increases most rapidly after 1950.

The role of vertical mixing in the ocean

When simple climate models such as the 1-D upwelling–diffusion model are used to simulate transient climatic change, the simulated rate of change depends on three parameters or sets of parameters: the radiative forcing (which we have just discussed), the climate sensitivity (which is the key unknown factor that we would like to constrain), and the mixed layer–deep ocean mixing coefficients. As discussed in Section 6.1, the two mixing coefficients in the upwelling–diffusion model are the vertical diffusion coefficient (k) and the upwelling velocity (w). Previous work with the upwelling–diffusion model (and with AOGCMs) has generally used a larger diffusion coefficient than is observed in the real ocean. In the upwelling–diffusion model, k is not a real diffusion coefficient, but rather is an effective diffusion coefficient that represents, among other things, the effects of quasi-horizontal mixing along gently slop-

ing isopycnal surfaces. As explained in Box 11.2, the appropriate diffusion coefficient in the upper 500–1000 m is different for the different properties (tracers) that are being diffused (such as temperature, radioactive carbon, or tritium). More importantly, the effective diffusion coefficient is significantly smaller for temperature than for other tracers. Thus, if k is determined from the mixing of radioactive carbon or tritium (which were both injected into the atmosphere and hence oceanic mixed layer through the atmospheric testing of nuclear bombs) and this k is applied to heat, then the ocean will absorb heat too quickly, causing a greater reduction in the rate of warming. This in turn permits a larger climate sensitivity while still simulating close to the observed temperature increase during the past century or so. If k is determined by fitting Eq. (11.4) to observations using a vertically uniform value of w, this will also result in too large a value for k in the upper 500–1000 m of the ocean, thereby also tending to cause the estimated climate sensitivity to be too large. This is because the true upwelling flux peaks at some mid-depth level, so the required k is smaller in the upper ocean than for uniform w. Direct observations and theory indicate that, below the mixed layer, k increases with increasing depth (Harvey and Huang, 1999).

Previous estimates of the permitted climate sensitivity based on the fit of simulated temperature changes to observations have used simple climate models with vertically uniform values of k on the order of $1 \, \mathrm{cm}^2 \, \mathrm{s}^{-1}$. This is much larger, in the upper ocean, than can be justified based on observations. Harvey and Huang (1999) developed an alternative upwelling–diffusion model in which k increases from a value of $0.15 \, \mathrm{cm}^2 \, \mathrm{s}^{-1}$ at the base of the mixed layer to a value of $1.0 \, \mathrm{cm}^2 \, \mathrm{s}^{-1}$ at the 2 km depth. As shown in the next subsection, this results in a smaller permitted climate sensitivity than was obtained before, given the key assumption in all such work that the effective transient climate sensitivity is the same as the equilibrium sensitivity.

Simulated climatic change during the past 140 years

Figure 11.19 shows the variation in global mean temperature from 1850 to 1995 as simulated using the climate model described in Harvey and Huang (1999) and the radiative forcings given in Fig. 11.17. The solar forcing of Lean *et al.* (1995) and the volcanic forcing of Sato *et al.* (1993) were used for these simulations. The simulations began in 1750, so that the trend by 1850 includes the effect of the forcings during the preceding century, but results are only shown since 1850 because estimates of the variation in global mean temperature begin about then. Results are shown for a steady-state climate response to a CO_2 doubling of 1.0°C, 2.0°C, and 3.0°C. Figure 11.19(a) shows the temperature variation when no aerosol cooling effect is assumed, while Figure 11.19(b) shows the results with offsets in 1990 equal to 20%, 40%, and 60% of the 1990 GHG radiative forcing for climate sensitivities of 1.0°C, 2.0°C, and 3.0°C, respectively. Best results are obtained for a climate sensitivity of 1.0°C with no aerosol forcing. Earlier studies (summarized in Santer *et al.*, 1996; their Section 8.4.1.3) implied that the historical temperature variations are consistent with a CO_2 doubling response of 1.5°C and 4.5°C, but, as noted above, the models used in these studies assumed a considerably stronger coupling between the mixed layer and deep ocean than used here.

One very notable discrepancy between the model-simulated temperature variation and the observed temperature variation is the large dip after 1883 in the model results. This is in response to the 1881 eruption of Mt Krakatoa, and subsequent eruptions extending to Mt Katmai in 1912. The peak cooling is modestly greater for larger climate sensitivity, but the greatest effect of higher climate sensitivity is in the long-term response. This behaviour was also found by Lindzen and Giannitsis (1998), using an even simpler model, and suggests a climate sensitivity near the lower end of the range considered here.

The following simple back-of-the-argument calculation demonstrates the reasonableness of the 1.0°C climate sensitivity in the absence of aerosol forcing: first, the radiative forcing by 1990 amounts to about 2.6 W m^{-2} due to GHGs plus 0.4 W m^{-2} due to solar variability, for a total of 3.0 W m^{-2}; second, this is 80% of the adjusted forcing for a CO_2 doubling; and third, the observed warming of 0.65 ± 0.05°C is about 80% of the instant-equilibrium response of 0.8°C that would occur for a climate sensitivity of 1.0°C for a CO_2 doubling. This is a reasonable response, in light of the model behaviour shown in Fig. 11.7.

Even though the estimated combined sulphate–black carbon aerosol is close to zero, it is possible that the aerosol forcing is still near the upper end of the range considered above. This is because the global mean indirect forcing could be in excess of 1.0 W m^{-2} according to many calculations, and because the SH forcing might be a large fraction of the NH forcing (Table 7.5). The indirect forcing in the SH could range from as little as 20% of the average NH

forcing to as much as 90%. If the SH:NH ratio is small, than the aerosol forcing cannot be very large, since otherwise the NH warming will be suppressed too much relative to the SH warming. This in turn implies a climate sensitivity of 1.0–2.0°C. However, if the aerosol forcing is largely an indirect forcing with comparable mean values in the NH and SH, then a larger climate sensitivity is permitted.

There are two other considerations that might permit a larger climate sensitivity. First, there could have been a natural cooling tendency over this period due to an internal oceanic oscillation that masked some of the warming due to increasing GHGs that would have otherwise occurred. Second, the effective climate sensitivity during the transient response, which is really what is being examined here, could be smaller than the equilibrium sensitivity. As discussed in Section 11.2, experiments with the Hadley Centre AOGCM give a much smaller transient sensitivity than the equilibrium sensitivity (1.0°C instead of 2.8°C). Thus, the true equilibrium climate sensitivity could still be within the upper part of the range considered in Fig. 11.19 even without aerosol cooling.

Simulated climatic change during the past 20 years

Hansen *et al.* (1997c) used the GISS AGCM combined with a variety of simple representations of the oceans to simulate climatic variations during the period 1979–1995. They were able to accurately simulate the observed changes in global mean surface, tropospheric, and stratospheric temperatures using forcings due to (a) the increase in well-mixed greenhouse gases during the period 1979–1995, (b) the decrease in stratospheric O_3, (c) the two climatically important volcanic eruptions (El Chichon and Mt Pinatubo), (d) solar variability, and (e) a constant forcing of 0.65 W m^{-2}, which represents the radiative disequilibrium in 1979 due to the fact that the climate at the start of the simulation was not fully adjusted to the preceding buildup of GHGs and aerosols. Plate 12 compares the model and observations (extended here to 1998) for the surface, troposphere, and stratosphere. Among the interesting findings resulting from this work are the following.

- Owing to the cooling effect of the El Chichon and Mt Pinatubo eruptions, there would have been no overall warming between 1979 and 1995 in spite of the continuing increase in GHG concentrations, were it not for the preceding buildup of GHGs and the resulting initial radiative disequilibrium.

Box 11.2 Isopycnal mixing and the effective vertical diffusion coefficient

Mixing in the oceans involves turbulent eddies. Vertical motions are resisted by the density stratification (the layering of less dense water over more dense water), while motions along surfaces of constant density – known as *isopycnal surfaces* – are not resisted at all by the density field. Consequently, the diffusion coefficient – which is the large-scale representation of small-scale turbulence – is several orders of magnitude larger for mixing oriented along isopycnal surfaces than for mixing across isopycnal surfaces (*diapycnal mixing*). The sloping isopycnal surfaces and associated perpendicular line therefore form the natural coordinate system for the representation of diffusion, rather than the vertical and horizontal directions.

Figure 11.2.1 illustrates an isopycnal surface with slope δ. The heat flux along this surface is $k_i \partial T / \partial s$, where k_i is the along-isopycnal diffusion coefficient, and $\partial T / \partial s$ is the temperature gradient on the isopycnal surface. The vertical component of this flux is $\delta/(1 + \delta^2) k_i \partial T / \partial s$. The vertical flux due to diapycnal mixing is equal to $1/(1 + \delta^2) k_n \partial T / \partial n$, where k_n is the diapycnal diffusion coefficient and $\partial T / \partial n$ is the temperature gradient perpendicular to the isopycnal surfaces. Thus, the total vertical heat flux is approximately given by

$$F_z = -\frac{1}{1 + \delta^2}\left[k_n \frac{\partial T}{\partial n} + \delta \frac{\partial T}{\partial s} \right] \qquad (11.2.1)$$

The effective diffusion coefficient, k_{eff}, is that coefficient which, when multiplied by the vertical temperature gradient, $\partial T / \partial z$, gives the same flux as in Eq. (11.2.1). Since $\partial T / \partial z$ and $\partial T / \partial n$ are essentially the same, it follows that

$$k_{eff} = -\frac{1}{1 + \delta^2}\left[k_n + \delta \left(\frac{\partial T / \partial s}{\partial T / \partial n} \right) \right] \qquad (11.2.2)$$

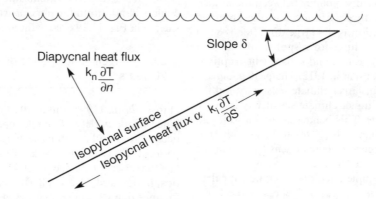

Fig. 11.2.1 Illustration of an isopycnal surface, with slope δ, and the diapycnal and isopycnal heat fluxes.

The constant-density surfaces in the ocean tend to be almost exactly parallel to constant-temperature surfaces. Hence, $\partial T / \partial s \approx 0$ and $k_{eff} \approx k_n$. For some tracers, such as radioactive carbon and tritium, there are significant gradients on isopycnal surfaces in middle and high-latitude regions because there has been much greater penetration of these tracers into the ocean at high latitudes than at low latitudes. This is illustrated in Fig. 11.2.2 for tritium. As a result, $K_{eff} >> K_n$, and the K_{eff} so derived is not applicable to temperature. For further analysis, see Harvey (1995) and Harvey and Huang (1999).

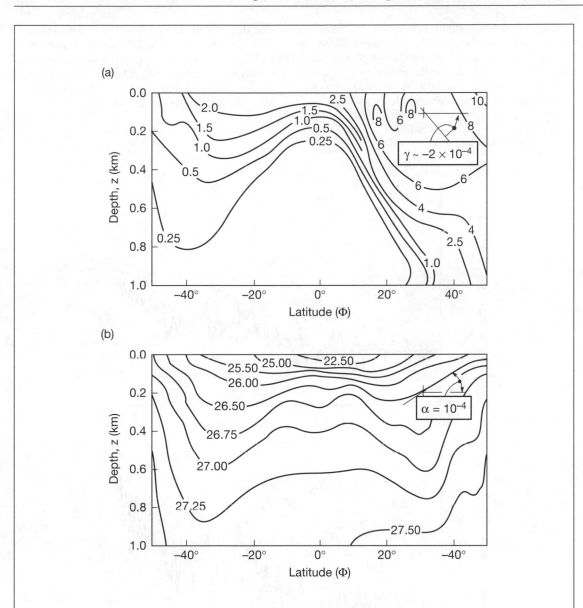

Fig. 11.2.2 (a) Latitude-depth contours of tritium concentration (in units of 10^{-15} moles tritium per mole of hydrogen in seawater) in the Atlantic basin in 1977 due to bomb testing. (b) Latitude-depth contours of Atlantic basin isopycnals in units of $\Delta\rho = \rho - \rho_0$. (kgm^{-3}), where $\rho_0 = 1000$ kgm^{-3} is a reference density. Reproduced from Hoffert and Flannery (1985).

- With the initial disequilibrium, the global mean surface warming is 0.1–0.2°C (depending on the ocean model used), which compares well with the observed warming during this period of 0.16°C.
- There is a maximum in the warming in the tropics in the upper troposphere in simulations without loss of stratospheric O_3, but this maximum disappears when O_3 loss is included – in accord with observations.

- The global mean stratospheric cooling between 1979 and 1995 due to the decrease in O_3 alone is 0.6°C, and 0.7°C when changes in well-mixed GHGs are included, both of which are less than the observed global mean cooling of 0.9°C given by the MSU data (Fig. 5.13(a)).

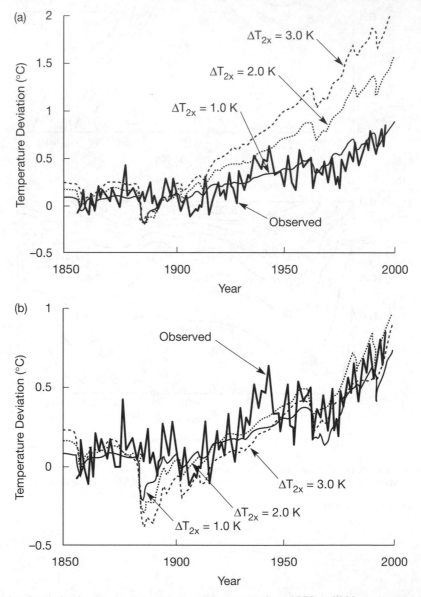

Fig. 11.19 Simulated variation in global mean surface air temperature from 1850 to 2000, and comparison with observations (to 1997), for the GHG, solar, and volcanic forcings given in Fig. 11.16, and using climate sensitivities of 1.0 K, 2.0 K, and 3.0 K for a CO_2 doubling. Results are given for (a) no aerosol cooling, and (b) aerosol coolings reaching 20%, 40%, and 60% of the 1990 GHG forcing by 1990 for climate sensitivities of 1.0 K, 2.0 K, and 3.0 K, respectively.

The equilibrium warming for a CO_2 doubling is 3.5°C for the version of the GISS AGCM used here. However, the good fit to observed temperature variations does not validate the high climate sensitivity, since there are relatively small differences in the initial transient response for models with large differences in sensitivity, as explained in Section 11.1 and illustrated in Fig. 11.7. Instead, the good agreement with observations serves to validate the processes affecting the vertical structure of temperature changes.

With regard to the cooling of the stratosphere, Ramaswamy *et al.* (1996) obtained a global mean stratospheric cooling of 0.6°C due to loss of ozone over the period 1979–1990, which is the same as the observed cooling during this period (see Fig. 5.13(a)). In their case, there was no additional cooling due to the increase in GHGs, as the computed heating effect of increasing SF_6 and CFCs balanced the cooling effect of increasing CO_2, CH_4, and N_2O (see Table 3.1). The discrepancy of 0.2°C in Hansen *et al.* (1997c) could be due to errors in the specified changes in stratospheric O_3 (which are highly uncertain, as discussed in Section 5.6), or due to the omission of the observed increase in stratospheric water vapour (see Section 5.5 and item 5 in Table 5.2). Since Hansen *et al.* (1997c) prescribed an increase in tropospheric O_3 (see Fig. 7.5), and increasing tropospheric O_3 also tends to cool the stratosphere (Section 10.1), the computed stratospheric cooling depends on the prescribed changes in both tropospheric and stratospheric O_3, although the latter dominates over the period 1979–1995. The stratospheric cooling and surface–troposphere radiative forcing due to the decrease in stratospheric O_3 are tightly coupled, since the radiative forcing is dominated by the decrease in downward infrared radiation caused by the cooling of the stratosphere (see Section 7.2 and Fig. 7.4). Hansen *et al.* (1997c) obtained a radiative forcing during 1979–1995 due to loss of stratospheric O_3 of –0.3 W m^{-2}. The fact they are also able to simulate close to the observed stratospheric cooling is evidence in support of the computed surface–troposphere radiative forcing. Thus, although the ability to simulate the observed changes in stratospheric temperature does not validate the model-computed surface troposphere climate sensitivity, it does help to constrain the sensitivity inferred from the comparison of model and observed surface temperature changes (as in the preceding section) by constraining one of the input forcings.

11.5   Summary

This chapter has extensively analysed the time-dependent or transient climatic response to increasing concentrations of GHGs. The salient points to emerge from this discussion are as follows.

- The time scale for the adjustment of surface temperature to heating perturbations depends on both the ocean mixing coefficients and the climate sensitivity. The interaction of these two factors is such that the relative difference in the simulated climatic warming for cases that differ in climate sensitivity is considerably smaller during the initial transient than during the later transient. This in turn makes it difficult to assess long-term climate sensitivity from the early stages of warming, given the uncertainties in the ocean mixing parameters.
- Changes in the intensity of the large-scale ocean circulation or in the average temperature at which deep water forms have a modest effect on the transient surface temperature response but have essentially no effect on the global mean surface temperature response in equilibrium. Such changes, however, have a dramatic effect on the vertical profile of deep ocean temperature change, both during the transient and in the new equilibrium.
- Transient simulations with coupled AOGCMs show substantially different regional patterns of climatic change than equilibrium simulations. Since cloud feedbacks depend critically on the changes in atmospheric winds and in vertical motions, which depend on regional patterns of climatic change, the net cloud feedback can be quite different during the transient than for an equilibrium simulation. In one case where this has been analysed, the effective climate sensitivity during the early transient is only one-third as large as the equilibrium sensitivity.
- Comparison of observed and model-simulated global- and hemispheric-scale temperature changes indicates that much of the warming during the past century is due to anthropogenic effects.
- The observed temperature variations of the last 120 years are best simulated, using the climate model of Harvey and Huang (1999), assuming a climate sensitivity of 1.0°C and no offsetting aerosol forcing. However, given the possibility that the effective transient climate sensitivity could be much smaller than the equilibrium climate sensitivity, an equilibrium climate sensitivity in line with other lines of evidence (2.0–3.0°C) cannot be ruled out.

CHAPTER 12

Sea level rise

In this chapter the physical processes that determine the rise in sea level that will accompany global warming are discussed, followed by a presentation of recent projections of sea level rise through to the year 2100. Most of the work on possible future sea level rise has focused on the short term (the 21st century). However, the sea level rise issue, perhaps more than any other issue in the climate system response to emissions of GHGs, involves very long time scales and very long lags between the driving factors and the ultimate response of the system. Because of these long lags, an irreversible commitment to substantial sea level rise (several metres) could be made centuries before it occurs. This chapter therefore closes with a discussion of the potential for very large increases in sea level over periods of many centuries to thousands of years.

12.1 Processes tending to cause a rise in sea level

Sea level will tend to increase in association with a warmer climate as a result of (i) thermal expansion of sea water, (ii) melting of small mountain glaciers and ice caps, (iii) melting of the Greenland ice cap, and (iv) changes in the mass balance of the Antarctic ice cap. In considering the response of the Antarctic ice cap, it is important to distinguish between the East Antarctic ice cap, which is land-based, and the West Antarctic ice cap, much of which is anchored below sea level and which partly consists of floating ice shelves. The processes involved in each of these contributors to sea level rise are discussed below.

Thermal expansion of sea water

As sea water warms, its density decreases. Thus, a given mass of ocean water will occupy a greater volume as the ocean warms, thereby tending to increase the average sea level. The extent of sea level rise depends on the vertical profile of warming within the deep ocean. As discussed in Section 11.1 and illustrated with the 1-D upwelling–diffusion model, quite different temperature warming profiles can occur in association with the same global mean surface warming, depending on (i) the change in the average temperature at which bottom water forms, and (ii) whether and to what extent the thermohaline overturning circulation changes in intensity. The change in the average temperature of bottom water formation depends on the amount of warming of the surface waters in the polar regions where bottom warm forms, and any change in the proportion of North Atlantic and Antarctic sources for bottom water. Since bottom water in the North Atlantic forms at about 2°C, while that in the Antarctic forms at about –2°C, a sharp drop in the rate of bottom water formation in the North Atlantic in combination with modest surface warming in polar regions could result in little or no change in the average temperature of bottom water formation. For the case in which the thermohaline overturning intensity is fixed and the average temperature of bottom water formation is constant, the steady-state profile of temperature change decreases exponentially with depth according to Eq. (11.5). If the average temperature of bottom water formation increases by the same amount as the global mean mixed layer, then the steady-state warming is the same at all depths (and equal to the surface warming). This gives a vertically integrated oceanic warming that is about seven times larger than in the first case. If the thermohaline circulation weakens significantly – as is expected during at least the initial transient response to global warming – then the upper part of the deep ocean (below the mixed layer) could warm by substantially more than the surface mixed layer. For a given average ocean warming, the rise in sea level depends on where the warming occurs. The expansion coefficient for sea water is greater the warmer the water, so the greatest sea level rise will occur if the warming is concentrated in those regions where the water is already the warmest, namely, the upper few hundred metres at low latitudes.

Glaciers and small ice caps

Mountain glaciers and small ice caps can be divided into an upper accumulation zone, where the annual accumulation of snowfall exceeds annual melting or *ablation,* and a lower ablation zone, where the annual ablation exceeds annual accumulation. For a steady-state glacier (i.e., a glacier with constant mass), the net accumulation in the accumulation zone exactly balances the net loss in the ablation zone, with an internal transfer of the surplus from the accumulation zone to the ablation zone through flow of the glacier. The boundary between the two zones is referred to as the *equilibrium line altitude (ELA)*. An increase in temperature or decrease in the annual snowfall tends to raise the ELA, while an increase in snowfall (which could occur as climate warms if total precipitation increases but temperatures remain sufficiently cold) will tend to lower the ELA. If the ELA rises above the height of the glacier, then the glacier will eventually completely disappear. Glaciers differ substantially in how sensitive they are to changes in the ELA, depending on the geometry of the ablation and accumulation zones. This is illustrated in

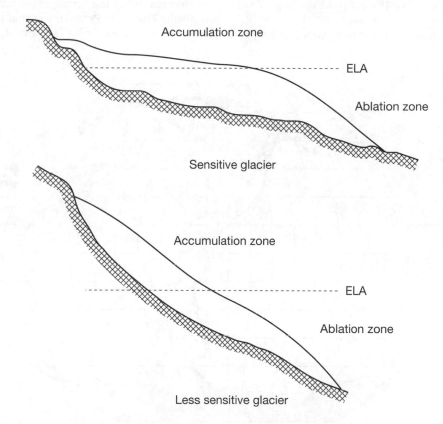

Fig. 12.1 Illustration of the dependence of glacier sensitivity to changes in ELA on the geometry of the ablation and accumulation zones.

Fig. 12.1. Different glaciers also differ dramatically in how quickly they will fully adjust to changes in ELA, with smaller glaciers tending to have a faster response (although Raper *et al.* (1996) provide an exception). However, the response time also depends on the rate of mass throughput; glaciers with large accumulation and ablation rates (per unit area), and hence a large mass throughout rate, will also tend to respond more quickly to changes in ELA. Figure 12.2 shows the distribution of mountain glaciers and small ice caps today. Complete melting of all mountain glaciers and small ice caps is estimated to raise sea level by 50 ± 10 cm (Warrick *et al.*, 1996).

Greenland and Antarctica

Greenland is of particular concern because of its large ice mass – sufficient to raise sea level by 7.4 m if it were to completely melt – and the fact that a regional warming of as little as 6°C appears to be sufficient to provoke the eventual melting of the entire ice mass (Letréguilly *et al.*, 1991). Figure 12.3 shows the distribution of summer surface air temper-

ature over Greenland. Today, July air temperatures over Greenland range from in excess of 5°C to somewhat colder than –10°C. Even though a 6°C warming would not provoke melting everywhere, the surface elevation would begin to drop in the coldest regions (and elsewhere), thereby provoking an increase in the areal extent of the ablation zone, further melting, and so on. The reduction in snow albedo as temperatures increase would constitute a further positive feedback. Simulations by Crowley and Baum (1995) indicate that, once the Greenland ice sheet disappeared, it would not reform, even if GHG concentrations returned to their pre-industrial values.

Models used to project changes in the mass of both the Greenland and Antarctic ice sheets generally assume that warmer temperatures lead to an increased rate of snowfall. This might not be true, at least in the case of the Greenland ice sheet. Cuffey and Clow (1997) found, based on the characteristics of ice cores from Greenland, that local temperatures over Greenland increased by about 20°C in going from the peak of the last ice age to the present interglacial (the Holocene Period), and that the rate of accumulation increased by a factor of 4. However,

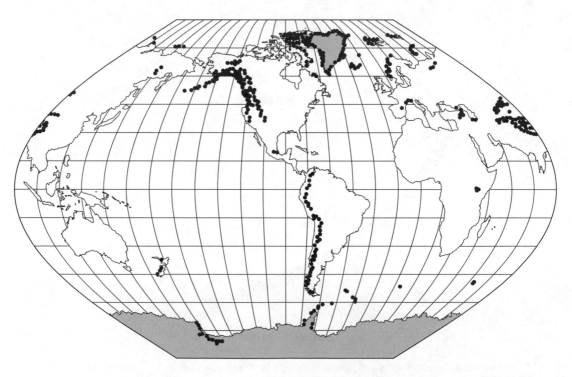

Fig. 12.2 Present-day distribution of mountain glaciers and small ice caps, reproduced from Warrick *et al.* (1996).

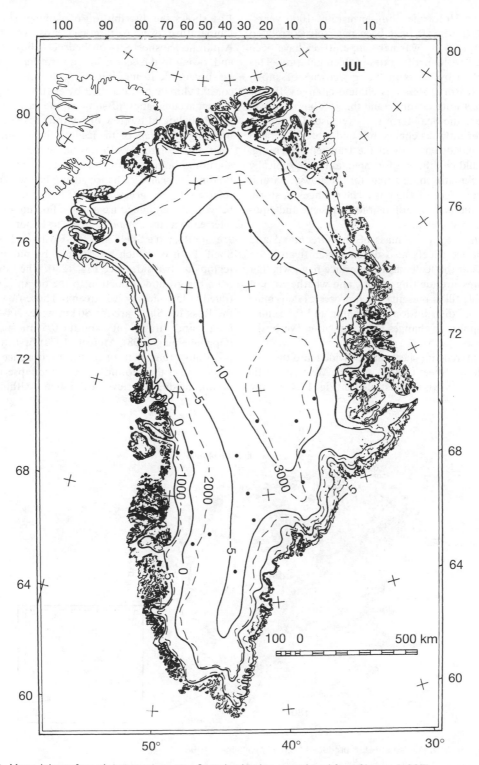

Fig. 12.3   Mean July surface-air temperature over Greenland today, reproduced from Ohmura (1987).

within the Holocene, long-term (500–1000 years) accumulation rates have been inversely correlated with temperature: warmer temperatures have been associated with smaller rates of accumulation. This is explained by shifts in the trajectories of snow-bearing storm systems as climate changes. If this applies to future changes (and this is a big if), it will exacerbate the near-term sea level rise. The way in which storm tracks change will undoubtedly depend on regional differences in the transient warming, which could be different for rapid, anthropogenically induced climatic change than for the slower, natural climatic changes of the past. The changes in storm tracks could also be different for transient and equilibrium climatic change.

The change in the mass of the Greenland and Antarctic ice sheets also depends significantly on processes at the ice/bedrock interface (e.g., whether or not pressure melting occurs, and whether or not deformable till is present). Such processes determine the regional distribution of thinning and thickening ice in response to changes in accumulation (Verbitsky and Oglesby, 1995).

The Antarctic ice sheet can be divided into the East Antarctic Ice Sheet (EAIS), and the West Antarctic Ice Sheet (WAIS), as illustrated in Fig. 12.4. Figure 12.5 shows the distribution of summer surface air temperature over Antarctica. Most of the East Antarctic ice sheet sits on bedrock above sea level and, owing to the very cold temperatures, would not be susceptible to melting along the edges until the coastal climate had warmed by 6–10°C. At present, the net accumulation of snow on the ice sheet will be largely balanced by flow of ice to the edges and the breaking or *calving* of the ice into the ocean. In contrast, most of the WAIS lies on bedrock below sea level, and large portions of the ice sheet consist of floating ice shelves. The ice sitting below sea level is referred to as *grounded ice*, and the boundary between the grounded ice and the floating ice shelf is referred to as the *grounding line*. The major ice shelves are the Ross Ice Shelf and the Filchner–Ronne Ice Shelf, both of which are fringed by mountainous regions on two sides (see Fig. 12.4). The flow of ice from the grounded portion to the ocean is concentrated in fast-moving ice streams. Those flowing into the Ross Ice Shelf are 30–80 km wide, 300–500 km long, and travel at about 0.5 km per year (Oppenheimer, 1998). Mercer (1978) first raised the possibility that melting of the floating ice shelves could cause the grounded ice to collapse into the ocean, raising sea level by 5–6 m. Although the

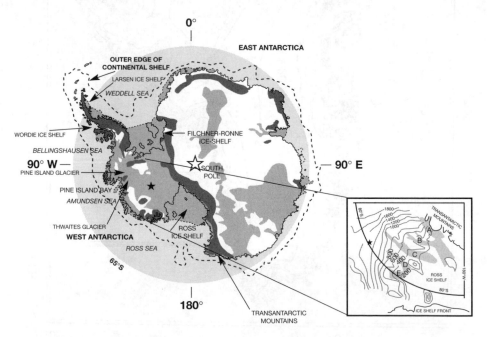

Fig. 12.4 The Antarctic ice sheet, reproduced from Oppenheimer (1998).

summer air temperature over the ice shelves might remain below the freezing point as the global climate warms, the ice shelves are thought to be particularly susceptible to melting as the seasonal sea ice around Antarctica retreats and the ocean water adjacent to and under the ice shelves begins to warm. The question of the stability of the WAIS is rather complex, and is discussed more fully in the following section.

12.2   Stability of the West Antarctic Ice Sheet

There is conflicting evidence concerning when the WAIS first formed: stratigraphic evidence from the Antarctic continental shelf suggests an inception 15–20 million years ago, while isotopic evidence from deep sea sediments suggests an inception 9 million years ago. It is unclear whether or not the WAIS has existed continuously since then. The initial concern

about the stability of the WAIS was based on the premise that the grounded portion of the ice sheet depends on back pressure from the floating ice shelves to prevent it from flowing into the ocean. However, the fact that outflow from the grounded ice is concentrated in wet-based ice streams diminishes the likelihood that the floating ice shelves are needed to stabilize the grounded ice according to Bentley (1998). This is supported by theoretical modelling work and by evidence indicating that the total outflow onto the Ross Ice Shelf (which takes half of the WAIS outflow) has been roughly constant during the past 1500 years in spite of major internal reorganizations.

Back-of-the-envelope calculations indicate that it would be very difficult for a rapid collapse of the WAIS to occur. The present-day outflow of the WAIS corresponds to a rise in sea level of 1 mm per year, were it not balanced by accumulation of snow on the ice sheet. To give a sea level rise of 0.5 m per century would require a fivefold increase in the rate of outflow (with no change in accumulation). Even if

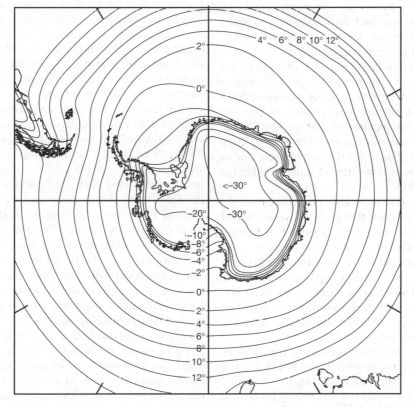

Fig. 12.5   Mean January surface air temperature over Antarctica today. Reproduced from Schwerdtfeger (1970).

hydrodynamic instability occurs, numerical models indicate that collapse of the WAIS would require *at least* 400 years (Thomas *et al.,* 1979), 1200 years (Bindschadler, 1997), or 1600–2400 years (MacAyeal, 1992). After reviewing the uncertainties and imperfections in existing numerical models, Oppenheimer (1998) concluded that "the early idea that WAIS is subject to hydrostatic instability and rapid grounding-line retreat seems unlikely, but the issue cannot be conclusively resolved" at present.

A key premise underlying concerns about the stability of the WAIS is that regional warming would lead to increased melting at the base of the floating ice shelves. The base of the floating ice shelves occurs at depths as great as 1500 m, and the basal melting that occurs today is driven by the fact that the melting point at the 1500 m depth is −2.9°C rather than −1.9°C, as at the surface. Dense surface water at −1.9°C forms in association with the ejection of a salty brine during the seasonal freezing of sea ice; this water is referred to as "High Salinity Shelf Water" (HSSW). HSSW flows along the underside of the floating ice shelves and, because it ends up warmer than the local melting point, is able to power substantial melting at the base of the ice shelves. Nicholls (1997) presents oceanographic data indicating that a reduction in the rate of formation of sea ice in winter (due to regional warming) would reduce the flux of HSSW, thereby reducing the rate of basal melting. Thus, a warmer climate would, at least initially, lead to an increase in the ice shelf mass, thereby tending to lower sea level. This is in addition to any lowering of sea level that would occur due to an increase in the accumulation of snow on either the EAIS or the WAIS.

A further premise underlying concerns about the stability of the WAIS is that the ocean water around Antarctica would warm. As discussed in Section 11.2, some AOGCMs show a slight *cooling* of the surface waters around Antarctica during the first several decades of the transient response to increasing atmospheric GHG concentrations. Of course, if this leads to an increase in the HSSW flux then, based on the arguments presented above, this could still lead to a thinning of the floating ice shelves!

In summary, the following conclusions can be drawn concerning the stability of the WAIS. First, it is not at all clear that the early stages of warming around Antarctica (whenever this should occur) will lead to melting of the floating ice shelves; it is entirely possible that the reduction in the rate of sea-sonal sea ice formation could reduce the flux of High Salinity Shelf Water beneath the floating ice shelves, thereby leading to a *reduction* in the rate of basal melting. Secondly, it may very well be that removal of the floating ice shelves would not trigger the collapse of the grounded portion of the ice sheet (loss of the floating ice shelves alone would have almost no effect on sea level). Third, if the WAIS ice sheet were to collapse, many hundred years to more than 1000 years would be required for this to occur.

12.3   Near-term projections (1990–2100)

The separate contributions of thermal expansion, mountain glaciers and small ice caps, and of Greenland and Antarctica, to changes in sea level have been computed using a variety of climate, glacier, and ice sheet models. Computation of the thermal expansion component requires simulating the downward penetration of heat into the oceans. It also requires having correctly computed the background temperature profile (prior to warming), since the thermal expansion for a given warming depends on the initial temperature. The downward penetration of heat directly affects the simulated surface temperature changes, with a greater downward penetration of heat leading to a slower rate of surface warming. The simulated surface warming is then used to drive models of mountain glaciers, small ice caps, the Greenland ice sheet, and the Antarctic ice sheet. The changes in glaciers and ice sheets projected over the next few hundred years do not, however, feed back onto the simulated change in regional or global mean surface temperature because the absolute change in areal extent is so small (so that any surface albedo–climate feedback is negligible, unlike the case of seasonal snowcover and sea ice). Thus, the glacier and ice sheet models can be run separately from the climate model. The glacier and ice sheet models can also be driven with idealized climatic change scenarios, in a stand-alone mode. When combining independent calculations of the sea level rise due to thermal expansion with that due to changes in ice mass, it should be kept in mind that the two will be inversely related to some extent, given that greater thermal expansion will tend to be associated with smaller transient warming of the surface climate, and hence smaller melting of glaciers.

Global mean sea level rise

Huybrechts and Oerlemans (1990) used a 3-D stand-alone model of the Antarctic ice sheet (including the floating ice shelves), and imposed varying degrees of climatic change. Accumulation was assumed to double for each 10 K warming. They found that for a warming over Antarctica of less than 5 K, there is a net decrease in sea level because the assumed increase in accumulation exceeds the increase in ablation. For warming greater than 5 K, a net sea level rise occurs. If local warming reaches 9 K by 2100 and is then held constant, Antarctica contributes a sea level rise of 20 cm by 2200 and 45 cm by 2300. These results are based on a very crude ice sheet model, and assume that no instability of the WAIS is triggered.

Van de Wal and Oerlemans (1997) used a 2-D (latitude–longitude) model of the Greenland ice sheet to investigate the sensitivity of the predicted changes in the Greenland ice mass to the way in which the model is formulated and to the choice of parameter values. As in the Antarctic ice sheet model, it is assumed that warmer temperatures lead to a greater rate of accumulation: in this case, accumulation is assumed to increase by 5% per degree Celsius of warming.

Loutre (1995) assessed the sensitivity of the Greenland ice sheet to CO_2 increases using a 2-D (latitude–height) climate model coupled to a 2-D (latitude–longitude) model of the Greenland ice sheet. She found that for a scenario in which the CO_2 concentration increases to 710 ppmv and is then held constant, the Greenland ice sheet completely disappears in 3000 years. Even if the CO_2 concentration returns to 350 ppmv 500 years from now, the Greenland ice sheet still disappears but takes over 5000 years to do so. Crowley and Baum (1995) conclude that the removal of the Greenland ice sheet would be irreversible, even if climate returned to pre-industrial conditions, since, in the absence of ice cover, summer temperatures would reach 10°C.

Two somewhat similar approaches have been used to project the contribution of mountain glaciers and small ice caps to future sea level rise. In one approach, developed by Wigley and Raper (1995), all the glaciers in the world are distributed into a series of "bins" representing different initial masses (expressed in terms of potential sea level rise). Within each bin, it is assumed that there is a range of temperature changes (ΔT^*) needed to raise the ELA to the top of the glacier; this range is prescribed. Finally, different glacier response time constants are applied for different ΔT^*s, such that glaciers with the smallest ΔT^* are assumed to respond most rapidly so as to maximize the computed effect on sea level. The whole scheme is somewhat *ad hoc,* owing to the lack of a comprehensive inventory of the world's glaciers. In the second approach, developed by Oerlemans and Fortuin (1992), all the world's glaciers and small ice caps are assigned to one of 100 different regions. For each region, a sensitivity of the glacier mass budget to temperature changes is prescribed that depends on the precipitation rate (glaciers with higher precipitation are assumed to be more sensitive to temperature changes than glaciers with a smaller precipitation rate). The feature common to both methods is that the calculated change in ice mass is performed over a distribution of glacier characteristics.

Table 12.1 Contribution to sea level rise (cm) from 1990 to 2100 due to various processes, as computed by different researchers. ΔT is the associated global mean warming from 1990 to 2100.

Reference	ΔT (°C)	Glaciers and small ice caps	Greenland	Antarctica	Thermal expansion	Total
Wigley and Raper (1995)	2.1	11.9–19.4	—	—	—	—
de Wolde et al. (1997)	2.2	12.3	7.2	−7.5	15.2	27
	2.5	14.6	10.4	−8.5	17.5	34
Gregory and Oerlemans (1998)	2.7	13.2	7.6	−7.9	22.5	35
	3.3	18.2	9.3	−9.7	29.9	48
Huybrechts and de Wolde (1999)	1.3	—	3.9	+0.7	—	—
	2.4	—	10.5	−3.0	—	—
	3.5	—	20.8	−6.4	—	—
Bryan (1996)	—	—	—	—	15 ± 5	—

Table 12.1 summarizes the projected (1990–2100) contributions to sea level rise from the thermal expansion of sea water, melting of mountain glaciers and small ice caps, and from changes in Greenland and Antarctica, as reported in several recently published studies. Also shown is the change in global mean surface air temperature during 1990–2100 associated with each set of results. The results of de Wolde *et al.* (1997) were obtained using a 2-D (latitude–height) energy balance climate model coupled to (i) the 2–D model of the Greenland ice sheet used by van de Wal and Oerlemans (1997), (ii) the 3-D Antarctic ice sheet model of Huybrechts and Oerlemans (1990), and (iii) the glacier and small ice cap model of Oerlemans and Fortuin (1992). The results of Gregory and Oerlemans (1998) were obtained using output from the Hadley Centre AOGCM to drive 2–D models of the Greenland and Antarctic ice sheets. They also used the distribution of mountain glaciers and small ice caps used by Oerlemans and Fortuin (1992). The results of Huybrechts and de Wolde (1999) were obtained by coupling the same 2-D energy balance climate model as used by de Wolde *et al.* (1997) to 3-D models of both the Greenland and Antarctic ice sheets. The smaller of the temperature change scenarios shown in Table 12.1 for both de Wolde *et al.* (1997) and Gregory and Oerlemans (1998) assume continued high emissions of sulphur, while the scenarios with greater temperature change assume no emissions of sulphur.

The various approaches yield surprisingly similar results owing, no doubt, to the fact that many of the critical underlying assumptions are similar (e.g., no collapse of the WAIS, snow accumulation increasing with temperature). The greatest uncertainty pertains to the response of the Antarctic ice cap. Both de Wolde *et al.* (1997) and Gregory and Oerlemans (1998) find that changes in the Antarctic and Greenland ice masses almost cancel out. This is supported by a study of the effect of a doubled CO_2 climate on the surface mass balance of Greenland and Antarctica by Ohmura *et al.* (1996) using a high-resolution AGCM. They found that the simulated *changes* in the surface mass balance on the two ice sheets largely cancel out. As noted in Section 5.8, one must assume a substantial contribution (5–14 cm) of Antarctica to sea level rise during the past century in order to explain the observed sea level rise. The results of de Wolde *et al.* (1997) and Gregory and Oerlemans (1998) do not include this possible background trend. In contrast, Huybrechts and de Wolde (1999) simulated the last two glacial cycles (beginning 250,000 years ago) prior to the present, so that the glacier ice mass "inertia" due to past climatic changes is reflected in their projections of future climatic change. They obtained a contribution of Antarctica to sea level rise during the past 100 years of 3.9 cm which, as noted in Section 5.8, closes about one-third to two-thirds of the gap between the observed sea level rise of 18 ± 1 cm and the computed rise due to melting of glaciers, thermal expansion of sea water, and direct anthropogenic effects (5–12 cm). For a constant climate, the projected effect of Antarctica to 2100 is a sea level rise of 5.0 cm. The sea level rises shown in Table 12.1 include this background trend plus the effect of a warming climate, and so the net result is a much smaller tendency for falling sea level than obtained by de Wolde *et al.* (1997) or Gregory and Oerlemans (1998).

The uncertainty in the model-computed background trend is quite large; it could be close to zero, or it could be twice as large. Given the aforementioned discrepancy between observed and computed sea level rise during the past century, the latter seems to be a distinct possibility. For this reason and because of the uncertainty in all of the computed components to sea level rise, it seems reasonable to broaden the original range (27–48 cm by 2100) to 25–70 cm by 2100. This range pertains to GHG concentrations stabilizing at the equivalent of a CO_2 doubling, (with a global mean warming of 2.2–3.3°C). Larger (smaller) warming would cause a larger (smaller) sea level rise by 2100.

Regional variations in sea level rise

There are marked regional variations in sea level today (by up to 1 m), which give rise to a non-zero slope of the sea surface. Surface ocean currents are in a balance between the Coriolis force due to horizontal motion, and the horizontal pressure gradient force created by the sea surface slope. The present-day variation in sea level is related to horizontal variations in the density of sea water (which in turn depend on horizontal variations in the temperature and salinity of sea water), and to horizontal transport of water in the mixed layer due to the applied wind stress (such transport is at right angles to the wind direction, as explained in any introductory oceanography textbook, and is referred to as Ekman transport).

Now suppose that a column of water experiences greater warming than the surrounding water. Sea level will increase more in the column experiencing greater warming, and one might expect water to flow out from the dome of warmer water until the sea level rise is the same everywhere. However, water cannot easily flow out from a dome of higher sea level because of the deflection of moving water perpendicularly to its motion by the Coriolis force. Instead, the strength of the currents encircling the dome will adjust themselves to the new sea surface slope, with little redistribution of water mass. The adjustment of the global distribution of sea level to non-uniform heating of the ocean occurs in two ways: (i) through a type of wave called a *Kelvin wave,* which propagates along coastlines and along the equator, and (ii) through another kind of wave called a *Rossby wave,* which propagates westward from the eastern boundaries of ocean basins. Hsieh and Bryan (1996) illustrate these processes through the presentation of a series of figures that illustrate the propagation of changes in sea level for two idealized, non-uniform distributions of ocean temperature change. Even in steady state (i.e., after the waves have had time to fully propagate), there will still be significant variations in the sea level caused by non-uniform warming of the oceans, because of compensating changes in ocean currents.

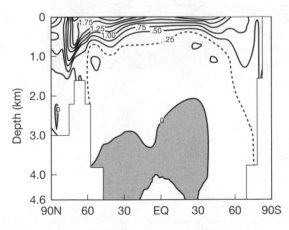

Fig. 12.6 Latitude–depth variation in the zonal mean ocean warming at year 70, as given by Bryan (1996) for a run of the GFDL AOGCM in which the atmospheric CO_2 concentration increases at a compounded rate of 1% per year.

Bryan (1996) computed the variation in sea level rise for a transient run of the GFDL AOGCM in which the atmospheric CO_2 concentration increases by 1% per year (the same experiment as reported by Manabe *et al.* (1991, 1992) and discussed in Section 11.2). Figure 12.6 shows the variation with latitude and depth of the zonally averaged temperature change in the ocean. The greatest warming occurs in the North Atlantic Ocean, due to the deep mixing of surface heat (this deep mixing also causes the minimum in surface air warming in this region, seen in Fig. 11.12). Figure 12.7 shows the regional variation in sea level rise averaged over years 70–79; only the contributions due to thermal expansion of sea water and changes in ocean currents are represented in this figure. The average sea level rise is 15 cm, but differs by in excess of 15 cm from one region to another. The greatest sea level rise occurs in the North Atlantic Ocean and along the east coast of North America, owing to the markedly greater warming in this region (seen in Fig. 11.12). However, the greater warming at depth in the North Atlantic Ocean than at lower latitudes reduces the horizontal density contrast, which drives the thermohaline overturning (recall Box 2.2). The thermohaline overturning and the associated poleward transport of heat thus decrease in strength. This in turn reduces the ocean warming at high latitudes and increases it at low latitudes, thereby reducing the difference in sea level rise. Thus, the main smoothing out of the horizontal variations in sea level rise occurs through a smoothing out of the temperature changes due to induced changes in ocean currents, rather than through a redistribution of ocean mass.

An interesting consequence of the non-uniform sea level rise is that the sea level rise averaged over all coastlines can be quite different from the global mean sea level rise (Hsieh and Bryan, 1996). This complicates the comparison of tidal gauge records of sea level rise with model-predicted sea level rise.

Finally, changes in atmospheric pressure and in ocean currents *induced by changes in atmospheric winds* also contribute to regional variations in sea level. Cui and Zorita (1998) find that these processes could cause the sea level rise along the Chinese coast to differ by a few centimetres from what it would otherwise be at the end of the next century. Note that ocean currents can play two very distinct roles in sea level rise: they can either inhibit the redistribution of water mass (and the smoothing out of variations in the change in sea level) when sea level rise is induced by thermal expansion, or provoke a redistribution of water mass (and thus *produce* regional variations in sea level) in response to a change in surface wind stress.

12.4   Prospects for long-term sea level rise

As noted in the introduction to this chapter, the sea level response to climatic change involves very long time scales and entails commitments that will endure for centuries after consumption of fossil fuels has ended. The sea level rise projected for the end of the 21st century – 25 to 70 cm – is likely to be only a small fraction of the sea level rise that will ultimately occur, even if GHG concentrations are rapidly stabilized during the early part of the 21st century.

Thermal expansion of sea water

Experiments with a 2-D ocean model by Harvey (1994) indicate that a 3.0°C global mean surface warming could be associated with an eventual sea level rise ranging from less than 0.5 m to almost 3.0 m, depending on how the temperature change varies with depth. The time to reach 50% of the final sea level rise ranged from about 200 to 1000 years, while the time required to reach 75% of the final sea level rise ranged from about 400 to 2400 years. The largest and smallest sea level rises were associated with the *shortest* response time, with intermediate amounts of sea level having the longest response time. The cases with large sea level rise involved a near-uniform warming with depth and little change in the oceanic circulation, while the cases with small sea level rise were associated with very limited deep ocean warming and major reorganizations of the oceanic circulation (such that the cold Antarctic bottom water increased in importance).

Rather different results were obtained by Mikolajewicz *et al.* (1990) using the Hamburg OGCM. For a mean ocean surface warming of 4.0°C, they obtained a steady-state sea level rise of 0.51 m, with 50% and 75% responses obtained after 80 and 210 years, respectively. This implies that the steady-state ocean temperature warming decreased with increasing depth in their model. In contrast, Manabe and Bryan (1985) found that the steady-state deep ocean warming in their AOGCM after a CO_2 increase was significantly larger than the mean surface warming. Since their surface climate sensitivity was comparable to that in Harvey (1994), this implies an even larger sea level rise than the 3 m rise obtained by Harvey (1994). Until the likely variation of ocean warming with depth can be constrained, there will be a large uncertainty in the long-term sea level rise due to thermal expansion.

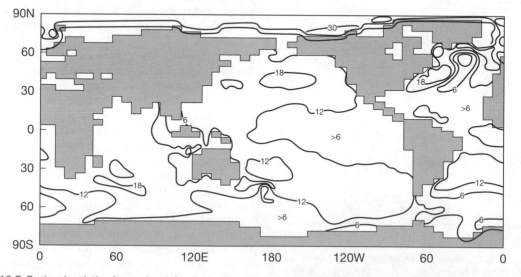

Fig. 12.7 Regional variation in sea level rise (cm), averaged over years 70–79 of the same experiment as in Fig. 12.6. Only the effects of thermal expansion of seawater and changes in ocean currents are shown here.

Greenland and Antarctica

Complete melting of Greenland, which could be triggered by as little as an equivalent CO_2 doubling, would lead to an increase in sea level by about 7 m, over a period of 1000–5000 years. Huybrechts and de Wolde (1999) find, using a high-resolution, 3-D model of the Greenland ice sheet, that melting of Greenland contributes 9–60 cm sea level rise per century over the period 2100–3000 for CO_2 concentrations stabilizing at two to eight times the present concentration (assuming a climate sensitivity for a CO_2 doubling of 2.2°C). The total contribution to sea level rise over the period 2000–3000 is 0.9–5.8 m. Collapse of the West Antarctic ice sheet, which could also be triggered by as little as an equivalent CO_2 doubling, would lead to a sea level rise of 5–6 m, over a time period of 1000–2000 years. The melting of East Antarctica, on the other hand, would require a regional warming on the order of perhaps 20°C, which is not likely except under scenarios of very high GHG emissions combined with moderate to high climate sensitivity.

Total sea level rise

Based on the above discussion, it can be seen that a sustained increase in GHG concentrations equivalent to a CO_2 doubling runs a significant risk of provoking an eventual sea level rise of up to 14–17 m. The component of this sea level rise due to melting of ice would be reduced by up to one-third by the tendency for the oceanic crust to sink under the weight of the additional water mass. This would occur on a time scale of several thousand years, which is comparable to the time required for large changes in the Greenland and Antarctic ice masses. Peak sea level rise may therefore be on the order of 10–13 m, which is still substantial. The risk is particularly large if high emission scenarios are combined with a climate sensitivity for a CO_2 doubling on the order of 3.0°C or larger, but is probably quite small if the climate sensitivity is on the order of 1.0°C (1.0–3.0°C being the range considered to be most likely, based on the evidence presented in Chapter 9).

12.5   Summary

Based on the evidence presented in this chapter, it is concluded that (i) global mean sea level is likely to rise by 25–70 cm by the year 2100, and (ii) a sustained increase in GHG concentrations equivalent to a doubling of atmospheric CO_2 runs a strong risk of an increase in sea level of 10–13 m over a period of several thousand years.

PART 3

THE SCIENCE–POLICY INTERFACE

The first two parts of this book have brought together information from the disparate disciplines that are needed to understand how the climate system responds to anthropogenic emissions of GHGs and aerosols and their precursors. We have examined the emission factors that link human activities to gas emissions, and have traced the sequence of events and processes involved in the carbon cycle and other biogeochemical cycles. These cycles determine the buildup in concentrations, from which one can calculate the radiative forcing. We then examined the transient and steady-state climate response to the projected radiative forcing, with an emphasis on those results that are independent of the details of any given simulation model. Lastly, we considered the effect on sea level rise. We have thus built an understanding of the processes involved, in nature, in translating human emissions of gases into physical impacts – changes in climate and changes in sea level at the regional level.

The development of a policy response to the prospect of global warming and rising sea level requires much more than the consideration of physical impacts. The anticipated changes in climate and sea level need to be translated into a wide array of social, economic, and biological impacts – changes in food production, in forests, in human welfare and well-being, in the health and vitality of natural ecosystems. Many of the impacts have costs that cannot be quantified in economic terms. The valuation of many impacts – particularly those involving loss of species and ecosystems – will differ from individual to individual, among different cultures, and probably from one generation to the next. Most of the economically quantifiable costs depend on the extent and ease of adaptation, which in turn depends on the nature of society and technology by the time

the changes occur. There is a wide divergence of opinion concerning the costs of climatic change, as reflected in reviews and discussions by Pearce *et al.* (1996) and Demeritt and Rothman (1998a,b).

The development of a policy response also requires consideration of the costs (and non-climatic benefits) of actions to reduce emissions of GHGs. In the case of energy-related emissions of GHGs, this entails improving the efficiency with which energy is used and transformed beyond the efficiency improvements that would occur anyway, increasing the provision of non-fossil fuel sources of energy beyond that which is expected to occur anyway, possibly reinjecting some CO_2 into the ground or deep into the ocean, and implementing broad-based policies to stabilize the human population at the lowest possible level. Many of these options and the associated cost estimates are reviewed in Ishitani *et al.* (1996), Kashiwagi *et al.* (1996), Michaelis *et al.* (1996), and Levine *et al.* (1996). Options to limit population growth, and related issues, are discussed by Bongaarts (1990, 1994). The reduction of land-use related emissions of GHGs entails developing sustainable forestry and agricultural practices (Brown *et al.*, 1996). In the case of agricultural emissions of methane and N_2O, this entails modifying current practices in ways that amount to increasing the efficiency with which fertilizers and animal feed are used (Mosier *et al.*, 1998a,b). As in the projected costs of climatic change, there is an enormous range of opinion concerning what the costs of these actions will be, with the anticipated costs being highly dependent on somewhat arbitrary assumptions concerning the ease with which new technologies can be introduced, the future cost of non-fossil fuel energy sources, and a whole array of social and economic factors (Repetto and Austin, 1997). Climatic change and efforts to limit emissions both entail risks (Harvey, 1996a,b). The development of a policy response to climatic change requires placing all the available information in a risk-minimizing framework.

Finally, the development of a policy response requires consideration of the politics of climatic change – the fact that an effective response requires coordinated, broadly based international action involving countries with widely differing economic and social structures and widely differing interests. One must take into account the *process of negotiating* effective international agreements to limit emissions of GHGs, the *structure* of such an agreement once in place, and the range of international *policy instruments* available to ensure effective and equitable international participation in any agreements to limit emissions of GHGs (Sebenius, 1991; Grubb, 1992).

It is far beyond the scope of this book to address any of the three issue areas identified above. Rather, in this concluding part, we shall use the tools developed in Parts One and Two to present current scientific understanding in a way that is useful for the development of limits on GHG emissions at the global scale and in the development of strategies to deal with the threat posed by unrestricted emissions of GHGs. We shall do this by using simple models to generate scenarios of climatic change for a range of possible future emissions of GHGs and aerosol precursors. Since climatic change, as it unfolds in reality, is likely to provide a number of surprises, Part Three and this book conclude with a discussion of the potential for surprises and the need to take this potential into account when developing a policy response.

CHAPTER 13

Scenarios of future climatic change

In order to explore the interface between science and policy, a series of policy-oriented questions will be addressed in this chapter. The purpose here is not to provide answers to specific policy questions, but rather to show how the scientific information presented in this book can be properly used as an *input* to the process of developing a policy response to the prospect of large, anthropogenically induced climatic change. Prior to doing this, we must first address the question "What is the appropriate scale of analysis?" That is, can useful policy insights be gained through the analysis of global-scale emissions and global-scale climatic changes, or must the policy analysis be grounded in changes at the regional level? Or is some mixture of the two scales of analysis most appropriate?

13.1 What are the appropriate scales of analysis?

Anticipated climate change is expected to vary greatly from region to region, and from season to season. First, the warming of the climate is expected to be largest at high latitudes and smallest at low latitudes. At high latitudes, the warming is expected to be greatest in winter (due to the thinning of sea ice) and smallest in summer. Elsewhere, the seasonal variation is expected to be much smaller, with greatest warming occurring in spring in some regions (due to earlier retreat of snowcover) and in summer in some other regions (due to reduced soil moisture). Within the broad latitudinal zones, however, there is expected to be much variability in the temperature and moisture response. The former was illustrated in Fig 10.1, which shows large variations in the projected equilibrium summer or winter temperature change for a CO_2 doubling at any given latitude.

The impacts of climatic change in any given region depend on the specific climatic changes that occur in that region. As noted above, local climatic changes can differ substantially from the globally averaged climatic change. The net global impact of climatic change will be given by the summation of regional impacts (modified by international trade, in the case of economic impacts), driven by regionally heterogeneous changes in climate. A consideration of climatic change at the regional level would therefore appear to be mandatory in the development of both national-scale and global-scale policy responses to the prospect of global warming. Indeed, this imperative has been driving the effort to develop better predictions of climatic change at the regional level.

There is a fundamental problem: the information that is most needed in order to assess the impacts of climatic change (namely, change at the regional level) is also the least reliable output from computer simulation models. This is evident from Fig. 10.1, where large differences between the different models can be seen in many regions. The predicted temperature change in a given region differs by up to a factor of 6. This is much larger than the difference in the predicted global mean temperature response among the models shown in Figs. 10.1, which ranges from 2.5°C to 4.0°C – less than a factor of 2 (among all the models shown in Table 9.6, the range in global mean warming is somewhat more than a factor of 2). It is a basic property of the climate system that the uncertainty in projected change is larger at the regional scale than at the global scale. This is because some of the differences between different models at the regional scale cancel out when summed globally. Since it is difficult to see how the uncertainty in the global mean temperature response can ever be reduced below a factor of 2 (at best), it is also unlikely that the uncertainty in the predicted climatic change will ever be reduced much below that seen in Fig. 10.1. This is especially the case for transient (time-evolving) climatic change, which introduces another source of uncertainty beyond that applicable to the equilibrium temperature changes shown in Fig. 10.1: the impact of regionally varying rates of uptake of heat by the oceans, and the repercussions of this for changes in ocean currents, atmospheric wind patterns, and regional cloud feedbacks. Changes in all of these could be quite different from the equilibrium changes, so several additional degrees of freedom (and associated uncertainty) are added to the transient response.

In light of these uncertainties, what is the value of assessments of the impact of climatic change at the regional level? Regional-scale projections of climatic change, and associated impact assessments, indicate the *risks* that climatic change poses for specific regions. They indicate the potential *magnitude* of local climatic change that could be associated with the generally much smaller, global mean climatic change. However, there is no scientific justification for developing policy responses at the national level based on the climatic changes and impacts within the jurisdiction under question, as given by any one climate model. Rather, the scientific understanding of anthropogenic climatic change can be used to support a *collective* (i.e., global-scale) policy response based on *generalized risks,* rather than nation-by-nation policy responses based on expected nation-by-nation impacts. In any case, the latter approach would not be effective, since emission reductions by one nation are of little value in reducing the buildup of GHGs unless all or most nations undertake similar actions.

This thinking is implicit in the wording of the United Nations Framework Convention on Climate Change (UNFCCC) and in the Kyoto Protocol, that was agreed to in December 1997. Nowhere do either of these agreements contemplate preventative policy responses based an anticipated impacts within the jurisdiction contemplating preventative responses. Rather, the UNFCCC refers to "common but differentiated responsibilities and respective capabilities" (Article 3.1).

In short, the role of regional-scale analysis is to indicate the level of risk, and the potential magnitude of local climatic change, associated with a given global mean warming. The global mean warming can therefore be thought of as an *index* of the risks posed by a particular scenario of GHG emissions, since regional impacts can be expected, on average, to increase with increasing global mean temperature.[1] This provides the justification for focusing on global-mean climatic change, as projected by globally averaged models, in assessing the merits of alternative scenarios of global-scale emissions.

13.2   Simulation methodologies

Two different methodologies are possible in analysing emissions and climatic changes at the global scale. First, the consequences for climate and sea level of a set of scenarios of future emissions of GHGs and aerosol precursors can be computed through a series of *forward calculations* – proceeding from emissions, to concentrations, to radiative forcing, to climatic change and sea level rise, in the same sequence as presented in Parts One and Two of this book. Second, a set of upper limits for the concentration of CO_2 and other GHGs can be adopted, involving eventual stabilization of concentrations, and the time-varying permitted emissions can be worked out using an *inverse calculation.* The mathematical procedure involved in the inverse calculation is outlined in Wigley (1991a). The procedure involves repeated calculations on each time step until no further change in the computed emission for that time step occurs, a procedure that would be prohibitively time-consuming if high-resolution models were used. The associated climatic change and sea level rise can be deduced using the usual forward calculations, given the prescribed variation in concentrations. Given a set of assumptions concerning the growth of the human population and of average per capita income (the product of which gives the growth of the global economy), the required rates of improvement in the overall efficiency with which energy is used and in the

---

[1] The global mean warming seems to be a good index of overall risks when the warming is increasing due to increasing forcing, with a roughly constant spatial pattern of warming. However, as discussed in Section 10.7, lower global mean warming due to a downward revision in the climate sensitivity may not imply reduced impacts in mid-latitude regions if the lower climate sensitivity is due to reduced warming in the tropics, since there would be a smaller increase in the moisture supply to mid-latitudes in this case.

rate of introduction of non-fossil energy sources can be computed, and the tradeoffs between these two options examined.

Using the forward approach, the following issues can be addressed: (i) the overall consequences of different scenarios of fossil fuel CO_2 emissions, (ii) the relative climatic benefits, and the timing of these benefits, when greater efforts are made to reduce emissions of different GHGs, and (iii) the tradeoffs involved when simultaneous reductions of emissions of GHGs and aerosol precursors occur.

The usefulness of the inverse approach stems from the fact that the UNFCCC declares its ultimate objective to be "stabilization of greenhouse gas concentrations in the atmosphere at a level that would prevent dangerous anthropogenic interference with the climate system". The UNFCCC does not specify what concentrations constitute "dangerous" interference, but the Convention does establish three broad objectives which are to be satisfied by whatever stabilization ceilings are adopted: (i) stabilization is to be achieved "within a time frame sufficient to allow ecosystems to adapt naturally to climate change"; (ii) it should "ensure that food production is not threatened"; and (iii) it should "enable economic development to proceed in a sustainable manner" (Article 2).

Simple climate models such as the 1-D upwelling–diffusion model can be combined with a globally aggregated model of the terrestrial biosphere, and used for both forward and inverse calculations. Such models are ideal tools for global-scale policy analysis, because the key parameters that govern the model response – such as the climate sensitivity, the intensity of ocean mixing, and the CO_2 fertilization effect – can be easily altered to reflect a wide range of possibilities concerning the behaviour of the real world. For each set of physical and biophysical parameters, a wide range of emission scenarios can be used. However, the number of cases that could be considered quickly multiplies, and simple models provide the only practical means for examining the cases of greatest interest. The use of simple models, however, restricts the analysis to the global scale, because such models are either globally averaged or have very coarse spatial resolution. However, we have already argued (in Section 11.1) that this is the most appropriate scale for the analysis of alternative emission scenarios.

13.3 Forward calculations using simple models

In this section, examples of the forward calculation approach are presented. This is followed in the next section by examples of inverse calculations.

Scenarios of future GHG and aerosol emissions

The projection of future GHG and aerosol emissions is fraught with enormous uncertainty, inasmuch as future emissions depend on a wide range of economic, sociological, and technical conditions. Not only are some of these conditions inherently unpredictable, but many of them – such as future levels of energy efficiency, lifestyles, and the form of new urban areas – are subject to deliberate choice. Projections of future GHG and aerosol emissions should therefore not be regarded as predictions, but rather as scenarios showing what the future would be like for specific sets of assumptions.

Figure 13.1 shows projections of fossil fuel emissions of CO_2 and total anthropogenic emissions of CH_4, N_2O, and SO_2 for a typical business-as-usual (BAU) scenario, a GHG emission stabilization scenario, and a GHG emission reduction scenario. Adoption of these scenarios does not imply that any one of them is necessarily feasible or desirable. The BAU CO_2 scenario used here is the IPCC IS92a scenario (Leggett et al., 1992), in which global fossil fuel CO_2 emissions increase from about 6.0 Gt C per year in the 1990s to 20 Gt C per year by 2100. Methane emissions increase from 263 Tg CH_4 per year to 524 Tg CH_4 per year, while N_2O emissions increase from 6.25 Tg N per year to 12.6 Tg N per year.

Future sulphur emissions will depend on the degree of concern over acid rain, since sulphur emissions are the single most important precursor to acid rain. If, as many economists project, the entire world becomes richer during the 21st century, then sulphur emissions can be expected to decrease even if CO_2 emissions rise dramatically, since wealthier people tend to demand – and can afford – a cleaner environment. As noted in Section 4.1, the technology already exists to decouple S and C emissions. Given that programmes are already in place to significantly reduce S emission in North America and Europe, and that current S emissions in Asia are already causing serious acid rain problems (Mohan and Kumar, 1998; Qian and Zhang, 1998), it is unlikely that future total global emissions will rise much above present levels, and they could very well decrease substantially. For all the CO_2 emission scenarios shown in Fig. 13.1, it is assumed that the S:C emission ratio decreases linearly to 25% of its year 2000 value by 2100. For the base case CO_2 emission scenario, absolute S emissions increase by about 50% before returning to the current level by 2100. If CO_2 emissions are reduced relative to the BAU scenario, then absolute S emissions fall more rapidly, but this is consistent with the greater overall concern over environmental issues that is implied by a CO_2 emission reduction scenario.

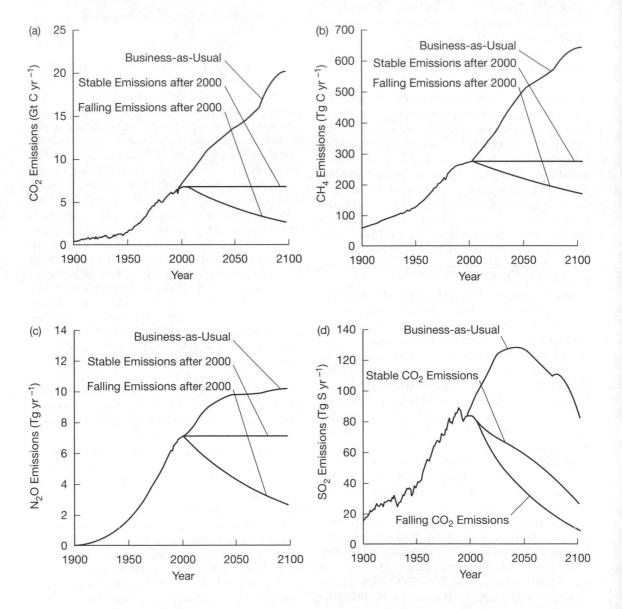

Fig. 13.1 (a) Fossil fuel emissions of CO_2, (b) total anthropogenic emissions of CH_4, (c) total anthropogenic emissions of N_2O, (d) and anthropogenic emissions of SO_2, for a typical business-as-usual scenario, a scenario in which GHG emissions are stabilized, and an aggressive GHG emission reduction scenario.

Scenarios of future GHG concentrations and radiative forcing

The coupled climate–carbon cycle model of Harvey and Huang (1999) is used here to project the buildup in the concentration of GHGs for the set of emission scenarios presented above. As noted in Section 8.4, there are major uncertainties concerning the present role of the terrestrial biosphere as a carbon sink, such that the current rate of uptake of anthropogenic CO_2 could lie between 0.5 and 4.5 Gt C yr^{-1}. As discussed in Section 8.9, the uncertainty in the current oceanic rate of uptake appears to be smaller, with most estimates lying between 1.4 and 2.6 Gt C yr^{-1}. Finally, as discussed in Section 4.2, there are substantial uncertainties concerning the current rate of emission of CO_2 due to deforestation and land degradation, and how these emissions varied in the past. In order to simulate the observed buildup of atmospheric CO_2 up to the present, a larger terrestrial biosphere sink is required the larger the assumed land use emission, for a given oceanic uptake. Thus, land use emissions of 0.5 and 1.5 Gt C yr^{-1} by 1990 require a terrestrial biosphere sink of 0.7 and 1.7 Gt C in 1990, respectively, given an oceanic sink in 1990 of 2.0 Gt C. The size of the terrestrial biosphere sink in turn can be altered by changing the value of the model parameters that govern the response of photosynthesis and respiration to increasing atmospheric CO_2 and temperature, within the range of observational uncertainty. Whatever the current CO_2 emission rate due to deforestation, the rate will decrease substantially within 20–30 years, if for no other reason than that there will be fewer forests left. However, since a greater rate of deforestation emissions today implies a greater responsiveness of net primary production (photosynthesis minus respiration) to increasing atmospheric CO_2, this results in smaller future projected CO_2 concentrations.

Figure 13.2(a) shows the buildup of atmospheric CO_2 for the three fossil fuel emission scenarios given in Fig. 13.1, in each case assuming that the deforestation emissions were 0.75 Gt C yr^{-1} in 1990 and decrease to zero by 2020. Figures 13.2(b) and (c) show the buildup of methane and N_2O, respectively, using the corresponding emission scenarios given in Fig. 13.1.

With regard to the forcing by halocarbons and changes in stratospheric ozone, emission scenarios similar to that used by Solomon and Daniel (1996)

are adopted. The buildup of halocarbons is computed using Eq. (8.13) (F and τ now pertaining to whatever gas is being considered) and the atmospheric lifetimes and radiative forcings as given in Table 7.1. The effective stratospheric chlorine concentration peaks around year 2000, as does the negative radiative forcing due to loss of stratospheric ozone. As shown by Solomon and Daniel (1996), the forcing due to halocarbons peaks at around 0.3 W m^{-2} in year 2000, but declines much more slowly (as CFC concentrations fall) than the recovery in stratospheric ozone. This is due to the fact that the HFCs have no ozone-depleting effect but are strong GHGs, and the heating effect due to the growth in their concentration largely compensates for the decrease in CFCs.

Figure 13.3 shows the variation in the forcing due to CO_2 alone due to all other GHGs, and due to aerosols, for the concentration scenarios given in Fig. 13.2. The radiative forcing due to the buildup of aerosols was computed by assuming (i) that the forcing in 1990 was 20%, 40%, or 60% of the 1990 GHG forcing, with one-third due to direct effects and two-thirds to indirect effects, (ii) that the direct forcing varies in direct proportion to the total S emission (given in Fig. 13.1(d)), and (iii) that the indirect forcing varies with the natural logarithm of the total emission.

Simulation of future rates of climatic change for surprise-free scenarios

The next step is to use the radiative forcings given in Fig. 13.3 to drive a climate model that takes into account the absorption of heat by the oceans. Here, the same quasi-one-dimensional upwelling–diffusion model that was used to compute the oceanic uptake of CO_2 is used for this purpose. Since all models, particularly the 1-D upwelling–diffusion model, are limited in their ability to simulate abrupt climatic changes or "surprises", the results obtained with the 1-D model will be referred to as "surprise-free" scenarios. Reality may very well include a number of surprises, either pleasant or unpleasant. The likelihood of surprises is likely to increase the greater and the faster warming occurs, and this should be kept in mind when viewing the results presented here. The prospects for surprises in the climate system response to increasing GHG concentrations, and their policy implications, are briefly discussed in Chapter 14.

The major uncertain parameter affecting future climatic changes, for a given emission scenario, is the climate sensitivity. Whatever the assumed climate sensitivity, the model should be constrained to roughly simulate the observed increase in temperature over the past 140 years. This in turn requires assuming a larger present-day aerosol cooling effect the larger the assumed climate sensitivity (otherwise, the larger the assumed climate sensitivity, the greater the simulated temperature change). As shown in Section 11.4, a reasonable simulation of the observed variation in global mean temperature during the past 140 years is obtained if the three aerosol forcing cases given above (whereby the forcing reaches 20%, 40%, or 60% of the GHG heating by 1990) are used in combination with climate sensitivities for a CO_2 doubling of 1°C, 2°C, and 3°C, respectively. Carbon dioxide doubling sensitivities of 1.0°C and 3.0°C represent reasonable upper and lower limits to the climate response, while 20% and 60% aerosol offsets (in the global mean) represent reasonable upper and lower limits to the aerosol forcings. The base-case ocean mixing parameter values from Harvey and Huang (1999) are used, and the thermohaline overturning is assumed to decrease by 10% per degree Celsius of mixed layer warming. This has negligible

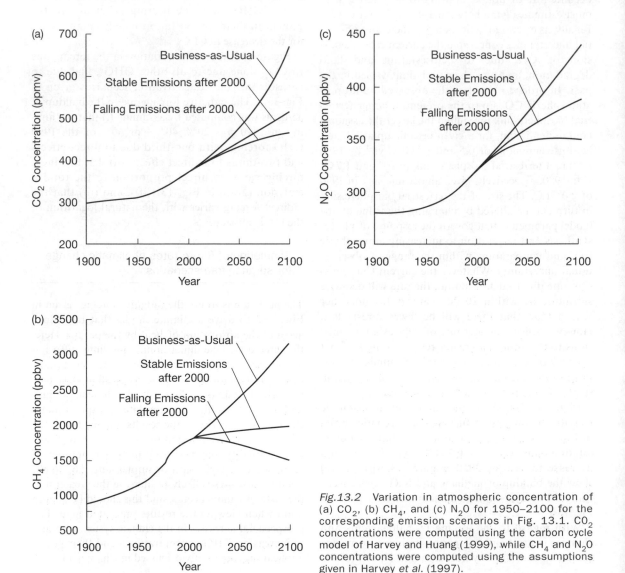

Fig.13.2 Variation in atmospheric concentration of (a) CO_2, (b) CH_4, and (c) N_2O for 1950–2100 for the corresponding emission scenarios in Fig. 13.1. CO_2 concentrations were computed using the carbon cycle model of Harvey and Huang (1999), while CH_4 and N_2O concentrations were computed using the assumptions given in Harvey et al. (1997).

effect on the CO_2 buildup but slightly slows down the transient temperature response compared to the case with fixed overturning intensity.

Figure 13.4 shows the projected change in global mean surface air temperature for the three climate sensitivities and for the three emission scenarios. For the case of low climate sensitivity (Fig. 13.4(a)), the impact of imposing constant or falling emissions after year 2000 is evident by 2010. For the case of high climate sensitivity (Fig. 13.4(c)), the negative aerosol forcing is assumed to be three times larger than for the case with 1°C climate sensitivity, and the

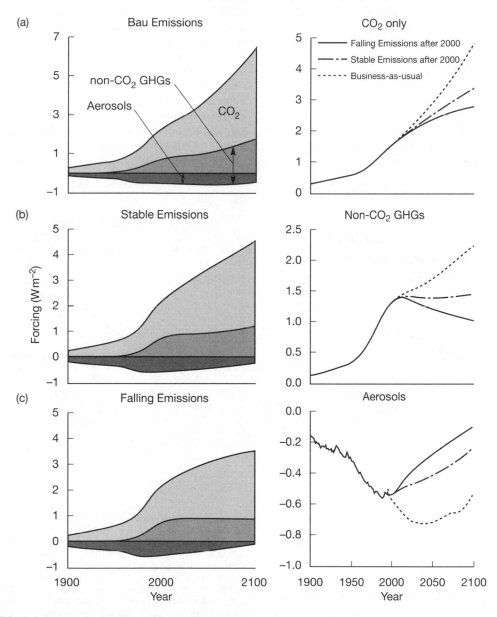

Fig. 13.3 Left Panels: Global mean radiative forcing due to aerosols, non-CO_2 greenhouse gases (GHGs), and CO_2 for (a) Business-as-usual GHG emissions, (b) stable GHG emissions, and (c) falling GHG emissions. Right Panels: Comparison of the global mean radiative forcing for the business-as-usual emission scenario, stable emissions scenario, and falling emissions scenario for (a) CO_2, (b) non-CO_2 GHGs, and (c) aerosols.

reduction in the aerosol cooling effect when CO_2 (and other GHG) emissions are constrained initially more than compensates for the reduced GHG heating. This is because the decrease in aerosol cooling occurs immediately as emissions are reduced, owing to the very short lifespan of aerosols in the atmosphere, whereas the impact of reduced emissions of CO_2 (and, to a lesser extent, most other GHGs) does not appear until after a few decades, when there is finally a noticeable difference in the cumulative buildup. As result, the transient warming for the CO_2 emission stabilization or reduction scenarios is initially slightly *larger* than for the BAU scenario. Reduced warming for these scenarios is not evident until 2040. These results are similar to those of Wigley (1991b), who was the first to show that simultaneous controls on CO_2 and SO_x emissions could initially cause greater climatic warming.

The above results appear to imply that there is little climatic benefit from reducing GHG emissions before several decades have passed, because there is a coupling in the scenarios presented above between CO_2 emissions and aerosol cooling (reduced CO_2 emissions lead to reduced S emissions and an immediate reduction in the cooling tendency). However, one can recast the apparent tradeoff in an entirely different light by postulating a series of increasingly stringent reductions in sulphur emissions, *independently* of what happens to GHG emissions. One can then examine the extent and the time frame over which increasingly stringent reductions in CO_2 alone, non-CO_2 GHGs, and all GHGs together can offset the extra warming tendency due to the reduction in sulphur emissions. The value in isolating CO_2 and non-CO_2 GHGs is that many of the latter have relatively short atmospheric lifetimes, so that reduced

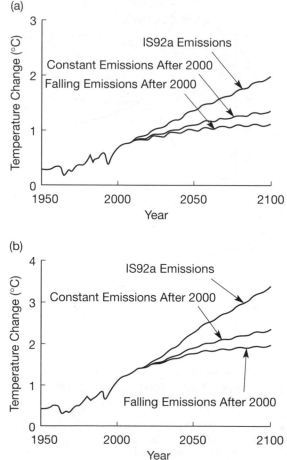

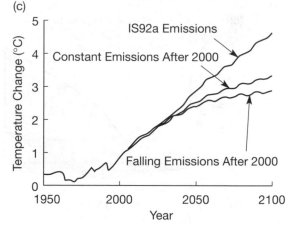

Fig. 13.4 Scenarios of future climatic change, corresponding to the radiative forcing variations given in Fig. 13.3, which in turn correspond to the emission scenarios given in Fig. 13.1. Results are given for (a) a climate sensitivity of 1.0°C for a CO_2 doubling and an aerosol cooling equal to 20% of the GHG heating in 1990, (b) a climate sensitivity of 2.0°C for a CO_2 doubling and an aerosol cooling equal to 40% of GHG heating in 1990, and (c) a climate sensitivity of 3.0°C for a CO_2 doubling and an aerosol cooling equal to 60% of the GHG heating in 1990. In all cases the S:C emission ratio decreases to 25% of its 1990 value by 2100.

warming due to reductions in their emissions will be seen sooner, thereby providing an opportunity to partially offset some of the near-term extra warming due to reduced sulphur emissions.

Figure 13.5 shows the results of a series of experiments designed to answer this question. The three solid curves in each panel assume the IS92a GHG emissions, but with different assumptions concerning sulphur emissions. Under BAU conditions, the S:C emission ratio is likely to fall substantially during the 21st century. To illustrate the effect of this anticipated reduction on global mean temperature change, a constant S:C emission ratio is adopted as the base case. The other two solid curves show the temperature change when the S:C ratio drops to 25% or 6.25% of the 1990 value by 2100; the former holds global S emissions roughly constant, and allows substantially more warming than for the (unrealistic) base case. The three dashed curves all assume the intermediate S emission scenario (in which S:C drops to 25% of the 1990 value), but increasingly stringent restrictions on GHG emissions or ozone precursors. The uppermost dashed curve shows the effect of assuming that the emission ratio of ozone precursors to CO_2 decreases to 6.25% of the 1990 level (i.e., that there is a substantial effort to address problems of ground-level ozone) and that CH_4 and N_2O emissions are stabilized. The middle dashed curves additionally assume stabilization of CO_2 emissions, while the lower dashed curves assume the scenarios of declining CO_2, CH_4, and N_2O emissions shown in Fig. 13.1.

The results shown in Fig. 13.5 show that if aerosols are currently offsetting only 20% of the GHG heating effect, then reductions in non-CO_2 GHGs alone can offset the extra heating effect (in the global mean) due to stabilization of global S emissions. For the case in which aerosols currently offset 60% of the GHG heating effect, merely stabilizing global S emissions results in substantially greater warming. With vigorous restrictions on emissions of all GHGs (and ozone precursors), there is still greater warming throughout the 21st century than if no restrictions are placed on GHG or S emissions. The good news is that, if reductions in S emissions are regarded as inevitable, then the extra climate warming can be largely nullified (in the global mean) by the end of the 21st century. Furthermore, temperatures have stabilized by that time for the scenario in which restrictions are placed on S and all GHG emissions, but continue to rise for the scenario with no emission restrictions (and would rise rapidly once S emissions decrease, as they must do eventually).

One could extend this analysis further to examine regional and seasonal temperature changes using models of intermediate complexity. Indeed, this is a case where analysis at the large regional scale is highly appropriate in spite of uncertainties in the regional effects for GHG increases alone, simply because the aerosol cooling effect is highly concentrated regionally. Nevertheless, the examples presented here serve to illustrate some of the tradeoffs involving multiple GHGs and S emissions.

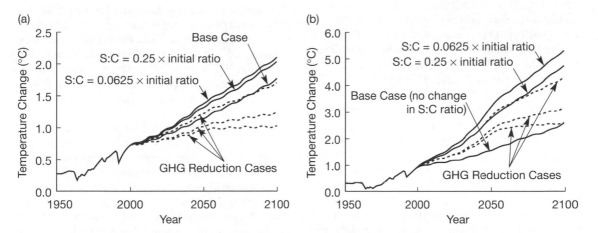

Fig. 13.5 Global mean temperature change for business-as-usual GHG emissions and a constant S:G emission ratio, or a ratio decreasing to 25% or 6.25% of the 1990 value by 2100 (solid lines); and for the same absolute S emissions as for the 25% case, but with increasingly stringent controls on GHG emissions (dashed lines). Results are given for a global mean equilibrium warming for a CO_2 doubling of (a) 1.0°C, and (b) 3.0°C.

13.4 Inverse calculations using simple models

The alternative to specifying emission scenarios and working forward to calculate GHG concentrations and thence climatic change, is to specify a variety of concentration pathways corresponding to stabilization at different concentrations, and to work out the permitted emissions. Past emissions and permitted future CO_2 emissions can be decomposed into the product of the following factors:

- population (P),
- economic output (dollars) per person ($/P$),
- average energy consumption (joules) per dollar of economic output (J/$),
- average CO_2 emission (kg) per joule of energy consumption (kg/J).

Thus,

$$\text{Emission} = P \times (\$/P) \times (J/\$) \times (kg/J) \qquad (13.1)$$

This equation makes it clear that future emissions depend on the future human population, the growth of per capita income, the kinds of economic activities undertaken and the efficiency with which energy is used (which together determine J/$), and the mix of energy sources used and the efficiency with which they are transformed into intermediate energy forms such as electricity and refined petroleum products, which together determine kg/J. Hoffert *et al.* (1998) performed this decomposition for the IPCC BAU scenario (IS92a), and the results are presented in Fig. 13.6. The BAU scenario assumes that the human population reaches 11.3 billion by 2100 (about double the present population) and that the average per capita income increases by a factor of 6, giving an increase in size of the global economy by a factor of 12. CO_2 emissions would grow by a comparable factor, were it not for the fact that the energy intensity of the global economy is assumed to decrease by 1% per year throughout the 21st century (each year, 1% less energy is required to produce a dollar of economic output than the year before). This is comparable to the rate of decrease that has occurred over the past century, and reflects both increases in the efficiency with which energy is used, and a shift from energy-intensive primary industries to less energy-intensive service industries. Finally, the carbon intensity of the energy supply system is assumed to decrease by about 25% as a result of an increase in

the proportion of non-carbon energy sources (nuclear power and/or renewable energy).

Figure 13.7 shows the concentration stabilization pathways for CO_2 of Wigley *et al.* (1996), with concentrations stabilizing at 450, 550, 650, and 750 pmv. Fig. 13.8(a) shows the permitted global fossil fuel CO_2 emissions, as computed using the carbon cycle model of Jain *et al.* (1996) and presented in Hoffert *et al.* (1998). Also given is the IPCC BAU scenario (the IS92a scenario), which is the same BAU scenario as used previously in this chapter. Stabilization of CO_2 at a concentration of 450 ppmv would still result in the equivalent of close to a CO_2 doubling from the pre-industrial concentration of 280 ppmv, given the radiative heating effect of increases in other GHGs that would also occur. Stabilization at 450 ppmv requires reducing emissions to about half the current emission rate by 2100. Also shown in Fig. 13.8 is the breakdown of emissions into emissions from natural gas, oil, and coal; the fuels are assumed to be used in this order, subject to the cap on total emissions (recall from Table 4.1 that the CO_2 emission per unit of fossil fuel energy is smallest for natural gas, followed by oil and then by coal). Fig. 13.8(b) shows the provision of primary power corresponding to each of the scenarios, while Fig. 13.8(c) shows the amount of carbon-free power that is required. This is given by the difference between the primary power required for the IS92a scenario and the total fossil fuel power that is permitted for the stabilization scenarios. The permitted fossil fuel primary power for the stabilization scenarios is computed by assuming that the same amount of natural gas and oil is used as for the IS92a scenario, and any remaining permitted emission is assigned to coal. Even for the BAU scenario, the required non-fossil energy supply reaches 11 TW (terawatts, 1 TW = 10^{12} W), which is equal to the total global energy supply in 1990. For the stabilization scenarios, an even larger non-fossil power supply is needed.

These results indicate that a significant expansion of non-fossil fuel energy supply – either nuclear energy or renewable energy – is implicit even in BAU scenarios, scenarios in which atmospheric CO_2 reaches several times the pre-industrial concentration. The required non-fossil energy supply can be reduced if the energy intensity of the global economy can be made to decrease at a faster rate than 1% per year. The tradeoff between the required non-fossil energy supply and the rate of decrease of energy intensity is shown in Fig. 13.9 for stabilization of CO_2 at 550 ppmv. More of one allows less of the

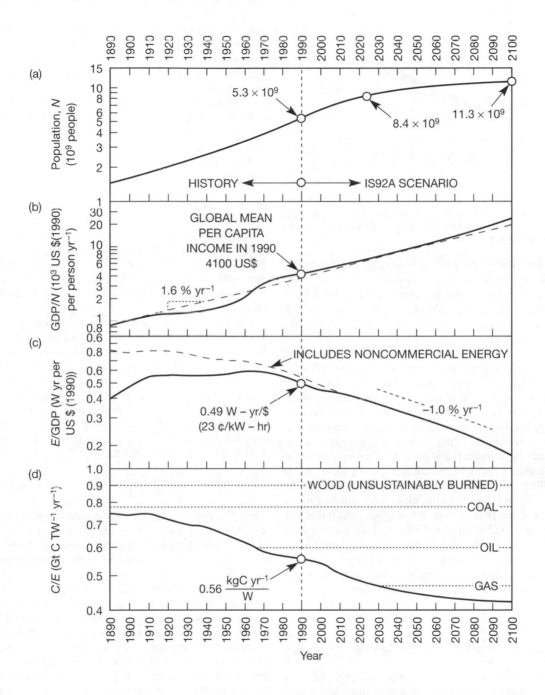

Fig. 13.6 Decomposition of past global fossil fuel carbon emissions, and future emissions under the IS92a scenario, according to (a) global population, (b) per capita income, (c) primary energy use per unit of GDP, and (d) carbon emission per unit of primary energy. The horizontal lines in (d) give the carbon intensity for individual fuels. Redrafted from Hoffert *et al.* (1998).

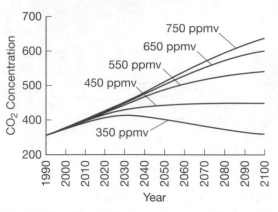

Fig 13.7 The CO_2 stabilization scenarios of Wigley *et al.* (1996). The concentrations at which CO_2 eventually stabilizes are shown. Produced based on data kindly provided in electronic form.

other. If the energy intensity were to improve by 2% per year rather than 1% per year, almost no expansion of non-fossil energy supply would be required.

Since the ultimate objective of the UNFCCC will almost certainly require *some* limitation of CO_2 emissions to below the IS92a scenario, there is already a clear policy implication, even in the absence of detailed scientific information: a major effort is required to ensure that non-fossil energy supply equal to at least the current total global energy supply is available a mere 50 years from now, or a major effort is required to accelerate the improvement in the efficiency with which energy is used. Since implementation of either of these options, to the extent required to stabilize atmospheric CO_2 at 550 ppmv, would be very difficult, and there could be further, unexpected difficulties, one can go further and state that strong efforts are *simultaneously* required on both fronts. This is an example of robust policy advice that can be derived based on the first-order relationships between the driving factors and climate change – relationships that are most clearly illustrated with the aid of simple models.

13.5 Global Warming Potential

The final topic under the subject of the science–policy interface that will be discussed here is that of the "Global Warming Potential" (GWP). The GWP represents an attempt to develop a single index for quantitatively comparing the climatic effects of equal emissions of different GHGs. The GWP is an attempt to simplify complex scientific information for purposes of policy analysis. However, as discussed below, the resultant index cannot be scientifically defended, and it is not necessary, given the ease with which climatic changes with different emission scenarios can be calculated (as exemplified by the examples given in Section 11.3).

As noted in Chapter 2 (Table 2.7), different GHGs differ in the radiative forcing per molecule and in the average lifespan of their molecules in the atmosphere. Thus, if equal quantities of two different gases are emitted at the same time, the relative amounts of the two gases will change over time. Hence, the relative radiative forcings – given by the ratio of the forcing per molecule times the number of molecules remaining in the atmosphere for the two gases – will also change over time, so it is not possible to uniquely specify the relative climatic effect of two different gases. However, if the concentration of both emission pulses decays to zero, the *integrated* forcing over all time arising from the emission pulse can be compared for the two gases.

The natural choice is to compare the integrated forcing of the gas in question with that for CO_2, which serves as the reference gas. There is, however, a fundamental problem: the concentration of a CO_2 emission pulse does not decay to zero (for all practical purposes). Thus, the integrated forcing over all time will be infinite. This problem is solved by performing the integration over some finite period of time called the *time horizon*. The *Global Warming Potential (GWP)* is thus defined as

$$\text{GWP}_i(T) = \frac{\int_0^T f_i(t) C_i(t) \mathrm{d}t}{\int_0^T f_r(t) C_r(t) \mathrm{d}t} \qquad (13.2)$$

where $f_i(t)$ and $f_r(t)$ are the heat-trapping abilities per unit mass, $C_i(t)$ and $C_r(t)$ are the amounts remaining in the atmosphere at time t, the subscripts i and r refer to the gas in question and to the reference gas, respectively, and T is the time horizon over which the integration is performed. The major drawback of the GWP is that it depends on the choice of T: for gases that are removed more quickly than CO_2, the GWP will be smaller the longer the time horizon, while for gases that are removed more slowly than the initial rate of removal of CO_2, the GWP will be larger the longer the time horizon. Whatever time horizon is chosen for the GWP, the computed GWP will not accurately reflect the relative warming effects of different gases for other time horizons.

The Kyoto Protocol places restrictions on a basket of six gases or groups of gases: CO_2, CH_4 (methane),

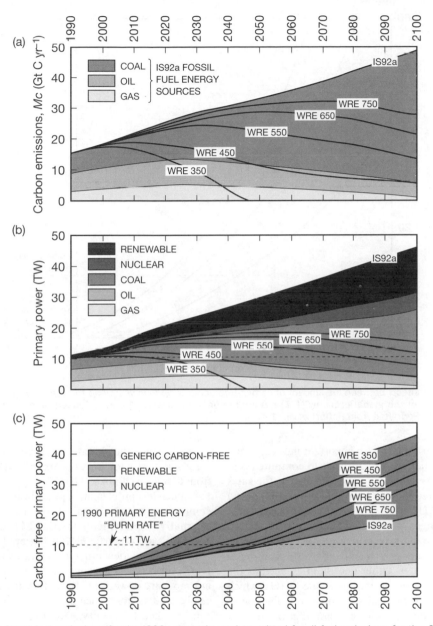

Fig. 13.8 (a) fossil fuel emissions for the IS92a scenario, and permitted fossil fuel emissions for the CO_2 concentration stablization pathways shown in Fig. 13.7. Also shown are the contributions of natural gas, oil and coal to the permitted CO_2 emissions. (b) the total primary power requirement, and (c) the carbon-free primary power requirement, corresponding to the IS92a and CO_2 stablization scenarios. Redrafted from Hoffert *et al.* (1998).

N_2O (nitrous oxide), SF_6 (sulphur hexafluoride), the HFCs (hydrofluorocarbons), and the PFCs (perfluorocarbons). The Kyoto Protocol will use GWPs as computed by the Intergovernmental Panel on Climate Change (IPCC) for a time horizon of 100 years.

There are a number of significant uncertainties associated with GWPs, that are more fully reviewed in Harvey (1993): (i) the direct radiative forcing of some gases (the HFCs in particular) is still uncertain by up to ±30%; (ii) some gases (the halocarbons and

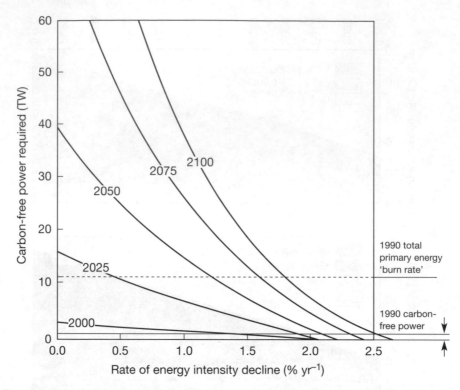

Fig. 13.9 The tradeoff between the amount of carbon-free power in various years and the rate of improvement in the energy intensity of the global economy (J/$) required in order to stabilize the CO_2 concentration at twice the pre-industrial value. Redrafted from Hoffert *et al.* (1998).

methane in particular) have important indirect radiative forcings that are very difficult to compute accurately; (iii) atmospheric lifetimes of some of the gases being compared with CO_2 are still uncertain and, in the case of CH_4, are expected to change over time; and (iv) changes in climate could cause important changes in the net rate of removal of CO_2 from the atmosphere, and this would alter the computed GWPs for all gases. Other uncertainties, such as the assumed future concentrations of the GHGs, which alter the $f_i(t)$ and $f_r(t)$, are less important. The use of the GWP as defined above contains a number of implicit assumptions as well. Most importantly, it is assumed that the climatic effect of a given GHG is directly proportional to its radiative forcing (that is, that the climate sensitivity is the same for all gases). As discussed in Section 3.7, this assumption is valid to within ±20% when comparing the GHGs that are covered by the Kyoto Protocol. The concept of a GWP is completely inapplicable to gases such as CO and NO_x (which affect climate through their role in the production of tropospheric ozone) and to

aerosols, since the climate sensitivity to ozone and aerosol radiative forcing appears to be quite different from that for well-mixed GHGs.

A much less problematic application of GWPs is in comparing the impact on future climatic change of alternative replacements for CFC-11 and CFC-12. The CFCs are powerful GHGs but are being phased out because of their impact on stratospheric ozone. HFC replacements have no effect on the ozone layer but most are powerful GHGs. In this case, CFC-11 can be used as a reference gas and the time horizon T can be taken to infinity. The difficulties involving CO_2 as a reference gas can thus be avoided. One may still choose to carry out the integration in Eq. (13.2) over a finite time horizon rather than over all time, if one is interested in near-time rates of warming rather than the integrated warming. Calculated values of GWPs using CFC-11 as a reference gas can be found in Table 2.8 of Schimel *et al.* (1996).

To summarize, the concept of GWP is an attempt to force the complexities of nature into a single number. This is not scientifically justifiable. Even if the heating effects and lifetimes of GHGs were accurately known, the GWP calculated for any one time horizon will not accurately reflect the relative climatic effects of gases for other time horizons. It is more scientifically justifiable to compare climatic changes, and rates of climatic change, for alternative emission scenarios using simple models that can perform forward calculations. In this way, alternative ways of achieving overall targets concerning maximum permitted climatic change can be compared. In addition, attention can be directed to the question of offsetting the inevitable acceleration of warming as sulphur emissions are reduced (as discussed in this chapter), rather than the question of how much reduction in CO_2 emissions can be avoided through reduced emissions of other gases. However, the political process has demanded a single number to intercompare different GHGs. This is because current international agreements have ignored scientific reality by focusing on a "basket" of GHGs rather than by framing gas-by-gas restrictions. The latter would eliminate the most problematic applications of GWPs altogether.

CHAPTER 14   The prospects for surprises

In the preceding chapters, projections have been presented of the buildup of GHG concentrations in response to anthropogenic emissions, of the gradual warming of the climate in response to the radiative heating caused by the increase in GHG concentrations, and of the long-term rise in sea level. These projections have been made with a variety of mostly simple models which, by their very design, preclude the possibility of abrupt changes in the rate of warming or in climatic parameters. The most sophisticated models – the coupled AOGCMs – allow for and indeed sometimes simulate abrupt (multidecadal time scale) changes in oceanic circulation, but the details and timing of such changes are highly dependent on uncertain model parameters, such as the subgrid-scale diffusion coefficients or the precipitation and evaporation fluxes obtained through coupling with the atmosphere. Furthermore, the way in which most such models are coupled probably distorts their ability to correctly respond to the kinds of things that could trigger abrupt changes, as will be explained later. Thus, the models used to project future climatic change either cannot simulate possible abrupt changes in the climate system, or where the possibility of simulating such changes exists, the results cannot be relied upon. All that can be said with high confidence is that the physics of the climate system permit several reorganizations of the atmosphere–ocean subcomponents, and were such reorganizations to occur, the effects would be profound.

Nature is likely to provide many "surprises" – abrupt and largely unanticipated changes in the climate system. Changes in ocean circulation represent only one way in which this might occur. The greater the amount of global warming that is allowed to occur, the more likely it would seem that surprises will occur. True surprises, by definition, cannot be predicted in advance. However, most so-called "surprises" in society (such as the accident at Chernobyl, the Challenger disaster, or the disaster at Bhopal) were not a surprise to at least some segments of society, and were predictable in advance. In spite of the rather pessimistic assessment given above of the ability of climate models to predict "surprises", there is much that is useful that can be said about the prospects for surprises in the climate system. An awareness of the possibilities for surprises underlines the need to build in a margin of error when developing a policy response to the risks posed by unrestrained emissions of GHGs. This chapter, on surprises, is therefore presented as the concluding chapter of this book.

14.1   A typology of surprise

It is instructive first to briefly consider a typology or classification of surprises. Such a typology is presented in Fig. 14.1, based on the work of Faber *et al.* (1994). A surprise can represent an outcome that was not known in advance, or an outcome that was known to be possible but either had, or was thought to have had, a low probability of occurring. The set of known outcomes can be accompanied by an acknowledgement that unknown outcomes may exist, or by a refusal to acknowledge that there could be unknown outcomes. Further distinctions are indicated in Fig. 14.1.

In this chapter we, of necessity, deal with known outcomes. These are either thought to be of low probability, or are known to be of non-negligible probability by the scientific community but not by society at large. In either case, their occurrence would be widely regarded by society as a "surprise".

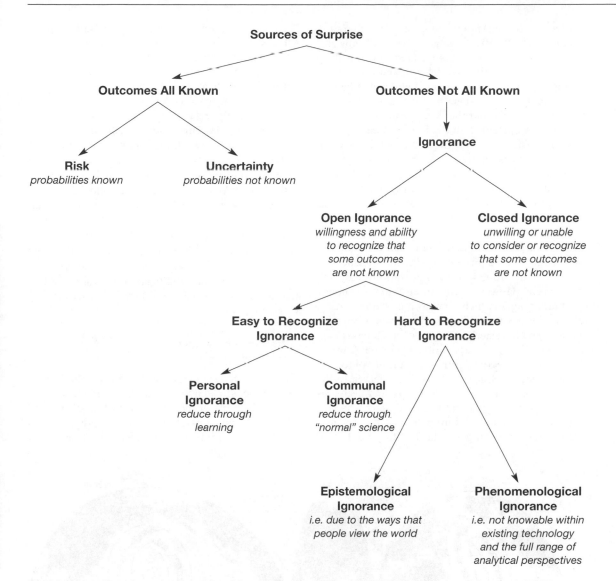

Fig. 14.1 A typology of surprise, reproduced from Schneider and Turner (1995) and based on Faber *et al.* (1994).

14.2 Abrupt reorganization of the oceanic circulation

The first potential surprise involves an abrupt reorganization of the oceanic circulation. The possibility of such occurring has in fact been anticipated for a long time, and the general way in which it could occur understood, although the timing and detailed nature of any changes in oceanic circulation remain uncertain. We shall discuss this potential surprise in terms of general principles, paleoclimatic evidence, and atmosphere–ocean climate model results.

Background principles

As explained in Box 2.2, the large-scale thermohaline overturning circulation is driven by the large-scale horizontal gradient of density in the ocean. The present-day, basin-averaged thermohaline overturning was illustrated in Fig. 11.2. Surface water flows to the North Atlantic Ocean, sinks at high latitudes, flows southward to the Antarctic region, and enters the Indian and Pacific basins, where it upwells. Additional surface water is injected into this deep flow by sinking in the Weddell Sea and other areas around Antarctica. The circuit is completed by surface currents from the Pacific into the Indian oceans between Australia and the Asian land mass, and by the Agulhas Current from the Indian Ocean, around southern Africa, and into the South Atlantic Ocean. This trajectory is referred to as the "conveyor belt", and is illustrated in Fig. 14.2. This is a highly idealized and simplified picture of the ocean circulation that neglects near-surface, wind-driven, and coastal circulations. Schmitz (1995) provides a recent review of the conveyor belt circulation.

All polar waters are deficient in salt compared to the oceanic average because of the net transport of water vapour from low to high latitudes by the atmosphere. However, the orientation of major mountain ranges (which serve as a moisture barrier) relative to the prevailing westerly winds in mid-latitudes and to easterly winds at low latitudes is such that there is more evaporation from the Atlantic Ocean than precipitation and runoff onto the Atlantic Ocean (Broecker, 1997). This increases the average salinity of the Atlantic Ocean and decreases the average salinity of the Pacific Ocean. The tendency for continuous salinification of the Atlantic Ocean and freshening of the Pacific and Indian Oceans is balanced by a net transfer of salt from the Atlantic Ocean to the other oceans by the conveyor belt circulation. Dense water is also continuously created in the North Atlantic Ocean, through the wintertime loss of heat, and this must be balanced by a net surface flow towards the North Atlantic Ocean of water of slightly less than average density.

The present-day circulation is such that it simultaneously rebalances both the salinity and the density budgets. Today, this balance is achieved with roughly equal rates of production of bottom water in the North Atlantic Ocean and around Antarctica. Any major perturbation in either the heat or the salt budget could therefore require a whole new circulation pattern in order to re-establish balance. In particular, both the north–south and east–west transfers of freshwater could be altered, thereby altering the regional salinification or freshening tendencies of the oceanic mixed layer and necessitating a new circulation.

Paleoclimatic evidence

There is abundant evidence that during the last ice age significant climatic changes, comparable to more than half of the difference between full glacial and full interglacial climates, occurred abruptly (within a few decades). Some of this evidence is reviewed by

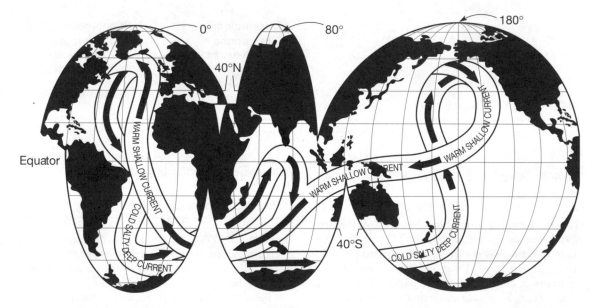

Fig. 14.2 The Conveyor-Belt circulation, from Schmitz (1995).

Broecker (1997). In contrast, the climate seems to have been quite stable during the Holocene period and during the preceding interglacial period (Adkins et al., 1997). Some rapid climatic changes have been documented during the Holocene, but these involved the transition from the remaining glacial conditions to full postglacial conditions around 8000 years ago (Stager and Mayewski, 1997).[1] This transition occurred within 200 years or less, and was marked by synchronous or near-synchronous changes in Greenland, Africa, and Antarctica.

Results from coupled atmosphere–ocean models

It is a robust result of 2-D and 3-D coupled atmosphere–ocean models that an at least partial collapse of the thermohaline overturning occurs during the transition to a warmer climate. This occurs irrespective of how the salinity budget changes, as a partial to total circulation collapse can be induced through thermal effects alone (Gough and Lin, 1992; Harvey, 1994). The reason is that even with uniform heating of the mixed layer, heat will penetrate more quickly into the deeper ocean where downwelling occurs than where upwelling occurs. In upwelling regions, greater surface warming and hence greater heat flux damping (to the atmosphere) occurs, so the depth-integrated oceanic warming is smaller than in downwelling regions. Since the downwelling regions are the regions with the greatest density, the greater heat uptake in the downwelling regions reduces the horizontal density contrast, which causes the thermohaline overturning to weaken. Once the thermohaline overturning weakens, the reduced poleward transport of salt from low latitudes leads to a further reduction in high-latitude density, thereby tending to further reduce the overturning. On the other hand, there is a reduction in the poleward transport of heat, which tends to increase high-latitude density and limit the collapse of the thermohaline overturning. The former seems to dominate, since in both 2-D and 3-D atmosphere–ocean models, a rapid reduction in the thermohaline circulation (by up to 90%) is triggered in response to surface warming (Stouffer et al., 1989; Washington and Meehl, 1989; Mikolajewicz et al., 1990; Harvey, 1994).

[1] Deglaciation itself began much earlier, depending on location, with warming beginning around 19,000 years ago in Antarctica and 14,500 years ago in Greenland.

The magnitude of the reduction, however, depends on the details of the model used. In some cases, transient reductions of only 30% occur. In the 2-D model of Stocker and Schmittner (1997), the occurrence of a near-complete shutdown depends on changes in both the heat and salinity budget, and also depends on the *rate* of increase in temperature. Slower rates of warming are less likely to lead to a near-total collapse of the thermohaline overturning. Nevertheless, some reduction can be expected irrespective of how slowly the climate warms, as long as the warming occurs over a shorter time scale than that required for near-complete mixing of heat within the Atlantic basin (hundreds of years), where significant overturning occurs today.

The surface branch of the thermohaline overturning in the NH consists of the Gulf Stream/North Atlantic Drift current system, which is responsible for the relatively warm climate of western Europe. A collapse of the thermohaline overturning in the North Atlantic Ocean could therefore cause a sudden cooling in western Europe that would be superimposed on the heating effect due to higher GHG concentrations. Results from two experiments have recently been published in which a near-total collapse of the thermohaline overturning in the North Atlantic occurred. In one case (Tziperman, 1997), the collapse provoked a mean annual cooling over the North Atlantic basin of up to 5°C but only 1–2°C along the west coast of Europe. In the other case (Schiller et al., 1997), a mean annual cooling of 5–15°C occurs along the west coast of Europe, with least cooling in the south. A cooling of this magnitude would locally overwhelm the heating effect of increasing GHGs.

In some stand-alone OGCMs, an initial interruption in convective mixing due to extra freshwater input at high latitudes leads to further freshening of the surface layer, owing to the fact that high-latitude convection no longer transports salt upward. The further freshening of the surface layer leads to a further weakening in convection, further freshening, and so on, until a complete collapse of convection occurs. This collapse, which can occur within 10 years and is accompanied by a collapse in the thermohaline overturning as well, is referred to as the "polar halocline catastrophe" (Zhang et al., 1993; Lenderink and Haarsma, 1994). In short, a partial to total collapse of the large-scale thermohaline overturning can occur through some combination of thermal and freshwater effects operating at a large scale, or through the localized initial suppression of convective mixing. In either case, a marked regional cooling tendency would be superimposed on the heating effect of increased GHG concentrations.

Shifts in the *locations* of oceanic convection can produce modestly different oceanic circulations for the same global-scale climatic conditions. For example, Rahmstorf (1995) reports oceanic circulations involving 16 Sv and 20 Sv of bottom water formation (Sv = Sverdrup, where 1 Sv = a volumetric flow of $10^6 \, m^3 \, s^{-1}$), associated with deep convective mixing in different parts of the North Atlantic basin. A cycle involving transitions from one circulation intensity to another occurs every 22 years for certain patterns of freshwater input to the ocean but not for others. Although this specific result depends on the particular model used, it does illustrate how, as the climate and the patterns of freshwater input change, there could be abrupt changes in the degree of interdecadal climate *variability* in certain regions.

A difficulty with AOGCMs is that large, fixed fluxes of heat and freshwater often need to be added to the ocean at individual grid points in order to prevent the coupled atmosphere–ocean model from drifting into an unrealistic simulated climate (e.g., the MPI model of Cubasch *et al.,* 1992; the GFDL model of Manabe *et al.,* 1991; the UKMO model of Murphy, 1995; and the CSIRO model of Gordon and O'Farrell, 1997). Sensitivity studies by Zaucker *et al.* (1994) show that the simulated oceanic circulation is very sensitive to errors in the simulated heat and moisture fluxes. Consequently, the simulated climate – and climatic change – will also be very sensitive to errors in the simulated fluxes. This is the reason why hard-wired fluxes are added. The hard-wired fluxes are as large as or larger than the freely predicted fluxes at many grid points, as illustrated in Schiller *et al.* (1997). However, unlike the freely predicted fluxes, the hard-wired fluxes cannot respond to changes in the atmospheric or oceanic state. This in turn suppresses a number of feedbacks that are important to the stability and behaviour of the coupled system (Nakamura *et al.,* 1994). Recently, some AOGCMs have been successfully run without these hard-wired fluxes (e.g., the NCAR model of Meehl *et al.,* 1993b; and the BMRC model of Colman *et al.,* 1995). However, most models, and most or all experiments investigating the stability of the atmosphere–ocean system using coupled AOGCMs, have used hard-wired fluxes.

Many of the instabilities of concern have been investigated using ocean-only models, with fixed mixed-layer temperature and prescribed air–sea water fluxes as boundary conditions. However, stabilizing feedbacks are strongly suppressed in such models. For this reason, instabilities such as the polar halocline catastrophe are now thought to be less likely in coupled AOGCMs and hence presumably less likely in reality than initially believed (Zhang *et al.,* 1993; Rahmstorf and Willebrand, 1995).

Summary

A consideration of the underlying forces driving the large-scale, thermohaline circulation in the oceans indicates that this circulation could be susceptible to an abrupt reorganization as the climate warms. Paleoclimatic data provide evidence of abrupt changes in climate during glacial periods but not during past periods as warm as or slightly warmer than at present. However, it is a robust result of both 2-D and 3-D coupled atmosphere–ocean models that the thermohaline circulation will undergo an at least partial collapse during the *transition* from the present climate to a warmer climate. This collapse is driven by the greater penetration of heat into the ocean at high latitudes than at low latitudes. A northward shift in the mid-latitude precipitation belt over the North Atlantic Ocean also contributes to this collapse, with the magnitude and timing of the circulation collapse dependent on the details of the precipitation shift and on the rate of warming. However, apart from this first-order effect of a warming climate on the oceanic circulation, nothing else can be said with great confidence because of the widespread use of hard-wired air–sea heat and freshwater fluxes in AOGCMs and because of the extreme sensitivity of the oceanic circulation in coupled models to small errors in the simulated fluxes.

14.3 Breakdown of the Arctic halocline and the abrupt disappearance of Arctic sea ice

The surface waters of the Arctic Ocean occur at a temperature near the freezing point of sea water (–1.8°C), but are underlain by relatively warm water (+0.5°C) that originates at mid-depth in the Atlantic Ocean. Normally, cold water would be more dense than warm water, so vertical mixing – and upward transfer of the heat from the Atlantic water – would be expected. However, the surface waters have unusually low salinity (32‰) while the deeper waters have higher salinity, close to the global average of about 35‰. There is thus a strong vertical salinity gradient or *halocline* in the upper part of the Arctic Ocean. Since the halocline keeps the surface waters less dense than the warm deeper waters, it inhibits vertical mixing and thereby insulates the surface from the warm waters below. Figure 14.3 shows the average vertical temperature and salinity variation in the Arctic Basin.

The Arctic halocline is maintained through the seasonal formation of sea ice over the continental shelves (Aagaard *et al.*, 1981). As sea water freezes to form sea ice, most of the salt is ejected (reducing the salinity of sea ice from 35‰ to 5‰). The salt-rich liquid that is ejected as sea ice forms is called *brine*. It sinks to the bottom of the shelf and slides off the shelf into the mid-depth Arctic Ocean, thereby maintaining the subsurface salinity maximum that is seen in Fig. 14.3.

As climate warms and the extent of winter sea ice formation over the continental shelves decreases, the salt pumping action will weaken. At some point a threshold is likely to be crossed, whereby intense vertical mixing can occur once the vertical density profile becomes unstable. This would pump large amounts of heat to the surface layer, thereby provoking a rapid retreat of the remaining perennial Arctic sea ice.

14.4   Abrupt increase in climate sensitivity

As discussed in Section 11.2, simulations with the Hadley Centre AOGCM indicate that the effective climate sensitivity during the initial transition to a warmer climate could be as little as one-third of the equilibrium climate sensitivity. This is due to the net cloud feedback being more negative or less positive during the transient compared to an equilibrium climatic change, in turn a result of different spatial patterns of climatic change. The analysis of the sensitivity studies of cloud feedback with the GISS AGCM, presented in Section 11.2, also indicates that cloud feedback could be much more negative during the initial transition to a warmer climate, than in an equilibrium warming. A rapid increase in the transient climate sensitivity, perhaps due to some reorganization of the ocean and atmospheric circulation, would result in a period of rapid climatic warming that would otherwise be unexpected.

14.5   Other possible abrupt changes

Listed below are a number of other ways in which "surprises" might occur.

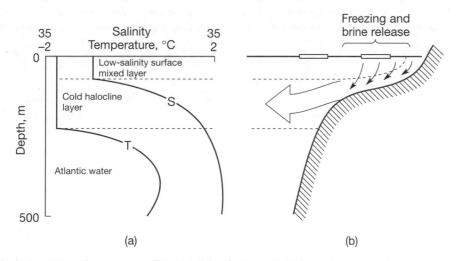

(a)   (b)

Fig. 14.3 Vertical variation of temperature (T) and salinity (S) in the Arctic Basin: (a) observed structure; (b) conceptual model of halocline maintenance. Reproduced from Aagaard *et al.* (1981).

Abrupt changes in precipitation patterns

Paleoclimatic data provide evidence of frequent and abrupt changes in precipitation patterns. Gasse and van Campo (1994) review the evidence for such abrupt changes in the tropics and subtropics during the Holocene Period. Were such changes to occur as the climate warms during the coming centuries, they too would appear as "surprises".

Collapse of the West Antarctic ice sheet

The possibility of a collapse of the West Antarctic Ice Sheet (WAIS) over a relatively short period of time (decades) has often been raised. However, as discussed in Section 12.2, it is not even clear that the WAIS is unstable. If it is unstable, there are compelling reasons for believing that it would take many centuries to collapse.

Large release of methane clathrate

Another frequently cited example of a non-linear feedback that could greatly amplify future warming is the destabilization of methane clathrates and the subsequent release of large amounts of CH_4. This possibility was extensively analysed by Harvey and Huang (1995), and although they concluded that a large amplification of future warming is unlikely, it cannot be ruled out at present. Other potential positive feedbacks involving the carbon cycle were discussed in Section 8.10, where it was pointed out that the conditions needed for all the potential positive feedbacks to occur simultaneously are to some extent mutually exclusive – that is, the conditions that favour one positive feedback either preclude another positive feedback or also favour an at least partially compensating negative feedback.

Abrupt dieback of forests

The possibility of a large dieback of forests due to climate zones shifting at a faster rate than that at which forest zones can migrate was discussed in Section 8.2. Such dieback could cause a large transient flux of CO_2 into the atmosphere. However, forests might be able to resist collapse for quite some time, but once one or more critical thresholds are crossed, rapid changes might occur.

14.6   Concluding comments

The climate system is sufficiently complex, climate models are sufficiently simple, and our understanding of the phenomenon of global warming and climatic change sufficiently rudimentary, that surprises of some sort can certainly be expected. This chapter has outlined some of the "surprises" that might occur in the future, but the true surprises wait to be seen.

Glossary

Ablation The process of melting at the surface of a glacier.

Ablation zone That portion of a glacier where annual ablation exceeds the annual accumulation of snow, so that there is a net loss of mass over the course of a year. When a glacier is in a steady (non-changing) state, the net loss will be exactly balanced by the inflow of mass to the ablation zone from the accumulation zone.

Accumulation zone That portion of a glacier where annual accumulation of snow exceeds the annual ablation, so that there is a net gain of mass over the course of a year. When a glacier is in a steady (non-changing) state, the net gain will be exactly balanced by the outflow of mass from the accumulation zone to the ablation zone.

Acid rain Rainfall with a greater than average acidity (lower pH) due to the presence of acids produced from anthropogenic emissions of sulphur and nitrogen oxides.

Adjustment time The time required for the perturbation in the amount of a gas in the atmosphere to decrease to 1/e of the amount initially injected, where the initial injection occurred as a sudden pulse. This will be the same as the lifespan or turnover time τ if the presence of the pulse does not increase the lifespan of the pre-existing gas.

Aerodynamic resistance The resistance to the movement of gases from the surface of leaves to the free air a few metres above the ground surface. It depends on the degree of turbulence of the air next to the ground, being smaller the greater the turbulence.

Aerosol Small airborne particles.

Albedo The fraction of solar radiation incident on a surface that is reflected.

Ammonification The process involved in the mineralization of organic nitrogen, in which it is converted to ammonia (NH_4^+).

Anthropogenic Of human origin.

Asymbiotic Used here to refer to nitrogen-fixing bacteria that are not living in a symbiotic relationship with a higher plant, but rather live freely in the soil. See also *symbiotic*.

Bathymetry Refers to the shape of an ocean basin, or the variation of depth with horizontal position.

Biological pump The process whereby carbon and nutrients are transferred from the surface waters to the deep ocean, due to the occurrence of photosynthesis near the ocean surface, the settling of some dead organic matter into the subsurface layer, and the decomposition of organic matter in deeper water.

Biomass Refers to living plant or animal matter.

Biosphere Refers to all living organisms – the living "sphere".

Black carbon Refers to soot aerosol – an aerosol consisting of highly absorbing carbon compounds that are produced as a result of the incomplete combustion of fossil fuels or biomass.

Brine A salt-rich solution produced by the ejection of salt during the freezing of sea water to produce sea ice.

Buffer factor The ratio of the percentage increase in the partial pressure of CO_2 in a layer of the ocean to the percentage increase in total dissolved carbon in that layer.

C_3 plant A plant that uses one of the two major pathways for photosynthesis, in which intermediate compounds with three carbon atoms are produced.

C_4 plant A plant that uses one of the two major pathways for photosynthesis, in which intermediate compounds with four carbon atoms are produced.

Calcium carbonate The material ($CaCO_3$) used to construct the skeletons in several major groups of marine micro-organisms.

Calving The breaking off into the ocean water of pieces of an ice sheet.

Carbonate (CO_3^{2-}) One of three forms in which inorganic carbon occurs in dissolved form in ocean water, the other two forms being as dissolved CO_2 and as bicarbonate (HCO_3^-).

CFCs Chlorofluorocarbons – a group of compounds built around one or more carbon atoms, and containing chlorine and fluorine.

Clausius–Clapeyron equation The concave upward relationship between the vapour pressure of water at saturation and temperature.

Climate sensitivity The ratio of the global mean equilibrium or steady-state temperature change to the radiative forcing that caused the temperature change, taking into account only the fast feedbacks.

CO_2 fertilization Refers to the tendency for higher atmospheric CO_2 concentrations to stimulate plant growth. This can occur as a result of the direct stimulation of photosynthesis, or an improvement in the efficiency of water use, which allows more plant growth for a given supply of water.

Convection The processes whereby parcels of air (in the atmosphere) or water (in the ocean) are exchanged vertically when the underlying air parcel becomes less dense than the overlying air due to heating, or the overlying water parcel becomes more dense than the underlying water due to cooling or salinification of the overlying parcel.

Coriolis force A force that varies in direct proportion to the speed of an object, and which acts perpendicularly to the right (left) of the motion in the NH (SH).

Cryosphere Refers to the cold sphere – glaciers, ice sheets, sea ice, and land snow cover.

Denitrification The process whereby nitrate (NO_3^-) or ammonium (NH_4^+) is converted to gaseous N_2 or N_2O.

Diapycnal mixing Mixing in the ocean in a direction perpendicular to isopycnal (constant-density) surfaces.

Diffusive mixing Refers to the net mixing due to small-scale eddies and turbulence. See also *diapycnal mixing* and *isopycnal mixing*.

Dissolved inorganic carbon The sum of dissolved CO_2, carbonate, and bicarbonate in the ocean.

Downregulation The tendency for the initial stimulation of photosynthesis caused by higher atmospheric CO_2 concentration to decrease over time.

e-folding time Following a sudden radiative heating perturbation, the time required for the temperature to reach to within $1 - 1/e$ or 63.2% of the final response.

Effective climate sensitivity The climate sensitivity that would occur in steady state if the strengths of the various feedback processes at a given point in the transient response were to persist.

El Niño A particular anomaly in the temperature and pressure patterns in the tropical Pacific Ocean, with worldwide effects, in which the eastern tropical Pacific is warmer than average and has lower surface pressure than average, while opposite changes occur in the western tropical Pacific.

Emissivity The ratio of the amount of radiation emitted by an object to the maximum permissible emission (the latter being given by the Stefan–Boltzmann relationship).

Equilibrium climatic change The climatic change that occurs after the atmosphere and ocean have fully adjusted to the imposed radiative forcing.

Equilibrium line altitude The altitude at which the transition from the ablation zone to the accumulation zone in a glacier occurs.

Equivalent CO_2 increase The increase in CO_2 concentration alone that has the same global mean radiative forcing as that arising from increases in a variety of well-mixed greenhouse gases.

Equivalent potential temperature The temperature that an air parcel would have if all its water vapour were condensed, and the parcel is brought to a pressure of 1000mb without exchanging heat with the surroundings.

Evapotranspiration The combination of evaporation of water directly from the surface of plants, and the transpiration or transfer of water from the interior of leaves to the exterior, largely through small openings called stomata.

External forcing The radiative forcing that results from some perturbation that is external to the climatic system, such as a change in solar luminosity, volcanic eruptions, or anthropogenic emissions of GHGs or aerosol precursors.

Extratropical cyclone A travelling synoptic-scale (1000 km) low-pressure system that occurs in middle and high latitudes, bringing with it rain.

Faculae Bright areas on the sun's surface, usually occurring in close proximity to sunspots. The presence of faculae contributes to an increase in the energy output from the sun.

Fast feedbacks Refers to those feedbacks in the climate system that operate over a period of days to months, and are used in the computation of climate sensitivity.

Flux A rate of flow of something.

General Circulation Model (GCM) A three-dimensional model of the atmosphere or ocean in which the motions at the scale of the model grid are simulated (along with many other processes).

Geostrophic balance A balance between the horizontal pressure gradient force and the horizontal component of the Coriolis force.

Global Warming Potential An single index that attempts to quantify the climatic impact of the same emission of a given greenhouse gas relative to that of a reference gas (usually CO_2), by taking into account the different heat-trapping abilities on a molecule-by-molecule basis and the different rates of removal of the two gases.

Greenhouse gas Any gas that emits and absorbs radiation in the infrared part of the electromagnetic spectrum, thereby tending to make the climate warmer than it would be in its absence.

Grounding line The boundary between the portion of an ice sheet that is on bedrock below sea level and the portion that is floating. Applicable to the West Antarctic ice sheet.

Growth respiration Respiration required to power the process of growth in a plant.

Halocarbons Compounds containing either chlorine, bromine, or fluorine, together with carbon. Such compounds can act as powerful greenhouse gases in the atmosphere. The chlorine- and bromine-containing halocarbons are also involved in depletion of the ozone layer.

Halocline A region in the ocean where there is a strong gradient in salinity. A halocline occurs in the upper part of the Arctic ocean, with relatively fresh water on top of saltier water.

Heat flux damping The increase in heat fluxes away from a surface (or layer) as the surface (or layer) warms up.

Heterotrophic respiration Respiration by the animals within an ecosystem.

Hydrochlorofluorocarbons (HCFCs) Compounds similar to the CFCs, except that one or more chlorine or fluorine atoms have been replaced with a hydrogen atom, thereby making them decompose much more readily.

Hydrofluorocarbons (HFCs) Carbon- and fluorine-containing compounds which, by virtue of the absence of chlorine or bromine, have no direct effect on stratospheric ozone, but do act as greenhouse gases.

Immobilization The process whereby nutrients that are released by mineralization are taken up by microbes and become unavailable for use by plants.

Impulse response The variation over time in the amount of a gas remaining in the atmosphere following a sudden injection.

Infrared radiation Electromagnetic radiation between wavelengths of $4.0\,\mu m$ and $50.0\,\mu m$, which is emitted by objects at temperatures typical of the Earth's surface and atmosphere.

Inverse problem Here, this refers to the problem of determining the magnitude and/or geographical distribution of the sources and sinks of a given gas based on the observed geographical and/or temporal variation in its concentration in the atmosphere.

Inversion A condition where temperature increases with increasing height in the atmosphere, which is opposite to the usual situation of decreasing temperature with increasing height.

Isopycnal mixing Mixing in the ocean along surfaces of constant density.

Isopycnal surface A set of adjoining points in 3-D space having the same density – a constant-density surface.

Isotope A variety of an element having a different number of neutrons in its nucleus and hence a different atomic mass.

Kyoto Protocol The agreement reached in Kyoto, Japan, on 10 December 1997 to limit emission of greenhouse gas by industrialized countries.

Lapse rate The negative of the rate of change of temperature with height. Since temperature normally decreases with increasing height, this gives a positive lapse rate. A greater lapse rate means a greater rate of decrease of temperature with increasing height.

Latent heat The heat that is required to melt ice or evaporate water, and that is released when freezing or condensation occurs.

Lithosphere The crust or outer layer of the Earth.

Maintenance respiration The background rate of respiration required to maintain living processes. See also *growth respiration* and *photorespiration*.

Mineralization The process whereby nutrients in organic matter are converted back to forms that can be readily used by plants or micro-organisms.

Mixed layer The surface layer of the ocean, that is well mixed by the action of wind-induced turbulence or the sinking of water parcels that become denser due either to surface cooling or to an excess of evaporation over precipitation plus runoff.

Monsoon The seasonal shift in wind patterns over south Asia and west Africa, whereby air flows from ocean to continent in summer and the reverse in winter. The summer monsoon brings rainfall to the affected regions.

Negative feedback A feedback in which the initial change (ΔA) provokes a change in some intermediate quantity (ΔB), and the change ΔB acting alone tends to provoke a further change in A that is opposite to the initial change. A negative feedback has a stabilizing effect.

Net Primary Production (NPP) The rate of photosynthesis by a plant minus the rate of growth and maintenance respiration.

Net radiation Absorbed solar radiation minus emitted infrared radiation.

Nitrification The process that converts NH_3 or NH_4^+ into NO_2^- or NO_3^-.

Nitrogen fixation The process that converts atmospheric N_2 into NH_3 or NH_4^+.

Nutrient Utilization Efficiency (NUE) An increase in the C:nutrient ratio in plant organic matter, such that less nutrients are required per unit of assimilated carbon. This can occur as a consequence of the biochemical downregulation of the photosynthetic response to higher CO_2, in which the concentration of the rubisco enzyme decreases.

Oxidation An element is oxidized when it gives up electrons to an electron acceptor or oxidizing agent, the most common electron acceptor being oxygen. It so doing it becomes chemically bound to the oxidizing agent. Biomass and fossil fuels are oxidized to produce CO_2 and H_2O.

Parameterization The technique of representing processes in a model that cannot be explicitly resolved at the resolution of the model (sub-grid scale processes) by relationships between the area-averaged effect of such sub-grid scale processes and the larger-scale variables (that are represented in the model).

Partial pressure The pressure due to a given gas acting alone.

Perfluorocarbons (PFCs) Carbon and fluorine compounds. Examples include CF_4 (carbon tetrafluoride or perfluoromethane) and C_2F_6 (perfluoroethane).

Photorespiration Respiration that is mediated by the same enzyme that catalyses photosynthesis (the rubisco enzyme) and that occurs only in the presence of sunlight. When an O_2 molecule attaches to the rubisco enzyme, photorespiration can occur, whereas when a CO_2 molecule attaches, photosynthesis can occur.

Planetary boundary layer The atmospheric layer next to the Earth's surface, in which there is intense vertical mixing.

Polar amplification Refers to the tendency, in computer climate models and in paleoclimatic data, for temperature changes at high latitudes to be greater than temperature changes at low latitudes when there is an overall warming or cooling of the climate.

Positive feedback A feedback in which the initial change (ΔA) provokes a change in some intermediate quantity (ΔB), and the change ΔB acting alone tends to provoke a further change in A that is in the same direction as the initial change. A positive feedback has a destabilizing effect.

Potential evapotranspiration The rate of evapotranspiration that would occur from a wet surface, or from a well-watered plant.

Precipitation efficiency The fraction of moisture transported upward by cumulus clouds that is quickly converted into rainfall.

Precursor A chemical that is emitted into the atmosphere and reacts with other chemicals to form a compound of interest. For example, CO, hydrocarbons, and NO_x are all precursors of tropospheric ozone, while SO_2 is a precursor for sulphate aerosol.

Radiative damping The increase in net emission of radiation to space in response to the warming of an object.

Radiative forcing The perturbation in the radiative balance at the tropopause in response to some externally applied perturbation (such as an increase in greenhouse gas concentrations other than water vapour).

Rainout The deposition of aerosols or other atmospheric constituents at the Earth's surface in rainwater.

Residence time The average length of time that a molecule spends in a reservoir before being removed.

Respiration The process of oxidizing or "burning" organic matter to provide energy, usually through the consumption of O_2. CO_2 is produced as a by-product of respiration. See also *growth respiration, heterotrophic respiration, maintenance respiration,* and *photorespiration.*

Senescence The process of ageing in a plant, characterized by a slowing down of physiological activity, decreasing growth rates, and increased sensitivity to abiotic and biotic stresses. This occurs at both the decadal and the seasonal time scales in deciduous trees, and is associated with leaf drop during autumn.

Sensible heat Heat that can be sensed or felt, related to the temperature of an object, gas, or fluid.

Silicates Minerals built around SiO_2 units.

Sink A reservoir into which a gas is absorbed and stored for a long period of time.

Snowline The boundary between the ablation zone and accumulation zone in a glacier.

Solar occultation A technique in which a satellite measures the amount of a gas (such as O_3) in the atmosphere by looking at the sun, when it is on the horizon, at two different wavelengths. One wavelength is where the gas absorbs radiation, and the other wavelength is one where it does not absorb.

Solubility pump The tendency for a net downward transfer of dissolved inorganic carbon in the ocean due to the fact that downwelling water is colder and therefore has a greater solubility than upwelling water.

Stomata Small openings on the surfaces of leaves that permit the passage of CO_2 into the interior of the leaf and the passage of water vapour from inside to outside the leaf.

Stomatal conductance The ease with which gases pass through the stomata. A greater conductance implies a smaller resistance.

Stomatal resistance The reciprocal of the stomatal conductance.

Stratosphere The layer of the atmosphere between a height of 10–17 and 50 km which contains the ozone "layer", and in which temperatures are either uniform with height (the lower stratosphere) or increase with increasing height (producing an inversion).

Sulphur hexafluoride (SF_6) An extremely powerful greenhouse gas used in the aluminium industry.

Sunspot Dark areas seen on the sun's surface, in some cases perceptible with the naked eye. The presence of sunspots contributes to a reduction in the energy output from the sun. See also *faculae.*

Surprise-free scenario Refers to the fact that the way the future unfolds in reality will almost certainly contain unexpected outcomes or "surprises" which, by definition, cannot be anticipated in the scenarios that are built using models. Such model-based projections therefore constitute surprise-free scenarios.

Symbiotic A mutually beneficial relationship between two different species.

System A set of components that interact with each other.

Thermohaline overturning The vertical overturning circulation in the oceans that is driven by horizontal differences in the density of water.

Transient climatic change The time-dependent climatic change during the transition from one equilibrium or statistically steady-state climate to another.

Tropopause The boundary between the troposphere and the stratosphere.

Troposphere The lowest layer of the atmosphere, in which weather disturbances occur. The troposphere varies in thickness from about 20 km at the equator to about 14 km at the poles.

Upwelling–diffusion model (UD model) A one-dimensional ocean model that is averaged horizontally (globally) but resolved vertically, and in which vertical mixing by diffusion and the upwelling of the thermohaline circulation are represented.

Water use efficiency (WUE) The ratio of carbon assimilation by photosynthesis to water loss through transpiration. WUE efficiency tends to increase as the atmospheric CO_2 concentration increases.

Well-mixed gases Gases whose lifetime in the atmosphere is sufficiently long (several years or more) that they have time to spread throughout the atmosphere before being removed. As result, they acquire a fairly uniform concentration within the atmosphere, the concentration being more uniform the longer the atmospheric lifetime and the more dispersed the sources and sinks.

References

Aagaard, K., Coachman, L.K. and Carmack, E. (1981) On the halocline of the Arctic Ocean. *Deep Sea Research*, **28A**, 529–545.

Acidifying Emissions Task Group. (1997) *Towards a National Acid Rain Strategy*, Environment Canada, Ottawa, 98 pages.

Adkins, J.F., Boyle, E.A., Keigwin, L. and Cortijo, E. (1997) Variability of the North Atlantic thermohaline circulation during the last interglacial period. *Nature*, **390**, 154–156.

Albritton, D.L., Derwent, R.G., Isaksen, I.S.A., Lal, M. and Wuebbles, D.J. (1995) Trace gas radiative forcing indices, in Houghton, J.T., Filho, L.G.M., Bruce, J., Lee, H., Callander, B.A., Haites, E., Harris, N. and Maskell, K. (eds), *Climate Change 1994: Radiative Forcing of Climate Change and an Evaluation of the IPCC IS92 Emission Scenarios*, Cambridge University Press, Cambridge, 205–231.

Andreae, M.O. (1995) Climatic effect of changing atmospheric aerosol levels, in Henderson-Sellers, A. (ed.), *Future Climates of the World: a Modelling Perspective*, World Survey of Climatology, Volume 16, Elsevier, Amsterdam, 347–398.

Angell, J.K. (1997) *Annual and Seasonal Global Temperature Anomalies in the Troposphere and Low Stratosphere, 1958–1996*, Carbon Dioxide Information Analysis Center, Report NDP-008, Oak Ridge National Laboratory, Oak Ridge, Tennessee.

Antoine, D. and Morel, A. (1995) Modelling the seasonal course of the upper ocean pCO_2 (i). Development of a one-dimensional model. *Tellus*, **47B**, 103–121.

Arakawa, A. and Schubert, W.H. (1974) Interaction of a cumulus cloud ensemble with the large-scale environment, I. *Journal of the Atmospheric Sciences*, **31**, 674–701.

Archer, D. (1996a) An atlas of the distribution of calcium carbonate in sediments of the deep sea. *Global Biogeochemical Cycles*, **10**, 159–174.

Archer, D. (1996b) A data-driven model of the global calcite lysocline. *Global Biogeochemical Cycles*, **10**, 511–526.

Archer, D., Kheshgi, H. and Maier-Reimer, E. (1997) Multiple timescales for neutralization of fossil fuel CO_2. *Geophysical Research Letters*, **24**, 405–408.

Atlas, R., Wolfson, N. and Terry, J. (1993) The effect of SST and soil moisture anomalies on GLA model simulations of the 1988 U.S. summer drought. *Journal of Climate*, **6**, 2034–2048.

Augustsson, T. and Ramanathan, V. (1977) A radiative–convective model study of the CO_2 climate problem. *Journal of the Atmospheric Sciences*, **34**, 448–451.

Bacastow, R. and Maier-Reimer, E. (1990) Ocean-circulation model of the carbon cycle. *Climate Dynamics*, **4**, 95–125.

Bakwin, P.S., Tans, P.P. and Novelli, P.C. (1994) Carbon monoxide budget in the Northern Hemisphere. *Geophysical Research Letters*, **21**, 433–436.

Baliunas, S. and Jastrow, R. (1990) Evidence for long term brightness changes of solar-type stars. *Nature*, **348**, 520–523.

Barnes, D.W. and Edmonds, J.A. (1990) *An Evaluation of the Relationship Between the Production and Use of Energy and Atmospheric Methane Emissions*, US Department of Energy, DOE/NBB-088P.

Barnett, T.P., Hegerl, G.C., Santer, B. and Taylor, K. (1998) The potential effect of GCM uncertainties and internal atmospheric variability on anthropogenic signal detection. *Journal of Climate*, **11**, 659–675.

Barnola, J.M., Pimienta, P. and Korotkevich, Y.S. (1991). CO_2-climate relationship as deduced from the Vostok ice core: a re-examination based on new measurements and on a re-evaluation of the air dating. *Tellus*, **43B**, 83–90.

Bates, T.S., Lamb, B.K., Guenther, A., Dignon, J. and Stoiber, R.E. (1992) Sulfur emissions to the atmosphere from natural sources. *Journal of Atmospheric Chemistry*, **14**, 315–337.

Battle, M. *et al.* (1996) Atmospheric gas concentrations over the past century measured in air from firn at the South Pole. *Nature*, **383**, 231–235.

Bazzaz, F.A. (1990) The response of natural ecosystems to the rising global CO_2 levels. *Annual Review of Ecology and Systematics*, **21**, 167–196.

Bazzaz, F.A., Miao, S.L. and Wayne, P.M. (1993) CO_2-induced growth enhancements of co-occurring tree species decline at different rates. *Oecologia*, **96**, 478–482.

Beck, L.L., Piccot, S.D. and Kirchgessner, D.A. (1993) Industrial sources, in Khalil, M. (ed.), *Atmospheric Methane: Sources, Sinks, and Role in Global Change*, Springer-Verlag, Berlin, 399–431.

Beersma, J.J., Rider, K.M., Komen, G.J., Kaas, E. and Kharin, V.V. (1997) An analysis of extra-tropical storms in the North Atlantic region as simulated in a control and $2 \times CO_2$ time-slice experiment with a high-resolution atmospheric model. *Tellus*, **49A**, 347–361.

Bekki, S. and Law, K.S. (1997) Sensitivity of the atmospheric CH_4 growth rate to global temperature changes observed from 1980 to 1992. *Tellus*, **49B**, 409–416.

Bekki, S., Law, K.S. and Pyle, J.A. (1994) Effects of ozone depletion on atmospheric CH_4 and CO concentrations. *Nature*, **371**, 595–597.

Benkovitz, C.M. *et al.* (1996) Global gridded inventories of anthropogenic emissions of sulfur and nitrogen. *Journal of Geophysical Research*, **101**, 29239–29253.

Bentley, C.R. (1998) Rapid sea-level rise from a West Antarctic ice-sheet collapse: a short-term perspective. *Journal of Glaciology*, **44**, 157–163.

Berner, R.A. (1990) Atmospheric carbon dioxide levels over Phanerozoic time. *Science*, **249**, 1382–1386.

Berner, R.A. (1994) Geocarb II: a revised model of atmospheric CO_2 over phanerozoic time. *American Journal of Science*, **294**, 56–91.

Berner, R.A., Lasaga, A.C. and Garrels, R.M. (1983) The carbonate–silicate geochemical cycle and its effect on atmospheric carbon dioxide over the past 100 million years. *American Journal of Science*, **283**, 641–683.

Berntsen, T.K., Isaksen, I.S.A., Myhre, G., Fuglestvedt, J.S., Stordal, F., Larsen, T.A., Freckleton, R.S. and Shine, K.P. (1997) Effects of anthropogenic emissions on tropospheric ozone and its radiative forcing. *Journal of Geophysical Research*, **102**, 28101–28126.

Betts, A.K. (1990) Greenhouse warming and the tropical water budget. *Bulletin of the American Meteorological Society*, **71**, 1464–1465.

Betts, R.A., Cox, P.M., Lee, S.E. and Woodward, F.I. (1997) Contrasting physiological and structural vegetation feedbacks in climate change simulations. *Nature*, **387**, 796–799.

Bhaskaran, B., Mitchell, J.F.B., Lavery, J.R. and Lal, M. (1995) Climatic response of the Indian subcontinent to doubled CO_2 concentrations. *International Journal of Climatology*, **15**, 873–892.

Billings, W.D., Luken, J.O., Mortensen, D.A. and Peterson, K.M. (1982) Arctic tundra: a source or sink for atmospheric carbon dioxide in a changing environment? *Oecologia*, **53**, 7–11.

Bindschadler, R. (1997) Actively surging West Antarctic ice streams and their response characteristics. *Annals of Glaciology*, **24**, 409–414.

Bintanja, R. (1995) *The Antarctic Ice Sheet and Climate*, PhD Thesis, Utrecht University, 200 pages.

Blanchard, D.C. (1983) The production, distribution, and bacterial enrichment of the sea-salt aerosol, in Liss, P.S. and Slinn, W.G.N. (eds), *Air–Sea Exchange of Gases and Particles*, NATO ASI Series 108, Reidel, Dordrecht, 407–454.

Boden, T.A., Kaiser, D.P., Sepanski, R.J. and Stoss, F.W. (eds) (1994) *Trends '93: A Compendium of Data on Global Change*. ORNL/CDIAC-65. Carbon Dioxide Information Analysis Center, Oak Ridge National Laboratory, Oak Ridge, Tennessee, 984 pages + appendices.

Boer, G.J., McFarlane, N.A. and Lazare, M. (1992) Greenhouse gas-induced climate change simulated with the CCC second-generation general circulation model. *Journal of Climate*, **5**, 1045–1077.

Bojkov, R.D. and Fioletov, V.E. (1995) Estimating the global ozone characteristics during the last 30 years. *Journal of Geophysical Research*, **100**, 16537–16551.

Bolker, B.M., Pacala, S.W., Bazzaz, F.A., Canham, C.D. and Levin, S.A. (1995) Species-diversity and ecosystem response to carbon dioxide fertilization – conclusions from a temperature forest model. *Global Change Biology*, **1**, 373–381.

Bonan, G.B. (1997) Effects of land use on the climate of the United States. *Climatic Change*, **37**, 449–486.

Bond, T.C., Charlson, R.J. and Heintzenberg, J. (1998) Quantifying the emission of light-absorbing particles: measurements tailored to climate studies. *Geophysical Research Letters*, **25**, 337–340.

Bongaarts, J. (1990) The measurement of unwanted fertility. *Population and Development Review*, **16**, 487–506.

Bongaarts, J. (1994) Population policy options in the developing world. *Science*, **263**, 771–776.

Borzenkova, I.I. (1992) *The Changing Climate During the Cenozoic*, Hydrometeoizdat, Saint Petersburg, 247 pages (in Russian; in English by Kluwer Academic).

Bottomley, M., Folland, C.K., Hsiung, J., Newell, R.E. and Parker, D.E. (1990) *Global Ocean Surface Temperature Atlas (GOSTA)*. Joint Meteorological Office/ Massachusetts Institute of Technology Project, HMSO, London, 20 + iv pages and 313 plates.

Boucher, O. and Anderson, T.L. (1995) General circulation model assessment of the sensitivity of direct climate forcing by anthropogenic sulfate aerosols to aerosol size and chemistry. *Journal of Geophysical Research*, **100**, 26117–26134.

Boucher, O. and Lohmann, U. (1995) The sulfate–CCN–cloud albedo effect. A sensitivity study with two general circulation models. *Tellus*, **47B**, 281–300.

Boutin, J.E.J. and Merlivat, L. (1991) Seasonal variation of the CO_2 exchange coefficient over the global ocean using satellite wind speed measurements. *Tellus*, **43B**, 247–255.

Branscome, L.E. and Gutowski, W.J. (1992) The impact of doubled CO_2 on the energetics and hydrologic processes of mid-latitude transient eddies. *Climate Dynamics*, **8**, 29–37.

Briffa, K.R., Jones, P.D., Schweingruber, F.H., Shiyatov, S.G. and Cook, E.R. (1995) Unusual twentieth-century summer warmth in a 1000-year temperature record from Siberia. *Nature*, **376**, 156–159.

Broccoli, A.J., Manabe, S., Mitchell, J.F.B. and Bengtsson, L. (1995) Comments on "Global climate change and tropical cyclones": Part II. *Bulletin of the American Meteorological Society*, **76**, 2243–2245.

Broecker, W.S. (1997) Thermohaline circulation, the Achilles Heel of our climate system: Will man-made CO_2 upset the current balance? *Science*, **278**, 1582–1588.

Brown, P.G. (1992) Climate change and the planetary trust. *Energy Policy*, **20**, 208–222.

Brown, S. *et al.* (1996) Management of forests for mitigation of greenhouse gas emissions, in Watson, R.T., Zinyowera, M.C. and Moss, R.H. (eds), *Climate Change 1995 – Impacts, Adaptations and Mitigation of Climate Change: Scientific–Technical Analysis*, Cambridge University Press, Cambridge, 771–797.

Brunke, E.-G., Scheel, H.E. and Seiler, W. (1990) Trends of tropospheric carbon monoxide, nitrous oxide and methane as observed at Cape Point, South Africa. *Atmospheric Environment*, **24**, 585–595.

Bryan, F. (1987) Parameter sensitivity of primitive equation ocean general circulation models. *Journal of Physical Oceanography*, **17**, 970–985.

Bryan, K. (1996) The steric component of sea level rise associated with enhanced greenhouse warming: a model study. *Climate Dynamics*, **12**, 545–555.

Bryant, E. (1997) *Climate Process and Change*, Cambridge University Press, New York, 209 pages.

Cane, M.A. *et al.* (1997) Twentieth-century sea surface temperature trends. *Science*, **275**, 957–960.

Cao, H.-X., Mitchell, J.F.B. and Lavery, J.R. (1992) Simulated diurnal range and variability of surface temperature in a global model for present and doubled CO_2 climates. *Journal of Climate*, **5**, 920–943.

Cerling, T.E., Ehleringer, J.R. and Harris, J.M. (1998) Carbon starvation, the development of C-4 ecosystems, and mammalian evolution. *Philosophical Transactions of the Royal Society of London, Series B*, **353**, 159–170.

Cess, R.D. *et al.* (1990) Intercomparison and interpretation of climate feedback processes in nineteen atmospheric general circulation models. *Journal of Geophysical Research*, **95**, 16601–16615.

Cess, R.D. *et al.* (1991) Interpretation of snow–climate feedback as produced by 17 general circulation models. *Science*, **253**, 888–892.

Cess, R.D. *et al.* (1993) Uncertainties in carbon dioxide radiative forcing in atmospheric general circulation models. *Science*, **262**, 1252–1255.

Cess, R.D. *et al.* (1996) Cloud feedback in atmospheric general circulation models: an update. *Journal of Geophysical Research*, **101**, 12791–12794.

Chan, J.C.L. and Shi, J. (1996) Long-term trends and inter-annual variability in tropical cyclone activity over the western North Pacific. *Geophysical Research Letters*, **23**, 2765–2767.

Chappellaz, J., Barnola, J.M., Raynaud, D., Korotkevich, Y.S. and Lorius, C. (1990) Ice-core record of atmospheric methane over the past 160,000 years. *Nature*, **345**, 127–131.

Charlson, R.J., Langner, J., Rodhe, H., Leovy, C.B. and Warren, S.G. (1991) Perturbation of the Northern Hemisphere radiative balance by backscattering from anthropogenic sulfate aerosols. *Tellus*, **43AB**, 152–163.

Charlson, R.J., Anderson, T.L. and McDuff, R.E. (1992) The sulfur cycle, in Butcher, S.S., Charlson, R.J., Orians, G.H. and Wolfe, G.V. (eds), *Global Biogeochemical Cycles*, Academic Press, London, 285–300.

Chase, T.N., Pielke, R.A., Kittel, T.G.F., Nemani, R.R. and Running, S.W. (1999) Simulated impacts of historical land cover changes on global climate. *Climate Dynamics* (in press).

Chen, C.-T., Roeckner, E. and Soden, B.J. (1996) A comparison of satellite observations and model simulations of column-integrated moisture and upper-tropospheric humidity. *Journal of Climate*, **9**, 1561–1585.

Chen, J. (1998) CO_2 emissions relief through blended cements, www.civil.nwu.edu/ACBM/Chen/CO2%20 emissions%20calculations.htm.

Chin, M., Jacob, D.J., Gardner, G.M., Foreman-Fowler, M.S. and Spiro, P.A. (1996) A global three-dimensional model of tropospheric sulfate. *Journal of Geophysical Research*, **101**, 18667–18690.

Chou, M.-D. (1990) Parameterizations for the absorption of solar radiation by O_2 and CO_2 with application to climate studies. *Journal of Climate*, **3**, 209–217.

Christidis, N., Hurley, M.D., Pinnock, M.D., Shine, K.P. and Wallington, T.J. (1997) Radiative forcing of climate change by CFC-11 and possible CFC replacements. *Journal of Geophysical Research*, **102**, 19597–19609.

Christopherson, R.W. (1992) *Geosystems: An Introduction to Physical Geography*, Macmillan, New York, 663 pages + appendices.

Christy, J.R. and Spencer, R.W. (1995) Assessment of precision in temperatures from the microwave sounding units. *Climatic Change*, **30**, 97–102.

Christy, J.R., Spencer, R.W. and McNider, R.T. (1995) Reducing noise in the MSU daily lower-tropospheric global temperature dataset. *Journal of Climate*, **8**, 888–896.

Christy, J.R., Spencer, R.W. and Braswell, W.D. (1997) How accurate are satellite thermometers? *Nature*, **389**, 342.

Christy, J.R., Spencer, R.W. and Lobl, E.S. (1998) Analysis of the merging procedure for the MSU daily temperature time series. *Journal of Climate*, **11**, 2016–2041.

Chuang, C.C., Penner, J.E., Taylor, K.E., Grossman, A.S. and Walton, J.J. (1997) An assessment of the radiative effects of anthropogenic sulfate. *Journal of Geophysical Research*, **102**, 3761–3778.

Chýlek, P. (1999) Effective medium approximations for heterogeneous particles, in Mishchenko, M. (ed.), *Light Scattering by Nonspherical Particles*, Academic Press (in press).

Chýlek, P., Ramaswamy, V. and Cheng, R.J. (1984) Effect of graphitic carbon on the albedo of clouds. *Journal of the Atmospheric Sciences*, **41**, 3076–3084.

Chýlek, P., Srivastava, V., Pinnock, R.G. and Wang, R.T. (1988) Scattering of electromagnetic waves by composite spherical particles: experiments and effective medium approximations. *Applied Optics*, **27**, 2396–2404.

Chýlek, P., Videen, G., Ngo, D., Pinnick, R.G. and Klett, J.D. (1995) Effect of black carbon on the optical properties and climate forcing of sulfate aerosols. *Journal of Geophysical Research*, **100**, 16325–16332.

Ciais, P. *et al.* (1995) Partitioning of ocean and land uptake of CO_2 as inferred by $\delta^{13}C$ measurements from the NOAA Climate Monitoring and Diagnostics Laboratory global air sampling network. *Journal of Geophysical Research*, **100**, 5051–5070.

Colman, R.A. and McAvaney, B.J. (1995) The sensitivity of the climate response of an atmospheric general circulation model to changes in convective parameterization and horizontal resolution. *Journal of Geophysical Research*, **100**, 3155–3172.

Colman, R.A. and McAvaney, B.J. (1997) A study of general circulation model climate feedbacks determined from perturbed sea surface temperature experiments. *Journal of Geophysical Research*, **102**, 19383–19402.

Colman, R.A., Power, S.B., McAvaney, B.J. and Dahni, R. (1995) A non-flux corrected transient CO_2 experiment using the BMRC coupled atmosphere/ocean. GCM *Geophysical Research Letters*, **22**, 3047–3050.

Colman, R.A., Power, S.B. and McAvaney, B.J. (1997) Nonlinear climate feedback analysis in an atmospheric general circulation model. *Climate Dynamics*, **13**, 717–731.

Cooke, W.F. and Wilson, J.J.N. (1996) A global black carbon aerosol model. *Journal of Geophysical Research*, **101**, 19395–19409.

Cox, S.J., Wang, W.-C. and Schwartz, S.E. (1995) Climate response to radiative forcings by sulfate aerosols and greenhouse gases. *Geophysical Research Letters*, **22**, 2509–2512.

Crowley, T.J. (1990) Are there any satisfactory geologic analogs for a future greenhouse warming? *Journal of Climate*, **3**, 1282–1292.

Crowley, T.J. and Baum, S.K. (1995) Is the Greenland ice sheet bistable? *Paleoceanography*, **10**, 357–363.

Crutzen, P.J. (1988) Variability in atmosphere–chemical systems, in Rosswall, T., Woodmansee, R.G. and Risser, P.G. (eds), *Scales and Global Change*, SCOPE 35, Wiley, Chichester, 81–108.

Crutzen, P.J. and Andreae, M.O. (1990) Biomass burning in the tropics: impact on atmospheric chemistry and biogeochemical cycles. *Science*, **250**, 1669–1678.

Crutzen, P.J. and Zimmermann, P.H. (1991) The changing photochemistry of the troposphere. *Tellus*, **43AB**, 136–151.

Cubasch, U., Hasselmann, K., Hock, H., Maier-Reimer, E., Mikolajewicz, U., Santer, B.D. and Sausen, R. (1992) Time-dependent greenhouse warming computations with a coupled ocean–atmosphere model. *Climate Dynamics*, **8**, 55–69.

Cuffey, K.M. and Clow, G.D. (1997) Temperature, accumulation, and ice sheet elevation in central Greenland through the last deglacial transition. *Journal of Geophysical Research*, **102**, 26383–26396.

Cui, M. and Zorita, E. (1998) Analysis of the sea-level variability along the Chinese coast and estimation of the impact of a CO_2-perturbed atmospheric circulation. *Tellus*, **50A**, 333–347.

Cunnington, W.M. and Mitchell, J.F.B. (1990) On the dependence of climate sensitivity on convective parameterization. *Climate Dynamics*, **4**, 85–93.

Dai, A. and Fung, I.Y. (1993) Can climate variability contribute to the "missing" CO_2 sink? *Global Biogeochemical Cycles*, **7**, 599–609.

Daniel, J.S., Solomon, S. and Albritton, D.L. (1995) On the evaluation of halocarbon radiative forcing and global warming potentials. *Journal of Geophysical Research*, **100**, 1271–1285.

de la Mare, W.K. (1997) Abrupt mid-twentieth-century decline in Antarctic sea-ice extent from whaling records. *Nature*, **389**, 57–60.

De Wolde, J.R., Bintanja, R. and Oerlemans, J. (1995) On thermal expansion over the last one hundred years. *Journal of Climate*, **8**, 2881–2891.

De Wolde, J.R., Huybrechts, P., Oerlemans, J. and van de Wal, R.S.W. (1997) Projections of global mean sea level rise calculated with a 2D energy-balance climate model and dynamic ice sheet models. *Tellus*, **49A**, 486–502.

Del Genio, A.D. (1993) Convective and large-scale cloud processes in GCMs, in Raschke, E. and Jacob, D. (eds), *Energy and Water Cycles in the Climate System*, Springer-Verlag, Berlin, 95–121.

Del Genio, A.D. (1996) GCM implications for mechanisms determining cloud and water vapor feedbacks, in Le Treut, H. (ed.), *Climate Sensitivity to Radiative Perturbations: Physical Mechanisms and their Validation*, NATO ASI Series, Vol. I 34, Springer-Verlag, Berlin, 107–125.

Del Genio, A.D., Lacis, A.A. and Ruedy, R.A. (1991) Simulations of the effect of a warmer climate on atmospheric humidity. *Nature*, **351**, 382–385.

Del Genio, A.D., Kovari, W. and Yao, M.-S. (1994) Climatic implications of the seasonal variation of upper tropospheric water vapor. *Geophysical Research Letters*, **21**, 2701–2704.

Del Genio, A.D., Yao, M.-S., Kovari, W. and Lo, K.K.-W. (1996) A prognostic cloud water parameterization for global climate models. *Journal of Climate*, **9**, 270–304.

Delworth, T., Manabe, S. and Stouffer, R.J. (1993) Interdecadal variations of the thermohaline circulation in a coupled ocean-atmosphere model. *Journal of Climate*, **6**, 1993–2011.

DeMaria, M. and Kaplan, J. (1994) Sea surface temperature and the maximum intensity of Atlantic tropical cyclones. *Journal of Climate*, **7**, 1324–1334.

Demeritt, D. and Rothman, D.D. (1998a) Comments on J.B. Smith (*Climatic Change* 32, 313–326) and the aggregation of climate damage costs. *Climatic Change*, **40**, 699–704.

Demeritt, D. and Rothman, D.S. (1998b) Figuring the costs of climate change: an assessment and critique. *Environment and Planning A* (in press).

Dentener, F.J. and Crutzen, P.J. (1993) Reaction of N_2O_5 on tropospheric aerosols: impact on the global distributions of NO_x, O_3, and OH. *Journal of Geophysical Research*, **98**, 7149–7163.

Dentener, F.J., Carmichael, G.R., Zhang, Y., Lelieveld, J. and Crutzen, P.J. (1996) Role of mineral aerosol as a reactive surface in the global troposphere. *Journal of Geophysical Research*, **101**, 22869–22889.

Dey, B. and Bhanu Kumar, O.S.R.U. (1982) An apparent relationship between Eurasian spring snow cover and the advance period of the Indian summer monsoon. *Journal of Applied Meteorology*, **21**, 1929–1932.

Diaz, S., Grime, J.P., Harris, J. and McPherson, E. (1993) Evidence of a feedback mechanism limiting plant response to elevated carbon dioxide. *Nature*, **364**, 616–617.

Dickinson, R.E. (1983) Land surface processes and climate–surface albedos and energy balance. *Advances in Geophysics*, **25**, 305–353.

Dickinson, R.E. (1984) Modelling evapotranspiration for three-dimensional global climate models, in Hanson, J.E. and Takahashi, T. (eds), *Climate Processes and Climate Sensitivity, Geophysical Monographs No. 29*, American Geophysical Union, 58–72.

Dickinson, R.E., Meleshko, V., Randall, D., Sarachik, E., Silva-Dias, P. and Slingo, A. (1996) Climate processes, in Houghton, J.T., Filho, L.G.F., Callander, B.A., Harris, N., Kattenberg, A. and Maskell, K. (eds), *Climate Change 1995: The Science of Climate Change*, Cambridge University Press, Cambridge, 193–227.

Dickson, R.R. (1984) Eurasian snow cover versus Indian monsoon rainfall – an extension of Hahn–Shukla results. *Journal of Climate and Applied Meteorology*, **23**, 171–173.

Ding, P. and Randall, D.A. (1998) A cumulus parameterization with multiple cloud base levels. *Journal of Geophysical Research*, **103**, 11341–11353.

Dixon, R.K., Brown, S., Houghton, R.A., Solomon, A.M., Trexler, M.C. and Wisniewski, J. (1994) Carbon pools and flux of global forest ecosystems. *Science*, **263**, 185–190.

Dlugokencky, E.J., Masarie, K.A., Lang, P.M. and Tans, P.P. (1998) Continuing decline in the growth rate of the atmospheric methane burden. *Nature*, **393**, 447–450.

DOE (US Department of Energy) (1992) *Electric Power Annual 1990*, DOE/EIA-0348(90), Washington, DC.

Douglas, B.C. (1991) Global sea level rise. *Journal of Geophysical Research*, **96**, 6981–6992.

Douglas, B.C. (1992) Global sea level acceleration. *Journal of Geophysical Research*, **97**, 12699–12706.

Douglas, B.C. (1995) Global sea level change: determination and interpretation. *Reviews of Geophysics* (supplement), 1425–1432.

Douglas, B.C. (1997) Global sea rise: a redetermination. *Surveys in Geophysics*, **18**, 279–292.

Douville, H., Royer, J.-F. and Mahfouf, J.-F. (1995) A new snow parameterization for the Météo-France climate model. *Climate Dynamics*, **12**, 21–35.

Drake, B.G., Gonzàlez-Meler, M.A. and Long, S.P. (1996a) More efficient plants: a consequence of rising atmospheric CO_2? *Annual Review of Plant Physiology and Plant Molecular Biology*, **48**, 609–639.

Drake, B.G., Muehe, M.S., Peresta, G., Gonzàlez-Meler, M.A. and Matamala, R. (1996b) Acclimation of photosynthesis, respiration, and ecosystem carbon flux of a wetland on Chesapeake Bay, Maryland to elevated atmospheric CO_2 concentration. *Plant and Soil*, **187**, 111–118.

Druffel, E.R.M., Williams, P.M., Bauer, J.E. and Ertel, J.R. (1992) Cycling of dissolved and particulate organic matter in the open ocean. *Journal of Geophysical Research*, **97**, 15639–15659.

Easterling, D.R. *et al.* (1997) Maximum and minimum temperature trends for the globe. *Science*, **277**, 364–367.

Ehleringer, J.R., Sage, R.F., Flanagan, L.B. and Pearcy, R.W. (1991) Climate change and the evolution of C_4 photosynthesis. *Trends in Ecology and Evolution*, **6**, 95–99.

EIA (Energy Information Administration) (1998) Appendix B: Characteristics of Coal Supply Contracts, in *Coal Industry Annual 1996* (obtainable from http://www.eia.doe.gov/cneaf/coal/cia/).

Eischeid, J.K., Baker, C.B., Karl, T.R. and Diaz, H.F. (1995) The quality control of long-term climatological data using objective data analysis. *Journal of Applied Meteorology*, **34**, 2787–2795.

Eisele, F.L., Mount, G.H., Tanner, D., Jefferson, A., Shetter, R., Harder, J.W. and Williams, E.J. (1997) Understanding the production and interconversion of the hydroxyl radical during the Tropospheric OH Photochemistry Experiment. *Journal of Geophysical Research*, **102**, 6457–6465.

Elliott, W.P. (1995) On detecting long-term changes in atmospheric moisture. *Climatic Change*, **31**, 349–367.

Elliott, W.P. and Angell, J.K. (1997) Variations of cloudiness, precipitable water, and relative humidity over the

United States: 1973–1993. *Geophysical Research Letters*, **24**, 41–44.

Elliott, W.P. and Gaffen, D.J. (1991) On the utility of radiosonde humidity archives for climate studies. *Bulletin of the American Meteorological Society*, **72**, 1507–1520.

Emanuel, K.A. (1991) A scheme for representing cumulus convection in large-scale models. *Journal of the Atmospheric Sciences*, **48**, 2313–2335.

Emanuel, K.A. (1995) Comments on "Global climate change and tropical cyclones": Part I. *Bulletin of the American Meteorological Society*, **76**, 2241–2243.

EPA (Environmental Protection Agency) (1998) Acid Rain Program – 1997 Compliance Report, Appendix B3 (obtainable from http://www.epa.gov/acidrain/cmprtpt97/ cr1997.htm).

Epstein, P.R. *et al.* (1998) Biological and physical signs of climate change: focus on mosquito-borne diseases. *Bulletin of the American Meteorological Society*, **79**, 409–417.

Etheridge, D.M., Pearman, G.I. and Fraser, P.J. (1992) Changes in tropospheric methane between 1841 and 1978 from a high accumulation-rate Antarctic ice core. *Tellus*, **44B**, 282–294.

Etheridge, D.M., Steele, L.P., Langenfelds, R.L., Francey, R.J., Barnola, J.-M. and Morgan, V.I. (1996) Natural and anthropogenic changes in atmospheric CO_2 over the last 1000 years from air in Antarctic ice and firn. *Journal of Geophysical Research*, **101**, 4115–4128.

Evans, J.L. and Allan, R.J. (1992) El Niño/Southern Oscillation modification to the structure of the monsoon and tropical cyclone activity in the Australasian region. *International Journal of Climatology*, **12**, 611–623.

Faber, M., Manstetten, R. and Proops, J.L.R. (1994) Humankind and the environment: an anatomy of surprise and ignorance. *Environmental Values*, **1**, 217–241.

Falkowski, P.G., Barber, R.T. and Smectacek, V. (1998) Biogeochemical controls and feedbacks on ocean primary production. *Science*, **281**, 200–205.

Fan, S., Gloor, M., Mahlman, J., Pacala, S., Sarmiento, J., Takahashi, T. and Tans, P. (1998) A large terrestrial carbon sink in North America implied by atmospheric and oceanic carbon dioxide data and models. *Science*, **282**, 442–446.

Fearnside, P.M. (1995) Hydroelectric dams in the Brazilian Amazon as sources of "greenhouse" gases. *Environmental Conservation*, **22**, 7–19.

Feichter, J., Lohmann, U. and Schult, I. (1997) The atmospheric sulfur cycle in ECHAM-4 and its impact on the shortwave radiation. *Climate Dynamics*, **13**, 235–246.

Field, C.B., Jackson, R.B. and Mooney, H.A. (1995) Stomatal responses to increased CO_2: implications from the plant to the global scale. *Plant, Cell and Environment*, **18**, 1214–1225.

Fischer, H., Werner, M., Wagenbach, D., Schwager, M., Thorsteinnson, T., Wilhelms, F. and Kipfstuhl, J. (1998) Little ice age clearly recorded in northern Greenland ice cores. *Geophysical Research Letters*, **25**, 1749–1752.

Fishman, J., Watson, C.E., Larsen, J.C. and Logan, J.A. (1990) Distribution of tropospheric ozone determined from satellite data. *Journal of Geophysical Research*, **95**, 3599–3617.

Folland, C.K. and Parker, D.E. (1989) Observed variations of sea surface temperature, in Schlesinger, M.E. (ed.), *Climate–Ocean Interaction*, NATO Workshop, Kluwer, Dordrecht, 21–52.

Folland, C.K. and Parker, D.E. (1995) Correction of instrumental biases in historical sea surface temperature data. *Quarterly Journal of the Royal Meteorological Society*, **121**, 319–367.

Folland, C.K., Parker, D.E. and Kates, F.E. (1984) Worldwide marine surface temperature fluctuations 1856–1981. *Nature*, **310**, 670–673.

Forster, P. M. de F. and Shine, K.P. (1997) Radiative forcing and temperature trends from stratospheric ozone changes. *Journal of Geophysical Research*, **102**, 10841–10855.

Forster, P.M. de F., Johnson, C.E., Law, K.S., Pyle, J.A. and Shine, K.P. (1996) Further estimates of radiative forcing due to tropospheric ozone changes. *Geophysical Research Letters*, **23**, 3321–3324.

Forster, P.M. de F., Freckleton, R.S. and Shine, K.P. (1997) On aspects of the concept of radiative forcing. *Climate Dynamics*, **13**, 547–560.

Fraedrich, K., Müller, K. and Kuglin, R. (1992) Northern hemisphere circulation regimes during extremes of the El Niño/Southern Oscillation. *Tellus*, **44A**, 33–40.

Frakes, L.A. (1979) *Climate Through Geologic Time*, Elsevier, Amsterdam, 310 pages.

Francey, R.J., Tans, P.P., Allison, C.E., Enting, I.G., White, J.W.C. and Troller, M. (1995) Changes in oceanic and terrestrial carbon uptake since 1982. *Nature*, **373**, 326–330.

Francois, L.M. and Walker, J.C.G. (1992) Modelling the Phanerozoic carbon cycle and climate: constraints from the $^{87}Sr/^{86}Sr$ isotopic ratio of seawater. *American Journal of Science*, **292**, 81–135.

Friis-Christensen, E. and Lassen, K. (1991) Length of the solar cycle: an indicator of solar activity associated with climate. *Science*, **254**, 698–700.

Gaffen, D.J. (1994) Temporal inhomogeneities in radiosonde temperature records. *Journal of Geophysical Research*, **99**, 3667–3676.

Gaffen, D.J. (1998) Falling satellites, rising temperatures? *Nature*, **394**, 615–616.

Gaffen, D.J., Elliott, W.P. and Robock, A. (1992) Relationships between tropospheric water vapor and surface temperature as observed by radiosondes. *Geophysical Research Letters*, **19**, 1839–1842.

Gas Research Institute (1997) Effect of methane emissions on global warming, Appendix B in *Methane Emissions from the Natural Gas Industry*, Gas Research Institute, Chicago, B2–B28.

Gasse, F. and van Campo, E. (1994) Abrupt post-glacial climate events in West Asia and North Africa monsoon domains. *Earth and Planetary Science Letters*, **126**, 435–456.

Gates, W.L. (1992) AMIP: The Atmospheric Model Intercomparison Project. *Bulletin of the American Meteorological Society*, **73**, 1962–1970.

Gates, W.L. *et al.* (1996) Climate models – evaluation, in Houghton, J.T., Filho, L.G.F., Callander, B.A., Harris, N., Kattenberg, A. and Maskell, K. (eds), *Climate Change 1995: The Science of Climate Change*, Cambridge University Press, Cambridge, 229–284.

Geller, L.S., Elkins, J.W., Lobert, J.M., Clarke, A.D., Hurst, D.F., Butler, J.H. and Myers, R.C. (1998) Tropospheric SF_6: observed latitudinal distribution and trends, derived emissions and interhemispheric exchange time. *Geophysical Research Letters*, **24**, 675–678.

Gifford, R.M. (1992) Interaction of carbon dioxide with growth-limiting environmental factors in vegetation productivity: implications for the global carbon cycle. *Advances in Bioclimatology*, **1**, 24–58.

Gifford, R.M. (1994) The global carbon cycle: a viewpoint on the missing sink. *Australian Journal of Plant Physiology*, **21**, 1–15.

Gordon, H.B. and O'Farrell, S.P. (1997) Transient climatic change in the CSIRO coupled model with dynamic sea ice. *Monthly Weather Review*, **125**, 875–907.

Gordon, H.B., Whetton, P.H., Pittock, A.B., Fowler, A.M. and Haylock, M.R. (1992) Simulated changes in daily rainfall intensity due to the enhanced greenhouse effect: implications for extreme rainfall events. *Climate Dynamics*, **8**, 83–102.

Gorham, E. (1991) Northern peatlands: role in the carbon cycle and probable responses to climatic warming. *Ecological Applications*, **1**, 182–195.

Gornitz, V. (1995) Monitoring sea level changes. *Climatic Change*, **31**, 515–544.

Gornitz, V., Rosenzweig, C. and Hillel, D. (1997) Effects of anthropogenic intervention in the land hydrologic cycle on global sea level rise. *Global and Planetary Change*, **14**, 147–161.

Gough, D.O. (1981) Solar interior structure and luminosity variations. *Solar Physics*, **74**, 21–34.

Gough, W.A. and Lin, C.A. (1992) The response of an ocean general circulation model to long time-scale surface temperature anomalies. *Atmosphere–Ocean*, **30**, 653–674.

Gregg, W.W. and Walsh, J.J. (1992) Simulation of the 1979 spring bloom in the mid-Atlantic bight: a coupled physical/biological/optical model. *Journal of Geophysical Research*, **97**, 5723–5743.

Gregory, J.M. and Oerlemans, J. (1998) Simulated future sea-level rise due to glacier melt based on regionally and seasonally resolved temperature changes. *Nature*, **391**, 474–476.

Gregory, J.M., Mitchell, J.F.B. and Brady, A.J. (1997) Summer drought in northern midlatitudes in a time-dependent CO_2 climate experiment. *Journal of Climate*, **10**, 662–686.

Groisman, P.Y., Karl, T.R., Knight, R.W. and Stenchikov, G.L. (1994) Changes of snow cover, temperature, and radiative heat balance over the Northern Hemisphere. *Journal of Climate*, **7**, 1633–1656.

Grossman, A.S., Grant, K.E., Blass, W.E. and Wuebbles, D.J. (1997) Radiative forcing calculations for CH_3Cl and CH_3Br. *Journal of Geophysical Research*, **102**, 13651–13656.

Grubb, M. (1992) *The Greenhouse Effect: Negotiating Targets*, Royal Institute of International Affairs, London, 56 pages.

Gutzler, D.S. (1992) Climatic variability of temperature and humidity over the tropical Western Pacific. *Geophysical Research Letters*, **19**, 1595–1598.

Haigh, J.D. (1994) The role of stratospheric ozone in modulating the solar radiative forcing of climate. *Nature*, **370**, 544–546.

Hall, C.A.S. and Uhlig, J. (1991) Refining estimates of carbon released from tropical land-use change. *Canadian Journal of Forest Research*, **21**, 118–131.

Hall, N.M.J., Hoskins, B.J., Valdes, P.J. and Senior, C.A. (1994) Storm tracks in a high-resolution GCM with doubled carbon dioxide. *Quarterly Journal of the Royal Meteorological Society*, **120**, 1209–1230.

Hansell, D.A. and Carlson, C.A. (1998) Deep-ocean gradients in the concentration of dissolved organic carbon. *Nature*, **395**, 263–266.

Hansen, J. and Lebedeff, S. (1987) Global trends of measured surface air temperature. *Journal of Geophysical Research*, **92**, 13345–13372.

Hansen, J., Russell, G., Rind, D., Stone, P., Lacis, A., Lebedeff, S., Ruedy, R. and Travis, L. (1983) Efficient three-dimensional global models for climate studies: Model I and II. *Monthly Weather Review*, **111**, 609–662.

Hansen, J., Sato, M. and Ruedy, R. (1995a) Long-term changes in the diurnal temperature cycle: implications about mechanisms of global .climate change. *Atmospheric Research*, **37**, 175–209.

Hansen, J., Wilson, H., Sato, M., Ruedy, R., Shah, K. and Hansen, E. (1995b) Satellite and surface temperature data at odds? *Climatic Change*, **30**, 103–117.

Hansen, J., Ruedy, R. and Sato, M. (1996a) Global surface air temperature in 1995: return to pre-Pinatubo level. *Geophysical Research Letters*, **23**, 1665–1668.

Hansen, J. *et al.* (1996b) A Pinatubo climate modeling investigation, in Fiocco, G., Duà, D. and Visconti, G. (eds), *The Mount Pinatubo Eruption: Effects on the Atmosphere and Climate*, Springer-Verlag, Berlin, 233–272.

Hansen, J., Sato, M. and Ruedy, R. (1997a) Radiative forcing and climate response. *Journal of Geophysical Research*, **102**, 6831–6864.

Hansen, J., Sato, M., Lacis, A. and Ruedy, R. (1997b) The missing climate forcing. *Philosophical Transactions of the Royal Society*, **352**, 231–240.

Hansen, J. *et al.* (1997c) Forcings and chaos in interannual to decadal climate change. *Journal of Geophysical Research*, **102**, 25679–25720.

Hansen, J.E., Sato, M., Lacis, A., Ruedy, R., Tegen, I. and Matthews, E. (1998) Climate forcings in the industrial era. *Proceedings of the National Academy of Sciences USA*, **95**, 12753–12758.

Hargreaves, D., Eden-Green, M. and Devaney, J. (1994) *World Index of Resources and Population*, Dartmouth, Brookfield, Vermont.

Harnisch, J., Borchers, R., Fabian, P. and Maiss, M. (1996) Tropospheric trends for CF_4 and C_2F_6 since 1982 derived from SF_6 dated stratospheric air. *Geophysical Research Letters*, **23**, 1099–1102.

Harris, R.N. and Chapman, D.S. (1997) Borehole temperatures and a baseline for 20th-century global warming estimates. *Science*, **275**, 1618–1621.

Harvey, L.D.D. (1988a) Climatic impact of ice age aerosols. *Nature*, **334**, 333–335.

Harvey, L.D.D. (1988b) A semi-analytic energy balance climate model with explicit sea ice and snow physics. *Journal of Climate*, **1**, 1065–1085.

Harvey, L.D.D. (1989) Effect of model structure on the response of terrestrial biosphere models to CO_2 and temperature increases. *Global Biogeochemical Cycles*, **3**, 137–153.

Harvey, L.D.D. (1992) A two-dimensional ocean model for long-term climatic simulations: stability and coupling to atmospheric and sea ice models. *Journal of Geophysical Research*, **97**, 9435–9453.

Harvey, L.D.D. (1993) A guide to global warming potentials. *Energy Policy*, **21**, 24–34.

Harvey, L.D.D. (1994) Transient temperature and sea level response of a two-dimensional ocean–climate model to greenhouse gas increases. *Journal of Geophysical Research*, **99**, 18447–18466.

Harvey, L.D.D. (1995) Impact of isopycnal diffusion on heat fluxes and the transient response of a two-dimensional ocean model. *Journal of Physical Oceanography*, **25**, 2166–2176.

Harvey, L.D.D. (1996a) Development of a risk-hedging CO_2-emission policy, Part I: Risks of unrestrained emissions. *Climatic Change*, **34**, 1–40.

Harvey, L.D.D. (1996b) Development of a risk-hedging CO_2-emission policy, Part II: Risks associated with measures to limit emissions, synthesis, and conclusions. *Climatic Change*, **34**, 41–71.

Harvey, L.D.D. (1999) Transient response of a coupled climate–carbon cycle model to changes in ocean circulation. *Journal of Geophysical Research* (submitted).

Harvey, L.D.D. and Huang, Z. (1995) An evaluation of the potential impact of methane-clathrate destabilization on future global warming. *Journal of Geophysical Research*, **100**, 2905–2926.

Harvey, L.D.D. and Huang, Z. (1999) A quasi-one-dimensional coupled climate–carbon cycle model: description, calibration, and sensitivity. *Journal of Geophysical Research* (submitted).

Harvey, L.D.D. and Schneider, S.H. (1985) Transient climatic response to external forcing on 10^0–10^4 year time scales, 1, Experiments with globally averaged coupled atmosphere and ocean energy balance models. *Journal of Geophysical Research*, **90**, 2191–2205.

Harvey, L.D.D., Gregory, J., Hoffert, M., Jain, A., Lal, M., Leemans, R., Raper, S., Wigley, T. and de Wolde, J. (1997) *An Introduction to Simple Climate Models Used in the IPCC Second Assessment Report*, Intergovernmental Panel on Climate Change, Technical Paper II, 50 pages.

Hättenschwiler, S., Miglietta, F., Raschi, A. and Körner, C. (1997) Thirty years of in situ tree growth under elevated CO_2: a model for future forest repsonses? *Global Change Biology*, **3**, 463–471.

Hauglustaine, D.A., Granier, C., Brasseur, G.P. and Mégie, G. (1994) The importance of atmospheric chemistry in the calculation of radiative forcing on the climate system. *Journal of Geophysical Research*, **99**, 1173–1186.

Haywood, J.M. and Ramaswamy, V. (1998) Global sensitivity studies of the direct radiative forcing due to anthropogenic sulfate and black carbon aerosols. *Journal of Geophysical Research*, **103**, 6043–6058.

Haywood, J.M., Roberts, D.L., Slingo, A., Edwards, J.M. and Shine, K.P. (1997) General circulation model calculations of the direct radiative forcing by anthropogenic sulfate and fossil-fuel soot aerosol. *Journal of Climate*, **10**, 1562–1577.

Haywood, J.M., Schwarzkopf, M.D. and Ramaswamy, V. (1998) Estimates of radiative forcing due to modeled increases in tropospheric ozone. *Journal of Geophysical Research*, **103**, 16999–17007.

Hegerl, G.C., Hasselmann, K., Cubasch, U., Mitchell, J.F.B., Roeckner, E., Voss, R. and Waszkewitz, J. (1997) Multi-fingerprint detection and attribution analysis of greenhouse gas, greenhouse gas-plus-aerosol and solar forced climate change. *Climate Dynamics*, **13**, 613–634.

Hegg, D.A. (1994) Cloud condensation nucleus–sulphate mass relationship and cloud albedo. *Journal of Geophysical Research*, **99**, 25903–25907.

Heimann, M. and Maier-Reimer, E. (1994) On the relations between the uptake of carbon dioxide and its isotopes by the ocean. *Global Biogeochemical Cycles*, **10**, 89–110.

Heimann, M. and Monfray, P. (1989) *Spatial and Temporal Variation of the Gas Exchange Coefficient for CO_2: 1. Data Analysis and Global Validation*, Max-Planck-Institut für Meteorologie, Report No. 31, Hamburg, 33 pages.

Hein, R., Crutzen, P.J. and Heimann, H. (1997) An inverse modelling approach to investigate the global atmospheric methane cycle. *Global Biogeochemical Cycles*, **11**, 43–76.

Held, I.M. and Suarez, M.J. (1974) Simple albedo feedback models of the icecaps. *Tellus*, **26**, 613–630.

Henderson-Sellers, A., McGuffie, K. and Gross, C. (1995) Sensitivity of global climate model simulations to increased stomatal resistance and CO_2 increases. *Journal of Climate*, **8**, 1738–1756.

Henderson-Sellers, A. *et al.* (1998) Tropical cyclones and global climate change: a post-IPCC assessment. *Bulletin of the American Meteorological Society*, **79**, 19–38.

Hennessy, K.J., Gregory, J.M. and Mitchell, J.F.B. (1997) Changes in daily precipitation under enhanced greenhouse conditions. *Climate Dynamics*, **13**, 667–680.

Hense, A., Krahe, P. and Flohn, H. (1988) Recent fluctuations of tropospheric temperature and water vapour content in the tropics. *Meteorology and Atmospheric Physics*, **38**, 215–227.

Hirakuchi, H. and Giorgi, F. (1995) Multiyear present-day and $2 \times CO_2$ simulations of monsoon climate over eastern Asia and Japan with a regional climate model nested in a general circulation model. *Journal of Geophysical Research*, **100**, 21105–21125.

Hoffert, M.I. and Covey, C. (1992) Deriving global climate sensitivity from paleoclimate reconstructions. *Nature*, **360**, 573–576.

Hoffert, M.I., Callegari, A.J. and Hseih, C.-T. (1980) The role of deep sea heat storage in the secular response to climatic forcing. *Journal of Geophysical Research*, **85**, 6667–6679.

Hoffert, M.I., Callegari, A.J. and Hseih, C.-T. (1981) A box-diffusion carbon cycle model with upwelling, polar bottom water formation and a marine biosphere, in Bolin, B. (ed.), *Carbon Cycle Modeling, SCOPE 16*, John Wiley and Sons, New York, 287–305.

Hoffert, M.I., and Flannery, B.P. (1985) Model projections of the time-dependent response to increasing carbon dioxide, in MacCracken, M.C. and Luther, F.M. (eds), *Projecting the Climatic Effects of Increasing Carbon Dioxide*, DOE/ER-0237, US Department of Energy, Washington, 149–190.

Hoffert, M.I., Frei, A. and Narayanan, V.K. (1988) Application of solar max acrim data to analysis of solar-driven climatic variability on earth. *Climatic Change*, **13**, 267–286.

Hoffert, M.I. *et al.* (1998) Energy implications of future stabilization of atmospheric CO_2 content. *Nature*, **395**, 881–884.

Holdsworth, G., Higucji, K., Zielinski, G.A., Mayewski, P.A., Wahlen, M., Deck, B., Ch_lek, P., Johnson, B. and Damiano, P. (1996) Historical biomass burning: late 19th century pioneer agriculture revolution in northern hemisphere ice core data and its atmospheric interpretation. *Journal of Geophysical Research*, **101**, 23317–23334.

Houghton, R.A. (1987) Biotic changes consistent with the increased seasonal amplitude of atmospheric CO_2 concentrations. *Journal of Geophysical Research*, **92**, 4223–4230.

Houghton, R.A. (1996) Terrestrial sources and sinks of carbon inferred from terrestrial data. *Tellus*, **48**, 420–433.

Houghton, R.A. (1998) The annual net flux of carbon to the atmosphere from changes in land use, 1850–1990. *Tellus* (in press).

Houghton, R.A., Hobbie, J.E., Melillo, J.M., Moore, B., Peterson, B.J., Shaver, G.R. and Woodwell, G.M. (1983) Changes in the carbon content of terrestrial biota and soils between 1860 and 1980: a net release of CO_2 to the atmosphere. *Ecological Monographs*, **53**, 235–262.

Hovine, S. and Fichefet, T. (1994) A zonally averaged three-basin ocean circulation model for climate studies. *Climate Dynamics*, **10**, 313–331.

Hoyt, D.V. and Schatten, K.H. (1993) A discussion of plausible solar irradiance variations, 1700–1992. *Journal of Geophysical Research*, **98**, 18895–18906.

Hsieh, W.W. and Bryan, K. (1996) Redistribution of sea level rise associated with enhanced greenhouse warming: a simple model study. *Climate Dynamics*, **12**, 535–544.

Hulme, M. (1991) An intercomparison of model and observed global precipitation climatologies. *Geophysical Research Letters*, **18**, 1715–1718.

Hulme, M., Zhao, Z.-C. and Jiang, T. (1994) Recent and future climate change in East Asia. *International Journal of Climatology*, **14**, 637–658.

Hungate, B.A., Holland, E.A., Jackson, R.B., Chapin III, F.S., Mooney, H.A. and Field, C.B. (1997) The fate of carbon in grasslands under carbon dioxide enrichment. *Nature*, **388**, 576–579.

Hurrell, J.W. and Trenberth, K.E. (1997) Spurious trends in satellite MSU temperatures from merging different satellite records. *Nature*, **386**, 164–167.

Hurrell, J.W. and Trenberth, K.E. (1998) Difficulties in obtaining reliable temperature trends: reconciling the surface and satellite microwave sounding unit records. *Journal of Climate*, **11**, 945–967.

Huybrechts, P. and de Wolde, J. (1999) The dynamic response of the Greenland and Antarctic ice sheets to multiple-century climatic warming. *Journal of Climate* (submitted).

Huybrechts, P. and Oerlemans, J. (1990) Response of the Antarctic ice sheet to future greenhouse warming. *Climate Dynamics*, **5**, 93–102.

Huybrechts, P., Letreguilly, A. and Reeh, N. (1991) The Greenland ice sheet and greenhouse warming. *Palaeogeography, Palaeoclimatology, Palaeoecology*, **89**, 399–412.

Idso, K.E. and Idso, S.B. (1994) Plant responses to atmospheric CO_2 enrichment in the face of environmental constraints: a review of the past 10 years' research. *Agricultural and Forest Meteorology*, **69**, 153–203.

IEA (International Energy Agency) (1991a) *Greenhouse Gases, Abatement and Control*, International Energy Agency, Paris.

IEA (International Energy Agency) (1991b) *Greenhouse Gas Emissions. The Energy Dimension*, International Energy Agency, Paris.

Ishitani, H. *et al.* (1996) Energy supply mitigation options, in Watson, R.T., Zinyowera, M.C. and Moss, R.H. (eds), *Climate Change 1995 – Impacts, Adaptations and Mitigation of Climate Change: Scientific–Technical Analysis*, Cambridge University Press, Cambridge, 587–647.

Jaffe, D.A. (1992) The nitrogen cycle, in Butcher, S.S., Charlson, R.J., Orians, G.H. and Wolfe, G.V. (eds), *Global Biogeochemical Cycles*, Academic Press, London, 263–284.

Jain, A.K., Kheshgi, H.S., Hoffert, M.I. and Wuebbles, D.J. (1995) Distribution of radiocarbon as a test of global carbon cycle models. *Global Biogeochemical Cycles*, **9**, 153–166.

Jain, A.K., Kheshgi, H.S. and Wuebbles, D.J. (1996) A globally aggregated reconstruction of cycles of carbon and its isotopes. *Tellus*, **48B**, 583–600.

Jaques, A.P. (1992) *Canada's Greenhouse Gase Emissions: Estimates for 1990*, Report EPS 5/AP/4, Environment Canada, Ottawa, 78 pages.

Jenkins, A., Helliwell, R.C., Swingewood, P.J., Sefton, C., Renshaw, M. and Ferrier, R.C. (1998). Will reduced sulphur emissions under the Second Sulphur Protocol lead to recovery of acid sensitive sites in UK? *Environmental Pollution*, **99**, 309–318.

Jiang, Y. and Yung, Y.L. (1996) Concentrations of tropospheric ozone from 1979 to 1992 over tropical Pacific South America from TOMS data. *Science*, **272**, 714–716.

Johns, T.C., Carnell, R.E., Crossley, J.F., Gregory, J.M., Mitchell, J.F.B., Senior, C.A., Tett, S.F.B. and Wood, R.A. (1997) The second Hadley Centre coupled ocean–atmosphere GCM: model description, spinup and validation. *Climate Dynamics*, **13**, 103–134.

Johnsen, S.J., Dansgaard, W. and White, J.W.C. (1989) The origin of Arctic precipitation under present and glacial conditions. *Tellus*, **41B**, 452–468.

Johnson, C., Henshaw, J. and McInnes, G. (1992) Impact of aircraft and surface emissions of nitrogen oxides on tropospheric ozone and global warming. *Nature*, **355**, 69–71.

Jonas, P.R., Charlson, R.J. and Rodhe, H. (1995) Aerosols, in Houghton, J.T., Filho, L.G.M., Bruce, J., Lee, H., Callander, B.A., Haites, E., Harris, N. and Maskell, K. (eds), *Climate Change 1994: Radiative Forcing of Climate Change and an Evaluation of the IPCC IS92 Emission Scenarios*, Cambridge University Press, Cambridge, 127–162.

Jones, A. and Slingo, A. (1996) Predicting cloud-droplet effective radius and indirect sulphate aerosol forcing using a general circulation model. *Quarterly Journal of the Royal Meteorological Society*, **122**, 1573–1595.

Jones, A., Roberts, D.L. and Slingo, A. (1994) A climate model study of indirect radiative forcing by anthropogenic sulphate aerosols. *Nature*, **370**, 450–453.

Jones, P.D. (1988) Hemispheric surface air temperature variations: recent trends and an update to 1987. *Journal of Climate*, **1**, 654–660.

Jones, P.D. (1994) Hemispheric surface air temperature variations: a reanalysis and an update to 1993. *Journal of Climate*, **7**, 1794–1802.

Jones, P.D., Raper, S.C.B., Bradley, R.S., Diaz, H.F., Kelly, P.M. and Wigley, T.M.L. (1986a) Southern Hemisphere surface air temperature variations, 1985–1984. *Journal of Climate and Applied Meteorology*, **25**, 161–179.

Jones, P.D., Raper, S.C.B., Bradley, R.S., Diaz, H.F., Kelly, P.M. and Wigley, T.M.L. (1986b) Northern Hemisphere surface air temperature variations, 1985–1984. *Journal of Climate and Applied Meteorology*, **25**, 1213–1230.

Jones, P.D., Wigley, T.M.L. and Wright, P.B. (1986c) Global temperature variations between 1861–1984. *Nature*, **322**, 430–434.

Jones, P.D., Groisman, P. Ya., Coughlan, M., Plummer, N., Wang, W.-C. and Karl, T.R. (1990) Assessment of urbanization effects in time series of surface air temperature over land. *Nature*, **347**, 169–172.

Jones, P.D., Wigley, T.M.L. and Farmer, G. (1991) Marine and land temperature data sets: a comparison and a look at recent trends, in Schlesinger, M.E. (ed.), *Greenhouse-gas-induced Climatic Change: A Critical Appraisal of Simulations and Observations*, Developments in Atmospheric Science, **19**, Elsevier, Amsterdam, 153–172.

Jones, P.D., Osborn, T.J. and Briffa, K.R. (1997) Estimating sampling errors in large-scale temperature averages. *Journal of Climate*, **10**, 2548–2568.

Kane, R.P. (1997) Relationship of El Niño–Southern Oscillation and Pacific sea surface temperature with rainfall in various regions of the globe. *Monthly Weather Review*, **125**, 1792–1800.

Karl, T.R. and Knight, R.W. (1998) Secular trends of precipitation amount, frequency, and intensity in the United States. *Bulletin of the American Meteorological Society*, **79**, 231–241.

Karl, T.R., Knight, R.W. and Christy, J.R. (1994) Global and hemispheric temperature trends: uncertainties related to inadequate sampling. *Journal of Climate*, **7**, 1144–1163.

Karl, T.R., Knight, R.W. and Plummer, N. (1995) Trends in high-frequency climate variability in the twentieth century. *Nature*, **377**, 217–220.

Karl, T.R., Knight, R.W., Easterling, D.R. and Quayle, R.G. (1996) Indices of climate change for the United States. *Bulletin of the American Meteorological Society*, **77**, 279–292.

Kashiwagi, T. *et al.* (1996) Industry, in Watson, R.T., Zinyowera, M.C. and Moss, R.H. (eds), *Climate Change 1995 – Impacts, Adaptations and Mitigation of Climate Change: Scientific–Technical Analysis*, Cambridge University Press, Cambridge, 649–677.

Kasibhatla, P., Chameides, W.L. and St John, J. (1997) A three-dimensional global model investigation of the seasonal variation in the atmospheric burden of anthropogenic sulfate aerosols. *Journal of Geophysical Research*, **102**, 3737–3759.

Kattenberg, A., Giorgi, F., Grassl, H., Meehl, G.A., Mitchell, J.F.B., Stouffer, R.J., Tokioka, T., Weaver, A.J. and Wigley, T.M.L. (1996) Climate models – projections of future climate, in Houghton, J.T., Filho, L.G.F., Callander, B.A., Harris, N., Kattenberg, A. and Maskell, K. (eds), *Climate Change 1995: The Science of Climate Change*, Cambridge University Press, Cambridge, 285–357.

Kaufmann, R.K. and Stern, D.I. (1997) Evidence for human influence on climate from hemispheric temperature relations. *Nature*, **388**, 39–44.

Keeling, C.D. and Whorf, T.P. (1998) Atmospheric CO_2 records from sites in the SIO air sampling network, in *Trends: A Compendium of Data on Global Change*, Carbon Dioxide Information Analysis Center, Oak Ridge National Laboratory, Oak Ridge, Tennessee.

Keeling, C.D., Whorf, T.P., Wahlen, M. and van der Plicht, J. (1995) Interannual extremes in the rate of rise of atmospheric carbon dioxide since 1980. *Nature*, **375**, 666–670.

Keeling, R.F., Piper, S.C. and Heimann, M. (1996) Global and hemispheric CO_2 sinks deduced from changes in atmospheric O_2 concentration. *Nature*, **381**, 218–221.

Keller, M., Jacob, D.J., Wofsy, S.C. and Harris, R.C. (1991) Effects of tropical deforestation on global and regional atmospheric chemistry. *Climatic Change*, **19**, 139–158.

Kellogg, W.W. and Zhao, Z.-C. (1988) Sensitivity of soil moisture to doubling of carbon dioxide in climate model experiments. Part I: North America. *Journal of Climate*, **1**, 348–366.

Kelly, P.M. and Wigley, T.M.L. (1992) Solar cycle length, greenhouse forcing and global climate. *Nature*, **360**, 328–330.

Kheshgi, H.S. and Lapenis, A.G. (1996) Estimating the accuracy of Russian paleotemperature reconstructions. *Palaeogeography, Palaeoclimatology, Palaeoecology*, **121**, 221–237.

Kiehl, J.T. and Ramanathan, V. (1982) Radiative heating due to increased CO_2: the role of H_2O continuum absorption in the 12–18 μm region. *Journal of the Atmospheric Sciences*, **39**, 2923–2926.

Kiehl, J.T., Schneider, T.L., Rasch, P.J., Barth, M.C. and Wong, J. (1999) Radiative forcing due to sulfate aerosols from simulations with the NCAR Community Climate Model (CCM3). *Journal of Geophysical Research* (submitted).

Kiladis, G.N. and Diaz, H.F. (1989) Global climatic anomalies associated with extremes in the Southern Oscillation. *Journal of Climate*, **2**, 1069–1090.

Kirchgessner, D.A., Piccot, S.D. and Winkler, J.D. (1993) Estimate of global methane emissions from coal mines. *Chemosphere*, **26**, 453–472.

Knutson, T.R. and Manabe, S. (1998) Model assessment of decadal variability and trends in the tropical Pacific Ocean. *Journal of Climate*, **11**, 2273–2296.

Knutson, T.R., Manabe, S. and Gu, D. (1997) Simulated ENSO in a global coupled ocean–atmosphere model: multidecadal amplitude modulation and CO_2 sensitivity. *Journal of Climate*, **10**, 138–161.

Knutson, T.R., Tuleya, R.E. and Kurihara, Y. (1998) Simulated increase of hurricane intensities in a CO_2-warmed climate. *Science*, **279**, 1018–1020.

Ko, M.K.W., Sze, N.D., Wang, W.-C., Shia, G., Goldman, A., Murcray, F.J., Murcray, D.G. and Rinsland, C.P. (1993) Atmospheric sulfur hexafluoride: sources, sinks, and greenhouse warming. *Journal of Geophysical Research*, **98**, 10499–10507.

Kogan, Z.N., Kogan, Y.L. and Lilly, D.K. (1996) Evaluation of sulfate aerosols indirect effect in marine stratocumulus clouds using observation-derived cloud climatology. *Geophysical Research Letters*, **23**, 1937–1940.

Körner, C. and Miglietta, F. (1994) Long term effects of naturally elevated CO_2 on Mediterranean grassland and forest trees. *Oecologia*, **99**, 343–351.

Krol, M., van Leeuwen, P.J. and Lelieveld, J. (1998) Global OH trend inferred from methylchloroform measurements. *Journal of Geophysical Research*, **103**, 10697–10711.

Kuo, H.L. (1974) Further studies of the parameterization of the influence of cumulus convection on large-scale flow. *Journal of the Atmospheric Sciences*, **31**, 1232–1240.

Lal, M. and Ramanathan, V. (1984) The effects of moist convection and water vapor processes on climate sensitivity. *Journal of the Atmospheric Sciences*, **41**, 2238–2249.

Lambert, S.J. (1995) The effect of enhanced greenhouse warming on winter cyclone frequencies and strengths. *Journal of Climate*, **8**, 1447–1452.

Lambert, S.J. (1996) Intense extratropical northern hemisphere winter cyclone events: 1899–1991. *Journal of Geophysical Research*, **101**, 21319–21325.

Landsea, C.W., Nicholls, N., Gray, W.M. and Avila, L.A. (1996) Downward trends in the frequency of intense Atlantic hurricanes during the past five decades. *Geophysical Research Letters*, **23**, 1697–1700.

Langner, J. and Rodhe, H. (1991) A global three-dimensional model of the tropospheric sulfur cycle. *Journal of Atmospheric Chemistry*, **13**, 225–263.

Lapenis, A.G. and Shabalova, M.V. (1994) Global climate changes and moisture conditions in the intracontinental arid zones. *Climatic Change*, **27**, 259–297.

Larsen, J.C., Chiou, E.W., Chu, W.P., McCormick, M.P., McMaster, L.R., Oltmans, S. and Rind, D. (1993) A comparison of the Stratospheric Aerosol and Gas Experiment II tropospheric water vapor to radiosonde measurements. *Journal of Geophysical Research*, **98**, 4897–4917.

Law, K.S. and Nisbet, E.G. (1996) Sensitivity of the CH_4 growth rate to changes in CH_4 emissions from natural gas and coal. *Journal of Geophysical Research*, **101**, 14387–14397.

Le Treut, H. and Li, Z.-X. (1991) Sensitivity of an atmospheric general circulation model to prescribed SST changes: feedback effects associated with the simulation of cloud optical properties. *Climate Dynamics*, **5**, 175–187.

Le Treut, H., Li, Z.X. and Forichon, M. (1994) Sensitivity of the LMD general circulation model to greenhouse forcing associated with two different cloud water parameterizations. *Journal of Climate*, **7**, 1827–1841.

Lean, J., Beer, J. and Bradley, R. (1995) Reconstruction of solar irradiance since 1610: implications for climate change. *Geophysical Research Letters*, **22**, 3195–3198.

Lee, W.-H., Iacobellis, S.F. and Somerville, R.C.J. (1997) Cloud radiative forcings and feedbacks: general circulation model tests and observational validation. *Journal of Climate*, **10**, 2479–2496.

Leggett, J., Pepper, W.J. and Swart, R.J. (1992) Emissions scenarios for the IPCC: an update, in Houghton, J.T., Callander, B.A. and Varney, S.K. (eds), *Climate Change 1992: The Supplementary Report to the IPCC Scientific Assessment*, Cambridge University Press, Cambridge, 69–95.

Lelieveld, J. and Crutzen, P.J. (1992) Indirect chemical effects of methane on climate warming. *Nature*, **355**, 339–341.

Lelieveld, J., Crutzen, P.J. and Dentener, F.J. (1998) Changing concentration, lifetime and climate forcing of atmospheric methane. *Tellus*, **50B**, 128–150.

Lenderink, G. and Haarsma, R.J. (1994) Variability and multiple equilibria of the thermohaline circulation associated with deep-water formation. *Journal of Physical Oceanography*, **24**, 1480–1493.

Letréguilly, A., Huybrechts, P. and Reeh, N. (1991) Steady-state characteristics of the Greenland ice sheet under different climates. *Journal of Glaciology*, **37**, 149–157.

Levine, M.D. *et al.* (1996) Mitigation options for human settlements, in Watson, R.T., Zinyowera, M.C. and Moss, R.H. (eds), *Climate Change 1995 – Impacts, Adaptations and Mitigation of Climate Change: Scientific–Technical Analysis*, Cambridge University Press, Cambridge, 713–743.

Levitus, S. (1982) *Climatological Atlas of the World Ocean*, NOAA Professional Paper 13, Washington, DC, 173 pages.

Levitus, S., Conkright, M.E., Reid, J.L., Najjar, R.G. and Mantyla, A. (1993) Distribution of nitrate, phosphate and silicate in the world oceans. *Progress in Oceanography*, **31**, 245–273.

Levy II, H., Kasibhatla, P.S., Moxim, W.J., Klonecki, A.A., Hirsh, J.I., Oltmans, S.J. and Chameides, W.L. (1997) The global impact of human activity on tropospheric ozone. *Geophysical Research Letters*, **24**, 791–794.

Lighthill, J., Holland, G., Gray, W., Landsea, C., Craig, G., Evans, J., Kurihara, Y. and Guard, C. (1994) Global climate change and tropical cyclones. *Bulletin of the American Meteorological Society*, **75**, 2147–2157.

Lindzen, R. (1990) Some coolness concerning global warming. *Bulletin of the American Meteorological Society*, **71**, 288–299.

Lindzen, R. and Giannitsis, C. (1998) On the climatic impli-

cations of volcanic cooling. *Journal of Geophysical Research*, **103**, 5929–5941.

Liousse, C., Penner, J.E., Chuang, C., Walton, J.J., Eddleman, H. and Cachier, H. (1996) A global three-dimensional model study of carbonaceous aerosols. *Journal of Geophysical Research*, **101**, 19411–19432.

Liss, P.S. and Merlivat, L. (1986) Air–sea gas exchange rates: introduction and synthesis, in Buat-Menard, P. (ed.), *The Role of Air–Sea Exchange in Geochemical Cycling*, Reidel, 113–127.

Liu, Z. (1998) The role of ocean in the response of tropical climatology to global warming: the west–east SST contrast. *Journal of Climate*, **11**, 864–875.

Logan, J.A. (1994) Trends in the vertical distribution of ozone: an analysis of ozonesonde data. *Journal of Geophysical Research*, **99**, 25553–25585.

Lohmann, U. and Feichter, J. (1997) Impact of sulfate aerosols on albedo and lifetime of clouds: a sensitivity study with the ECHAM4 GCM. *Journal of Geophysical Research*, **102**, 13685–13700.

Lohmann, U. and Roeckner, E. (1995) Influence of cirrus cloud radiative forcing on climate and climate sensitivity in a general circulation model. *Journal of Geophysical Research*, **100**, 16305–16323.

Lohmann, U., Feichter, J., Chuang, C. and Penner, J.E. (1999) Predicting the number of cloud droplets in the ECHAM GCM. *Journal of Geophysical Research* (in press).

Loutre, M.-F. (1995) Greenland ice sheet over the next 5000 years. *Geophysical Research Letters*, **22**, 783–786.

Luo, Y., Jackson, R.B., Field, C.B. and Mooney, H.A. (1996) Elevated CO_2 increases belowground respiration in California grasslands. *Oecologia*, **108**, 130–137.

Luther, F.M. and Ellingson, R.G. (1985) Carbon dioxide and the radiation budget, in MacCracken, M.C. and Luther, F.M. (eds.), *Projecting the Climatic Effects of Increasing Carbon Dioxide*, US Department of Energy, DOE/ER-0237, 25–55.

Luxmore, R.J., Wullschleger, S.D. and Hanson, P.J. (1993) Forest responses to CO_2 enrichment and climate warming. *Water, Air, and Soil Pollution*, **70**, 309–323.

Lynch-Stieglitz, M. (1994) The development and validation of a simple snow model for the GISS GCM. *Journal of Climate*, **7**, 1842–1855.

MacAyeal, D.R. (1992) Irregular oscillations of the West Antarctic ice sheet. *Nature*, **359**, 29–32.

Machida, T., Nakazawa, T., Fujii, Y., Aoki, S. and Watanabe, O. (1995) Increase in the atmospheric nitrous oxide concentration during the last 250 years. *Geophysical Research Letters*, **22**, 2921–2924.

Maier-Reimer, E., Mikolajewicz, U. and Winguth, A. (1996) Future ocean uptake of CO_2: interactions between ocean circulation and biology. *Climate Dynamics*, **12**, 711–721.

Manabe, S. and Bryan, K. (1985) CO_2-induced change in a coupled ocean–atmosphere model and its paleoclimatic implications. *Journal of Geophysical Research*, **90**, 11689–11707.

Manabe, S. and Stouffer, R.J. (1994) Multiple-century response of a coupled ocean–atmosphere model to an increase of atmospheric carbon dioxide. *Journal of Climate*, **7**, 5–23.

Manabe, S. and Stouffer, R.J. (1997) Climate variability of a coupled ocean–atmosphere–land surface model: implications for the detection of global warming. *Bulletin of the American Meteorological Society*, **78**, 1177–1185.

Manabe, S., Smagorinsky, J. and Strickler, R.F. (1965) Simulated climatology of a general circulation model with a hydrologic cycle. *Monthly Weather Review*, **93**, 769–798.

Manabe, S., Stouffer, R.J., Spelman, M.J. and Bryan, K. (1991) Transient responses of a coupled ocean–atmosphere model to gradual changes of atmospheric CO_2. Part I: Annual mean response. *Journal of Climate*, **4**, 785–818.

Manabe, S., Spelman, M.J. and Stouffer, R.J. (1992) Transient responses of a coupled ocean–atmosphere model to gradual changes of atmospheric CO_2. Part II: Seasonal response. *Journal of Climate*, **5**, 105–126.

Mann, M.E., Bradley, R.S. and Hughes, M.K. (1998) Global-scale temperature patterns and climate forcing over the past six centuries. *Nature*, **392**, 779–787.

Marenco, A. and Said, F. (1989) Meridional and vertical ozone distribution in the background troposphere (70 N–60 S; 0–12 km altitude) from scientific aircraft measurements during the Stratoz III experiment (June 1984). *Atmospheric Environment*, **23**, 185–200.

Marenco, A., Gouget, H., Nédélec, P., Pagés, J.P. and Karcher, F. (1994) Evidence of a long-term increase in tropospheric ozone from the Pic du Midi series: consequences: positive radiative forcing. *Journal of Geophysical Research*, **99**, 16617–16632.

Marland, G., Boden, T.A., Andres, R.J., Brenkert, A.L. and Johnston, C. (1998) Global, regional, and national CO_2 emissions, in *Trends: A Compendium of Data on Global Change*, Carbon Dioxide Information Analysis Center, Oak Ridge National Laboratory, Oak Ridge, Tennessee.

Marshall, S., Oglesby, R.J., Larson, J.W. and Saltzman, B. (1994) A comparison of GCM sensitivity to changes in CO_2 and solar luminosity. *Geophysical Research Letters*, **21**, 2487–2490.

Martin, J.H. and Fitzwater, S.E. (1992) Dissolved organic carbon in the Atlantic, Southern and Pacific oceans. *Nature*, **356**, 699–700.

Maslanik, J.A., Serreze, M.C. and Barry, R.G. (1996) Recent decreases in Arctic summer ice cover and linkages to atmospheric circulation changes. *Geophysical Research Letters*, **23**, 1677–1680.

Mauzerall, D.L. *et al.* (1998) Photochemistry of biomass burning plumes and implications for tropospheric ozone over the tropical South Atlantic. *Journal of Geophysical Research*, **103**, 8401–8423.

McConnaughay, K.D.M., Bassow, S.L., Berntson, G.M. and Bazzaz, F.A. (1996) Leaf senescence and decline of end-of-season gas exchange in five temperate deciduous tree species grown in elevated CO_2 concentrations. *Global Change Biology*, **2**, 25–33.

McGuire, A.D., Melillo, J.M., Joyce, L.A., Kicklighter, D.W., Grace, A.L., Moore III, B. and Vorosmarty, C.J. (1992) Interactions between carbon and nitrogen dynamics in estimating net primary productivity for potential vegetation in North America. *Global Biogeochemical Cycles*, **6**, 101–124.

McKane, R.B., Rastetter, E.B., Melillo, J.M., Shaver, G.S., Hopkinson, C.S., Fernandes, D.N., Skole, D.L. and Chomentowski, W.H. (1995) Effects of global change on carbon storage in tropical forests of South America. *Global Biogeochemical Cycles*, **9**, 329–350.

Meehl, G.A. and Washington, W.M. (1988) A comparison of soil-moisture sensitivity in two global climate models. *Journal of the Atmospheric Sciences*, **45**, 1476–1492.

Meehl, G.A. and Washington, W.M. (1990) CO_2 climate sensitivity and snow–sea-ice albedo parameterization in an atmospheric GCM coupled to a mixed-layer ocean model. *Climatic Change*, **16**, 283–306.

Meehl, G.A. and Washington, W.M. (1993) South Asian summer monsoon variability in a model with doubled atmospheric carbon dioxide concentration. *Science*, **260**, 1101–1103.

Meehl, G.A. and Washington, W.M. (1995) Climate change from increased CO_2 and direct and indirect effects of sulphate aerosols. *Geophysical Research Letters*, **23**, 3755–3758.

Meehl, G.A. and Washington, W.M. (1996) El Niño-like climate change in a model with increased atmospheric CO_2 concentrations. *Nature*, **382**, 56–60.

Meehl, G.A., Branstator, G.W. and Washington, W.M. (1993a) Tropical Pacific interannual variability and CO_2 climate change. *Journal of Climate*, **6**, 42–63.

Meehl, G.A., Washington, W.M. and Karl, T. (1993b) Low-frequency variability and CO_2 transient climate change. Part 1. Time-averaged differences. *Climate Dynamics*, **8**, 117–133.

Meehl, G.A., Washington, W.M., Erickson III, D.J., Briegleb, B.P. and Jaumann, P.J. (1996) Climate change from increased CO_2 and direct and indirect effects of sulfate aerosols. *Geophysical Research Letters*, **23**, 3755–3758.

Meehl, G.A., Boer, G.J., Covey, C., Latif, M. and Stouffer, R.J. (1997) Intercomparison makes for a better climate model. *EOS*, **78**, 445–451.

Melillo, J.M., McGuire, A.D., Kicklighter, D.W., Moore III, B., Vorosmarty, C.J. and Schloss, A.L. (1993) Global climate change and terrestrial net primary production. *Nature*, **363**, 234–240.

Melillo, J.M., Prentice, I.C., Farquhar, G.D., Schulze, E.-D. and Sala, O.E. (1996) Terrestrial biotic responses to environmental change and feedbacks to climate, in Houghton, J.T., Filho, L.G.F., Callander, B.A., Harris, N., Kattenberg, A. and Maskell, K. (eds), *Climate Change 1995: The Science of Climate Change*, Cambridge University Press, Cambridge, 445–481.

Mercer, J.H. (1978) West Antarctic ice sheet and CO_2 greenhouse effect: a threat of disaster. *Nature*, **271**, 321–325.

Michaelis, L. *et al.* (1996) Mitigation options in the transportation sector, in Watson, R.T., Zinyowera, M.C. and Moss, R.H. (eds), *Climate Change 1995 – Impacts, Adaptations and Mitigation of Climate Change: Scientific–Technical Analysis*, Cambridge University Press, Cambridge, 679–712.

Mikolajewicz, U., Santer, B.D. and Maier-Reimer, E. (1990) Ocean response to greenhouse warming. *Nature*, **345**, 589–593.

Miller, K.G., Fairbanks, R.G. and Mountain, G.S. (1987) Tertiary oxygen isotope synthesis, sea level history, and continental margin erosion. *Paleoceanography*, **2**, 1–19.

Miller, R.L. and Del Genio, A.D. (1994) Tropical cloud feedbacks and natural variability of climate. *Journal of Climate*, **7**, 1388–1402.

Milly, P.C.D. (1992) Potential evaporation and soil moisture in general circulation models. *Journal of Climate*, **5**, 209–226.

Minschwaner, K., Carver, R.W., Briegleb, B.P. and Roche, A.E. (1998) Infrared radiative forcing and atmospheric lifetimes of trace species based on observations from UARS. *Journal of Geophysical Research*, **103**, 23243–23253.

Misra, P.K., Bloxam, R., Fung, C. and Wong, S. (1989) Non-linear response of wet deposition to emissions reduction: a model study. *Atmospheric Environment*, **23**, 671–687.

Mitchell, J.F.B. and Ingram, W.J. (1992) Carbon dioxide and climate: mechanisms of changes in cloud. *Journal of Climate*, **5**, 5–21.

Mitchell, J.F.B. and Johns, T.C. (1997) On modifications of global warming by sulfate aerosols. *Journal of Climate*, **10**, 245–267.

Mitchell, J.F.B. and Warrilow, D.A. (1987) Summer dryness in northern mid-latitudes due to increased CO_2. *Nature*, **330**, 238–240.

Mitchell, J.F.B., Wilson, C.A. and Cunnington, W.M. (1987) On CO_2 climate sensitivity and model dependence of results. *Quarterly Journal of the Royal Meteorological Society*, **113**, 293–322.

Mitchell, J.F.B., Manabe, S., Meleshko, V. and Tokioka, T. (1990) Equilibrium climate change – and its implications for the future, in Houghton, J.T., Jenkins, G.J. and Ephraums, J.J. (eds), *Climate Change: The IPCC Scientific Assessment*, Cambridge University Press, Cambridge, 131–172.

Mitchell, J.F.B., Johns, T.C., Gregory, J.M. and Tett, S.F.B. (1995) Climate response to increasing levels of greenhouse gases and sulphate aerosols. *Nature*, **376**, 501–504.

Mohan, M. and Kumar, S. (1998) Review of acid rain potential in India: future threats and remedial measures. *Current Science*, **75**, 579–593.

Monteith, J.L. (1965) Evaporation and environment. *Symposium of the Society of Experimental Biology*, **XIX**, 205–234.

Montzka, S.A., Butler, J.H., Myers, R.C., Thompson, T.M., Swanson, T.H., Clarke, A.D., Lock, L.T. and Elkins, J.W. (1996) Decline in the tropospheric abundance of halogen from halocarbons: implications for stratospheric ozone depletion. *Science*, **272**, 1318–1322.

Mopper, K., Zhou, X., Kieber, R.J., Kieber, D.J., Sikorski, R.J. and Jones, R.D. (1991) Photochemical degradation of dissolved organic carbon and its impact on the oceanic carbon cycle. *Nature* **353**, 60–62.

Morantine, M. and Watts, R.G. (1990) Upwelling diffusion climate models: analytical solutions for radiative and upwelling forcing. *Journal of Geophysical Research*, **95**, 7563–7571.

Mosier, A.R., Duxbury, J.M., Freney, J.R., Heinemeyer, O., Kinami, K. and Johnson, D.E. (1998a) Mitigating agricultural emissions of methane. *Climatic Change*, **40**, 39–80.

Mosier, A.R., Duxbury, J.M., Freney, J.R., Heinemeyer, O. and Kinami, K. (1998b) Assessing and mitigating N_2O emissions from agricultural soils. *Climatic Change*, **40**, 7–38.

Murphy, J.M. (1995) Transient response of the Hadley Centre coupled ocean–atmosphere model to increasing carbon dioxide. Part III: Analysis of global-mean response using simple models. *Journal of Climate*, **8**, 496–514.

Murphy, J.M. and Mitchell, J.F.B. (1995) Transient response of the Hadley Centre coupled ocean–atmosphere model to increasing carbon dioxide. Part II: Spatial and temporal structure of response. *Journal of Climate*, **8**, 57–80.

Myers, N. (1989) *Deforestation Rates in Tropical Forests and their Climatic Implications*, Friends of the Earth, London.

Myhre, G. and Stordal, F. (1997) Role of spatial and temporal variations in the computation of radiative forcing and GWP. *Journal of Geophysical Research*, **192**, 11181–11200.

Myhre, G., Highwood, E.J., Shine, K.P. and Stordal, F. (1998a) New estimates of radiative forcing due to well mixed greenhouse gases. *Geophysical Research Letters*, **25**, 2715–2718.

Myhre, G., Stordal, F., Restad, K. and Isaksen, I.S.A. (1998b) Estimation of the direct radiative forcing due to sulfate and soot aerosols. *Tellus*, **50B**, 463–477.

Myneni, R.B., Keeling, C.D., Tucker, C.J., Asrar, G. and Nemani, R.R. (1997) Increased plant growth in the northern high latitudes from 1981–1991. *Nature*, **386**, 698–702.

Mysak, L.A. and Manak, D.K. (1989) Arctic sea-ice extent and anomalies, 1953–1984. *Atmosphere–Ocean*, **27**, 376–405.

Najjar, R.G., Sarmiento, J.L. and Toggweiler, J.R. (1992) Downward transport and fate of organic matter in the ocean: simulations with a general circulation model. *Global Biogeochemical Cycles*, **6**, 45–76.

Nakamura, M., Stone, P.H. and Marotzke, J. (1994) Destabilization of the thermohaline circulation by atmospheric eddy transports. *Journal of Climate*, **7**, 1870–1882.

Nakicenovic, N., Grübler, A., Ishitani, H., Johansson, T., Marland, G., Moreira, J.R. and Rogner, H.-H. (1996) Energy primer, in Watson, R.T., Zinyowera, M.C., Moss, R.H. and Dokken, D.J. (eds), *Climate Change 1995 – Impacts, Adaptations and Mitigation of Climate Change: Scientific–Technical Analysis*, Cambridge University Press, Cambridge, 75–92.

National Acid Precipitation Assessement Program (NAPAP) (1987). *Volume II: Emissions and Control*, Washington D.C.

Nedoluha, G.E., Bevilacqua, R.M., Gomez, R.M., Siskind, D.E., Hicks, B.C., Russell III, J.M. and Connor, B.J. (1998) Increases in middle atmospheric water vapor as observed by the Halogen Occultation Experiment and the ground-based Water Vapor Millimeter-wave Spectrometer from 1991 to 1997. *Journal of Geophysical Research*, **103**, 3531–3543.

Nevison, C.D., Weiss, R.F. and Erickson, D.J. (1995) Global oceanic emissions of nitrous oxide. *Journal of Geophysical Research*, **100**, 15809–15820.

Newell, R.E. (1978) *The Global Circulation of the Atmosphere*, Royal Meteorological Society, Bracknell, England.

Newman, M.J. and Rood, R.T. (1977) Implications of solar evolution for the Earth's early atmosphere. *Science*, **198**, 1035–1037.

Nicholls, K.W. (1997) Predicted reduction in basal melt rates of an Antarctic ice shelf in a warmer climate. *Nature*, **388**, 460–462.

Nicholls, N., Gruza, G.V., Jouzel, J., Karl, T.R., Ogallo, L.A. and Parker, D.E. (1996) Observed climate variability and change, in Houghton, J.T., Filho, L.G.F., Callander, B.A., Harris, N., Kattenberg, A. and Maskell, K. (eds), *Climate Change 1995: The Science of Climate Change*, Cambridge University Press, Cambridge, 133–192.

Nimitz, J.S. and Skaggs, S.R. (1992) Estimating tropospheric lifetimes and ozone-depletion potentials of one- and two-carbon hydrofluorocarbons and hydrochlorofluorocarbons. *Environmental Science and Technology*, **26**, 739–744.

Norby, R.J., Gunderson, C.A., Wullschleger, S.D., O'Neill, E.G., and MacCracken, M.K. (1992) Productivity and compensatory response of yellow-poplar trees in elevated CO_2. *Nature*, **357**, 322–324.

Novelli, P.C., Masarie, K.A., Tans, P.P. and Lang, P.M. (1994) Recent changes in atmospheric carbon monoxide. *Science*, **263**, 1587–1590.

Novelli, P.C., Masarie, K.A. and Lang, P.M. (1998) Distributions and recent changes of carbon monoxide in the lower troposphere. *Journal of Geophysical Research*, **103**, 19015–19033.

O'Dowd, C.D., Lowe, J.A., Smith, M.H., Davison, B., Hewitt, C.N. and Harrison, R.M. (1997) Biogenic sulphur emissions and inferred non-sea-salt-sulphate cloud condensation nuclei in and around Antarctica. *Journal of Geophysical Research*, **102**, 12839–12854.

Oechel, W.C., Hastings, S.J., Vourlitis, G., Jenkins, M., Riechers, G. and Grulke, N. (1993) Recent change of Arctic tundra ecosystems from a net carbon dioxide sink to a source. *Nature*, **361**, 520–523.

Oechel, W.C., Cowles, S., Grulke, N., Hastings, S.J., Lawrence, B., Prudhomme, T., Riechers, G., Strain, B., Tissue, D. and Vourlitis, G. (1994) Transient nature of CO_2 fertilization in Arctic tundra. *Nature*, **371**, 500–503.

Oerlemans, J. (1994) Quantifying global warming from retreat of glaciers. *Science*, **264**, 243–245.

Oerlemans, J. and Fortuin, J.P.F. (1992) Sensitivity of glaciers and small ice caps to greenhouse warming. *Science*, **258**, 115–117.

Oglesby, R.J. and Saltzman, B. (1992) Equilibrium climate statistics of a general circulation model as a function of atmospheric carbon dioxide. Part I: Geographic distributions of primary variables. *Journal of Climate*, **5**, 66–92.

Ohmura, A. (1987) New temperature distribution maps for Greenland. *Zeitschrift für Gletscherkunde und Glazialgeologie*, **23**, 1–45.

Ohmura, A., Wild, M. and Bengtsson, L. (1996) A possible change in mass balance of Greenland and Antarctic ice sheets in the coming century. *Journal of Climate*, **9**, 2124–2135.

Oltmans, S.J. and Hofmann, D.J. (1995) Increase in lower-stratospheric water vapour at a mid-latitude Northern Hemisphere site from 1981 to 1994. *Nature*, **374**, 146–149.

Oort, A.H. and Liu, H. (1993) Upper-air temperature trends over the globe, 1958–1989. *Journal of Climate*, **6**, 292–307.

Oppenheimer, M. (1998) Global warming and the stability of the West Antarctic Ice Sheet. *Nature*, **393**, 325–332.

Oram, D.E., Sturges, W.T., Penkett, S.A., McCulloch, A. and Fraser, P.J. (1998) Growth of fluoroform (CHF_3, HFC-23) in the background atmosphere. *Geophysical Research Letters*, **25**, 35–38.

Orr, J.C. (1993) Accord between ocean models predicting uptake of anthropogenic CO_2. *Water, Air, and Soil Pollution*, **70**, 465–481.

Osborn, T.J. and Wigley, T.M.L. (1994) A simple model for estimating methane concentrations and lifetime variations. *Climate Dynamics*, **9**, 181–193.

Overland, J.E., Walter, B.A., Curtin, T.B. and Turet, P. (1995) Hierarchy and sea ice mechanics: a case study from the Beaufort Sea. *Journal of Geophysical Research*, **100**, 4559–4571.

Pan, W., Tatang, M.A., McRae, G.J. and Prinn, R.G. (1997) Uncertainty analysis of direct radiative forcing by anthropogenic sulfate aerosols. *Journal of Geophysical Research*, **102**, 21915–21924.

Pan, W., Tatang, M.A., McRae, G.J. and Prinn, R.G. (1998) Uncertainty analysis of indirect radiative forcing by anthropogenic sulfate aerosols. *Journal of Geophysical Research*, **103**, 3815–3823.

Papasavva, S., Tai, S., Illinger, K.H. and Kenny, J.E. (1997) Infrared radiative forcing of CFC substitutes and their atmospheric reaction products. *Journal of Geophysical Research*, **102**, 13643–13650.

Parker, D.E. (1994) Effets of changing exposure of thermometers at land stations. *International Journal of Climatology*, **14**, 1–31.

Parker, D.E. and Cox, D.I. (1995) Towards a consistent global climatological rawinsonde data-base. *International Journal of Climatology*, **15**, 473–496.

Parker, D.E. and Folland, C.K. (1991) Worldwide surface temperature trends since the mid-19th century, in Schlesinger, M.E. (ed.), *Greenhouse-gas-induced Climatic Change: A Critical Appraisal of Simulations and Observations*, Developments in Atmospheric Science, **19**, Elsevier, Amsterdam, 173–193.

Parker, D.E., Folland, C.K. and Jackson, M. (1995) Marine surface temperature: observed variations and data requirements. *Climatic Change*, **31**, 559–600.

Pastor, J. and Post, W.M. (1988) Response of northern forests to CO_2-induced climate change. *Nature*, **334**, 55–58.

Pearce, D.W., Cline, W.R., Achanta, A.N., Fankhauser, S., Pachauri, R.K., Tol, R.S.J. and Vellinga, P. (1996) The social costs of climate change: greenhouse damage and the benefits of control, in Bruce, J.P., Lee, H. and Haites, E.F. (eds), *Climate Change 1995 – Economic and Social Dimensions of Climate Change*, Cambridge University Press, Cambridge, 179–224.

Peltier, W.R. and Tushingham, A.M. (1989) Global sea level rise and the greenhouse effect: might they be connected? *Science*, **244**, 806–810

Peng, L., Chou, M.-D. and Arking, A. (1982) Climate studies with a multi-layer energy balance model. Part I: Model description and sensitivity to the solar constant. *Journal of the Atmospheric Sciences*, **39**, 2639–2656.

Peng, T.-H., Takahashi, T., Broecker, W.S. and Olafsson, J. (1987) Seasonal variability of carbon dioxide, nutrients and oxygen in the northern North Atlantic surface water: observations and a model. *Tellus*, **39B**, 439–458.

Penman, H.L. (1948) Natural evaporation from open water, bare soil and grass. *Proceedings of the Royal Society of London*, **A193**, 120–145.

Pham, M., Müller, J.-F., Brasseur, G.P., Granier, C. and Mégie, G. (1996) A 3D model study of the global sulphur cycle: contributions of anthropogenic and biogenic sources. *Atmospheric Environment*, **30**, 1815–1822.

Philander, S.G., Pacanowski, R.C., Lau, N.-C. and Nath, M.J. (1992) Simulation of ENSO with a global atmospheric GCM coupled to a high-resolution, tropical Pacific Ocean GCM. *Journal of Climate*, **5**, 308–329.

Pinnock, S., Hurley, M.D., Shine, K.P., Wallington, T.J. and Smyth, T.J. (1995) Radiative forcing of climate by hydrochlorofluorocarbons and hydrofluocarbons. *Journal of Geophysical Research*, **100**, 23227–23238.

Plöchl, M. and Cramer, W. (1995) Coupling global models of vegetation structure and ecosystem processes. *Tellus*, **47B**, 240–250.

Pollard, D. and Thompson, S.L. (1994) Sea-ice dynamics and CO_2 sensitivity in a global climate model. *Atmosphere–Ocean*, **32**, 449–467.

Poorter, H. (1993) Interspecific variation in the growth response of plants to an elevated ambient CO_2 concentration. *Vegetatio*, **104/105**, 77–97.

Portmann, R.W., Solomon, S., Fishman, J., Olson, J.R., Kiehl, J.T. and Briegleb, B. (1997) Radiative forcing of the Earth's climate system due to tropical tropospheric ozone production. *Journal of Geophysical Research*, **102**, 9409–9417.

Post, W.M., Pastor, J., King, A.W. and Emanuel, W.R. (1992) Aspects of the interaction between vegetation and soil under global change. *Water, Air, and Soil Pollution*, **64**, 345–363.

Prabhakara, C., Iacovazzi Jr, R., Yoo, J.-M. and Dalu, G. (1998) Global warming deduced from MSU. *Geophysical Research Letters*, **25**, 1927–1930.

Prather, M., Ibrahim, A.M., Sasaki, T. and Stordal, F. (1992) Future chlorine–bromine loading and ozone depletion, in *Scientific Assessment of Ozone Depletion: 1991*, World Meteorological Organization, Geneva.

Prather, M., Derwent, R., Ehhalt, D., Fraser, P., Sanhueza, E. and Zhou, X. (1995) Other trace gases and atmospheric chemistry, in Houghton, J.T., Filho, L.G.M., Bruce, J., Lee, H., Callander, B.A., Haites, E., Harris, N. and Maskell, K. (eds), *Climate Change 1994: Radiative Forcing of Climate Change and an Evaluation of the IPCC IS92 Emission Scenarios*, Cambridge University Press, Cambridge, 73–126.

Princiotta, F.T. (1991) Pollution control for utility power generation, 1990–2020, in Tester, J.W., Wood, D.O. and Ferrari, N.A. (eds), *Energy and the Environment in the 21st Century*, MIT Press, Cambridge, MA, 631–650.

Prinn, R.G., Weiss, R.F., Miller, B.R., Huang, J., Alyea, F.N., Cunnold, D.M., Fraser, P.J., Hartley, D.E. and Simmonds, P.G. (1995) Atmospheric trends and lifetime of CH_3CCl_3 and global OH concentrations. *Science*, **269**, 187–192.

Qian, J.J. and Zhang, K.M. (1998) China's desulfurization potential. *Energy Policy*, **26**, 345–351.

Quay, P.D., Tilbrook, B. and Wong, C.S. (1992) Oceanic uptake of fossil fuel CO_2: carbon-13 evidence. *Science*, **256**, 74–79.

Rahmstorf, S. (1995) Bifurcations of the Atlantic thermohaline circulation in response to changes in the hydrological cycle. *Nature*, **378**, 145–149.

Rahmstorf, S. and Willebrand, J. (1995) The role of temperature feedback in stabilizing the thermohaline circulation. *Journal of Physical Oceanography*, **25**, 787–805.

Ramanathan, V. (1977) Interactions between ice-albedo, lapse-rate and cloud-top feedbacks: an analysis of the nonlinear response of a GCM climate model. *Journal of the Atmospheric Sciences*, **34**, 1885–1897.

Ramanathan, V., Cicerone, R.J., Singh, H.B. and Kiehl, J.T. (1985) Trace gas trends and their potential role in climate change. *Journal of Geophysical Research*, **90**, 5547–5566.

Ramaswamy, V., Schwarzkopf, M.D. and Randel, W.J. (1996) Fingerprint of ozone depletion in the spatial and temporal pattern of recent lower-stratospheric cooling. *Nature*, **382**, 616–618.

Randall, D.A. *et al.* (1994) Analysis of snow feedbacks in 14 general circulation models. *Journal of Geophysical Research*, **99**, 20757–20771.

Randall, D.A. *et al.* (1996) A revised land surface parameterization (SiB2) for atmospheric GCMs. Part III: The greening of the Colorado State University general circulation model. *Journal of Climate*, **9**, 738–763.

Randall, D.A. *et al.* (1998) Status of and outlook for large-scale modelling of atmosphere–ice–ocean interactions in the Arctic. *Bulletin of the American Meteorological Society*, **79**, 197–219.

Raper, S.C.B., Briffa, K.R. and Wigley, T.M.L. (1996) Glacier change in northern Sweden from AD500: a simple geometric model of Storglaciären. *Journal of Glaciology*, **42**, 341–351.

Rasch, P.J., Barth, M.C., Kiehl, J.T., Schwartz, S.E. and Benkowitz, C.M. (1999) A description of the global sulfur cycle and its controlling processes in the NCAR CCM3. *Journal of Geophysical Research* (submitted).

Rastetter, E.B., Ryan, M.G., Shaver, G.R., Mclillo, J.M., Nadelhoffer, K.J., Hobbie, J. and Aber, J.D. (1991) A general biogeochemical model describing the responses of the C and N cycles in terrestrial ecosystems to changes in CO_2, climate, and N deposition. *Tree Physiology*, **9**, 101–126.

Rastetter, E.B., McKane, R.B., Shaver, G.R. and Melillo, J.M. (1992) Changes in C storage by terrestrial ecosystems: how C–N interactions restrict responses to CO_2 and temperature. *Water, Air, and Soil Pollution*, **64**, 327–344.

Rennó, N.O., Emanuel, K.A. and Stone, P.H. (1994a) Radiative–convective model with an explicit hydrologic cycle 1. Formulation and sensitivity to model parameters. *Journal of Geophysical Research*, **99**, 14429–14441.

Rennó, N.O., Emanuel, K.A. and Stone, P.H. (1994b) Radiative–convective model with an explicit hydrologic cycle 2. Sensitivity to large changes in solar forcing. *Journal of Geophysical Research*, **99**, 17001–17020.

Repetto, R. and Austin, D. (1997) *The Costs of Climate Protection: A Guide for the Perplexed*, World Resources Institute, Washington, 51 pages.

Reynolds, R.W. (1988) A real-time global sea surface temperature analysis. *Journal of Climate*, **1**, 75–86.

Rind, D. (1987) The doubled CO_2 climate: impact of the sea surface temperature gradient. *Journal of the Atmospheric Sciences*, **44**, 3235–3268.

Rind, D. (1988) The doubled CO_2 climate and the sensitivity of the modeled hydrologic cycle. *Journal of Geophysical Research*, **93**, 5385–5412.

Rind, D., Goldberg, R., Hansen, J., Rosenzweig, C. and Ruedy, R. (1990) Potential evapotranspiration and the likelihood of future drought. *Journal of Geophysical Research*, **95**, 9983–10004.

Rind, D., Healy, R., Parkinson, C. and Martinson, D. (1995) The role of sea ice in $2 \times CO_2$ climate model sensitivity. Part I: The total influence of sea ice thickness and extent. *Journal of Climate*, **8**, 449–463.

Robock, A. and Free, M.P. (1995) Ice cores as an index of global volcanism from 1850 to the present. *Journal of Geophysical Research*, **100**, 11549–11567.

Robock, A. and Mao, J. (1995) The volcanic signal in surface temperature observations. *Journal of Climate*, **8**, 1086–1103.

Rodhe, H. and Crutzen, P. (1995) Climate and CCN. *Nature*, **375**, 111.

Roelofs, G.-J., Lelieveld, J. and Dorland, R. (1997) A three-dimensional chemistry/general circulation model simulation of anthropogenically derived ozone in the troposphere and its radiative climate forcing. *Journal of Geophysical Research*, **102**, 23389–23401.

Roelofs, G.-J., Lelieveld, J. and Ganzeveld, L. (1998) Simulation of global sulfate distribution and the influence on effective cloud drop radii with a coupled photochemistry–sulfur cycle model. *Tellus*, **50B**, 224–242.

Rosa, L.P. and Schaeffer, R. (1995) Global warming potentials: the case of emissions from dams. *Energy Policy*, **23**, 149–158.

Ryan, M.G. (1991) Effects of climate change on plant respiration. *Ecological Applications*, **1**, 157–167.

Sagan, C., Toon, O.B. and Pollak, J.B. (1979) Anthropogenic albedo changes and the Earth's climate. *Science*, **206**, 1363–1368.

Sahagian, D.L., Schwartz, F.W. and Jacobs, D.K. (1994) Direct anthropogenic contributions to sea level rise in the twentieth century. *Nature*, **367**, 54–57.

Salathé, E.P., Chesters, D. and Sud, Y.C. (1995) Evaluation of the upper-tropospheric moisture climatology in a general circulation model using TOVS radiance observations. *Journal of Climate*, **8**, 2404–2414.

Sandroni, S., Anfossi, D. and Viarengo, S. (1992) Surface ozone levels at the end of the nineteenth century in South America. *Journal of Geophysical Research*, **97**, 2535–2539.

Santer, B.D. *et al.* (1996) A search for human influences on the thermal structure of the atmosphere. *Nature*, **382**, 39–46.

Sarmiento, J.L. and Le Quéré, C. (1996) Oceanic carbon dioxide uptake in a model of century-scale global warming. *Science*, **274**, 1346–1350.

Sarmiento, J.L. and Sundquist, E.T. (1992) Revised budget for the oceanic uptake of anthropogenic carbon dioxide. *Nature*, **356**, 589–593.

Sarmiento, J.L., Orr, J.C. and Siegenthaler, U. (1992) A perturbation simulation of CO_2 uptake in an ocean general circulation model. *Journal of Geophysical Research*, **97**, 3621–3645.

Sarmiento, J.L., Slater, R.D., Fasham, M.J.R., Ducklow, H.W., Toggweiler, J.R. and Evans, G.T. (1993) A seasonal three-dimensional ecosystem model of nitrogen cycling in the North Atlantic euphotic zone. *Global Biogeochemical Cycles*, **7**, 417–450.

Sato, M., Hansen, J.E., McCormick, M.P. and Pollack, J.B. (1993) Stratospheric aerosol optical depths, 1850–1990. *Journal of Geophysical Research*, **98**, 22987–22994.

Sato, N., Sellers, P.J., Randall, D.A., Schneider, E.K., Shukla, J., Kinter III, J.L., Hou, Y.-T. and Albertazzi, E. (1989) Effects of implementing the simple biosphere model in a general circulation model. *Journal of the Atmospheric Sciences*, **46**, 2757–2782.

Schiller, A., Mikolajewicz, U. and Voss, R. (1997) The stability of the North Atlantic thermohaline circulation in a coupled ocean–atmosphere general circulation model. *Climate Dynamics*, **13**, 325–347.

Schimel, D. *et al.* (1996) Radiative forcing of climate change, in Houghton, J.T., Filho, L.G.F., Callander, B.A., Harris, N., Kattenberg, A. and Maskell, K. (eds), *Climate Change 1995: The Science of Climate Change*, Cambridge University Press, Cambridge, 65–131.

Schlesinger, M.E. (1985) Appendix A: Analysis of results from energy balance and radiative convective models, in MacCracken, M.C. and Luther, F.M. (eds), *Projecting the Climatic Effects of Increasing Carbon Dioxide*, US Department of Energy, DOE/ER-0237, 281–319.

Schlesinger, M.E. and Ramankutty, N. (1992) Implications for global warming of intercycle solar irradiance variations. *Nature*, **360**, 330–333.

Schlesinger, M.E. and Ramankutty, N. (1994) An oscillation in the global climate system of period 65–70 years. *Nature*, **367**, 723–726.

Schlesinger, W.H. (1991) *Biogeochemistry: An Analysis of Global Change*, First Edition, Academic Press, San Diego, 443 pages.

Schlesinger, W.H. (1997) *Biogeochemistry: An Analysis of Global Change*, Second Edition, Academic Press, San Diego, 588 pages.

Schmetz, J. and van de Berg, L. (1994) Upper tropospheric humidity observations from Meteosat compared with short-term forecast fields. *Journal of Geophysical Research*, **21**, 573–576.

Schmitz, W.J. (1995) On the interbasin-scale thermohaline circulation. *Reviews of Geophysics*, **33**, 151–173.

Schneider, S.H. and Turner, B.L. (1995) Anticipating Global Change Surprise, in Hassol, S.J. and Katzenburger, J. (eds), Elements of Change 1994, Aspen Global Change Institute, Aspen, Colorado.

Schöenwise, C.-D., Ullrich, R., Beck, F. and Rapp, J. (1994) Solar signals in global climatic change. *Climatic Change*, **27**, 259–281.

Scholes, R.J., Ward, D.E. and Justice, C.O. (1996) Emissions of trace gases and aerosol particles due to vegetation burning in southern hemisphere Africa. *Journal of Geophysical Research*, **101**, 23677–23682.

Schult, I., Feichter, J. and Cooke, W.F. (1997) Effect of black carbon and sulfate aerosols on the global radiation budget. *Journal of Geophysical Research*, **102**, 30107–30117.

Schwerdtfeger, W. (1970) The climate of the Antarctic, in Orvig, S. (ed.), *Climates of the Polar Regions*, World Survey of Climatology, Volume 14, Elsevier, Amsterdam, 253–355.

Seager, R. and Murtugudde, R. (1997) Ocean dynamics, thermocline adjustment, and regulation of tropical SST. *Journal of Climate*, **10**, 521–534.

Sebenius, J.K. (1991) Designing negotiations toward a new regime: the case of global warming. *International Security*, **15**, 110–148.

Seitz, F. (1994) *Global Warming and Ozone Hole Controversies: A Challenge to Scientific Judgment*, George C. Marshall Institute, Washington, 34 pages.

Sellers, P.J., Mintz, Y., Sud, Y.C. and Dalcher, A. (1986) The design of a simple biosphere model (SiB) for use within General Circulation Models. *Journal of the Atmospheric Sciences*, **43**, 505–531.

Sellers, P.J. *et al.* (1996a) A revised land surface parameterization (SiB2) for atmospheric GCMs. Part I: Model formulation. *Journal of Climate*, **9**, 676–705.

Sellers, P.J. *et al.* (1996b) A revised land surface parameterization (SiB2) for atmospheric GCMs. Part II: The generation of global fields of terrestrial biophysical parameters from satellite data. *Journal of Climate*, **9**, 706–737.

Sellers, P.J., *et al.* (1996c) Comparison of radiative and physiological effects of doubled atmospheric CO_2 on climate. *Science*, **271**, 1402–1406.

Sellers, P.J. *et al.* (1997) Modeling the exchanges of energy, water, and carbon between continents and the atmosphere. *Science*, **275**, 502–509.

Senior, C.A. and Mitchell, J.F.B. (1993) Carbon dioxide and climate: the impact of cloud parameterization. *Journal of Climate*, **6**, 393–418.

Shindell, D.T., Rind, D. and Lonergan, P. (1998) Increased polar stratospheric ozone losses and delayed eventual recovery owing to increasing greenhouse-gas concentrations. *Nature*, **392**, 589–592,

Shine, K.P. and Henderson-Sellers, A. (1985) The sensitivity of a thermodynamic sea ice model to changes in surface albedo parameterization. *Journal of Geophysical Research*, **90**, 2243–2250.

Shine, K.P., Derwent, R.G., Wuebbles, D.J. and Morcrette, J.-J. (1990) Radiative forcing of climate, in Houghton, J.T., Jenkins, G.J. and Ephraums, J.J. (eds), *Climate Change: The IPCC Scientific Assessment*, Cambridge University Press, Cambridge, 41–68.

Shine, K.P., Fouquart, Y., Ramaswamy, V., Solomon, S. and Srinivasan, J. (1995) Radiative forcing, in Houghton, J.T., Filho, L.G.M., Bruce, J., Lee, H., Callander, B.A., Haites, E., Harris, N. and Maskell, K. (eds), *Climate Change 1994: Radiative Forcing of Climate Change and an Evaluation of the IPCC IS92 Emission Scenarios*, Cambridge University Press, Cambridge, 163–203.

Siegenthaler, U. and Joos, F. (1992) Use of a simple model for studying oceanic tracer distributions and the global carbon cycle. *Tellus*, **44B**, 186–207.

Siegenthaler, U. and Oeschger, H. (1987) Biospheric CO_2 emissions during the past 200 years reconstructed from deconvolution of ice core data. *Tellus*, **39B**, 140–154.

Siegenthaler, U. and Sarmiento, J.L. (1993) Atmospheric carbon dioxide and the ocean. *Nature*, **365**, 119–125.

Sinha, A. and Allen, M.R. (1994) Climate sensitivity and tropical moisture distribution. *Journal of Geophysical Research*, **99**, 3707–3716.

Sinha, A. and Shine, K.P. (1994) A one-dimensional study of possible cirrus cloud feedbacks. *Journal of Climate*, **7**, 158–173.

Smith, D.M. (1998) Recent increase in the length of the melt season of perennial Arctic sea ice. *Geophysical Research Letters*, **25**, 655–658.

Smith, I.N., Dix, M. and Allan, R.J. (1997) The effect of greenhouse SSTs on ENSO simulations with an AGCM. *Journal of Climate*, **10**, 342–352.

Soden, B.J. (1997) Variations in the tropical greenhouse effect during El Niño. *Journal of Climate*, **10**, 1050–1055.

Soden, B.J. and Bretherton, F.P. (1994) Evaluation of water vapor distribution in general circulation models using satellite observations. *Journal of Geophysical Research*, **99**, 1187–1210.

Soden, B.J. and Fu, R. (1995) A satellite analysis of deep convection, upper-tropospheric humidity, and the greenhouse effect. *Journal of Climate*, **8**, 2333–2351.

Soden, B.J. and Lanzante, J.R. (1996) An assessment of satellite and radiosonde climatologies of upper-tropospheric water vapor. *Journal of Climate*, **9**, 1235–1250.

Solanki, S.K. and Fligge, M. (1998) Solar irradiance since 1874 revisited. *Geophysical Research Letters*, **25**, 341–344.

Solomon, A.M. (1986) Transient response of forests to CO_2-induced climate change: simulation modeling experiments in eastern North America. *Oecologia*, **68**, 567–579.

Solomon, S. and Daniel, J.S. (1996) Impact of the Montreal Protocol and its amendments on the rate of change of global radiative forcing. *Climatic Change*, **32**, 7–17.

Somerville, R. and Remer, L. (1984) Cloud optical thickness feedbacks in the CO_2 climate problem. *Journal of Geophysical Research*, **89**, 9668–9672.

Spencer, R.W. and Braswell, W.D. (1997) How dry is the tropical free troposphere? Implications for global warming theory. *Bulletin of the American Meteorological Society*, **78**, 1097–1106.

Spiro, P.A., Jacob, D.J. and Logan, J.A. (1992) Global inventory of sulfur emissions with 1°×1° resolution. *Journal of Geophysical Research*, **97**, 6023–6036.

Stager, J.C. and Mayewski, P.A. (1997) Abrupt early to mid-Holocene climatic transition registered at the equator and the poles. *Science*, **276**, 1834–1836.

Stammerjohn, S.E. and Smith, R.C. (1997) Opposing Southern Ocean climate patterns as revealed by trends in regional sea ice coverage. *Climatic Change*, **37**, 617–639.

Stenchikov, G.L. and Robock, A. (1995) Diurnal asymmetry of climatic response to increased CO_2 and aerosols: forcings and feedbacks. *Journal of Geophysical Research*, **100**, 26211–26227.

Stephenson, D.B. and Held, I.M. (1993) GCM response of northern winter stationary waves and storm tracks to increasing amounts of carbon dioxide. *Journal of Climate*, **6**, 1859–1870.

Stern, D.I. and Kaufmann, R.K. (1996a) Estimates of global anthropogenic methane emissions 1860–1993. *Chemosphere*, **33**, 159–176.

Stern, D.I. and Kaufmann, R.K. (1996b) *Estimates of Global Anthropogenic Sulfate Emissions 1860–1993*, CEES Working Paper 9602, Center for Energy and Environmental Studies, Boston University, Boston, MA.

Stocker, T.F. and Schmittner, A. (1997) Influence of CO_2 emission rates on the stability of the thermohaline circulation. *Nature*, **388**, 862–865.

Stocker, T.F., Broecker, W.S. and Wright, D.G. (1994) Carbon uptake experiment with a zonally averaged global ocean circulation model. *Tellus*, **46B**, 103–122.

Stouffer, R.J., Manabe, S. and Bryan, K. (1989) Interhemispheric asymmetry in climate response to a gradual increase of atmospheric CO_2. *Nature*, **342**, 660–662.

Stouffer, R.J., Manabe, S. and Vinnikov, K.Ya. (1994) Model assessment of the role of natural variability in recent global warming. *Nature*, **367**, 634–636.

Stouffer, R.J., Hegerl, G. and Tett, S. (1999) A comparison of surface air temperature variability in three 1000-year coupled ocean–atmosphere model integrations. *Journal of Climate* (submitted).

Suhre, K., Andreae, M.O. and Rosset, R. (1995) Biogenic sulfur emissions and aerosols over the tropical South Atlantic 2. One-dimensional simulation of sulfur chemistry in the marine boundary layer. *Journal of Geophysical Research*, **100**, 11323–11334.

Sun, D.-Z. and Held, I.M. (1996) A comparison of modeled and observed relationships between interannual variations of water vapor and temperature. *Journal of Climate*, **9**, 665–675.

Sun, D.-Z. and Lindzen, R.S. (1993) Distribution of tropical tropospheric water vapor. *Journal of the Atmospheric Sciences*, **50**, 1643–1660.

Sun, D.Z. and Oort, A.H. (1995) Humidity–temperature relationships in the tropical troposphere. *Journal of Climate*, **8**, 1974–1987.

Suppiah, R. (1995) The Australian summer monsoon: CSIRO9 GCM simulations for $1 \times CO_2$ and $2 \times CO_2$ conditions. *Global and Planetary Change*, **11**, 95–109.

Suppiah, R. and Hennessy, K.J. (1998) Trends in total rainfall, heavy rain events and number of dry days in Australia, 1910–1990. *International Journal of Climatology*, **18**, 1141–1164.

Syncrude Canada (1998) *Environmental Progress Summary*, http://www.syncrude.com/5_env/progress_98/so2.htm.

Takahashi, T., Broecker, W.S. and Bainbridge, A.E. (1981) The alkalinity and total carbon dioxide concentration in the world oceans, in Bolin, B. (ed.), *Carbon Cycle Modelling*, SCOPE 16, John Wiley, Chichester, 271–286.

Tans, P.P., Fung, I.Y. and Takahashi, T. (1990) Observational constraints on the global atmospheric CO_2 budget. *Science*, **247**, 1431–1438.

Tans, P.P., Berry, J.A. and Keeling, R.F. (1993) Oceanic $^{13}C/^{12}C$ observations: a new window on oceanic CO_2 uptake. *Global Biogeochemical Cycles*, **7**, 353–368.

Tarasick, D.W., Wardle, D.I., Kerr, J.B., Bellefleur, J.J. and Davies, J. (1995) Tropospheric ozone trends over Canada: 1980–1993. *Geophysical Research Letters*, **22**, 409–412.

Tegen, I. and Fung, I. (1995) Contribution to the mineral aerosol load from land surface modification. *Journal of Geophysical Research*, **11**, 18707–18726.

Tegen, I., Hollrig, P., Chin, M., Fung, I., Jacob, D. and Penner, J. (1997) Contribution of different aerosol species to the global aerosol extinction optical thickness: estimates from model results. *Journal of Geophysical Research*, **102**, 23895–23915.

Tett, S.F.B. (1995) Simulation of El Niño–Southern Oscillation-like variability in a global AOGCM and its response to CO_2 increase. *Journal of Climate*, **8**, 1473–1502.

Tett, S.F.B., Mitchell, J.F.B., Parker, D.E. and Allen, M.R. (1996) Human influence on the atmospheric vertical temperature structure: detection and observations. *Science*, **274**, 1170–1173.

Thomas, R.H., Sanderson, T.J.O. and Rose, K.E. (1979) Effect of climate warming on the West Antarctic ice sheet. *Nature*, **277**, 355–358.

Tiedtke, M. (1989) A comprehensive mass flux scheme for cumulus parameterization in large-scale models. *Monthly Weather Review*, **117**, 1779–1800.

Timbal, B., Mahfouf, J.-F., Royer, J.-F., Cubasch, U. and Murphy, J.M. (1997) Comparison between doubled CO_2 time-slice and coupled experiments. *Journal of Climate*, **10**, 1463–1469.

Toumi, R., Bekki, S. and Law, K.S. (1994) Indirect influences of ozone depletion on climate forcing by clouds. *Nature*, **372**, 348–351.

Townsend, A.R., Vitousek, P.M. and Holland, E.A. (1992) Tropical soils could dominate the short-term carbon cycle feedbacks to increased global temperatures. *Climatic Change*, **22**, 293–303.

Trenberth, K.E. and Hoar, T.J. (1996) The 1990–1995 El Niño–Southern Oscillation event: longest on record. *Geophysical Research Letters*, **23**, 57–60.

Trenberth, K.E. and Hoar, T.J. (1997) El Niño and climatic change. *Geophysical Research Letters*, **24**, 3057–3060.

Trenberth, K.E. and Hurrell, W.K. (1997) How accurate are satellite thermometers? Reply. *Nature*, **389**, 342–343.

Trenberth, K.E., Houghton, J.T. and Meira Filho, L.G. (1996) The climate system: an overview, in Houghton, J.T., Filho, L.G.F., Callander, B.A., Harris, N., Kattenberg, A. and Maskell, K. (eds), *Climate Change 1995: The Science of Climate Change*, Cambridge University Press, Cambridge, 51–64.

Tziperman, E. (1997) Inherently unstable climate behaviour due to weak thermohaline ocean circulation. *Nature*, **386**, 592–595.

United Nations (1994) Protocol to the 1979 Convention on Long-Range Transboundary Air Pollution on Further Reductions of Sulphur Emissions. *International Legal Materials*, **33**, 1540–1555.

Van de Wal, R.S.W. and Oerlemans, J. (1997) Modelling the short-term response of the Greenland ice-sheet to global warming. *Climate Dynamics*, **13**, 733–744.

Van Dorland, R., Dentener, F.J. and Lelieveld, J. (1997) Radiative forcing due to tropospheric ozone and sulfate aerosols. *Journal of Geophysical Research*, **102**, 28079–28100.

Van Minnen, J.G., Goldewijk, K.K. and Leemans, R. (1996) The importance of feedback processes and vegetation transition in the terrestrial carbon cycle. *Journal of Biogeography*, **22**, 805–814.

Vaughan, D.G. and Doake, C.S.M. (1996) Recent atmospheric warming and retreat of ice shelves on the Antarctic Peninsula. *Nature*, **379**, 328–331.

VEMAP Members (1995) Vegetation ecosystem modeling and analysis project: comparing biogeography and biogeochemistry models in a continental-scale study of terrestrial ecosystem responses to climate change and CO_2 doubling. *Global Biogeochemical Cycles*, **9**, 407–437.

Verbitsky, M. and Oglesby, R.J. (1995) The CO_2-induced thickening/thinning of the Greenland and Antarctic ice sheets as simulated by a GCM (CCM1) and an ice sheet model. *Climate Dynamics*, **11**, 247–253.

Verbitsky, M. and Saltzman, B. (1995) Behavior of the East Antarctic ice sheet as deduced from a coupled GCM/ice-sheet model. *Geophysical Research Letters*, **22**, 2913–2916.

Vinnikov, K. Ya. and Yeserkepova, I.B. (1991) Soil moisture: empirical data and model results. *Journal of Climate*, **4**, 66–79.

Vinnikov, K., Groisman, P.Ya., Lugina, K.M. and Golubev, A.A. (1987) Variations in North Hemisphere mean surface air temperature over 1881–1985. *Meteorology and Hydrology*, **1**, 45–53 (in Russian).

Vinnikov, K., Groisman, P.Ya. and Lugina, K.M. (1990) Empirical data on contemporary global climate changes (temperature and precipitation). *Journal of Climate*, **3**, 662–677.

Vinnikov, K. Ya., Robock, A., Stouffer, R.J. and Manabe, S. (1996) Vertical patterns of free and forced climate variations. *Geophysical Research Letters*, **23**, 1801–1804.

Vose, R.S., Schmoyer, R.L., Steurer, P.M., Pterson, T.C., Heim, R., Karl, T.R. and Eischeid, J. (1992) *The Global Historical Climatology Network: Long-term Monthly Temperature, Precipitation, Sea Level Pressure, and Station Pressure Data*, Carbon Dioxide Information and Analysis Center, Oak Ridge National Laboratory, Oak Ridge, Tennessee, Report ORNL/CDIAC-53, NDP-041.

Waelbroeck, C., Monfray, P., Oechel, W.C., Hastings, S. and Vourlitis, G. (1997) The impact of permafrost thawing on the carbon dynamics of tundra. *Geophysical Research Letters*, **24**, 229–232.

Walsh, J.E. (1995) Long-term observations for monitoring of the cryosphere. *Climatic Change*, **31**, 369–394.

Wang, W.-C., Dudek, M.P., Liang, X.-Z. and Kiehl, J.T. (1991) Inadequacy of effective CO_2 as a proxy in simulating the greenhouse effect of other radiatively active gases. *Nature*, **350**, 573–577.

Wang, Y., Jacob, D.J. and Logan, J.A. (1998a) Global simulation of tropospheric O_3–NO_x–hydrocarbon chemistry 1. Model formulation. *Journal of Geophysical Research*, **103**, 10713–10725.

Wang, Y., Logan, J.A. and Jacob, D.J. (1998b) Global simulation of tropospheric O_3–NO_x–hydrocarbon chemistry 2. Model evaluation and global ozone budget. *Journal of Geophysical Research*, **103**, 10727–10755.

Wang, Y., Jacob, D.J. and Logan, J.A. (1998c) Global simulation of tropospheric O_3–NO_x–hydrocarbon chemistry 3. Origin of tropospheric ozone and effects of nonmethane hydrocarbons. *Journal of Geophysical Research*, **103**, 10757–10767.

Warrick, R.A., Le Provost, C., Meier, M.F., Oerlemans, J. and Woodworth, P.L. (1996) Changes in sea level, in Houghton, J.T., Filho, L.G.F., Callander, B.A., Harris, N., Kattenberg, A. and Maskell, K. (eds), *Climate Change 1995: The Science of Climate Change*, Cambridge University Press, Cambridge, 359–405.

Washington, W.M. and Meehl, G.A. (1989) Climate sensitivity due to increased CO_2: experiments with a coupled atmosphere and ocean general circulation model. *Climate Dynamics*, **4**, 1–38.

Washington, W.M. and Meehl, G.A. (1993) Greenhouse sensitivity experiments with penetrative cumulus convection and tropical cirrus albedo effects. *Climate Dynamics*, **8**, 211–223.

Washington, W.M. and Meehl, G.A. (1996) High-latitude climate change in a global coupled ocean–atmosphere–sea ice model with increased CO_2. *Journal of Geophysical Research*, **101**, 12795–12801.

Watson, A. (1994) Air–sea gas exchange and carbon dioxide, in Heimman, M. (ed.), *The Global Carbon Cycle*, Springer-Verlag, Berlin, 397–412.

Watterson, I.G. (1997) The diurnal cycle of surface air temperature in simulated present and doubled CO_2 climates. *Climate Dynamics*, **13**, 533–545.

Watterson, I.G., O'Farrell, S.P. and Dix, M.R. (1997) Energy and water transport in climates simulated by a general circulation model that includes dynamic sea ice. *Journal of Geophysical Research*, **102**, 11027–11037.

Weaver, A.J. and Sarachik, E.S. (1991) Evidence for decadal variability in an ocean general circulation model: an advective mechanism. *Atmosphere–Ocean*, **29**, 197–231.

Weaver, A.J., Sarachik, E.S. and Marotzke, J. (1991) Internal low frequency variability of the ocean's thermohaline circulation. *Nature*, **353**, 836–838.

Wentz, F.J. and Schabel, M. (1998) Effects of orbital decay on satellite-derived lower-tropospheric temperature trends. *Nature*, **394**, 661–664.

Wetherald, R.T. and Manabe, S. (1988) Cloud feedback processes in a general circulation model. *Journal of the Atmospheric Sciences*, **45**, 1397–1415.

Wetherald, R.T. and Manabe, S. (1995) The mechanisms of summer dryness induced by greenhouse warming. *Journal of Climate*, **8**, 3096–3108.

Whalen, S.C. and Reeburgh, W.S. (1990) Consumption of atmospheric methane by tundra soils. Nature, 346, 160–162.

Whiting, G.J. and Chanton, J.P. (1993) Primary production control of methane emission from wetlands. *Nature*, **364**, 794–795.

Wigley, T.M.L. (1989) Possible climate change due to SO_2-derived cloud condensation nuclei. *Nature*, **339**, 365–367.

Wigley, T.M.L (1991a) A simple inverse carbon cycle model. *Global Biogeochemical Cycles*, **5**, 373–382.

Wigley, T.M.L. (1991b) Could reducing fossil-fuel emissions cause global warming? *Nature*, **349**, 503–506.

Wigley, T.M.L. and Raper, S.C.B. (1990) Natural variability of the climate system and detection of the greenhouse effect. *Nature*, **344**, 324–327.

Wigley, T.M.L. and Raper, S.C.B. (1995) An heuristic model for sea level rise due to the melting of small glaciers. Geophysical Research Letters, 22, 2749–2752.

Wigley, T.M.L., Richels, R. and Edmonds, J.A. (1996) Economic and environmental choices in the stabilization of atmospheric CO_2 concentrations. *Nature*, **379**, 242–245.

Wild, M., Ohmura, A. and Cubasch, U. (1997) GCM-simulated surface energy fluxes in climate change experiments. *Journal of Climate*, **10**, 3093–3110.

Williams, L.D. (1979) An energy balance model of potential glacierization of Northern Canada. *Arctic and Alpine Research*, **11**, 443–456.

Wingenter, O.W., Wang, C.J.-L., Blake, D.R. and Rowland, F.S. (1998) Seasonal variation of tropospheric methyl bromide concentrations: constraints on anthropogenic input. *Geophysical Research Letters*, **25**, 2797–2800.

Woodruff, S.D., Slutz, R.J., Jenne, R.J. and Steurer, P.M. (1987) A comprehensive ocean–atmosphere data set. *Bulletin of the American Meteorological Society*, **68**, 1239–1250.

World Meteorological Organization (1994) *Scientific Assessment of Ozone Depletion: 1994*, World Meteorological Organization Global Ozone Research and Monitoring Project – Report No. 37, Geneva.

World Meteorological Organization (1998) *Scientific Assessment of Ozone Depletion: 1998*, World Meteorological Organization Global Ozone Research and Monitoring Project – Report No. 44, Geneva.

World Resources Institute (1998) World Resources 1998–99, World Resources Institute, Washington, DC, 369 pages.

Wright, D.G. and Stocker, T.F. (1991) A zonally averaged ocean model for the thermohaline circulation, I, Model development and flow dynamics. *Journal of Physical Oceanography*, **21**, 1713–1724.

Wullschleger, S.D., Ziska, L.H. and Bunce, J.A. (1994) Respiratory responses of higher plants to atmospheric CO_2 enrichment. *Physiologia Plantarum*, **90**, 221–229.

Zak, D.R., Pregitzer, K.S., Curtis, P.S., Teeri, J.A., Fogel, R. and Randlett, D.L. (1993) Elevated atmospheric CO_2 and feedback between carbon and nitrogen cycles. *Plant and Soil*, **151**, 105–117.

Zaucker, F., Stocker, T.F. and Broecker, W.S. (1994) Atmospheric freshwater fluxes and their effect on the global thermohaline circulation. *Journal of Geophysical Research*, **99**, 12443–12457.

Zhang, C. (1993) Large-scale variability of atmospheric deep convection in relation to sea surface temperature in the Tropics. *Journal of Climate*, **6**, 1898–1913.

Zhang, M.H., Hack, J.J., Kiehl, J.T. and Cess, R.D. (1994) Diagnostic study of climate feedback processes in atmospheric general circulation models. *Journal of Geophysical Research*, **99**, 5525–5537.

Zhang, S., Greatbatch, R.J. and Lin, C.A. (1993) A reexamination of the polar halocline catastrophe and implications for coupled ocean–atmosphere modeling. *Journal of Physical Oceanography*, **23**, 287–299.

Zhang, Y. and Wang, W.-C. (1997) Model-simulated northern winter cyclone and anticyclone activity under a greenhouse warming scenario. *Journal of Climate*, **10**, 1616–1634.

Zhao, Z.-C. and Kellogg, W.W. (1988) Sensitivity of soil moisture to doubling of carbon dioxide in climate model experiments. Part II: The Asian monsoon region. *Journal of Climate*, **1**, 367–378.

Zuo, Z. and Oerlemans, J. (1997) Contribution of glacier melt to sea-level rise since AD 1865: a regionally differentiated calculation. *Climate Dynamics*, **13**, 835–845.

Zwiers, F.W. and Kharin, V.V. (1998) Changes in the extreme of the climate simulated by CCC GCM2 under CO_2 doubling. *Journal of Climate*, **11**, 2200–2222.

Web sites used

Listed below are the Web sites from which data were downloaded for use in this book, or to which the reader can refer for up-to-date information.

Chapter 1

Paleoclimatic indicators from ice cores and marine sediments were obtained from the Paleoclimatology Website of the US National Oceanographic and Atmospheric Administration. The home page is *http://www.ngdc.noaa.gov/paleo*.

Chapter 4

National GHG emission inventories prepared under the United Nations Framework Convention on Climate Change can be obtained from *http://www.unfccc.de/*.

GHG emission estimates can also be obtained from the Pacific Institute for Studies in Development, Environment, and Security: *http://www.globalchange.org/*.

Information on sulphur emissions associated with various industrial activities can be obtained from the Energy Information Administration of the US Department of Energy *(http://www/eia/ doe/gov/cneaf/coal/cia/)*, the acid rain program of the US Environmental Protection Agency *(http://www.epa.gov/acidrain/)*, the natural gas industry *(http://naturalgas.com/environment/acid/html)*, and Syncrude Canada *(http://www.syncrude.com/5 env/progress 98/so2.htm)*.

Data on variations through time in GHG concentrations and emissions can be obtained from the Carbon Dioxide Information and Analysis Center (CDIAC) of the US Department of Energy *(http://cdiac.esd.ornl.gov)*.

Estimates of the historical variation of global sulphur emissions can be obtained from the Centre for Resource and Environment Studies of the Australian National University *(http://cres.anu.edu.au/~dstern/ anzsee/datasite.html)*.

Chapter 5

Grid point and hemispherically and globally averaged data on changes in sea surface or surface air temperature during the past century or so can be obtained from the Climate Research Institute of the University of East Anglia *(http://www.cru.uea.ac.uk)*. Much of the same data can be obtained from the UK Meteorological Office *(http://www.meto.gov.uk)*. An independent analysis of surface air temperature is available from the Goddard Institute for Space Studies *(http://www.giss.nasa.gov)*.

The variation in tropospheric temperature as obtained from a carefully chosen set of 63 radiosonde stations can be obtained from the Carbon Dioxide Information and Analysis Center *(http://cdiac.esd.ornl.gov/ftp/ndp008)*.

The variation in stratospheric and tropospheric temperature as determined from the satellite-based Microwave Sounding Unit (currently without correction for the drift in satellite altitude) can be obtained via FTP from 198.116.60.11 *(wind.atmos.uah.edu)*.

Chapter 7

Data on the concentrations of halocarbon gases can be obtained from the Nitrous Oxides and Halocompounds (NOAH) group of the US National Oceanographic and Atmospheric Administration *(http://www.cmdl.noaa.gov.noah)*.

Chapter 11

Data on AGCMs and AOGCMs can be obtained from the web site of the Climate Model Intercomparison Project (CMIP): *http://www-pcmdi.llnl.gov/cmip*. Results of transient climatic change simulations from seven (as of this writing) AOGCMs are available at the IPCC Data Distribution Centre (*http://ipcc-ddc.cru.uea.ac.uk*). The data for each model consist of monthly averaged, grid point values of daily mean, minimum, and maximum temperatures; precipitation; winds; and other variables for every month from 1860 to 2099. For each model, results are given for one control run, eight runs with GHG forcing only, and eight runs with GHG plus aerosol forcing.

Overall

In addition to the sites listed above, a good starting point for information (including recent research results) and data pertaining to climatic change is the *Global* Change Master Directory (*http://gcmd.gsfc.nasa.gov*), Another convenient starting point for information on climatic change is the Carbon Dioxide Information and Analysis Center (*http://cdiac.esd.ornl.gov*), mentioned above. Finally the United States Environmental Protection Agency global warming web site (*http://www.epa.gov/global warming*) contains links to about 100 climate-related web sites, including research centres, business associations, the environment ministries of some other governments, and some non-governmental organizations.

Index